"*The Routledge Handbook of Disaster Response and Recovery* highlights not only the critical importance of response to and recovery from disasters, but the diversity of approaches to these processes and the ways in which they are studied and seen by people around the world. Understanding the scope with which different experts, people, and cultures comprehend and move through these spaces is absolutely essential to improving our progress through them, and works like this propel our ability to discuss these issues through varied lenses."

Jennifer Trivedi, *Disaster Research Center, University of Delaware*

"*The Routledge Handbook of Disaster Response and Recovery* represents a significant contribution to understandings of these two phases of the emergency management cycle. Spanning topical areas from 'nuts and bolts' concepts to real-world and applied situations, this Handbook provides a comprehensive accounting of current challenges and opportunities for response and recovery. The editors have curated a volume worthy of not only a place on the bookshelf, but one that will be pulled down again and again as a reference manual for those interested in diving into the challenges and solutions presented within."

Christopher T. Emrich, *School of Public Administration, University of Central Florida*

THE ROUTLEDGE HANDBOOK OF DISASTER RESPONSE AND RECOVERY

The Routledge Handbook of Disaster Response and Recovery covers the two post-disaster stages of the disaster management cycle and presents an extensive and cutting-edge overview of their many considerations.

Organized into two parts, Response and Recovery, this handbook details the history, theories, methods, debates, and emerging issues in the stages of response and recovery. Using a transdisciplinary approach, the myriad topics examined in this handbook include search and rescue, myths related to disaster response, technological methods for response, recovery among vulnerable populations, and the intersection of disasters and mental health. Contributors discuss these issues both globally as well as country- and disaster-specific.

This book is an essential guide and reference not only for scholars engaged in disaster research, but also for undergraduate and graduate students, policy makers, disaster managers, international and supranational agencies, and humanitarian and volunteer organizations engaged in disaster management.

Bimal Kanti Paul is a professor of Geography and Geospatial Sciences at Kansas State University who specializes in various aspects of environmental hazards. He has contributed to the field by integrating the social sciences with the physical sciences to address environmental concerns more holistically. Paul has compiled an outstanding publication record, including seven books on environmental hazards as single or co-author. Recently, Stanford University identified Paul among the top 2% of researchers in the world. He was also editor of the *Geographical Review* (2013–2015), book review editor of *The Professional Geographer* (2011–2012), and is a fellow of the American Association of Geographers.

Luke Juran is a professor in the Department of Geography and at the Virginia Water Resources Research Center at Virginia Tech. Juran's research focuses on the human ecology of water and disaster reconstruction with an emphasis on the implementation of community-level water and sanitation systems after disasters. Juran's research investigates issues related to social vulnerability, water quality, and access to water with particular attention paid to gender and human-environment interactions. In addition to ongoing projects in India, Juran has conducted and supervised research in Bangladesh, Costa Rica, Mauritius, Mexico, Nepal, Sri Lanka, Tanzania, and the United States.

THE ROUTLEDGE HANDBOOK OF DISASTER RESPONSE AND RECOVERY

Edited by Bimal Kanti Paul and Luke Juran

Routledge
Taylor & Francis Group
LONDON AND NEW YORK

Designed cover image: Photograph by Luke Juran

First published 2025
by Routledge
4 Park Square, Milton Park, Abingdon, Oxon OX14 4RN

and by Routledge
605 Third Avenue, New York, NY 10158

Routledge is an imprint of the Taylor & Francis Group, an informa business

British Library Cataloguing-in-Publication Data
A catalogue record for this book is available from the British Library

ISBN: 978-1-032-54032-0 (hbk)
ISBN: 978-1-032-54034-4 (pbk)
ISBN: 978-1-003-41483-4 (ebk)

DOI: 10.4324/9781003414834

Typeset in Times New Roman
by KnowledgeWorks Global Ltd.

CONTENTS

FIGURES

TABLES

BOXES

CONTRIBUTORS

Prabin Acharya University of Alberta, Canada

Mokbul M. Ahmad Asian Institute of Technology, Thailand

Sanam K. Aksha University of Central Florida, USA

Sweta Baniya Virginia Tech, USA

Shirish Bhat University of Florida, USA

Grace Blazek Kansas State University, USA

Stephen T. Buckman Clemson University, USA

Sherilyn D. Burris Liberty University, USA

David A. Torres Castro National Agency for Cultural Heritage, Mexico

Anand Chand Fiji National University, Fiji Islands

Subarna Chatterjee Kansas State University, USA

Charles D. S. Chishiri University of the Free State, South Africa

Matthew Clarke Deakin University, Australia

Chandra Pauline Daniel Global Goals Ambassador, United Nations

BreAnne Danzi University of South Dakota, USA

Kobel Dubique Zanmi Lasante, Haiti

Magdalena Gil Pontificia Universidad Católica de Chile and National Research Center for Integrated Disaster Risk Management, Chile

Rob Grace Brandeis University, USA

Yetta Gurtner James Cook University, Australia

Joshua Hallwright Deakin University, Australia

Nathan Heazlewood Eagle Technology Group Ltd, New Zealand

Tran Thi Minh Hien Van Lang University, Vietnam

Jennifer A. Horney University of Delaware, USA

Gifty Immanuel Harvard University, USA

Md Sariful Islam Massachussets Institute of Technology, USA

Luke Juran Virginia Tech, USA

Emily R. Kaplan New York Medical College, USA

Shakil Bin Kashem Kansas State University, USA

Jessica T. Kelly University of South Dakota, USA

Md. Shohel Khan Noakhali Science and Technology University, Bangladesh

Hyung Jin Kim Kansas State University, USA

David King James Cook University, Australia

Rico Kongsager University College Copenhagen, Denmark

Madison Kueny Kansas State University, USA

Oriane Longerstaey Brown University, USA

Saadia Majeed The Department of Energy, Environment & Climate Action, Australia

Barsha Manandhar University of South Florida, USA

Mohammad Yunus Naseri Virginia Tech, USA

Thomas Evan O'Shea University of Salford, UK

Bimal K. Paul Kansas State University, USA

Shitangsu K. Paul University of Rajshahi, Bangladesh

Munshi K. Rahman Georgia Southern University, USA

Avantika Ramekar Kansas State University, USA

Rachel Routley Deakin University, Australia

Naxhelli Ruiz-Rivera Universidad Nacional Autónoma de México, Mexico

Abdulhadi A. Al Ruwaithi Umm Al-Qura University, Saudi Arabia

Sarah E. Scales University of Delaware, USA

Laila Shahzad University College London, UK

Julia Sherry Clemson University, USA

Saeidah Sobhaninia Arizona State University, USA

Mitchel Stimers Park University, USA

Wendy Stout Virginia Tech, USA

Cara Taubman New York City Health + Hospitals and Harlem Hospital, USA

Shehan Tawsif Shahjalal University of Science and Technology, Bangladesh

Jessy L. Thomas University of South Dakota, USA

Elizabeth Wagemann Universidad Diego Portales, Chile

Nafeeul Alam Walee Georgia Southern University, USA

INTRODUCTION

Bimal Kanti Paul and Luke Juran

Disaster Management Cycle

The disaster management cycle is a conceptual framework that seeks to categorize the stages or phases of disasters. Traditionally, the cycle comprises four overlapping temporal stages, two of which take place before an extreme event (mitigation and preparedness) and two that occur after an event (response and recovery) (Paul, 2011; Coppola, 2020). These stages represent useful typologies that have implications for planning and policy, scholarly research, and other activities to support disaster management. The cycle illustrates the ongoing activities through which various stakeholders collectively seek to reduce potential fatalities, injuries, and physical damage (McEntire, 2021). Disaster management also involves measures to recover from the impacts of disasters as quickly as possible and to prevent disasters from occurring (or at least mitigate their effects) in the next iteration of the cycle. Therefore, each stage demands careful attention among relevant actors—for example, government agencies, first responders, affected populations, nongovernmental organizations (NGOs), and the business community among many others—so that efforts are coordinated, proactive, and sustainable.

The *Handbook of Disaster Response and Recovery* covers the two post-disaster stages of the cycle by presenting a comprehensive and cutting-edge overview of their many considerations. The handbook draws upon the foundations and history of response and recovery and, in doing so, critiques and advances the theories, methods, and debates surrounding the two stages. The objective is to discuss major ongoing and emerging issues related to the stages that are confronted by supranational, international, national, local, and NGO entities as well as businesses, communities, and households. Harnessing a multidisciplinary approach, the topics included in the handbook are explored both globally as well as country- and disaster-specific in order to address the various scales and contexts in which disaster response and recovery are conducted.

Disaster Response and Recovery

Among the four traditional stages, response is the most time sensitive and refers to all activities during and immediately following an extreme event, including efforts to save lives through search and rescue (SAR) operations, treating injuries, and preventing further damage. The primary goals

DOI: 10.4324/9781003414834-1

are to get disaster survivors out of harm's and to help to provide for their immediate needs until more permanent and sustainable solutions can be deployed. Such needs include temporary shelter, water, sanitation, and food. For injuries, responders provide informal and formal medical attention and triage care; for deaths, responders recover bodies and perform proper rituals (PAHO, 2004; Gupta, 2009; Teahen, 2012). Response also involves restoring utilities, initial repairs to critical infrastructure, closure of roads and bridges, cleanup of debris, and damage assessments (Sawalha, 2020; McEntire, 2021). Finally, response should aim to facilitate and prepare society for a speedy recovery.

As the emergency situation is brought under control, the recovery stage of the disaster management cycle begins. The goal of recovery is to return the economy, infrastructure, and livelihoods of survivors to pre-disaster levels or preferably better (Philips, 2015; Paul, 2022). Recovery encompasses activities such as repairing and reconstructing houses, commercial establishments, and public buildings and infrastructure. There is no discrete point at which response changes into recovery and subsequently into long-term sustainable development, and in fact they often overlap.

There are many opportunities during the recovery period to reduce vulnerability, enhance disaster risk reduction (DRR), and address societal inequities through a 'window of opportunity' that often exists. This term refers to the margin of time during the recovery period when positive measures can be more easily undertaken due to an influx of resources, enhanced coordination, and increased political and social solidarity that often accompanies the post-disaster context (Juran, 2015; Paul, 2022). Recovery through a 'build back better' (BBB) approach was adopted as part of Priority 4 of the United Nations Sendai Framework for Disaster Risk Reduction 2015–2030 (Fernandez & Ahmed, 2019). The phrase seeks to popularize efforts to leave affected populations better off than they were before the disaster. However, some have argued that 'better' is vague and subjective (i.e., better than what?) and thus calls to amend the phrase to 'build back safer' have risen in order to place the focus squarely on enhancing safety and security from disasters. Increasingly, recovery is linked with the goal of fostering resilience, as stressed in the Sendai Framework (UNDRR, 2015). While disaster recovery is largely unplanned, it encompasses a key stage in which numerous policies and programs are developed and implemented. Depending on the scale of impact, availability of resources, and level of economic development and political stability, and the recovery stage typically lasts for several years to more than a decade (Ruszczyk, 2018).

As indicated, sustainable developmental considerations should contribute to all aspects of the disaster management cycle, particularly during recovery through the generation of more livable, integrated, and environmentally sensitive human settlements and economies. When this is achieved, societies have greater capacity and resilience when confronting future disasters and future recovery can occur more rapidly. Conversely, recovery processes that merely recreate or perpetuate past vulnerabilities can lead to increased or even compounding susceptibility to extreme events (Davis & Alexander, 2016; Juran, 2018).

The Handbook, Book Structure, and Chapter Summaries

The Routledge Handbook of Disaster Response and Recovery has a total of 36 chapters, including the introductory and concluding chapters. The handbook is divided into Part I on disaster response and Part II on disaster recovery. Each part contains 17 chapters subdivided into four thematic sections.

Chapters in this edited volume have been contributed by a cohort of established and emerging authors who are highly engaged in the multidisciplinary field of hazards and disaster management. Contributors not only represent a diversity of stages in their careers but also a variety of professional sectors (e.g., academia, government institutions, supranational entities, NGOs, first responders, and health care) that each possess valuable expertise and approaches. In total, 58 authors from 17 countries—Australia, Chile, Denmark, Fiji, Haiti, Mexico, New Zealand, Pakistan, South Africa, and Vietnam among others—have coalesced to share their knowledge, experiences, and best practices from the global north and south as well as from various cultural and gender identity perspectives. Additionally, the handbook incorporates a multiplicity of geographies and disaster types through its investigation of small and large nations, high- and low-income nations, and hydrometeorological, geophysical, and biological disasters. Many chapters include primary data or comprise original case studies in a conscious effort to impart new knowledge and advance the practical processes of response and recovery.

Part I: Disaster Response

The 17 chapters in Part I are categorized into four sections, each with a specific theme. Section I surveys key elements and activities of disaster response, and thus Chapter 1 by Avantika Ramekar sets out to briefly describe the major activities that generally comprise the response stage. Many of these activities are analyzed in detail in subsequent chapters of the handbook. The next chapter, by Oriane Longerstaey, examines SAR, including fundamentals of SAR activities, types of SAR (e.g., ground, urban, confined space, maritime, and mountain), and the use of aids such as canines, robots, and unmanned aerial vehicles (UAVs). The chapter ends with two brief case studies on the 2010 lightning strike in the Grand Tetons (USA) and the 2018 Tham Luang cave rescue in Thailand.

Many people in extreme events require immediate medical attention to stabilize their conditions and reduce their suffering. In Chapter 3, Emily Kaplan and Cara Taubman discuss the importance of emergency medical care. The chapter begins by introducing organizations that provide medical care in the aftermath of disasters both in the United States and internationally. Other topics include the coordination of emergency medical care at the local and national levels, medical care leadership and oversight both organizationally and from a clinical perspective, and medical care standards for disaster situations. Additionally, some individuals may face post-traumatic stress disorder (PTSD) or mental health issues. This important topic is broached in Chapter 3 and investigated in detail in Chapter 33. Disasters damage our built environment, knock down trees and structures, and move massive amounts of sediment, organic matter, and solid waste. Therefore, effective clearance of debris is necessary to facilitate emergency response, promote public health and safety, and enable post-disaster recovery. The objective of S. D. Burris in Chapter 4 is to discuss these critical aspects of debris removal, including protocols for hazardous substances and the criticality of maintaining a robust debris management plan.

Section II focuses on channeling disaster response efforts. In Chapter 5, Subarna Chatterjee examines flash appeals as a tool used to quickly secure funds when major disasters occur in low- and middle-income countries. Flash appeals are generally prepared by the United Nations Office for the Coordination of Humanitarian Affairs (UNOCHA) and/or UN Resident Coordinator within three to seven days of an emergency. Chatterjee outlines the scope and contents of flash appeals, revisions to flash appeals, the financial tracking service (FTS) used to monitor flash appeals, and limitations of the flash appeal instrument. The objective of Chapter 6, by Madison Kueny et al., is to highlight how Post-Disaster Needs Assessments (PDNAs) are utilized to foster disaster recovery. The authors detail the purpose, methods, responsible parties, limitations, and potential improvements of PDNAs. Similar to flash appeals, PDNAs are deployed for major disasters, typically in lesser developed countries, and they are headed by external actors.

In Chapter 7, Munshi Khaledur Rahman and others explore the numerous variables and determinants that affect flows of humanitarian aid after major global disasters. Examples of accepting and rejecting disaster aid, competition among donor states, and gaps between pledged amounts and actual amounts disbursed are provided. The chapter allows practitioners to better understand the rationale that governs the provision of aid, which has potential to inform post-disaster mitigation and vulnerability reduction. The final chapter of Section II, by Shakil Bin Kashem and Hyung Jin Kim, discusses the role social networks increasingly play in disaster response. Like many chapters in the handbook, Chapter 8 is equally relevant to both response and recovery. Methodological approaches for social network analysis (SNA) are assessed as is the significance of social capital. With the near ubiquitous utilization of technologies and online social networks, opportunities for linking offline and online networks are posited.

Section III of the handbook centers on disaster relief and humanitarian aid. In Chapter 9, Julia Sherry and Kobel Dubique begin by outlining the roles, strengths, and challenges of NGOs in disaster assistance. Given their functions as providers and distributors of humanitarian aid, the chapter also discusses NGO collaboration with governments and stakeholders, advocacy and fundraising, and involvement in long-term disaster recovery. Sherry and Dubique identify agility, flexibility, and independence as unique attributes that NGOs can leverage in times of disaster. Chapter 10, by Rachel Routley et al., describes the role domestic and foreign military forces play in disaster response and humanitarian relief. Militaries are able to quickly deploy personnel, assets, and assist in SAR, emergency medicine, and debris cleanup due to their specialized equipment and expertise. After raising concerns on the participation of military forces (e.g., coordination with the local government and other militaries), the chapter presents case studies on Pakistan and Australia before concluding with five key principles to improve the effectiveness of military participation in the post-disaster setting.

Chapter 11 assesses the intersection of diplomacy and disaster aid. Rob Grace offers an overview of various facets of disaster diplomacy, including humanitarian negotiations, the domestic politics of disaster response, (geo)political dynamics that influence international disaster response, and the role non-state armed groups play in this sphere. Another dimension of the diplomacy-disaster aid nexus aims to advance domestic and international agendas. This can be attempted through the medium of humanitarian aid and the deployment of high-level engagement to achieve geostrategic and security-related goals, economic gains, and to compel actors to pursue humanitarian objectives.

The following two chapters present original case studies on disaster relief in Bangladesh and volunteerism in Australia, respectively. First, Shitangsu K. Paul and coauthors analyze flows of disaster relief in Bangladesh as well as the institutional frameworks and distribution systems that manage its distribution. Paul et al. harness empirical data gathered from different parts of the country after different disasters (e.g., floods and cyclones). Despite improvements in disaster management efficiency, the authors identify several issues, such as timeliness of aid, coordination among stakeholders, and biases in relief distribution. In Chapter 13, Sweta Baniya and Badal Pokharel narrate the diverse activities of Sikh Volunteers, a volunteer-based NGO in Australia, during recent 2022 floods in New South Wales. The case study narrates how Sikh Volunteers contributed to disaster response efforts, and their actions are interpreted through lenses of Millora's (2020) typology of volunteering and humanitarian principles of the United Nations (UNHCR, 2019).

The last section (Section IV) on disaster response explores critical issues and visions that should be considered, such as biases and myths, the potential for outbreaks of disease, and the increased use of technologies in disaster response. First, in Chapter 14, Emily R. Kaplan et al. discuss biases in relief distribution and decision making. After identifying populations and communities that commonly experience enhanced vulnerability, the authors investigate types of biases (e.g., implicit and explicit) as well as methods for individuals and institutions to mitigate biases. It is here, with a heavy heart, we must share that Chandra Pauline Daniel, co-author of Chapter 14, passed away in the short time since producing this strong contribution. Daniel's work in the sphere of global health and disasters—most recently as a United Nations Global Goals Ambassador for the Sustainable Development Goals—will certainly be missed, but the knowledge and positive changes that she created will live on.

In Chapter 15, Thomas Evan O'Shea draws upon recent and historical examples to scrutinize the origins of myths related to disaster response. O'Shea not only demonstrates how myths, assumptions, and beliefs can serve to extend the impacts of disasters far past the initial event but also how they can be surmounted to foster capacity and resilience. The relationship between disasters and the outbreak of infectious diseases has been well established. Chapter 16 by Laila Shahzad details how extreme events can give rise to increased rates of communicable and vector-borne diseases, thereby posing a secondary emergency. Population displacement, overcrowding in temporary shelters, stressed healthcare systems, and hazardous environmental conditions are just a few of the factors that can facilitate the presence and transmission of pathogens in the post-disaster setting. Different modes of transmission (e.g., water, air, and vectors), underlying causes of disease outbreaks in low- and middle-income countries, and strategies to prevent the spread of diseases are also provided.

The final chapter in Section IV, Chapter 17, focuses on the use of technologies in disaster response. Wendy Stout and Nathan Heazlewood survey current and emerging technologies, including discussions on the use of geographic information science (GIS), remote sensing, and UAVs. Challenges and requirements related to digital infrastructure, data collection, and management are provided along with recommendations on how to keep up-to-date with technological advances. Stout and Heazlewood speak in lay language throughout the chapter in order to address a general audience.

Part II: Disaster Recovery

Part II on disaster recovery also has 17 chapters divided into four thematic sections. Similar to Part I on response, the first section (Section V) examines key elements and activities of disaster recovery. In Chapter 18, Rico Kongsager discusses recovery goals and objectives before outlining the major activities that typically unfold during disaster recovery. Kongsager divides these activities and processes into pre-disaster, short-term recovery, and long-term recovery—in particular noting the significance of being proactive and establishing resilience and long-term sustainability. In Chapter 19, Abdulhadi A. Al Ruwaithi details the evolution of the disaster recovery cycle and contends that a more comprehensive characterization of recovery must be adopted that accounts for physical as well as the non-physical components and outcomes. Several conceptual models are critiqued, and recovery frameworks of the United States and Australia are reviewed. Ultimately, Al Ruwaithi argues that recovery models necessarily oversimplify reality and thus they must be holistic, flexible, and informed by both theory and empirical evidence.

Chapter 20 provides a review of the outcomes of disaster recovery. Yetta Gurtner and David King stress that recovery should begin with a recovery plan. In this sense, outcomes are the products or end result of the recovery planning process. Unfortunately, recovery can create disparities, resulting in 'winners' and 'losers,' and post-disaster population loss is a dynamic that must often be managed. The authors contend that the affected community should be the focus of recovery, and they note that recovery may significantly diverge from initial plans as time passes and community needs evolve. In Chapter 21, Mitchel Stimers merges the concepts of 'sustainable development' and 'disaster recovery' to address the concept of sustainable disaster recovery (SDR). SDR emphasizes holistic, long-term resilience to disasters that simultaneously seeks to produce positive socio-environmental change. This involves a range of strategies, including the utilization of renewable energy sources, implementation of 'green' infrastructure, and land use that promotes social equity and ecosystem services. SDR concepts are increasingly used in conjunction with DRR measures, but challenges to successful SDR exist, such as monitoring and evaluation of outcomes, collaboration among diverse stakeholders, and prioritizing SDR across all stages of the disaster management cycle.

Section VI concentrates on theoretical approaches and methodologies for disaster recovery. Chapter 22 commences this discussion by surveying disaster recovery models. Elizabeth Wagemann and Magdalena Gil review several frameworks in public policy as well as recovery guidelines and manuals. The authors culminate by suggesting an adaptation of Davis and Alexander's (2016) sectoral model of recovery by adding a set of goals to be achieved for each sector. Wagemann and Gil ultimately argue that, while models oversimplify real-world systems, the value lies in their ability to support decision making. Continuing with critiques of recovery models, Chapter 23 appraises the well-known Kates-Pijawka model (Kates & Pijawka, 1977) against current-day understandings of recovery. Stephen T. Buckman and Saeideh Sobhanini detail the four phases of the traditional Kates-Pijawka model (i.e., emergency response, restoration, reconstruction (I), and reconstruction (II)) before updating the instrument to better suit today's needs without abandoning the spirit of the original model. The newly conceived model accounts for the affected location's pre-disaster status and level of resilience, and it also addresses issues of equity and social and emotional recovery that have come to the forefront in recent years.

In Chapter 24, Saadia Majeed charts the evolution of recovery studies and engages the discourse on recovery duration. Beginning in the 6th century with the reconstruction of Hagia Sophia in Turkey and continuing to the recent COVID-19 pandemic, Majeed demonstrates how recovery has transformed over time to incorporate different foci, including community-led recovery and nature-based solutions (NbS). Majeed identifies the complex factors that influence the duration of disaster recovery and proposes a nuanced recovery timeline that considers both operational and survivor perspectives. Naxhelli Ruiz-Rivera examines social scientific methods for recovery research in Chapter 25. After reviewing the relevant literature, two approaches are investigated: normative approaches, and frameworks that are founded on guidelines and principles. Each approach treats disaster recovery and its components in a different way, and each considers why and how societies change when confronted with disaster impacts. Four types of research designs are reviewed for each approach. Ethical considerations, longitudinal designs, SNA, and narrative techniques are reviewed for normative approaches, and PDNAs, non-economic loss and damage (NELD) assessments, forensic approaches, and auditing methodologies are reviewed for recovery frameworks.

Section VII contains five case studies. Beginning with small island developing states (SIDS), Anand Chand and Luke Juran provide insight on household and community roles during disasters in the South Pacific. The authors highlight the importance of social protection mechanisms, particularly social capital, reciprocity, and the nexus of community, traditional leadership, and faith. Given extreme isolation along with limited state resources and built infrastructure, the authors argue that formal response to disasters is hampered without the roles assumed by households and communities. Disaster recovery is widely viewed as a 'window of opportunity' to institute positive societal change. In Chapter 27, Mokbul Morshed Ahmad and Tran Thi Minh Hien illustrate various opportunities that arose during the COVID-19 pandemic in Ho Chi Minh City, Vietnam.

Despite the deaths, suffering, and restrictions to mobility, COVID-19 helped to fashion more adaptable supply chains, improved healthcare systems, digital advancements in education, and avenues for poverty reduction through remote work. Furthermore, increased public health awareness and a spirit of community engagement and cooperation emerged. Collectively, the authors claim that these transformations helped to create a more resilient post-pandemic society.

In Chapter 28, the sources and distribution of aid for post-disaster housing in Nepal is scrutinized by Barsha Manandhar. The chapter demonstrates how, after the 2015 Gorkha earthquake, powerful stakeholders shaped the distribution of financial assistance for housing. Inequities after the 2015 Gorkha earthquake are examined further by Sanam K. Aksha et al. in Chapter 29, who argue that vulnerable and marginalized communities were overlooked during the recovery period. The authors argue that the Government of Nepal missed the opportunity to both introduce resilient infrastructure and create a more just society. The last case study, Chapter 30, employs computer modeling to inform disaster recovery. Mohammad Yunus Naseri and Luke Juran employ the Height Above Nearest Drainage (HAND) model to map flood inundation extent in Hope Mills (USA) using Hurricane Matthew as a reference event. HAND-derived results are compared to actual inundation extents produced using field-verified high-water mark data, and avenues to improve the HAND model are posited.

Section VIII, the last section of the handbook, surveys four ongoing and emerging issues. Businesses sustain varying degrees of damage from disasters, and a considerable number of businesses are unable to reopen their doors. Thus, in Chapter 31, Charles D. S. Chishiri and Luke Juran discuss factors associated with business recovery, such as firm size, sector, infrastructure and supply chain disruptions, and pre-disaster financial conditions. A conceptual framework is also presented that gauges the spatio-temporal implications of disasters upon a business vis-à-vis its financial and market profile. A major lesson is that recovery is not only a function of the disaster, but more so the pre-disaster conditions and vulnerabilities of the business itself. Thus, the factors that have the greatest impact on post-disaster business failure are often the same as those that influence failure in non-disaster scenarios. Disasters also impact our heritage buildings, including historical sites, monuments, artifacts, and vernacular architecture. David A. Torres engages this subject in Chapter 32 by stressing the importance of preserving cultural assets. Torres calls for a paradigm shift that prioritizes cultural infrastructure in response and recovery. The approach should be grounded in inclusivity, equity, and human rights, and plans to safeguard cultural heritage should be in place in advance of the disaster.

In Chapter 33, BreAnne A. Danzi et al. tackle the difficult and increasingly important issue of mental health. Some disaster survivors develop psychiatric disorders, such as depression, anxiety, PTSD, and substance abuse problems. The chapter identifies several risk and protective factors that help to predict psychological functioning after disasters, including demographic characteristics, disaster exposure specifics, and aspects of the recovery environment. Danzi et al. advocate for a consideration of these factors and an integrated structure for post-disaster mental health service provision that utilizes Psychological First Aid and cognitive behavioral approaches. A final current and emerging issue that must be dissected is how the COVID-19 pandemic, a biological hazard/disaster, affected response, and recovery. In Chapter 34, Sarah E. Scales and others discuss how COVID-19 related deaths, sickness, burnout, and turnover affected the emergency response workforce. This left emergency operation center, frontline healthcare, and public health workers less prepared for pending disasters and to mitigate continued excess morbidity and mortality from the virus. Scales et al. maintain that the recent global pandemic exposed deep-seated inequalities in social determinants of health as well as the need to improve support systems for first responders and public health workers during complex emergencies.

Conclusion

The disaster management cycle is a framework that categorizes the phases of disasters into response, recovery, mitigation, and preparedness. This conceptualization has functional implications for planning and policy, theoretical applications in scholarly research, and helps to classify the numerous activities required to support disaster management in a way that still allows for overlap and interplay. That being said, this volume critiques and updates the concept of a traditional disaster management cycle in order to account for the complex factors associated with our contemporary global system.

In the *The Routledge Handbook of Disaster Response and Recovery*, we have attempted to produce a comprehensive volume on the two (primarily) post-disaster stages of response and recovery. Response involves tending to the immediate and short-term needs of disaster affected populations, while recovery seeks to return society to its pre-disaster status at minimum. Compared to response, recovery is protracted and can take many years. For example, recovery after the 2011 Tōhoku triple disaster (earthquake, tsunami, and Fukushima Daiichi nuclear plant meltdown) in Japan is yet to be completed. The disaster destroyed over 120,000 homes along the northeastern coast of Honshu, and the Japanese Reconstruction Agency is still working to reconstruct housing. As of February 2024, nearly 30,000 people were still occupying temporary accommodations, the majority of those in Fukushima Prefecture (Japan Data, 2024).

Response and recovery can be conducted in ways that are more sensitive to both human societies and the natural environment. While the effects of disasters are unpleasant and unwanted, the post-disaster context represents an opportunity to affect positive change by implementing principles of sustainable development and DRR. If we desire, stronger and more resilient systems can be created in terms of both our built and social environments. This text advocates for the heavy task of attempting to do so.

References

Coppola, D.P. 2020. *Introduction to International Disaster Management*, 4th edition. Boston, MA: Elsevier.

Davis, I., and D. Alexander. 2016. *Recovery from Disaster*. London: Routledge.

Fernandez, G., and I. Ahmed. 2019. "Build Back Better" Approach to Disaster Recovery: Research Trends since 2006. *Progress in Disaster Science* 1: 100003. http://dx.doi.org/10.1016/j.pdisas.2019.100003.

Gupta, K. 2009. Cross Cultural Analysis of Responses to Mass Fatalities following 2009 Cyclone Aila in Bangladesh and India. *Quick Response Report #216*. Boulder, CO: Natural Hazards Center, University of Colorado at Boulder.

Japan Data. 2024 (March 11). The State of Recovery in Tohoku 13 Years after 3/11. https://www.nippon.com/en/japan-data/h01917/.

Juran, L. 2015. A Framework for Integrating Water Projects in Disaster Reconstruction Processes in the International Context. *Public Works Management & Policy* 20(4): 305–321.

Juran, L. 2018. Disaster, Development, and Water: The Reconstruction and (Re)fabrication of Hazardous Waterscapes in Post-Tsunami India. In *The Asian Tsunami and Post-Disaster Aid*, edited by Reddy, S., pp. 79–99. Singapore: Springer Nature.

Kates, R.W., and D. Pijawka. 1977. From Rubble to Monument: The Pace of Reconstruction. In *Reconstruction Following Disaster*, edited by Haas, J.E., R.W. Kates, and M.J. Bowden, pp. 1–23. Cambridge, MA: M. I. T. Press.

McEntire, D. 2021. *Disaster Response and Recover: Strategies and Tactics for Resilience, Third Edition*. Danvers, MA: John Wiley.

Millora, C. 21st Century Volunteering Practices. 6 July 2020, https://www.unv.org/publications/21st-century-volunteering-practices.

PAHO (Pan American Health Organization). 2004. *Management of Dead Bodies in Disaster Situations*. Washington, DC: PAHO.

Paul, B.K. 2011. *Environmental Hazards and Disasters: Contexts, Perspective and Management*. Hoboken. NJ: Wiley-Blackwell.
Paul, B.K. 2022. Recovery. In *Routledge Handbook of Environmental Hazards and Society*, edited by McGee, T., and E. Penning-Rowsell, pp. 425–441. London: Routledge.
Philips, B.D. 2015. *Disaster Recovery*, 2nd edition. Boca Raton, FL: CRC Press.
Ruszczyk, H.A. 2018. The Earthquake and Ideas Lying Around. In *Evolving Narratives of Hazard and Risk: The Gorkha Earthquake, Nepal 2015*, edited by Bracken, L., H. A. Ruszczyk, and T. Robinson, pp. 125–139. Gewerbestrasse: The Palgrave Macmillan.
Sawalha, I.H. 2020. A Contemporary Perspective on the Disaster Management Cycle. *Foresight* 22(4): 469–482. https://doi.org/10.1108/FS-11-2019-0097.
Teahen, P.R. 2012. *Mass Fatalities: Managing the Community Response*. Boca Raton, FL: CRC Press.
UNDRR (United Nations Office for Disaster Risk Reduction. 2015. Sendai Framework for Disaster Risk Reduction 2015-2030. https://www.unisdr.org/we/inform/publications/43291.
UNHCR (United Nations High Commissioner for Refugees). Humanitarian Principles. 22 July 2019, https://emergency.unhcr.org/protection/protection-principles/humanitarian-principles.

PART I

Disaster Response

Section I

Disaster Response: Key Elements and Activities

1
KEY ELEMENTS AND ACTIVITIES OF DISASTER RESPONSE

Avantika Ramekar

Introduction

Disasters are exceptional events. Whether natural disasters such as earthquakes or human-made disasters such as terrorist attacks, disasters create emergency situations that disrupt social, political, economic, and physical order. They create situations in which basic life necessities, such as food, water, shelter, protection, and medical help, are needed (Narayanan & Ibe, 2012). To ensure swift action, quick coordination and reliable communication are required. Therefore, prior knowledge, preparedness, and planning can facilitate efficient delivery of services to those in dire need.

A typical disaster management cycle consists of the organization and management of resources before, during, and after an extreme event (Paul, 2011). Broadly, it can be divided into four continuums: response, recovery, mitigation, and preparedness. Learning from each of these phases feeds the following phase to be better equipped for the next disaster. This chapter will focus on the key elements and activities of disaster response stage.

The response phase of disaster management begins immediately after the disaster strikes and involves several steps (Todd & Todd, 2011). It encompasses the early part of relief in which the goal is to save lives, minimize damage to property, and enhance effectiveness of recovery in the shortest possible time. The magnitude and the type of event determine the response timeframe. The stage includes both immediate response, also known as relief, and medium-term response that moves toward the recovery phase of the disaster management cycle (Paul, 2011). Poor proficiency of institutions and administrations, along with poor interagency collaborations, worsens human suffering. Often inadequacy is associated with developing countries; however, there have been instances of institutional or administrative failures during natural disasters even in developed nations (Paul, 2011).

Organizations or agencies that are involved in the response stage are public, private, humanitarian, not-for-profit, and individuals from affected and non-affected areas or countries. This support mainly depends on the magnitude of the disaster and may also be impacted by geopolitical relations between political entities. More importantly, the design of disaster response activities should be as simple and realistic as possible (Todd & Todd, 2011). An important aspect of disaster response is coordination and communication (Paul, 2011).

Disaster response begins with search and rescue (SAR) operations. This involves a series of steps, including establishment of communication infrastructure, performing SAR operations, and

DOI: 10.4324/9781003414834-4

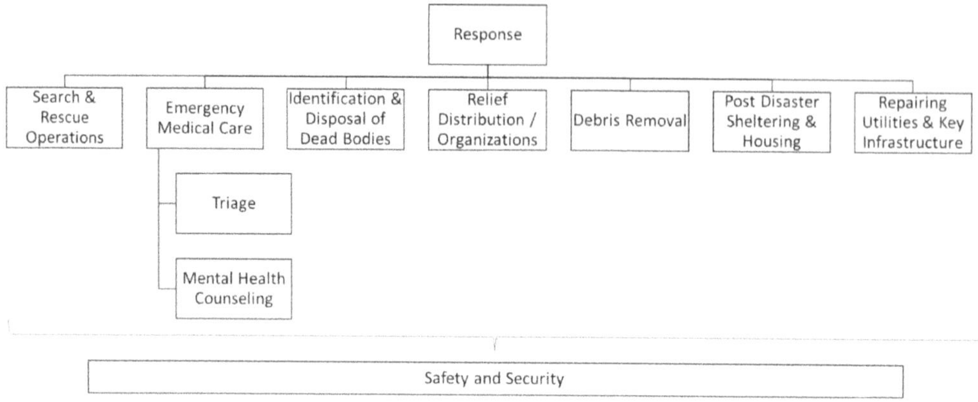

Figure 1.1 Action Steps for Disaster Response

Source: Modified from Paul (2011).

providing any needed first aid services (Narayanan & Ibe, 2012). These relief operations require a reliable communication network. These networks can be divided into two, according to Narayanan and Ibe (2012), disaster mitigation networks and disaster relief networks. A disaster mitigation network is used for pre-disaster planning, while the disaster relief network is a network that provides emergency support to disaster survivors and to crew members who are helping the survivors (Narayanan & Ibe, 2012). Broadly speaking, disaster response can be divided into the following action steps noted in Figure 1.1 and described in the following sections.

Search and Rescue (SAR) Operations

Tracking and identifying the locations of disaster survivors are two of the operations that take place in an SAR operation (Narayanan & Ibe, 2012). SAR operations are often characterized by time constraints—any delay can result in dramatic consequences, including potential human losses, especially if the environmental conditions are challenging (Waharte & Trigoni, 2010).

The scale of the disaster determines who participates first: local or regional individuals. However, it is the neighbors and neighboring communities that are first to respond (Paul, 2011). Over time, social networks such as personal, corporate, and religious as well as local and regional political units get involved in response and extend a hand in the relief work. Delayed response is the time gap between the occurrence of the event and the response time. The first 12 hours post-natural disaster are the most crucial (Paul, 2011). Therefore, prompt mobilization of emergency resources at both the personal and community levels is important. Later, national, foreign, armed forces, and international players may get involved.

In the past, dogs, human signs, and acoustic signals were primarily used for SAR. These techniques worked where accessibility to affected areas by humans and animals was relatively easy. In more challenging environments, the use of unmanned aerial vehicles (UAVs) can provide critical support to SAR operations. Other technologies such as sensors, wireless communication, robotics, computing, ground penetrating radar, and audio-visual technology have assisted SAR operations (Waharte & Trigoni, 2010). Additionally, global positioning system (GPS) technology is used in conjunction with other available technologies. However, there can be limitations in using technologies. For instance, GPS is useful in outdoor SAR operations, while radio

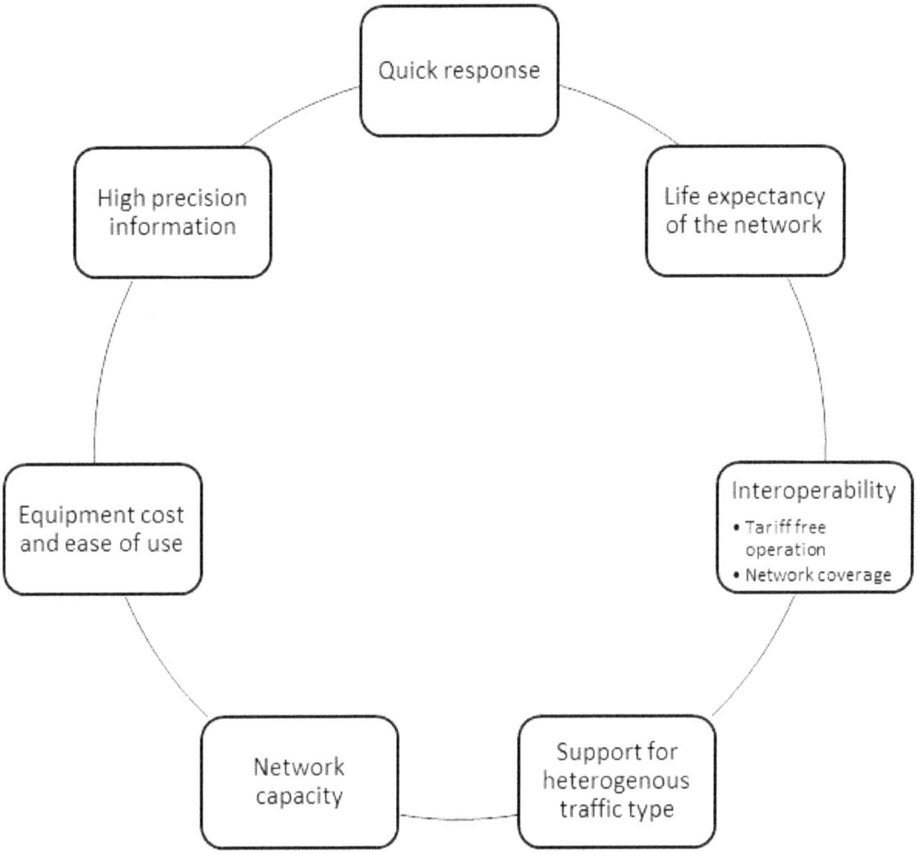

Figure 1.2 Key Requirements for Search and Rescue Operations

Source: Modified from Narayanan and Ibe (2012).

frequency identification (RFID)-based schemes are useful in indoor office applications (Waharte & Trigoni, 2010). Moreover, if the technology is dependent on cellular networks, availability of the network in the affected area will affect the capability of SAR (Waharte & Trigoni, 2010). The key requirements for efficient SAR operations are summarized in Figure 1.2.

Emergency Medical Care

Emergency medical care developed over 30 years ago. Individuals participating in emergency medical care during disaster management have a strong history of leadership, including preparedness planning, coordinated communication, triage assessment, patient transportation and care, resource management, and hazardous material response (Catlett et al., 2011). The emergency medical services role in disaster response falls into five categories: recognition of the event, notification, mobilization, response, and demobilization (Catlett et al., 2011). The key roles of these services are mass-casualty triage, on-scene treatment, communication, evacuation, coordination of patient transport, and patient tracking (Catlett et al., 2011). The goal is to obtain the best possible outcome for the greatest number of people. The first 72 hours following a natural disaster serve as a crucial window for emergency response—the way this window is utilized affects later outcomes (Paul, 2011; Hamilton et al., 2022).

Geographically speaking, populations in low- and middle-income countries are most vulnerable to natural disasters. Additionally, poor or substandard infrastructure and communication systems can result in delayed response operations and the availability of medical equipment and personnel. This in turn impedes the ability to cope with a sudden surge in demand of trauma care (Hamilton et al., 2022). In such cases, immediate response is provided by residents and local rescue efforts. Successively, national emergency services or national medical teams are usually dispatched to impacted areas. Depending on the scale of the disaster and the capacity of local and national authorities to deal with the emergent situation, the international community may offer various forms of assistance, including but not limited to emergency medical care (Hamilton et al., 2022). In some jurisdictions, emergency medical personnel may also take leadership roles during disaster response and be a part of command staff or regional or national assets (Catlett et al., 2011).

Establishment of an incident command system (ICS) was developed in the United States more than 30 years ago in the wake of a series of uncontrolled wildfires that devastated the state of California. This was before multi-institutional collaboration was available for such a large-scale event and as a result disaster response was chaotic, uncoordinated, and led to operational problems (O'Neill, 2005). By 1981, US federal officials had moved from ICS to a National Interagency ICS Management System (NIIMS). In 2004, NIIMS was revamped into the National Incident Management System (NIMS), which established standard organizational tools for disaster response for all incidences within the American National Emergency response, including state and local agencies. Over the years, the ICS structure, or some variation of it, has been adopted internationally (O'Neill, 2005). For example, integrating and coordinating first responders and multi-jurisdiction agencies at the time of disaster has given rise to the Hospital Emergency Incident Command System (HEICS). This has been an effective way of incorporating the basic knowledge and understanding of ICS principles and structure as essential for all individuals participating in disaster response (O'Neill, 2005).

Mental health issues may result from a range of negative outcomes for disaster survivors. Mental health diseases such as acute stress disorder (ASD), post-traumatic stress disorder (PTSD), depression, anxiety, panic, complicated bereavement, anger, and arrested or regressed childhood development may be initiated or exaggerated among individuals (Ruzek et al., 2004). Given the adverse emotional impact of a disaster not just on the survivors but also on medical staff, healthcare workers also benefit from increased training in self-care and stress management. Despite its importance, integration of mental health aspects of disaster response is not a feature of most current emergency medical disaster readiness plans. Such an integration will require collaboration between medical and mental healthcare providers. Ideally, medical emergency medicine preparation should establish relations with local emergency response teams and agencies, train hospital staff in mental and physical health, and include medicine planning efforts in regional planning efforts prior to the disaster (Ruzek et al., 2004).

Identification and Disposal of Dead Bodies

Natural disasters are associated with mass fatality. Since 1990, over 1.6 million fatalities globally have been caused by natural disasters (Ward et al., 2020). Circumstances of death, direct and indirect causes of death, location of death, activity at the time of death, awareness of the disaster, and attempt to take safety measures are some challenges that vary based on the type of disaster (Paul, 2021). It forms the foundation for identifying the medical cause and manner of death and provides important clues to coroners or medical examiners to certify the deceased (Paul, 2021).

The identification of disaster victims is not always easy and/or reliable, and in some cases it may be impossible. Therefore, for accuracy purposes, disaster victim identification (DVI) capacity and standards are required. For this purpose, the International Police Organizations (Interpol) resolution was created for DVI and is the only internationally recognized legislature functioning under international law (Brough et al., 2015). There are a total of 187 member countries of Interpol, and they recommend following common procedures to identify victims (Brough et al., 2015). This begins with each member nation establishing and managing one or more permanent DVI teams that are responsible for pre-planning and training of key personnel. The post-mortem intervention as described in the DVI protocol includes three vital steps: recovery and examination of bodies to establish post-mortem evidence from the deceased, search for ante-mortem information, and comparison of ante-mortem and post-mortem data (Brough et al., 2015).

After disasters, the fear of diseases has encouraged communities, local authorities, and governments to rapidly dispose of the bodies of disaster victims, sometimes without identification (Morgan & de Goyet, 2005; Suwalowska et al., 2021). Hasty disposal of bodies can result from misinformation regarding the spread of diseases (Morgan & de Goyet, 2005; Paul, 2021; Suwalowska et al., 2021). Repulsion to the smell of decay and fear of the spread of infectious diseases also hasten the disposal of disaster deaths. Corpses are often perceived as 'an inconvenience with the potential to pose a practical risk to the living' (Suwalowska et al., 2021, 2). Several studies have shown that the fear of infection and the logistics of managing a large number of fatalities force local authorities and community members to bury the dead in mass graves or to cremate them (Paul, 2021; Suwalowska et al., 2021). However, there remain large gaps in addressing ethical and social aspects of managing death. This may cause tensions, especially when foreign volunteers, tourists, untrained individuals, or lack of cultural knowledge are involved (Morgan & de Goyet, 2005). For instance, there are differences in terms of care of the body, timeliness of burial, and other funerary rites that have tremendous social and psychological importance for the families of the victims (Suwalowska et al., 2021). However, sometimes given restrictive policies, the chaos of crisis, or an overwhelming number of fatalities, it is impossible to perform the rites required by the culture or religion despite all best efforts (Suwalowska et al., 2021). This compounds distress for families and communities impacted by the event and demonstrates the necessity of disaster practitioners to always harness an ethical lens.

There are several stakeholders involved in the management of dead bodies, including family members, communities, community and civic organizations, religious leaders, and traditional healers. Additionally, those involved include last responders such as burial teams, pathologists, coroners, medical and national authorities, global organizations, and non-governmental organizations (Suwalowska et al., 2021). While perspectives of these stakeholders may overlap, often their roles and priorities are different, which can give rise to tensions. Moreover, media outlets also tend to gravitate toward sensational news about the number of lives lost, timely disposal of the dead, and (real or perceived) mishandling of or cultural taboos related to the victims (Suwalowska et al., 2021).

Relief Distribution/Organization

Relief distribution network in disaster response is a part of disaster management or emergency management. Broadly speaking, relief distribution efforts can be divided into short- and long term. Relief distribution networks follow a sequence of decisions and tasks once an emergency alert is given. These include design of the system to be followed and selection of the logistics centers, shelters, and distribution centers that will be used to support the operations. Distribution centers are normally located in an accessible area that is proximal to the disaster area and receive relief goods from external parties such as regional, national, international, and humanitarian entities. Decisions such as the location and allocation of resources and transportation should have already been planned in the preparedness phase but may have to be revised depending on the actual situation at the scene (Maghfiroh & Hanaoka, 2020). Two challenges to optimizing relief distribution are location and network design problems combined with transportation and routing problems (Anaya-Arenas et al., 2014; Fikar et al., 2016).

Debris Removal

Removal of debris in a timely manner can be challenging as communities may be overwhelmed with other demanding post-disaster activities. Depending upon the nature and severity of the disaster, the quantity of debris generated will vary (Brown & Milke, 2016). For instance, in a given community, a high impact disaster can generate up to five to ten times debris volumes of the annual waste normally generated by a community (Kim et al., 2014).

The type of debris depends on the nature of the disaster. For instance, debris caused by floods differs from debris caused by an earthquake or tornado. Typically, the type of debris generated includes vegetation/biomass, municipal solid waste, construction and demolition debris, vehicles, appliances, and electronics, and household hazardous waste (Table 1.1). Each type of waste may contain or be contaminated with toxic or hazardous constituents. In the short term, removal of debris is critical to facilitate response (Luther, 2008). However, with limited capacities after disasters, the debris removal process is often unable to be completed in the desired time, which can delay response and SAR operations and even impact access to medical aid (Kim et al., 2014). In other words, a poor management system for cleanup efforts can exacerbate disaster response issues, resulting in increased loss, damages, and slow recovery of the region and its communities.

In the United States, the debris removal process is undertaken in alliance with the city's waste management system. Establishing a management system during the pre-disaster management process is a complex balancing act. It involves balancing stakeholder desires, community needs, environmental factors, and political will. Adding a disaster to this equation adds to the complexity of management (Brown & Milke, 2016).

Table 1.1 Typical Debris Types and Their Constituents

Debris type	*Constituents of debris*
Municipal solid waste (MSW)	Personal belongings and general household trash
Natural	Soil, silt, rocks, and sediments
Putrescibles	Animal carcasses (pets and farm animals), rotten or spoiled fruits, vegetables, seafoods, and meats
White goods	Refrigerators, freezers, air conditioners, washers, dryers, stoves, water heaters, and dishwashers
Household hazardous waste	Oil, pesticides, paints, and cleaning agents
Construction and demolition	Roads, bridges, cinder blocks, building materials (asphalt, drywall, plaster), and roofing materials
Vegetation	Trees, branches, shrubs, logs, and roots
Vehicles	Cars, trucks, fuels, motor oil, batteries, and tires
Electronic waste	Computers, televisions, printers, stereos, and telephones

Source: Compiled from Luther (2008, 2011).

In times of disaster, a large volume of waste can quickly overburden existing landfill space (Luther, 2011). The likely best choice for most regions will involve diverting as much debris as possible from area landfills by burning, recycling, compositing, or participating in other ways of waste reduction. However, if proper pre-disaster planning was not considered, it can complicate the process (Luther, 2011).

Post-Disaster Sheltering and Housing

Disasters often result in mass displacement. According to Peacock et al. (2018), approximately 19.2 million people were displaced in 2016 alone, including 63,000 in the United States, 91,000 in Mexico, 2 million in Nepal, and 3.5 million in China. Quarantelli (1995) points out that the terms 'shelter' and 'house' are often used interchangeably, adding to confusion. To remedy this, he suggests using four different terms to reduce confusion, namely, emergency sheltering, temporary sheltering, temporary housing, and permanent housing. Briefly, these terms can be described as follows:

Emergency sheltering is a place where individuals shelter themselves for short periods of time before a hazard or immediately after impact. It is spontaneous and based on locational convenience and immediacy of need. In countries where hazard warning systems and evacuation protocols are well established, pre-hazard emergency sheltering is common (Quarantelli, 1995; Peacock et al., 2018).

Temporary sheltering is a place where survivors can stay for a longer period of time (though still short term) while waiting for it to be safe to return to permanent residences. These shelters require more preparation by not-for-profit agencies and the government. This is a publicly designated mass shelter (Quarantelli, 1995; Paul, 2011).

Temporary housing provides disaster survivors a place where they can stay for weeks to months to even years. Compared to temporary sheltering, at temporary housing, routine day-to-day household activities are reestablished (Quarantelli, 1995). Individuals in temporary housing are waiting to move to permanent housing, which may be their original home (now restored) or some type of alternate permanent home (Quarantelli, 1995; Peacock et al., 2018). In the United States, temporary housing arrangements after a disaster are funded by Federal Emergency Management Agency (FEMA) or the United States Department of Housing and Urban Development (HUD) through cash grants for temporary rental housing or the provision of manufactured homes (Peacock et al., 2018).

Permanent housing is when survivors return to their repaired or rebuilt original home/location or move into new quarters in the community. In either case, the move involves occupying permanent residential faculties. Unlike temporary sheltering and housing where the distinction can be less clear, there is a sharp difference between permanent housing (Quarantelli, 1995). While temporary housing can transition to permanent housing when displaced households cannot return to or refuse to return to their pre-disaster home, complications can arise due to housing quality and lack of long-term planning (Quarantelli, 1995; Peacock et al., 2018).

Reestablishing housing is critical for the response and recovery process whether it is for an individual, family, community, or entire region. Research has shown that reestablishing permanent housing is critical for carrying out normal daily activities and domestic functions and for reestablishing livelihood (Juran, 2018; Peacock et al., 2018). Delays in permanent housing can stymie the return to all aspects of life. However, the process of reestablishing permanent housing is complex and multidimensional as one moves from the individual level to a community and regional level. Further complexity is added when various forms of housing are considered, such as private, rented, single family, multi family, high end, and affordable housing (Peacock et al., 2018). Not having proper housing can impact a whole community, as employees leave, consumers are lost, and social networks are broken; hence, housing recovery is fundamental. For example, housing issues after Hurricane Katrina led to an out-migration from the impact zone, job-loss and disrupted workplaces, and financial crises within affected communities (Zottarelli, 2008).

Repairing Utilities and Key Infrastructure

Terms such as infrastructure systems, critical infrastructure, and lifelines refer to networks and systems in the built environment that provide essential goods and services for social and economic activities (Juran, 2015; Chang, 2016). Therefore, resilience against infrastructure failure is essential for ensuring the health and safety of communities during and following a disaster (Schweikert et al., 2019). Moreover, resilient infrastructure is important for rapid response to recovery efforts. For example, even if roads are damaged but still serviceable, routes might become evacuation routes or even help in quick movement of SAR teams, other responders, and relief supplies (Oh et al., 2012). Utility and infrastructure damage can be for aboveground, underground, and on-ground utility and infrastructure components. These include systems such as roads, bridges, electricity, gas, and water, wastewater, and waste treatment facilities (Oh et al., 2012). Post-disaster, the repair of these systems to pre-disaster levels is very important for the livelihood of citizens and maintaining the economy of the region. For example, during the 2010 Haiti earthquake, the fatalities and loss of property increased significantly due to weak infrastructure systems (Oh et al., 2012). Damaged infrastructure impeded the relief effort of emergency-related agencies by delaying their ability to reach affected areas (Oh et al., 2012).

The type of natural disaster determines the kind of loss to utility and infrastructure. For instance, hydrometeorological disasters have a direct influence on the vulnerability of road infrastructure such as bridges, culverts, and floodways (Gajanayake et al., 2020). Destruction of such infrastructure can exacerbate the consequences of the disaster on both temporal and spatial scales. For example, the state of Queensland, Australia, was severely affected by flooding events in 2010–2011 and spent 6.8 billion Australian dollars rebuilding road infrastructure (Gajanayake et al., 2020).

According to the National Oceanic and Atmospheric Administration (NOAA), the United States has experienced 376 weather and climate disasters since 1980, where the total cost exceeds $2.65 trillion (NOAA, 2019). Many of these events impacted urbanized areas, and researchers and policymakers have called for efforts to make cities and interconnected urban regions more disaster resilient (Chang et al., 2014; Loken et al., 2020). Rural areas support a small population compared to urban areas, although they house the agriculture sector and critical transportation lines. Lack of economic diversity and geographical remoteness tends to make rural communities less resilient to disasters than urban regions (Loken et al., 2020). Differences between urban and rural characteristics subject them to different levels and types of utility and infrastructure losses. For example, after the 2015 twin Nepal earthquakes, urban and rural centers experienced disparity in aid provision in which remote districts felt isolated. This was not only due to greater populations in urban areas but also due to greater infrastructure losses.

Understanding how natural hazards impact society in terms of economic cost, recovery time, and damage to critical infrastructure is essential for developing robust approaches to increasing resilience (Schweikert et al., 2019). Particularly, identification of specific vulnerabilities at a pre-disaster stage allows for better communication, planning, and situation-based intervention. This is an important component of the recovery process and a critical opportunity to 'build back better' for the government.

Safety and Security during the Response Stage

Security is a crucial component during all stages of disaster response (O'Neill, 2005). Without appropriate security, no phase of disaster response can proceed safely or effectively. The first step in any disaster response is to ensure the security of survivors and first responders. Generally, security concerns are divided into scene and hospital concerns. At the scene of the event, local police normally provide security. Depending on the size of the event, as the event evolves, additional security may be required from the state or national forces, or even the military (O'Neill, 2005). Identification of authorized personnel is required to restrict the impacted area to trained personnel to reduce further injuries. Additionally, security personnel at the scene must be able to manage volunteers, traffic, and roadways so that emergency personnel and rescue vehicles can gain easy and organized access. In the United States, prior to the September 11, 2001 attacks, disaster policies and programs tended to be reactive and were concentrated on solving problems after the situation revealed them rather than focusing on proactive risk and vulnerability assessments. After 9/11, new laws such as the Homeland Security Act of 2002, creation of new agencies such as the Department of Homeland Security and state and local homeland security offices, and new frameworks for managing disasters and catastrophic events were established. For instance, the San Antonio Chapter of the American Red Cross serves 21 county areas and has several branches. The local chapter is now responsible for health and safety training, disaster response services, support services to military members and their families, and education programs that facilitate other programs in partnership with Red Crescent (Palomo-Gonzalez & Rahm, 2008).

Security at hospitals and healthcare institutions is also a security concern. Providing security in healthcare institutions is considered the responsibility of that institution (O'Neill, 2005). Therefore, the role of security officers needs to be clearly defined, especially in disaster situations which are different from normal circumstances. Since most local and regional disaster response teams rely on hospitals to provide security, security officers need to be trained, should know the plan, and should participate actively in disaster drills. This reiterates the importance of multi-agency cooperation in disaster response plans and the need to reevaluate these plans from time to time (O'Neill, 2005).

Institutionalized Factors Impacting Disaster Response

Gender

Natural disasters are gender neutral, but the impacts of the disasters are revealed differently across genders. Members of different genders face disproportionate levels of exposure and vulnerability to natural hazards, and these are influenced by gender relations and discrimination in each society (Juran, 2012). Gender inequality arises from the expected roles of men and women in a society, where women (generally and especially in low-income countries) are in a disadvantaged position due to restricted access to literacy, employment, and societal status; control over economic resources; and decision-making rights and opportunities. Thus, it is common for more women to die in a disaster. However, data also show that men account for 70% of flood-related deaths in Europe and the United States, although this can partly be explained due to the overrepresentation of men in SAR operations (Erman et al., 2021).

A gender-sensitive approach in disaster response has been lacking in disaster management planning. Understanding gendered evacuation patterns and behaviors can help in rescue operations and even emergency care. Gender-sensitive planning recognizes the different needs of men and women in disaster situations (Medury, 2008). Women assume gendered responsibilities in disaster situations, such as providing care to family members and arranging for food, water, and fuel. For instance, in Bangladesh, a country that experiences serial flooding, women are required to prepare and store food during pre-disaster days. Similarly, in Sri Lanka, women are regular managers of food and water and manage them during difficult conditions. However, it should not be assumed that only women are tasked with domestic and natural resource-related work. In the aftermath of floods in Punjab, Pakistan, male members managed rehabilitation work in agriculture and livestock caring (Medury, 2008). Women also contributed to and shared in the responsibility of recovery work by preparing mud bricks and performing construction work.

Economically, disasters also have different effects for men and women, with women at a disadvantage (Juran, 2012; Erman et al., 2021). For example, in less developed countries, agriculture is a large and important sector of the economy. Thus, female employment is often high in this sector. Post-disaster, domestic responsibilities increase and the burden often falls upon women, meaning that they miss out on incoming-generating opportunities. Moreover, lack of banking knowledge and access to financial services means that women's assets are less protected than men. A gender-sensitive approach, through strategies such as gender mainstreaming, helps to identify differential vulnerabilities, capacities, and coping strategies for both men and women in crisis (Medury, 2008).

Populations with Special Needs

Given the complex, multifaceted nature of disaster management, response, and recovery, it is not surprising that requirements of special needs populations are often overlooked. Special needs populations are often individuals with physical or mental limitations, sociocultural isolation, medical or chemical dependence, homeless, and the frail and elderly (Penner & Wachsmuth, 2008).

From the disaster response point of view, special needs populations should be aware and informed of resources in cases of disaster. Moreover, emergency warning systems should be designed for and adapted for visual, hearing, and cognitive impairments and be repeated multiple times and through different media (Penner & Wachsmuth, 2008). Given the diverse abilities and mobilities of special needs populations, clear and timely information must be provided on evacuation and modified forms of response assistance may be required. Knowledge about and access to the location of transportation, food, water, shelter, medical care, prescription drug replacements, and housing vouchers are crucial (Penner & Wachsmuth, 2008). From a planning perspective, relief and rescue workers must be trained and interagency coordination is required for evacuation, relocation, and sheltering special needs populations when a disaster occurs.

Class, Caste, Race, and Ethnicity

Possessing and acting with cross-cultural knowledge can lead to the development of more robust and effective disaster preparation procedures (Roth, 1970). The consequences of disasters are often influenced by social structures and hierarchies, the prevailing governance system, and the history of decades of institutional arrangements (Jha, 2015). This means that in most disaster situations, certain groups of society may be more vulnerable than others.

The relationship between class, caste, and economy is interwoven so intimately that it is impossible to understand it without thinking of mutual impact (Ray-Bennett, 2009). Though mostly outlawed, the caste system in India, Sri Lanka, and Nepal continues to be a barrier for economic prosperity and equality within the social structure of society (Rao, 2010). The caste system has become the basis of prejudiced and discrimination-filled systems, both formal and informal. While ostensibly outlawed, the caste system remains embedded in the fabric of South Asia today. For instance, after massive floods in the Kosi region of Bihar, India in 2008, it was reported that Dalit and low caste populations were excluded at the time of boarding boats during evacuation and rescue operations (Jha, 2015). In the relief camps, influential people from dominant castes managed to get a bigger share of the relief material. Meanwhile, outside of the relief camps, Dalits were denied the use of hand-pumps to obtain water and some upper caste individuals had tied barbed wires around the handpumps to restrict access (Jha, 2015). Similarly, after twin earthquakes in Nepal in 2015, field observations from disaster response and relief activities showed that more criticisms of lack of help were registered from Dalits compared to other castes.

Research shows that interactions between relief personnel and racial and ethnic groups may bring forth preexisting social problems and structural stratification (Fothergill et al., 1999). For example, after Hurricane Fredric in the United States, power to black areas was restored only after it was first restored in white areas. Additionally, African American communities received less emergency shelter, ice, food, and overall assistance (Fothergill et al., 1999). Another example is after the 1989 Loma Prieta earthquake in the United States, where economically well-off areas received more volunteers and visits from the mayor compared to economically disadvantaged communities. Moreover, Anglo volunteers made racist remarks in low-income emergency shelters, and the mayor paid no visits to such locations (Fothergill et al., 1999).

Geopolitics

Geopolitics in disaster response and recovery is often covered under disaster diplomacy. Disasters can provide a catalyst for transformation between geopolitical parties, typically with three outcomes: short term, long term, and the opposite desired effect (Maciver, 2012). However, there are sometimes barriers to the 'window of opportunity' to transform diplomatic relations, which can be donor bias, donor fatigue, both physical and psychic distance, and the 'CNN Effect' (where media sensationalizes a disaster for only a short time) (Maciver, 2012; Mittelman & Dow, 2018). Here, two examples are provided involving geopolitics that produced divergent outcomes, one between international governments and one based on an intra-national case.

The relationship between Turkey and Greece has been strained for a long time. Goodwill between the two nations was challenged when a war almost broke out in 1974 when Turkey invaded Cyprus. The contentious relationship between the countries began to change only after 1999 when an earthquake hit Turkey in August, and then in September Greece was hit by another earthquake. Quick disaster response from each of these countries to the other fostered goodwill among government leaders and the general population (Maciver, 2012). In this case, disaster diplomacy opened a way for better relations between the two countries.

A second example, one of intra-national governmental disaster diplomacy, is within the state of Sri Lanka. Sri Lanka has a history rife with tensions and civil violence that stretches back years due to colonial influences and, especially, the relationship between the Tamil population (Liberation Tigers of Tamil Eelam, also known as the Tamil Tigers) and the rest of the Sri Lankan people (Maciver, 2012). Devastating floods in 2003 resulted in the Tamil Tigers donating aid to affected population as a sign of goodwill. However, when the 2004 Indian Ocean tsunami hit Sri Lanka, the national government refused aid from the Tamil Tigers as well as restricted outside organizations' access to Tamil areas. These actions shattered the hope for peace and helped to solidify existing divisive forces in the country (Maciver, 2012). Hence, disaster diplomacy failed. Due to the complexities, one of the biggest potentials for disaster diplomacy lies in local or micro-level cases where personal interactions and rapport can take root.

Conclusion

Several actors are involved in the disaster response stage, from local to international and from public to private to humanitarian entities. Depending on the type, magnitude, and extent of the disaster, priorities of the 'to-do-list' will differ. Post-disaster response activities include containing further life loss, damage, and disruption; addressing survivors' immediate needs; making the transition to recovery easier; and building trust among disaster practitioners and the affected populations (Tierney & Oliver-Smith, 2012). Planning that takes gender dynamics, class, race, and cultural and social sensitivities into account can help make disaster response more efficient without exacerbating existing tensions. Knowledge presented in this chapter can help in the design of improved and more comprehensive disaster response strategies and plans.

References

Anaya-Arenas, A. M., Renaud, J., & Ruiz, A. (2014). Relief distribution networks: A systematic review. *Annals of Operations Research, 223*(1), 53–79. https://doi.org/10.1007/s10479-014-1581-y.

Brough, A. L., Morgan, B., & Rutty, G. N. (2015). The basics of disaster victim identification. *Journal of Forensic Radiology and Imaging, 3*(1), 29–37. https://doi.org/10.1016/j.jofri.2015.01.002.

Brown, C., & Milke, M. (2016). Recycling disaster waste: Feasibility, method and effectiveness. *Resources Conservation and Recycling, 106*, 21–32. https://doi.org/10.1016/j.resconrec.2015.10.021.

Catlett, C. L., Jenkins, J. L., & Millin, M. G. (2011). Role of emergency medical services in disaster response: Resource document for the National Association of EMS physicians position statement. *Prehospital Emergency Care, 15*(3), 420–425. https://doi.org/10.3109/10903127.2011.561401.

Chang, S. E. (2016). "Socioeconomic Impacts of Infrastructure Disruptions" in *Oxford Research Encyclopedia of Natural Hazard Science*, 1–26. New York. https://doi.org/10.1093/acrefore/9780199389407.013.66.

Chang, S. E., Mcdaniels, T., Fox, J., Dhariwal, R., & Longstaff, H. (2014). Toward disaster-resilient cities: Characterizing resilience of infrastructure systems with expert judgments. *Risk Analysis, 34*(3), 416–434. https://doi.org/10.1111/risa.12133.

Erman, A., De Vries Robbe, S. A., Thies, S. F., Kabir, K., & Maruo, M. (2021). *Gender Dimensions of Disaster Risk and Resilience*. Washington, DC: International Bank for Reconstruction and Development/The World Bank. https://doi.org/10.1596/35202.

Fikar, C., Gronalt, M., & Hirsch, P. (2016). A decision support system for coordinated disaster relief distribution. *Expert Systems with Applications, 57*, 104–116. https://doi.org/10.1016/j.eswa.2016.03.039.

Fothergill, A., DeRouen Darlington, J. A., & Maestas, E. G. M. (1999). Race, ethnicity and disasters in the United States: A review of the literature. *Disasters, 23*(2), 156–173. https://doi.org/10.1111/1467-7717.00111.

Gajanayake, A., Zhang, G., Khan, T., & Mohseni, H. (2020). Postdisaster impact assessment of road infrastructure: State-of-the-art review. *Agricultural Economics, 21*(1). https://doi.org/10.1061/(ASCE)NH.1527-6996.0000343.

Hamilton, A. R. L., Södergård, B., & Liverani, M. (2022). The role of emergency medical teams in disaster response: A summary of the literature. *Natural Hazards, 110*(3), 1417–1426. https://doi.org/10.1007/s11069-021-05031-x.

Jha, M. K. (2015). Liquid disaster and frigid response: Disaster and social inclusion. *International Social Work, 58*(5), 704–716.

Juran, L. (2012). The gendered nature of disasters: Women survivors in post-tsunami Tamil Nadu. *Indian Journal of Gender Studies, 19*(1), 1–29.

Juran, L. (2015). A framework for integrating water projects in disaster reconstruction processes in the international context. *Public Works Management & Policy*, *20*(4), 305–321.
Juran, L. (2018). "Disaster, Development, and Water: The Reconstruction and (Re)Fabrication of Hazardous Waterscapes in Post-Tsunami India" in *The Asian Tsunami and Post-Disaster Aid*, 79–99. Edited by Sunita Reddy. Singapore: Springer Nature.
Kim, J., Deshmukh, A., & Hastak, M. (2014). *Selecting a Temporary Debris Management Site for Effective Debris Removal*. May, 214–218. https://doi.org/10.5703/1288284315362.
Loken, A., Wittich, C. E., Brito, L., & Saifullah, M. K. (2020). Digital reconnaissance and performance assessment of rural infrastructure for 2018 natural hazards. *Journal of Performance of Constructed Facilities*, *34*(4), 1–14. https://doi.org/10.1061/(asce)cf.1943-5509.0001460.
Luther, L. (2008). *Disaster Debris Removal after Hurricane Katrina: Status and Associated Issues*. Washington, DC: Congressional Research Service. https://sgp.fas.org/crs/misc/RL33477.pdf.
Luther, L. (2011). *Disaster Debris Management: Requirements, Challenges, and Federal Agency Roles*. Washington, DC: Congressional Research Service, 16. https://www.humanitarianlibrary.org/sites/default/files/2020/08/R44941.pdf.
Maciver, C. (2012). Disaster diplomacy: A brief review. *SAI – From Vision to Results*, *12*, 1–13. http://sai-dc.com/wp-content/uploads/2014/07/Disaster-Diplomacy-FINAL-Lit-Review.pdf.
Maghfiroh, M. F. N., & Hanaoka, S. (2020). Multi-modal relief distribution model for disaster response operations. *Progress in Disaster Science*, *6*. https://doi.org/10.1016/j.pdisas.2020.100095.
Medury, U. (2008). "Toward Disaster Resilient Communities: A New Approach for South Asia and Africa" in *Disaster Management Handbook*. Edited by Jack Pinkowski. Boca Raton, FL: CRC Press Taylor & Francis Group.
Mittelman, R., & Dow, D. (2018). Biases in charitable giving to international humanitarian aid: The role of psychic distance. *Journal of Macromarketing*, *38*(4), 383–399. https://doi.org/10.1177/0276146718796899.
Morgan, O., & de Goyet, C. D. V. (2005). Dispelling disaster myths about dead bodies and disease: The role of scientific evidence and the media. *Revista Panamericana de Salud Publica/Pan American Journal of Public Health*, *18*(1), 33–36. https://doi.org/10.1590/S1020-49892005000600006.
Narayanan, R. G. L., & Ibe, O. (2012). A joint network for disaster recovery and search and rescue operations. *Computer Network*, *56*(14), 3347–3373. https://doi.org/10.1016/j.comnet.2012.05.012.
NOAA (National Oceanic and Atmospheric Administration). (2019). "US billion-dollar weather climate disaster." Accessed February 19, 2024. https://www.ncei.noaa.gov/access/billions/.
O'Neill, P. A. (2005). The ABC's of disaster response. *Scandinavian Journal of Surgery*, *17*, 259–266.
Oh, E. H., Deshmukh, A., & Hastak, M. (2012). Criticality assessment of lifeline infrastructure for enhancing disaster response. *Natural Hazards Review*, *14*(2), 98–107. https://doi.org/10.1061/(asce)nh.1527-6996.0000084.
Palomo-Gonzalez, S., & Rahm, D. (2008). "Hurricanes Katrina and Rita: The Critical Role of the Nonprofit Community in the San Antonio Disaster Response" in *Disaster Management Handbook*. Edited by Jack Pinkowski. CRC Press Taylor & Francis Group.
Paul, B. K. (2011). *Environmental Hazards and Disasters: Contexts, Perspectives and Management*. Chichester: West Sussex; Hoboken, NJ: Wiley-Blackwell.
Paul, B. K. (2021). *Disaster Deaths Trends, Causes and Determinants*. New York: Routledge.
Peacock, W. G., Nicole, D., Yang, Z., & Shannon, V. Z. (2018). "Post-Disaster Sheltering, Temporary Housing and Permanent Housing Recovery" in *Handbooks of Sociology and Social Research*, 569–594. https://doi.org/10.1007/978-3-319-63254-4_27.
Penner, S. J., & Wachsmuth, C. (2008). "Disaster Management and Populations with Special Needs" in *Disaster Management Handbook*. Edited by Jack Pinkowski. Boca Raton, FL: CRC Press Taylor & Francis Group.
Quarantelli, E. L. (1995). Patterns of sheltering and housing in US disasters. *Disaster Prevention and Management*, *4*(3), 43–53. https://doi.org/10.1108/09653569510088069.
Rao, J. (2010). The caste system: Effects on poverty in. *Global Majority E-Journal*, *1*(2), 97–106.
Ray-Bennett, N. S. (2009). The influence of caste, class and gender in surviving multiple disasters: A case study from Orissa, India. *Environmental Hazards*, *8*(1), 5–22. https://doi.org/10.3763/ehaz.2009.0001.
Roth, R. (1970). Cross-cultural perspectives on disaster response. *American Behavioral Scientist*, *13*(3), 440–451. https://doi.org/10.1177/000276427001300314.

Ruzek, J. I., Young, B. H., Cordova, M. J., & Flynn, B. W. (2004). Integration of disaster mental health services with emergency medicine. *Prehospital and Disaster Medicine*, *19*(1), 46–53. https://doi.org/10.1017/S1049023X00001473.

Schweikert, A., Nield, L., Otto, E., & Deinert, M. (2019). Resilience and critical power system infrastructure: Lessons learned from natural disasters and future research need. *Natural Hazards Review*, *20*(1). https://doi.org/10.1596/1813-9450-8900.

Suwalowska, H., Amara, F., Roberts, N., & Kingori, P. (2021). Ethical and sociocultural challenges in managing dead bodies during epidemics and natural disasters. *BMJ Global Health*, *6*(11), 1–10. https://doi.org/10.1136/bmjgh-2021-006345.

Tierney, K., & Oliver-Smith, A. (2012). Social dimensions of disaster recovery. *International Journal of Mass Emergencies & Disasters*, *30*(2), 123–146. https://doi.org/10.1177/028072701203000210.

Todd, D., & Todd, H. (2011). "Natural Disaster Response: Lessons from Evaluations of the World Bank and Others" in *Evaluation Brief; 16*. Washington, DC: World Bank. https://openknowledge.worldbank.org/entities/publication/c803240f-8236-5e09-abbf-c62e8f6ad81d.

Waharte, S., & Trigoni, N. (2010). Supporting search and rescue operations with UAVs. *Proceedings – EST 2010 – 2010 International Conference on Emerging Security Technologies, ROBOSEC 2010 – Robots and Security, LAB-RS 2010 – Learning and Adaptive Behavior in Robotic Systems*, 142–147. https://doi.org/10.1109/EST.2010.31.

Ward, P. J., Blauhut, V., Bloemendaal, N., Daniell, E. J., De Ruiter, C. M., Duncan, J. M., Emberson, R., Jenkins, F. S., Kirschbaum, D., Kunz, M., Mohr, S., Muis, S., Riddell, A. G., Schäfer, A., Stanley, T., Veldkamp, I. E. T., & Hessel, W. C. (2020). Review article: Natural hazard risk assessments at the global scale. *Natural Hazards and Earth System Sciences*, *20*(4), 1069–1096. https://doi.org/10.5194/nhess-20-1069-2020.

Zottarelli, L. K. (2008). Post-Hurricane Katrina employment recovery: The interaction of race and place. *Social Science Quarterly*, *89*(3), 592–607. https://doi.org/10.1111/j.1540-6237.2008.00550.x.

2
SEARCH AND RESCUE (SAR)

Oriane Longerstaey

Introduction

What comes to mind when thinking of search and rescue (SAR)? Some people may remember the massive search for victims after the terrorist attacks on September 11, 2001 in the United States, or the more recent search for the Thai soccer team stuck in an underwater cave. However, SAR operations can take on many forms and rely on the same fundamental principles. The most important aspect is preparedness and practice. This chapter will cover the fundamentals of SAR in the United States at the federal and local level, as well as involvement with international operations. This chapter also discusses different strategies for SAR by type of terrain and environment.

Fundamentals of Search and Rescue

National Search and Rescue Plan of the United States

The National Search and Rescue Plan of the United States is devised by the Federal government and coordinates SAR operations domestically and internationally. Its goal is to identify all available resources for SAR and distribute responsibilities among participants. Participants include the Department of Homeland Security, United States Coast Guard (USCG), and Federal Emergency Management Agency (FEMA), Department of Transportation (Federal Aviation Agency [FAA], Maritime Administration), Department of Defense, Department of Commerce, Department of the Interior, National Park Service, Department of State, Federal Communications Commission, and National Aeronautics and Space Agency.

The National Search and Rescue Plan covers many types of SAR services. It coordinates multiple stages of maritime, aeronautical, and land SAR operations. Agencies perform assessment of conditions, provision of assistance, transportation of survivors to safety, and recover human remains. It does not cover the recovery of military operations, salvage operations, or the provision of assistance during events that disrupt the government. The National Park Service has jurisdiction over services on their lands and should assist nearby authorities. FEMA manages the National Urban Search and Rescue (US&R) Response System. All participants are encouraged to work cooperatively as needed. The Plan designates which participants are responsible for international

DOI: 10.4324/9781003414834-5

representation. The following agencies are responsible for representing the United States in SAR operations to partner organizations:

- The USCG reports to the International Maritime Organization (IMO);
- The FAA reports to the International Civil Aviation Organization (ICAO);
- National Oceanographic and Atmospheric Agency (NOAA) reports to the International Cospas-Sarsat Programme; and
- With invitation from NOAA and ICAO, the United States Air Force (USAF) and USCG provide experts to join working groups.

The United States may join in SAR operations internationally, either for U.S. citizens or in support of its humanitarian goals. Participants support the SAR operations of other countries within international law. If assisting with foreign SAR, participants need to consider several factors such as expertise, legal authority, applicable procedures, collaboration with other participants, and confirmation that they are partnering with legitimate international partners.

The Plan describes arrangements for SAR partnerships between the United States and foreign countries. However, it is important that these arrangements align with SAR principles, U.S. law, and guidance from the Department of State. The Plan delineates Search and Rescue Regions (SRR) within the United States. These support communication, infrastructure, and efficient SAR operations. These cover territory over land, sea, and air. Each SRR will have an SAR coordinator (SC) responsible for services and the Rescue Coordination Center (RCC).

National Urban Search and Rescue Response System

The National US&R System was established after the Loma Prieta earthquake in 1989. It functions to organize local emergency services for federal response. There are 28 sponsoring agencies that maintain the task forces that can be deployed.

FEMA Teams

FEMA is responsible for the National US&R Response. There are 28 task forces that can be deployed to a disaster area at any time. The task forces need to be able to deploy within 6 hours. Teams may be readied in anticipation of a disaster and can be activated if local agencies are overwhelmed, if a state government requests assistance, or if the President activates disaster assistance.

Each Type 1 Task Force has at least 70 members with a breadth of expertise. One of these teams can be divided into two Type 3 Task Forces so they can work in 12-hour shifts non-stop or deployed independently if a Type 1 Task Force is not required. The Type 1 Task Force can perform SAR operations in structures and water environments, emergency medical care, needs assessment, management of utilities, evaluation of hazardous materials, and stabilization of damaged structures. At least 14 of the 70 members can work in water environments and may be deployed as a Water Rescue Mission Ready Package. This team includes different types of boats and personal protective equipment (PPE).

Types of Search and Rescue

Although the fundamentals of SAR remain the same, there are considerations and methods depending on the terrain and situation. Searcher and search area safety are paramount in all rescue operations. The first step of SAR is reconnaissance to understand the situation and collect data to support safe and effective decision-making. Additional considerations in any environment are establishing modes of communication, such as emergency signals in more chaotic situations, and safe locations to bring victims. Searchers should utilize appropriate tracking techniques for all search types, including looking for signs such as discarded clothing, broken branches, and footprints that can be compared to known facts about the missing person.

Ground Rescue

There are different search techniques for ground rescue. The first step is identifying the last place the individual was seen to create a maximum search area based on possible distance traveled on foot or by car (KEM, nd). Then, a containment team should be placed along the borders of this area in a way that maximizes visibility.

A *bastard search* will focus on searching areas the lost individuals usually frequent, relying on prior knowledge of their usual patterns. Key aspects to this type of search include having this prior knowledge and a coordinating speedy response, so it is important to gather information quickly and communicate efficiently. For example, interviewing parents and friends of the missing person can be crucial to identifying key information.

A *hasty search* relies on a well-trained team of searchers entering the search area and quickly working through obvious hazards to find missing or injured persons and identifying tracks or other clues. This search will move quickly and require good communication between team members.

A *grid search* is typically used after other methods have failed as it requires significant manpower. Unlike the bastard search or hasty search, a grid search is much slower. The search area is divided into a grid and searchers are each assigned to sections of the grid. Overlap of sections will improve coverage between searchers is an important consideration. Each sector will have its own probability of area (POA), which is how likely it is that the subject is in that area. This differs from the probability of detection (POD), which is the probability that a search team will find the subject if in that area.

A *choke point search* is a unique method that depends on the terrain of the search area. Searchers identify parts of the search area that all lead to a certain point. For example, rivers or cliffs that would make a crossing difficult and would lead a missing person to naturally avoid them and head to a different point. Searchers are stationed at that point.

Finally, a *track trap search* tries to capture evidence of human movement. These trap tracks can be natural or manmade. For example, searchers will look for footprints in sand either on naturally occurring sandy areas or intentionally place sand in areas that will later be checked.

Urban Rescue

Types of Urban Structures

Understanding types of construction is critical in urban SAR. There are six main types of buildings that SAR personnel should be familiar with when faced with structural collapses (FEMA, 2020). Light frame construction is most commonly seen in low- and mid-rise residential and commercial buildings. These structures are made of lumber and metal in repetitive framing structures. Heavy wall construction are typically larger buildings that have heavy, fire-resistant exterior walls made of stone or concrete (FEMA, 2020). Heavy floor construction is seen in mid- to high-rise buildings or industrial facilities where the floors are cast-in-place concrete. Steel frame construction as a rigid steel frame making up the roof and floors and used mostly in high-rise structures (Empact International, 2023). Pre-engineered buildings can be constructed of moth metal or concrete in modular systems.

Urban Search and Rescue Techniques

US&R operations begin with reconnaissance to evaluate the incident and determine resources needed. Teams assess structures for hazards and instability and determine one of two primary strategies to safely perform the rescue. Teams can perform a hasty search, described above, or a primary search, which is a rapid search of the structures most likely to have survivors. After the primary search, a slower, and more methodical secondary search will be performed to search every space in a structure.

For each of these strategies, hazards are marked for future teams to identify whether a structure is accessible and safe, significantly damaged, or not safe for search, using a standard marking strategy (FEMA, 2020). Marking is done using bright colors and in obvious locations. These markings should also indicate if a building is completely or incompletely assessed and communicates the status and number of victims and survivors in the structure.

Confined Space Rescue

Confined space rescue techniques may or may not include entry into the space. For example, rescues can be performed with ropes, pulley systems, and harnesses, limiting the needs for rescuers to enter the confined space, as these can be very dangerous. The first step is to identify possible hazards such as toxic gases, dangerous mechanisms, temperatures, chemicals (KOSFM, nd). This should guide the appropriate choice in PPE. Then, the team should contact the person(s) in the confined space. Understanding their physical and mental condition may be helpful in choosing a rescue technique depending on whether the victims can actively participate in rescue efforts. The primary tools used in rescue are ropes, harnesses, tripods, and winches. Knowing the capacity of each tool is critical.

A vertical space rescue should utilize point attachments for lifting devices. However, it is important to ensure that the anchor used to secure the equipment or rescue is structurally sound. The Occupational Safety and Health Administration guidelines recommend structural anchors such as beams, stairwells, or columns. Other team members should be collaborating to watch the rigging and rescuer for safety.

Trench Rescue

Trench rescue operations are a subset of confined space rescue, where a primary concern is additional collapse. There are many factors within and outside rescuer control that further destabilize a trench (City of Phoenix, nd). It is particularly important to establish a hazard zone and shut down nearby roads as vehicle vibration can cause more collapse. It is standard to place two ladders into the trench prior to shoring up the trench and creating a safe zone. Then, victims should be identified, packaged, and removed from the trench. It may be necessary to remove water, dirt, and other debris from the patients.

Maritime Rescue

Swift Water

Compared to other settings, the search phase in swift water SAR may be relatively simple. The principles are similar to others discussed earlier in the chapter. The first step is identifying the last place the missing person(s) was seen, the Point Last Seen (PLS), to define the search area (International Technical Rescue Association [ITRA], 2020). Then, rescuers can proceed with a hasty search, usually downstream of the PLS, relying on understanding of currents.

Swift water rescue can be very dangerous. Swift water is not limited to geographic areas, as flooding can create swift water anywhere and rescuers must be trained. Rescuer safety is paramount.

Rescuers should wear life jackets, helmets, and never be tied with ropes, only with quick-release systems (Hatt, 2018). The first steps include evaluating the water level, direction of movement, and potential hazards. One rescuer should be stationed upstream to identify hazards or potential victims for the downstream team.

If the victim can collaborate with rescuers, teams on shore will throw ropes toward swimmers, who can grab the ropes and be pulled back to shore. Ropes are usually stuffed into throw bags to make it easier to throw to victims. This rope can be coiled after first use to make a second throw. Otherwise, a rescuer is attached to a quick-release rope system and enters the water to grab the victim before both are pulled back to shore.

Another shore-based rescue involves placing a stabilization rope across the water so victims can grab the rope as they move downstream. Then snag lines are placed lower on the victims to perform the rescue (ITRA, 2020). There are other shore-based cinch systems that can be used depending on whether rescuers are all on one side of the water and if the victim is close to the shore.

Flat Water

Like in swift water rescue, the first step-in flat-water rescue is to determine the PLS and mark that location using static objects on shore. Then, place a marker buoy at that point and have human searchers look in shallow waters and boat-based searchers using sonar and probing tools in deeper waters (GFD, 2019). Shallow water searchers should work together and work in straight lines, sweeping their feet across the bottom (ARC, 2016). Submerged searches are limited to trained personnel for safety concerns. Once the victim is located, visual contact should be always maintained. Rescuers can also utilize boats to approach and rescue the victim.

Flat Ice

Before attempting any rescue, rescuers must consider the characteristics and strength of ice involved, as there will be variations within the same body of water. Conditions that affect ice strength and thickness are age (old ice is weaker than new ice), wind, snow, or slush, overlying water, daily temperatures, and movement of water between the ice. It is difficult to establish the strength of the ice solely on visual inspection. Rescuers must identify possible hazards and access to the area. Rescuers should balance risks to rescuers with the benefits of the rescue and determine acceptable levels of risk to perform the rescue (MDFS, nd).

Rescue teams assess how many victims are involved and what their physical and mental situations are. They should also mark the victims' locations using static objects in case the victim becomes submerged (Green, 2016). As seen in swift water rescue, initial strategies involve the victim in the rescue. If the victim can engage with rescuers, rescuers should instruct them in safe self-rescue, such as lying flat on the ice and moving slowly across the ice to maximize weight distribution while being aware of noises around them that could indicate the ice breaking. If this is unsuccessful, rescuers can throw different types of objects such as robes, ladders, poles with a flotation device at the end (MDFS, nd). Once the victim can hold the end of the tool, they can be pulled back to shore. Inflatable boats or small flat-bottom boats can help rescuers contact the victim. Rescuers can slide the boat over the ice, and if there is any breaking of the ice, they can enter the boat and make their way to the victim.

Finally, the last option is a go rescue. A rescuer will move over the ice with appropriate equipment and precautions, attached to a non-locking rope system. Proper equipment includes dry suits, life jackets, ice awls, and safety spikes. When the rescuer reaches the victim, the rescuer will tether themselves to the victim, and the shore-team will pull them both back to shore (MDFS, nd).

Mountain Rescue

Rescues in mountainous terrains have other unique methods and considerations. Rope rescues may be used to gain access to and remove patients from locations with difficult access. There has been an increase in avalanche accidents over the years, and most injuries and deaths involve those participating in recreational activities (Strapazzon et al., 2021). There is high risk of death when people are completely buried, primarily from asphyxia.

SAR operations in mountainous areas are well-suited to using mixed equipment, both land and air based, manned and unmanned. A common technique for mountain SAR is the probabilistic search technique, based on the probability of finding lost individuals within the overall search area. Certain groups have proposed creating a network combined of unmanned ground vehicles, human rescuers, short- and long-range unmanned air vehicles to hasten information acquisition and rescue.

Avalanche Operations

Avalanche rescue is a subset of mountain rescue. It is critical to locate lost individuals buried in snow quickly as many die within the first 30 minutes and often are not buried very deep (Bianco et al., 2021). Often, victims will wear beacons that send signals between 40 and 80 meters (Arvesen, 2021). Rescuers can use receiving devices to detect these signals, even if the victim is buried. When doing a beacon search, all rescuers should set their beacons to receive mode. The first step is a signal search during which rescuers should search the area moving no further away from each other than the range of the beacon. Once the beacon has been detected, rescuers move into a coarse search along a curved path until they are within a short distance of the beacon, using the arrows on the screen. When rescuers are within 3 meters of the beacon, they will transition to a fine search under the snow.

If there are multiple victims, it will be more difficult to identify individual beacons. For closely buried beacons, rescuers will initially use a standard quadrant search. Once a signal has been found, they can use the *circle technique*, walking in smaller and smaller circles until the beacons have been located. An alternative method is the *split point method*. When two beacon signals are detected at the same time, searchers will mark that point and work around it using distance reads for each beacon to differentiate between locations. The last alternative for many signals in a small area is the *micro search and strip method*. It is a fine search in which rescuers will move back and forth across the area where the beacon was detected in lines of no more than 3 to 4 meters (AAI, 2020).

Once the victim has been found, they must be extricated from the snow. Given the weight of snow, rescuers should use aluminum shovels. The snow should be thrown downhill, and if there are multiple rescuers, others can clear the excess snow. Try and clear the victim's face first. If victims are not wearing beacons, SAR becomes infinitely more challenging and the likelihood of rescue decreases significantly. Rescuers must then rely on visual inspection of the area, look for likely burial sites, and probe them individually (AAI, 2020).

A crucial part of avalanche SAR is having a medical team ready at the time of extrication given how critically ill these patients can be after burial. They will often be hypoxic, with elevated levels of carbon dioxide. This is worse in cases of wetter snow where oxygen movement is more difficult (Strapazzon et al., 2021). It is also important to consider traumatic injuries in these patients. Trained helicopter teams can expedite medical care and transfer to higher level facilities.

Additional Aids

K9 Teams

Canine search teams add a unique skillset to an SAR operation due to dogs' sense of smell. Dogs are an estimated 100–1,000 times more capable than humans of detecting odors (Cengiz, 2022). Dogs can identify human scents and aid with both survivor retrieval and human remain detection (FEMA Canine, 2023). Each Type 1 US&R task force has at least four canine teams. The human remains teams are brought in after the search for survivors has concluded.

Robotics and Drones

Drones, or unmanned aerial vehicles (UAV), are becoming integral to SAR operations, extending current capabilities. They can search difficult or more dangerous terrain quicker and utilize additional technology such as thermal imaging (DHS, 2020). The addition of artificial intelligence (AI) to these devices will assist with real-time detection and facilitate rescue coordination (Rizk et al., 2021).

In addition to UAVs, robots are playing a larger role in SAR. There are many different types of robots used in SAR, each specifically suited to a situation. For example, snake robots have multiple joints and can work in small spaces, carry different types of equipment, and move over different types of terrain, including underwater caves (Li et al., 2022). The addition of robots with specialized skills such as firefighting and bomb disposal could limit dangerous exposure for rescuers (Huffington Post, 2017).

Case Studies

2010 Lightning Strike in the Grand Tetons

On July 21, 2010, a lightning storm struck the Grand Teton in Wyoming and trapped 17 climbers in three different groups, leading to the largest rescue on that mountain (Teton County SAR, 2019). The majority of the climbers suffered some injuries, and one hiker died after a fall. Fortunately, the first group of climbers was able to reach the Jenny Lake Climbing Rangers at 12:25PM and started coordinating SAR operations (Hallberg, 2020). The two other groups eventually made contact as well. The locally contracted helicopter flew rescue personnel and equipment to the Lower Saddle of the mountain at 11,600 feet (National Parks Traveler, 2010). From that point, rangers deployed to find the hikers, provide emergency medical care, and facilitate evacuation. Rescuers were transported under the helicopters via the 'short-haul' method, where rangers hang off of ropes attached to the bottom of the helicopter. This effort required the coordination of roughly 90 rescue personnel from federal, state, and local agencies in the setting of additional storm activity (Hallberg, 2020). Sixteen hikers were successfully rescued off the mountain.

2018 Tham Luang Cave Rescue

On June 23, 2018, rising waters trapped 13 members of a soccer team—12 boys and their coach—inside an underwater cave of about 2 kilometers in length (Davies, 2021). Their families alerted authorities that night, who activated the Thai Navy Seals and national police among other rescue teams (BBC, 2018). Efforts were complicated by persistent rainfall, difficulty pumping out water from the cave, and lack of cave-diving experience by rescue teams. Specialized international rescuers arrived on June 28th and found the soccer team on July 2nd. However, it had taken very experienced divers to enter this space, so the next challenge was extracting 13 untrained individuals before they ran out of air. One of the rescuers, Dr. Richard Harris, an anesthesiologist, recommended sedating the children so that the divers could move them through the cave more easily (Davies, 2021). The divers had to be trained to inject the medication and monitor them through the dive. Although the rescue efforts proved successful with all 13 team members being rescued, one retired Thai Navy diver, Saman Kunam, died during rescue operations (TIME, 2022). This case illustrates key concepts in SAR operations: coordinating multiple agencies (national and international), adapting to difficult environments, utilizing specialized rescuers, and patience. The rescue took two weeks during which teams practiced various methods to extract the boys and their coach, troubleshot falling oxygen levels, and addressed nutritional needs and other potential adversities (BBC, 2018).

Conclusion

The keys to a successful SAR operation are preparedness, training, and effective communication. Organizations must practice their skills in coordination with other groups they may need to coordinate with at the local, state, and/or federal level. Events that require SAR are tense situations. Teams must be equipped to rely on their training to minimize the effects of stress and uncertainty. The principles remain the same regardless of terrain or environment, and the priority remains the safety of rescuers, as adding victims to the search will complicate efforts. Rescue plans and teams should identify particular skills that may be needed, and potential aids such as drones, robots, or K9.

References

"*11 Robotic Applications for Search and Rescue*." HuffPost Contributor, 2017. https://www.huffpost.com/entry/11-robotic-applications-for-search-and-rescue_b_5a173c9ae4b0bf1467a845c4. Accessed 20 Feb. 2024.

American Avalanche Institute (AAI). "Avalanche Rescue." 2020. https://www.americanavalancheinstitute.com/.

American Red Cross (ARC). "Water Rescue Skills." 2016. www.nps.gov/subjects/aviation/upload/nps_short-haul_ops_plan.pdf.

Arvesen, Amelia. "Know How to Use Your Avalanche Transceiver." *Outside Online*, 2 Mar. 2021. https://www.outsideonline.com/outdoor-gear/tools/know-how-use-your-avalanche-transceiver/

Bianco, Giulio Maria, Romeo Giuliano, Gaetano Marrocco, and Franco Mazzenga. "LoRa System for Search and Rescue: Path-Loss Models and Procedures in Mountain Scenarios." *IEEE Internet of Things Journal*, vol. 8, no 3, 2021, pp. 1985–99. https://doi.org/10.1109/JIOT.2020.3017044

Cengiz, Sevil. "Possible Hazard and Risks that Search and Rescue (SAR) Dogs May Face in CBRN Incidents." *Gümüşhane University Journal of Health Sciences*, vol. 11, no. 1, 2022, pp. 300–307. https://doi.org/10.37989/gumussagbil.1003880.

City of Phoenix. "Trench Rescue Operations." nd. https://www.phoenix.gov/firesite/Documents/074768.pdf. Accessed 26 Nov. 2023.

Davies, Dave. "Documentary Follows the Divers Who Risked It All in the Thailand Cave Rescue." *NPR*, 11 Oct. 2021. https://www.npr.org/2021/10/11/1043363760/thailand-cave-rescue-documentary-film

Department of Homeland Security (DHS). "Snapshot: First Responders Assess Drones for Search and Rescue Missions." 2020. https://www.dhs.gov/science-and-technology/news/2020/04/02/snapshot-first-responders-assess-drones-search-and-rescue-missions

Empact International. "Urban Search & Rescue: Understanding the 5 Building Classifications." 2023. https://www.empact.ngo/post/usar-buildingtypes
Federal Emergency Management Agency (FEMA). "Urban Search and Rescue." 2020. https://www.fema.gov/emergency-managers/national-preparedness/frameworks/urban-search-rescue
FEMA Canine. "National US&R Response System." 2023. https://femacanine.org/
Gettysburg Fire Department. "Water Rescue Manual." 2019. https://www.gettysburgfd.com/files/doc/6/GFD%20Water%20Rescue%20Manual.pdf
Green, Glen. "Assisted Ice Rescue Techniques." 2016. https://www.philacanoe.org/resources/Documents/cold%20water%20safety/Assisted%20Ice%20Rescue%20Techniques.pdf
Hallberg, Tom. "10 Years after Traumatic Lightning Storm, Walker Returns to Grand." *Jackson Hole News & Guide*, 26 Aug. 2020. https://www.jhnewsandguide.com/news/features/10-years-after-traumatic-lightning-storm-walker-returns-to-grand/article_a805885a-2797-5f99-93c0-7ea5043020fa.html
Hatt, Kerri. "Swift Water Rescue Safety and Techniques." *EMS1*, 21 Feb. 2018. https://www.ems1.com/rescue/articles/swift-water-rescue-safety-and-techniques-9a4W6uRFLYlzgkzv/
International Technical Rescue Association (ITRA). "Swiftwater Rescue Course Manual." 2020. https://itra.international/wp-content/uploads/2020/05/SwiftwaterRescueKauffman1807.pdf
Kansas Office of the State Fire Marshall. "Confined Space Rescue Awareness." nd. https://www.firemarshal.ks.gov/DocumentCenter/View/169/Confined-Space-Rescue-Awareness-PDF. Accessed 26 Nov. 2023.
Kentucky Emergency Management. "Search and Rescue Field Search Methods." nd. https://kyem.ky.gov/Who%20We%20Are/Documents/SAR%20Field%20Search%20Methods.pdf. Accessed 26 Nov. 2023.
Li, Dongfang, et al. "Snake Robots Play an Important Role in Social Services and Military Needs." *The Innovation*, vol. 3, no. 6, Oct. 2022, p. 100333. *PubMed Central*. https://doi.org/10.1016/j.xinn.2022.100333.
One Dead, 16 Injured Following Lightning From Thunderstorm in Grand Teton National Park. 2010. https://www.nationalparkstraveler.org/2010/07/one-dead-16-injured-following-lightning-thunderstorm-grand-teton-national-park6291. Accessed 20 Feb. 2024.
Rizk, Mostafa., Slim, Fatima., Baghdadi, Amer., & Diguet, Jean-Philippe. (2022, September). Towards real-time human detection in maritime environment using embedded deep learning. In *International Conference on System-Integrated Intelligence* (pp. 583–593). Cham: Springer International Publishing.
Giacomo Strapazzon, Hannes Gatterer, Marika Falla, Tomas Dal Cappello, Sandro Malacrida, Rachel Turner, Kai Schenk, Peter Paal, Markus Falk, Jürg Schweizer, Hermann Brugger (2021). Hypoxia and hypercapnia effects on cerebral oxygen saturation in avalanche burial: A pilot human experimental study. *Resuscitation*, vol. 158, p. 175–182.
"'Thai Cave Rescue' and the True Story Filmmakers Can't Stop Retelling." *TIME*, 22 Sept. 2022. https://time.com/6215763/thai-cave-rescue-netflix-true-story/
"'The Fine Line' Podcast Relives the Largest Rescue Effort Ever on the Grand Teton." *Teton County Search and Rescue*, 16 Aug. 2019. http://www.tetoncountysar.org/latest-news/2019/8/15/the-fine-line-podcast-relives-the-largest-rescue-effort-ever-on-the-grand-teton
The Full Story of Thailand's Extraordinary Cave Rescue. 13 July 2018. www.bbc.com, https://www.bbc.com/news/world-asia-44791998

3
EMERGENCY MEDICAL CARE

Emily Kaplan and Cara Taubman

Introduction

The Emergency Event Database, a joint initiative between the Centre for Research on the Epidemiology of Disasters (CRED) and the World Health Organization (WHO), documented 387 catastrophic events including manmade and natural disasters worldwide in 2022—almost one per day globally (CRED, 2023). These disasters resulted in 30,704 deaths (CRED, 2023). During disasters, coordinated systems provide a wide range of emergency medical care services including: medical oversight; immediate first aid; land, air, and sea medical transport; forward care units such as mobile trauma units; specialty medical care services mobile health units for long duration incidents; and care coordination. A broad knowledge of disasters, the role of emergency management in disaster medical care, and a 'whole community' approach are necessary to purposefully approach emergency medical care in disaster scenarios. Given this background, this chapter:

- Introduces organizations providing medical care responders both in the United States and internationally,
- Describes coordination of medical care at the local and national levels to ensure care delivery and medical supply chain management in the United States,
- Outlines medical care leadership and oversight both organizationally and from a clinical perspective,
- Provides an overview of medical care team capacities for different types of medical care response and clinical care provided in disasters,
- Identifies organizations in the United States responsible for coordinating of medical care at the national and regional level,
- Describes standards of medical care under normal and disaster conditions,
- Reviews special considerations during disaster including mental health, pharmaceutical, medical device considerations, and public health.

Phases of Disaster Response

Emergency medical care in disasters is integrated in each phase of disaster response. Preparedness activities for emergency medical care are embedded in risk assessment, planning, training and education, stockpiling, community engagement, and interagency coordination. Response activities

 DOI: 10.4324/9781003414834-6

include immediate medical care, triage, and prioritization as well as the expansion of traditional medical care activities to novel environments. During recovery, emergency medical care ensures continued treatment of affected populations, manages chronic conditions, and addresses health issues arising directly from the disaster. In mitigation, responders and clinicians ensure appropriate community health planning to reduce the long-term health-related burdens of the disaster and enhance the resilience of medical care delivery systems.

Disasters

There exist multiple ways to define disaster and multiple terms to describe or denote disaster. For the purpose of this chapter, a useful definition of disaster is provided by CRED: 'A situation or event which overwhelms local capacity, necessitating a request for national or international level for assistance; an unforeseen event that causes great damage, destruction and human suffering' (CRED, 2023). A range of medical conditions from acute injury to chronic illnesses may occur during and/or after a disaster event.

Whole Communities in Emergency Medical Care

The 'whole community' is involved in disaster medical care and therefore a non-exhaustive list of those involved in medical care is lay responders, first responders, public safety personnel, and licensed clinicians. Federal, state, local, non-governmental organizations (NGOs), health care businesses, voluntary associations, and international organizations are also involved (US Department of Homeland Security, 2011). Each responder is integrated into emergency disaster care. National and international organizations (public, private, and NGOs) work jointly to provide medical care, supply chain services, data informatics, and finance systems to support the expansion of care services.

Domestic vs. International Availability of Emergency Medical Care

Emergency medical care access varies globally. Countries such as the United States, France, England, and Australia have emergency medical systems with robust resources, including ambulance and hospital systems, capable of providing medical care domestically and internationally under normal and disaster circumstances. Emergency care access is not necessarily universal. Even within nations with strong medical systems, access in the face of disaster may require national responses. In the United States, 4.5 million people live in areas without ready access to ambulance services, particularly in rural areas (Jonk, 2023). In India, while hospitals may have sufficient workforces, most hospitals lack clinicians dedicated to Emergency Departments or with a specialization in Emergency Medicine (Misra et al., 2023). Furthermore, nearly half the world's population lacks access to essential health care services (World Health Organization (WHO), 2017b).

Medical Care Emergency Management

Under normal conditions, hospitals and state, federal, and emergency medical services engage in the routine daily operations of their respective services. The services are bound by federal, state, and local regulations provided by entity-specific mission objectives, be it providing mobile integrated health care (MIH), patient transportation among facilities, routine response within a 9-1-1 system, primary care, surgical services, or tactical combat care. These services have memorandums of understanding (MOU) with a variety of entities that agree to participate, or are required by policy or legislation to participate, in the event of disaster.

During disaster in the United States, public, private, and NGO actors work with the National Incident Management System (ICS). It is a management system in which each service is responsible for their respective area and expertise within a unified command (UC) structure. UC ensures coordination, communication, and cooperation to carry out efficient response. Internationally, nation-specific coordination systems are adopted (i.e., the Coordinated Incident Management System (CIMS) as used in New Zealand) (NEMA (National Emergency Management Agency), 2020). Among nations, the United Nations (UN) uses the 'Cluster System' to coordinate UN and relief organizations, including health, via similar mechanisms to the United States such as the On-Site Operations Coordination Center (OSOCC) (UNOCHA (United Nations, Office for Coordinated Affairs), n.d.).

Medical Care Leadership

At the federal level in the United States, the Secretary of United States Health and Human Services (HHS) leads all responses to public health emergencies (i.e., the disaster setting within the National Response Framework (NRF)). The NRF includes all Emergency Support Functions (ESF), coordinating structures, support annexes, and descriptions of how support is organized. Within each, departments utilize the ICS framework within unified command. At the state level, state Departments of Health (DOH) are responsible for public health and in leadership and coordination for public health emergencies. Secretaries or Commissioners of Departments of Health may designate state health officers, often physicians, to be the primary individuals responsible for health policy and state public health programs. Internationally, the Executive Director of the WHO Health Emergencies Programme leads response to public health emergencies and outbreaks.

Medical Control

Medical control refers to the oversight of operations, systems, and clinicians by qualified physicians. Medical control is involved in planning, response, and recovery. Medical control also provides real-time direction for clinicians as well as indirect guidance through the creation of policies and protocols that govern clinical activities and standards of practice. During normal operations, medical control is provided by physicians dedicated to the operations of respective organizations (Baker and Cole, 2022). During disaster, medical control may occur under federal or state operational commands.

Coordination of Medical Care

State, national, and international disaster medical care response all require concerted coordination. In the United States, the Secretary of HHS coordinates national public health response and recovery actions through the Office of the Assistant Secretary for Preparedness and Response (ASPR) consistent with internal MOUs, policies, and procedures under ESF #8 (US Department of Homeland Security, 2008) (see Figure 3.1). It includes the structure for coordination interagency support for public health, medical, mental health services, and mass fatality management. HHS leverages the resources of multiple offices to respond to health-related emergencies including but not limited to the Centers for Disease Control and Prevention (CDC), Agency for Toxic Substances and Disease Registry (ATSDR), Food and Drug Administration (FDA), and Health Resources and Services Administration (HRSA). Under ASPR, the Technical Resource

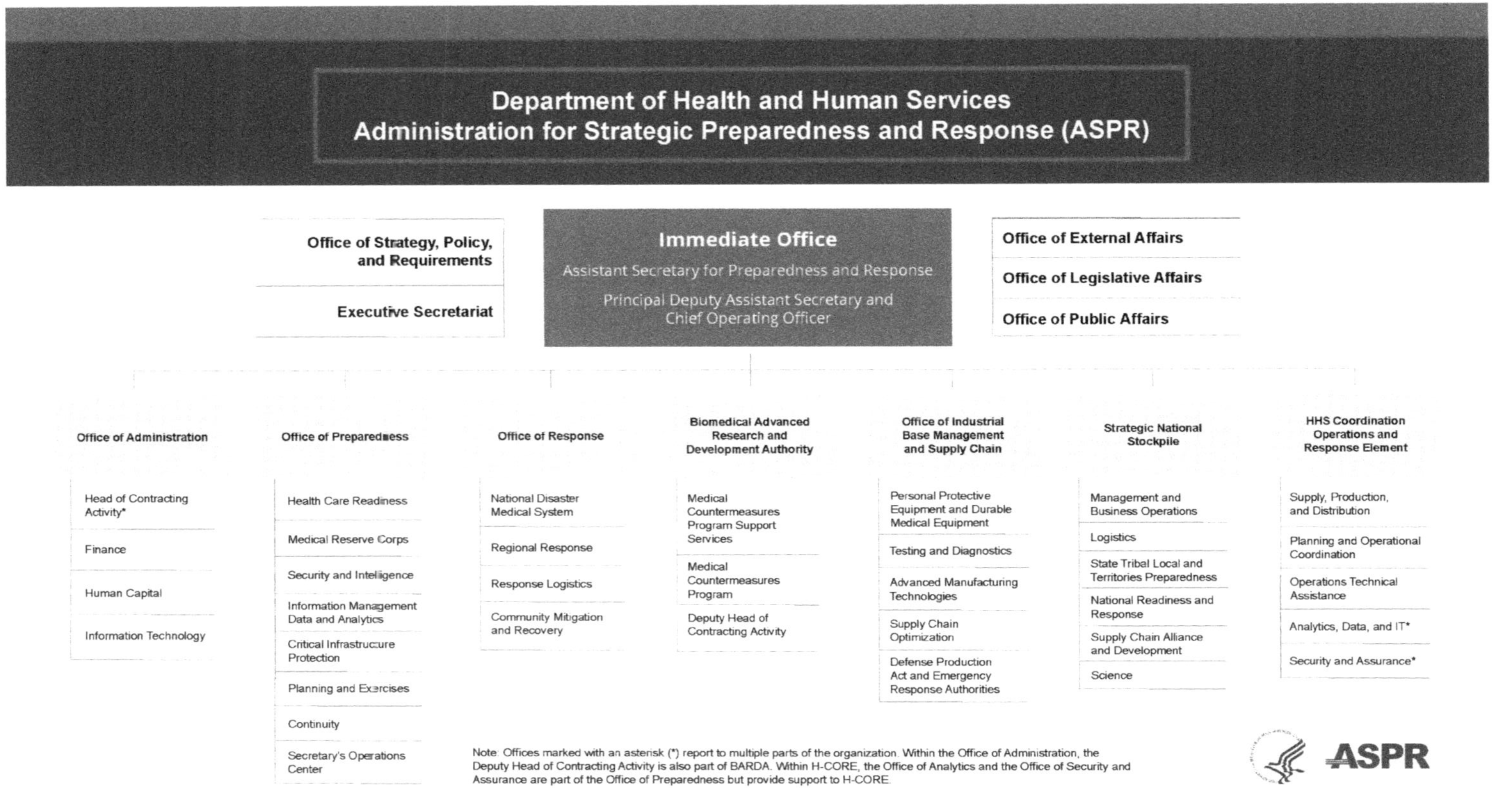

Figure 3.1 Administrative chart of Administration for Strategic Preparedness and Response (ASPR) of the US Department of Health and Human Services (HSS)

Source: Department of Health and Humar Services Administration for Strategic Preparedness and Response (ASPR), https://www.hhs.gov/about/agencies/orgchart/aspr/index.html

Center and Information Exchange (ASPR-TRACIE) is an online information and technical assistance hub for disaster response participants. Participation in and use of TRACIE ensures the sharing of current information for alignment of services (HHS, *ASPR-TRACIE*, n.d.). Within the Department of Health and Human Services, the Surgeon General oversees the US Public Health Commissioned Corps (USPHS). USPHS comprises health professionals representing a wide variety of fields from emergency medicine to virology. In disasters, USPHS professionals may be embedded in response teams.

In the United States, regional partnerships are essential to provide medical and health care across a 3.79 million square mile country. The Regional Disaster Health Response System (RHRDS) is an ASPR pilot program among four regions. RHRDS partnerships build on care coalitions and well-established trauma systems to expand access to specialty clinical care and medical care (HHS, *Regional Disaster*, n.d.).

National Disaster Medical Systems

In the United States, the National Disaster Medical System (NDMS), a response operations system within ASPR, expands capacities at a national level to ensure the provision of personnel, medical supplies, and national coordination of hospitals. NDMS resources may be requested on a state's request. NDMS coordinated operations are facilitated by the 'Health Care Facility/Partner Memorandum of Agreement for Definitive Medical Care,' also known as the 'Four Partners Agreement.' The Four Partners Agreement is among: (1) HHS; (2) Department of Defense (DoD); (3) Department of Homeland Security (DHS); and (4) the Department of Veterans Affairs. The agreement ensures coordination among the services in a public health emergency and to be present at locations when those areas are at risk of a public health emergency. Under Emergency Support Function (ESF) #8, within HHS, the Emergency Management Group (EMG) coordinates response by ASPR along with other agencies. Joint Special Operations Command (JSOC), a joint unit subordinate to the United States Special Operations Command (USSCOM), provides military special operations support such as medical training and response (US Department of Defense, n.d.).

Internationally, disaster management systems, including medical systems, are constructed for nations and among nations. The European Union, for example, uses the EU Civil Protection Mechanism which brings together the resources of EU member and participating states. Within each participating or member state, state response plans are prepared including for medical response. For example, the Danish Emergency Management Agency is responsible for advanced first aid emergency medical care. The UN coordinates response under the Emergency Response Framework to deploy Emergency Medical Teams (WHO, 2017a).

International Medical Care Teams and Support

Both government agencies and NGOs support emergency care operations domestically and internationally. NGOs coordinate, or work in partnership with, several disaster relief organizations including the WHO and the UN. Emergency Medical Teams participating with WHO are classified according to their mobility and ability to provide specialized care (World Health Organization (WHO), n.d.):

- Type 1 Mobile: Teams that provide acute trauma and medical care, refer patients to additional services, and provide community-based primary care. Daylight hours only. The team is mobile and works in various environments.

- Type 1 Fixed: Teams that provide acute trauma and medical care, referrals, and primary care in a set, fixed, outpatient facility.
- Type 3 Inpatient Surgical Emergency Care: Teams that provide acute trauma and medical care, referrals, and primary care in a set, fixed, outpatient facility. Type three teams are capable of complex care and intensive care.
- Specialized Care Teams: Provide specialty care services such as surgery, maternal and child care, outbreak, and interhospital technical support.

During some disasters, the UN Cluster System is employed. The Cluster System has been used since 2005 and was developed for non-refugee-based humanitarian emergencies and creates a system of accountability for services delivered. UN agencies, NGOs, and other international organizations are organized within 11 clusters: Health, Water Sanitation and Hygiene, Food Security, Logistics, Nutrition, Protection, Shelter, Camp Coordination, Early Recovery, Education, and Emergency Telecommunications.

A non-exhaustive list of NGO medical care organizations and governmental medical organizations is provided in Tables 3.1 and 3.2, respectively. For organizations that maintain an Emergency Medical Team with the UN, the team type(s) is included.

Table 3.1 Non-Governmental Organizations (NGOs) Providing Medical Care, Location, Types of Care Provided, and WHO Emergency Medical Team Types

Medical Care Agency	*Country*	*Activities*
AmeriCares	United States	Prestaging of medical supplies and providing medical response, long-term medical care activities, and support during recovery. WHO Type 1 provider.
Aspen Medical	Australia	Specialized care including surgical and outbreak. Air ambulance. Technical hospital support and medical supply. WHO Specialized Care (Aspen Medical).
Doctors Without Borders/ Médecins Sans Frontières (MSF)	Multiple MSF associations registered by country	Provides doctors, nurses, and technical expertise in various specialties, including pediatrics, surgery, and logistics worldwide (MSF). Six operational centers are located in Belgium, France, Netherlands, Spain, and Switzerland, with other offices worldwide.
Heart to Heart	United States	Humanitarian aid including medical and support teams, medicine, and hygiene kits (Heart to Heart). WHO Type 1 Mobile provider.
International Medical Corps	United States	Emergency medical care including emergent response and specialty care services such as maternal and newborn health. Services available during response and recovery (International Medical Corps.). WHO Type 1 Fixed and Mobile.

(Continued)

Table 3.1 (Continued)

Medical Care Agency	*Country*	*Activities*
International Federation of Red Cross and Red Crescent Societies (IFRC)	Multiple associations registered by country	Provides first aid, first aid training, and humanitarian relief in more than 190 countries. Deploy health personnel including physicians, nurses, and paramedical professionals during response and recovery including actives such as psychological first and dignified burial (IFRC).
MedShare	United States	Provides infectious disease control and prevention specialties and material, maternal and child health services, biomedical equipment training and repair, and primary care. Medical support volunteers. Technical support in health system administration (MedShare).
Médecins du Monde (Doctors of the World)	Multiple associations registered by country	Supports humanitarian crises and vulnerable populations with a focus on human rights. Active in approximately 30 countries providing physicians, nurses, psychological support, and medical support services (Doctors of the World).
Medical Teams International	Multiple associations registered by country	Faith-based organization providing hygiene and health kits, medical professionals, crisis health, and mental health care (Medical Teams International).
Mercy Malaysia	Malaysia	Provides medical teams, medical care operational support, and humanitarian care in Malaysia. WHO Type 1 Fixed.
Polish Center for International Aid (PCPM)	Poland	Provides medical support, including paramedical system support, and humanitarian aid (PCPM). WHO Type 1 Fixed.
Save the Children	Multiple associations registered by country	Provides child-focused global humanitarian response. Creates safe spaces for children and orphans in disasters. Frontline workers may engage in or support vaccination services, train health care workers, and establish child-friendly mobile medical care facilities (Save the Children).
Sauveters Sans Frontiers (SSF)	Multiple associations registered by country	Supports more than 800 medical volunteer physicians, nurses, psychologists, and paramedical personnel internationally for disaster response (SSF). WHO Type 1 Mobile.
Team Rubicon	United States	Veteran-led organization providing incident management, emergency medicine, and hazard mitigation throughout the disaster cycle (Team Rubicon). WHO Type 1 Mobile.
UK-MED	United Kingdom of Great Britain and Northern Ireland	Provides medical humanitarian aid including training and preparedness activities and medical response (UK-MED). WHO Type 1 Fixed.

Source: n.d. https://www.who.int/emergencies/partners/emergency-medical-teams/emt-global-classified-teams.

Table 3.2 Governmental Organizations Providing Medical Care, Location, and WHO Emergency Medical Team Types

Governmental Organization	*Country*	*Medical Team Types*
AUSMAT (NCCTRC)	Australia	WHO Type 2, Type 1, Type 1 Mobile and Specialized Care (Surgical, Outbreak)
All-Russian Centre for Disaster Medicine (Zaschita)	Russian Federation	WHO Type 2
Barbados Defense Force (BDF)	Barbados	WHO Type 1 Fixed
Central Airmobile Rescue Team of EMERCOM	Russian Federation	WHO Type 2
Caja Costarricense Seguro Social (CCSS)	Costa Rica	WHO Type 1 Fixed
EMT Regione Piermont	Italy	WHO Type 2
ESCRIM	France	WHO Type 2
FEMAT	Fiji	WHO Type 1 Fixed
Guangdong Hospital	China	WHO Type 2
INEM	Portugal	WHO Type 1 Fixed
Israeli Defense Force (IDF)	Israel	WHO Type 3
Japan Disaster Relief Team JDRT (JICA)	Japan	WHO Type 2; Type 1 Fixed and Mobile; Specialized Care (Surgical and Hemodialysis)
Korea Disaster Relief	Korea	WHO Team Type 1
Macao Health Bureau	China	WHO Type 1 Fixed
Ministerio Salud Pública Ecuador	Ecuador	Type 2 (2); Specialized Care (Surgical)
New Zealand Medical Assistance Team (NZMAT)	New Zealand	WHO Type 1; Fixed and Mobile
Norwegian EMT	Norway	WHO Type 1 Fixed and Mobile
SAMUR Protección Civil	Spain	WHO Type 1 Fixed
Shanghai East Hospital	China	WHO Type 2
Sichuan	China	WHO Type 3
Spanish Technical Aid Response Team (START)—Spanish Agency for International Development (AECID)	Spain	WHO Type 2
Swiss Humanitarian Aid	Switzerland	WHO Specialized Care (Reproductive, Maternal, Newborn, and Child Health)
Thailand EMT	Thailand	WHO Type 1 Fixed
Tianjin	China	WHO Type 2
United Kingdom of Great Britain and Northern Ireland EMT	United Kingdom Great Britain and Northern Ireland	WHO Type 2
UME	Turkey	WHO Type 2

Source: Adapted with permission from World Health Organization (WHO), n.d. https://www.who.int/emergencies/partners/emergency-medical-teams/emt-global-classified-teams.

US-Based Medical Care Teams

NDMS medical teams provide disaster medical assistance (DMAT), trauma and critical care teams, victim identification, mortuary, and veterinary response teams. DMATs and their respective team members provide prehospital triage, general emergency medical care, general primary care, hospital decompression, support patient movement, and mass prophylaxis (e.g. vaccinations). Trauma and Critical Care Teams (TCCT) provide operative and emergency care and are deployed as scalable teams. These teams provide critical, operative, and emergency care and may augment or establish medical facilities (HHS, *Trauma and Critical Care Teams*, n.d.). To further increase capacity, in development is the National Emergency Telemedicine Network, which will provide clinical oversight and care management during widespread disasters and facilitate access to care in remote or austere environments.

US Armed Services Support in Disaster Medical Care

USSCOM is responsible for providing coherent and coordinated military special operations, including the provision of medical support internationally. The US Army, Navy, and Air Force respectively maintain medical commands responsible for preparedness, training, and response. Specialties such as dentistry, nursing, paramedical professions, staff corps of scientists, and researchers as well as hospital administrators are represented in respective commands. The US Armed Services provide a wide range of medical care and support globally, including but not limited to medical evacuation; field hospitals and hospital ships; emergency medical treatment including doctors, nurses, and medics; logistical support for transportation and distribution of supplies; public health support augmenting capacity of public health agencies and services; and engineering support to repair or rebuild health care facilities and infrastructure in affected areas.

Under normal conditions, the Posse Comitatus Act (1878) generally restricts actions of the US Armed Services (i.e., Army, Navy, Air Force, and Marine Corps) from engaging in domestic law enforcement activities. In disasters, several federal statutes, such as the National Defense Authorization Act, enable the DoD to provide military medical personnel and support domestically and to rapidly scale aid. The Coast Guard, under the Department of Homeland Security, may, for example, be moved under the command of the US Navy to augment and coordinate services. The John Warner National Defense Authorization Act (2007) allows the President to use the military to aid in responding to incidents involving weapons of mass destruction (WMD), including chemical, biological, and nuclear (CBRN) events. The act, passed following the September 11th terrorist attacks on the World Trade Centers, grants authority for the military to provide support to civilian authorities.

States may use the National Guard, a reserve component of the US Armed Forces, to provide medical care and medical infrastructure due to their dual role as both state and federal military forces. As the Commander-In-Chief of the National Guard unit with a state, a state governor may deploy the National Guard for state-specific missions including public health emergencies. When the National Guard is federalized by an act of Congress, the National Guard is called into service for national defense purposes augmenting medical care resources.

Interstate Coordination of Medical Care

Within nations, medical care coordination across states and localities is necessary. In the United States, the Emergency Management Assistance Contract (EMAC) aids across states and territories when a disaster is declared by a state or territory. Utilization of National Guard resources also occurs under EMAC. Interstate assistance may include provision of ambulances and medical supplies (EMAC (Emergency Management Association Compact), 2023).

Hospitals and Health Care Coalitions

In the United States, ASPR maintains the Hospital Preparedness Program (HPP). The HPP provides leadership and funding in preparedness and response medical care. Funding to recipients is managed through cooperative agreements (CoAg). Recipients of funding under a CoAg agreement must adhere to the strategies, activities, and fiscal requirements of the HPP. The HPP program areas include but are not limited to interoperable communications systems, bed tracing, mass fatality management, biosurveillance, medical surge planning, trauma care during medical surges, continuity of operations planning, and planning for and coordinating health care evacuation and response (Assistant Secretary for Preparedness and Response (ASPR), 2016). HPP supports the development and sustainability of 326 Health Care Coalitions (HCCs) comprised of 42,128 members across the nation (Assistant Secretary for Preparedness and Response (ASPR), 2016). During the COVID-19 pandemic, for example, hospitals worked with coalition partners to increase the bed capacity. An HCC is a group of individual health care and response organizations (e.g., hospitals and EMS services) in a delineated geographic location. HHCs have a variety of governance structures, however, all engage community partners and stakeholders to ensure a state of readiness and coordination in response.

Patient Tracking Systems

On October 29, 2012, Hurricane Sandy, a Category 1 hurricane, struck the northeast coast of the United States resulting in a storm surge and widespread damage in New York and New Jersey. More than 6,400 patients were evacuated. In New York, one facility transported patients without records, ambulances transported patients to the wrong facilities, and patients were admitted to units without the proper facilities for the patient's condition (ACEP (American College of Emergency Physicians), 2015).

Integral to coordinated medical care are patient tracking systems. Patient tracking refers to the process of monitoring and managing the movement, status, and information of patients within and among health care facilities. The process includes entry into the health care system, each transition within the system, all the way through final disposition and exit from the system (Tavakoli et al., 2017). Patient tracking is often one of the most challenging activities inhibiting medical care delivery in disasters due to a lack or disruption of patient tracking and health informatics systems. Failure to track patients may result in lack of access to care, delayed medical care, poor health outcomes, 'lost' patients within a facility (Wolinsky et al., 2003), and delayed or absent reunification. Among countries, there exists a wide disparity in patient tracking abilities, from absent to technologically advanced. Many systems are available for patient tracking and ideally are interoperable within and among regional systems. Mobile solutions include a variety of features such as barcode patient wristbands and scanners (E-FINDS), digital photograph integration with a patient record, and child-parent medical record linking (Institute of Medicine, 2015).

Emergency Medical Services

Uniquely, emergency medical services (EMS) in the United States, such as ambulance and 9-1-1 systems, are not under a single federal agency. Various aspects of EMS services are respectively under the Department of Transportation (DOT) and Health and Human Services (HHS) and regulated under a plethora of agencies (Cordi and Goldstein, 2023). EMS systems in the United States vary substantially at the state and local levels. For example, standards of practice, typical types of care, and procedures performed vary. Prehospital providers in the United States perform basic and advanced medical care under the direction of licensed medical professionals. In some systems only basic level care may be provided while in other systems, physicians may actively participate in direct clinical care. Under normal conditions, EMS hold multiple roles including response within the 9-1-1 system and MIH inclusive of preventative and primary care activities.

In response to disaster, FEMA contracts with private industry EMS providers to scale ground ambulances, air ambulances (e.g. helicopters and fixed-wing jets), transport, as well as the provision of paratransit vehicles. During disasters, EMS providers may be leveraged to support a wide variety of activities from vaccination to chronic care patient management and support.

Internationally, EMS may be nationalized such as the National Health Service (NHS) service in the United Kingdom. Varying levels of licensed clinicians may be integrated into direct-care ambulance emergency responses, such as the inclusion of physicians in France (Adnet and Lapostolle, 2004). Scope of practice among providers cannot be readily compared internationally due to wide disparities in training and licensure.

Licensing, Credentialing, and Privileges

In normal operations, states and varied agencies license and certify clinicians. Clinicians in one state may not practice in another without obtaining licensure. At the local level, clinicians certified or credentialed by an agency or facility may only engage in practice under the credentialing and permissions granting institution. During a disaster, executive orders may be issued by a state governor to suspend normal licensing procedures and expand scopes of practice (Iserson, 2020). At the local level, for hospitals accredited by The Joint Commission, once a hospital's Emergency Operations Plan has been activated, an expedited process for granting privileges may occur (HHS, 2017).

Standards of Care and Practice

Standards of Care

Standards of care is a legal term. Standards of care are the minimal actions a reasonable person takes in a given situation (Moffett and Moore, 2021). Whether standards of care are met is determined by administrative agencies and legal courts. In consideration of whether a standard of care is met, the body with legal authority will often refer to clinical practice guidelines set by professional associations and convening panels of experts. 'Altered standards of care' may be accepted in a disaster setting. In low-resource environments, including disasters, the typical standard of care may not be practical. Under normal circumstances, when arriving on scene with multiple victims, an EMT would likely tend to a patient who is severely bleeding or without a pulse first. However, in a disaster, these patients may be passed over, and attention is paid to other victims who have a higher chance of survival.

Standards of Practice

Standards of practice are the written guidelines for clinical practice. Standards of practice during disaster are frequently modified. During the COVID-19 pandemic there were an insufficient number of ventilators for patients requiring breathing support. Hospitals with anesthesia gas machines

(AGMs) used the AGMs, an off-label use, to provide breathing support in place of ventilators. Under normal conditions, the use of ventilators for these patients is the standard of care and therefore AGM use would not be allowed. In the disaster setting, under altered standards of practice, AGM use was considered an acceptable alternative in the absence of ventilators (Haina, 2020).

Triage

Disaster medical care requires triage, sorting by acuity, of patients. Under normal operations patients are provided medical care to maximize each patient's health. Given limited resources in disaster medical care, medical care interventions are allocated to do the most good for the most people. Medical procedures, such as elective surgeries, may be forgone in disaster medical care. Similarly, the most ill may not receive care in favor of improving the survival of more individuals with less acute illnesses and injuries. There are multiple well-recognized systems (e.g., simple triage and rapid treatment (START) and sort, assess, lifesaving interventions, transport/treatment (SALT)) to assist providers in medical triage (Bazyar et al., 2019).

Use of Telehealth in Disasters

Telemedicine has been available since the 1950s with the use of radios to provide information allowing for diagnosis. While technology and capabilities have advanced significantly in the decades since, it took the COVID-19 pandemic for widespread adoption in the United States (Kichloo et al., 2020; Maese et al., 2020). The first application of telehealth in disasters was NASA's use of satellites to provide communications to aid victims (Litvak et al., 2022). Telemedicine use in disasters varies based on location and disaster type. It has been used in response to various natural disasters as well as in conflict and is most often used in the response and recovery phases of the disaster cycle (Ajami and Lamoochi, 2014; Salehinejad et al., 2020). Telehealth is a potentially powerful tool that can amplify the workforce available to respond to a disaster. However, work is still needed to fully understand its applications in disasters in both high and low-resourced locations.

Special Planning Considerations

Pharmaceutical and Device Management and Access

The US Federal Drug Agency (FDA) and its international counterparts (e.g., European Medicines Agency) are responsible for ensuring the safety of drugs and medical devices, often a time-consuming process, prior to approval for use. In a US-based disaster, multiple federal regulations allow for accelerated approval (Novack, 2022). Additionally, Expedited Use Authorizations (EUAs) may be issued in response to a CBRN threat (HHS, 2017). An EUA allows the FDA to authorize use of unapproved medical products and devices. During the COVID-19 pandemic, an EUA was issued by the Secretary of HHS which allowed the FDA to quickly provide provisional approval for use of newly developed vaccines.

Disasters reduce access to pharmaceuticals used in daily life. People may no longer have access to their pharmacies, particularly if they have been displaced or are not covered by health insurance. The Emergency Prescription Program (EPAP), authorized under the Robert T. Stafford Disaster Relief and Emergency Assistance Act (Stafford Act), assists people in a disaster area without insurance to obtain necessary pharmaceuticals and medical devices (Assistant Secretary for Preparedness and Response (ASPR), 2016; HHS, *The Emergency Preparedness Assistance Program*, n.d.).

Strategic National Stockpile

The Strategic National Stockpile (SNS) is a repository of medications and medical materiel used to support response to CBRN events (Neumeister and Gray, 2021). The SNS is collaboratively maintained by the US Department of Homeland Security (DHS) and US CDC. Management of the SNS requires inclusion of multiple disciplines including clinical care, pharmacy, and medical logistics. The SNS program manages separate inventory stocks for vaccines, antitoxins, antibiotics, and materiel. The SNS is supported and continually updated via ongoing research and drug discovery managed by the Biomedical Advanced Research and Development Agency (BARDA). Material provided includes but is not limited to personal protective equipment (PPE), ventilators, and Federal Medical Stations (FMS). FMS are caches of material and pharmaceuticals that may be resourced during a federal declared emergency to turn predetermined locations into temporary medical centers. Assets of the SNS are rapidly available to cities, towns, and states via push packages, managed inventory, and CHEMPACK programs. Push packages are prepacked stocks of pharmaceuticals made available within 12-hours of request by a state. Similarly, the CHEMPACK program manages stockpiles of nerve agent antidotes (e.g., atropine and 2-PAM) strategically placed around the country, primarily at hospitals and fire stations (HSS, *CHEMPAK*, n.d.). CHEMPACKs are remotely monitored as access results in automatic notification to the CDC.

Special Medical Equipment

Over 2 million people worldwide are dependent on dialysis machines due to end stage kidney disease (Chronic Disease Research Group (CDRG), 2024). The medical condition makes people dependent on dialysis and dialysis machines as they filter waste products and excess fluid from blood. People dependent on dialysis are particularly vulnerable to the effects of disasters as this therapy is administered on a strict schedule and missed sessions have significant and possibly deadly effects on a patient's health. Dialysis is also used as a treatment for patients whose kidneys are acutely injured, such as after crush injuries which can occur in natural disasters such as earthquakes. After an earthquake in Marmara, Turkey in 1999, 477 victims required dialysis (Sever et al., 2002). The development of a surge and flex plan for special medical equipment such as dialysis machines is an important aspect of disaster response as they are not included in SNS stockpiles.

Public Health Crises in Disaster Response

Disasters that result in overcrowding, breakdown of waste-management systems, scarcity of potable water, and food scarcity cause a host of issues that require medical care and intervention. Public health and medical professionals provide guidance to create safe environments and prevent the spread of disease. In the event of outbreaks, medical care teams then intervene. Cholera, caused by the bacterium *Vibrio cholerae*, is a common diarrheal outbreak infection where there is poor sanitation in the disaster setting. It is preventable with, among other measures, good hand hygiene and boiling water, and two vaccines are currently available for prophylactic care. Treatment includes oral rehydration and antibiotics may shorten the course of disease (Harris et al., 2012). Laboratory services, as part of clinical and biosurveillance teams, are particularly important to include in medical care and outbreak response.

Mental Health Care in Disaster Response

Most people affected by a disaster experience mental distress; for some, the effects may last a lifetime. Those with mental health illness are also more likely to experience exacerbation of illness following a disaster (WHO, *Mental Health,* n.d.; Makwana, 2019). Mental health care resources should be incorporated in response for both disaster victims and responders. A range of resources are available including support from community social support, peer-to-peer programs, professional organizations, governmental organizations (e.g., Substance Abuse and Mental Health Services Administration), Center for the Study of Traumatic Stress (CSTS), and the National Child & Traumatic Stress Network (NCTSN).

Summary

Quick access to medical care can make the difference between healthy lives lived, protracted illness and recovery, or death. Medical care is therefore embedded in all phases of the disaster response cycle. Medical care during disasters is inclusive of emergency management; coordination of resources; medical oversight; preventative, primary, chronic-illness, and specialty care services; and supply chain management. Key to ensuring medical care is information sharing and coordination across responding agencies. Coalition participation maximizes the resources and capacities of response systems during periods of patient surges. To meet disaster specific needs, medical response teams from many agencies deploy a range of resources. The medical services provided during disaster differ from normal operations due to disaster scale and scarcity of resources. As recovery from disaster is often protracted, medical services may be provided over several years during the recovery period. Medical care providers work closely with and among national and international governmental and non-governmental agencies. Integrating medical care in disaster response ensures, restores, and improves health and health care for both individuals and communities.

References

ACEP (American College of Emergency Physicians). *Lessons Learned from Hurricane Sandy and recommendations for Improved Healthcare and Public Health Response and Recovery for Future Catastrophic Events*. 2015. https://www.acep.org/siteassets/uploads/uploaded-files/acep/by-medical-focus/disaster/lessons-learned-from-hurricane-sandy-webpage.pdf

Adnet, Frédéric, and Frédéric Lapostolle. "International EMS Systems: France." *Resuscitation*, vol. 63, no. 1, 2004, pp. 7–9. doi:10.1016/j.resuscitation.2004.04.001

Ajami, Sima, and Parisa Lamoochi. "Use of Telemedicine in Disaster and Remote Places." *Journal of Education and Health Promotion*, vol. 3, no. 3, May. 2014, p. 26.https://doi:10.4103/2277-9531.131886

Aspen Medical. *"What We Do."* Aspenmedical. 2022. https://www.aspenmedical.com/what-we-do. Accessed 23 November, 2023.

Assistant Secretary for Preparedness and Response (ASPR). *2017-2022 Healthcare Preparedness and Response Capabilities*. ASPR. Department of Health and Human Services. 2016.

Baker, Japheth, and John Cole. "EMS Medical Oversight of Systems." *StatPearls*, StatPearls Publishing, 26 September 2022.

Bazyar, Jafar, Mehrdad Farrokhi, and Hamidreza Khankeh. "Triage Systems in Mass Casualty Incidents and Disasters: A Review Study With A Worldwide Approach." *Open Access Macedonian Journal of Medical Sciences*, vol. 7,no. 3, 12 Feb. 2019, pp. 482–494. doi:10.3889/oamjms.2019.119

Chronic Disease Research Group (CDRG). *Kidney Disease*. Retrieved 10 January. Chronic Disease Research Group. 2024, https://www.cdrg.org/chronic-diseases/nephrology/?gclid=Cj0KCQiAwvKtBhDrARIsAJj-kTjKh8vd9Di002WSu520iU33y23SlFMW08uxRSpMVD1HP_-s53Ym488aAkcdEALw_wcB

Congress. John Warner National Defense Authorization Act for Fiscal Year 2007. U.S. Government Publishing Office, 31 Dec 2020, https://www.govinfo.gov/app/details/COMPS-10387

Cordi, Heidi P, and Scott Goldstein. "EMS Federal Regulations." *StatPearls*, StatPearls Publishing, 14 August 2023.

CRED (Centre for Research on the Epidemiology of Disasters). 2022 Disaster in Numbers. University of Louvain. 20 March 2023. https://www.cred.be/sites/default/files/2022_EMDAT_report.pdf

Doctors of the World. "*What We Do*." https://doctorsoftheworld.org/what-we-do/. Accessed 2 December, 2023.

EMAC (Emergency Management Association Compact). "Emergency Medical Services." NEMA, 2023. https://www.emacweb.org/index.php/learn/learn-about-emac-your-discipline/emergency-medical-services. Accessed December, 2023.

"Emergency Management - Requirements for Granting Privileges During a Disaster." *The Joint Commission*, 2023, https://www.jointcommission.org/standards/standard-faqs/hospital-and-hospital-clinics/emergency-management-em/000002275/. Accessed 3 December 2023.

Haina, Kent M K Jr. "Use of Anesthesia Machines in a Critical Care Setting During the Coronavirus Disease 2019 Pandemic." *A&A Practice*, vol. 14, no. 7, 2020. https://doi:10.1213/XAA.0000000000001243

Harris, Jason B, Regina C LaRocque, Firdausi Quadri, Edward T Ryan, and Stephan B Calderwood. "Cholera." *Lancet (London, England)*, vol. 379, no. 9835, (2012, pp. 2466–2476. doi:10.1016/S0140-6736(12)60436-X

Heart to Heart. "*Disaster Response*." https://www.hearttoheart.org/disaster-response. Accessed 15 November, 2023.

He, Shuhan, Sammer Marzou, Adi Balk, Tehnaz Boyle, and Jarone Lee. The Telehealth Advantage: Supporting Humanitarian Disasters With Remote Solutions. *American Journal of Disaster Medicine*, vol. 17, no. 2, 2022, pp. 95–99. https://doi:10.5055/ajdm.2022.0423. PMID: 36494880; PMCID: PMC10316149.

Institute of Medicine. *Regional Disaster Response Coordination to Support Health Outcomes: Summary of a Workshop Series*. Washington, DC: The National Academies Press. 2015. https://doi.org/10.17226/21713.

International Federation of Red Cross and Red Crescent Societies (IFRC). *"Emergency Health."* Accessed 3 December, 2023. https://www.ifrc.org/our-work/health-and-care/emergency-health

Iserson, Kenneth V. "Augmenting the Disaster Healthcare Workforce." *The Western Journal of Emergency Medicine*, vol. 21, no. 3, 13 Apr. 2020, pp. 490–496, doi:10.5811/westjem.2020.4.47553

Jonk, Yvonne, Carly Milkowski, Zacharia Croll, and Karen Pearson. *Ambulance Deserts: Geographic Disparity of Provision of Ambulance Services [Chartbook]*. University of Southern Maine, Muskie School, Maine Rural Health Research Center. University of Southern Maine, Muskie School, Maine Rural Health Research Center. 2023.

Kichloo, Asim, Michael Albost, Kirk Dettloff, Fara Wani, Zain El-Amir, Jagmeet Singh, Michael Aljadah, Raja Chandra Chakinala, Ashok Kumar Kanugula, Shantanu Solanki, and Savneek Chugh. "Telemedicine, the Current COVID-19 Pandemic and the Future: A Narrative Review and Perspectives Moving Forward in the USA." *Family Medicine and Community Health*, vol. 8, no. 3, 2020, p. e000530. doi:10.1136/fmch-2020-000530

Lameire, Norbert. "Renal Replacement Therapies in the Aftermath of the Catastrophic Marmara Earthquake." *Kidney International*, vol. 62, no. 6, 2002, pp. 2264–71. https://doi:10.1046/j.1523-1755.2002.00669.x

Litvak, Mark, Katherine Miller, Tehnaz Boyle, Racher Bedenbaugh, Christina Smith, David Megurerdichian, David Reisman, Paul Biddinger, Adam Licurse, and Eric Goralnick. Telemedicine Use in Disasters: A Scoping Review. *Disaster Medicine and Public Health Preparedness*, vol. 16, no. 2. 2022, pp. 791–800. PubMed, https://doi:10.1017/dmp.2020.473.

Maese, John, Donna Seminara, and Anita Szerszen. Perspective: What a Difference a Disaster Makes: The Telehealth Revolution in the Age of COVID-19 Pandemic. *American Journal of Medical Quality*, vol. 35, no. 5. 2020, pp. 429–31. https://doi:10.1177/1062860620933587

Makwana, Nikunj. "Disaster and Its Impact on Mental Health: A Narrative Review." *Journal of Family Medicine and Primary Care*, vol. 8, no. 10, 31 Oct. 2019, pp. 3090–3095, doi:10.4103/jfmpc.jfmpc_893_19

Malaysia, Mercy. *Annual Report 2022*. https://www.mercy.org.my/wp-content/uploads/2023/07/MERCY_Malaysia_AR2022_full_report.pdf

Medicines Sans Frontiers (MSF). "*How We're Run.*" https://www.msf.org. Accessed 15 November, 2023.
MedShare. *"Our Programs*." https://www.medshare.org/programs/all-programs. Accessed 3 December, 2023.
Misra, Asit, Dolly Yadav, and Tamorish Kole. "Emergency Care in India Beyond 75 Years of Independence – Problems and Solutions." *Journal of Global Health*, vol. 13, 14 Apr. 2023, p. 03015, doi:10.7189/jogh.13.03015
Moffett, Peter, and Gregory Moore. "The Standard of Care: Legal History and Definitions: The Bad and Good News." *The Western Journal of Emergency Medicine*, vol. 12, no. 1, 2021, pp. 109–112.
NDMS (National Disaster Medical System). National Disaster Medical System Health Care Facility/Partner Memorandum of Agreement for Definitive Medical Care, n.d.. https://www.phe.gov/Preparedness/responders/ndms/definitive-care/Documents/ndms-moa-07112018.pdf. Accessed 3 December.
NEMA (National Emergency Management Agency). Coordinated Incident Management System (CIMS) third edition. 2020. https://www.civildefence.govt.nz/resources/coordinated-incident-management-system-cims-third-edition. Accessed 4 December, 2023.
Neumeister, Shondra M, and Joshua P Gray. "The Strategic National Stockpile: Identification, Support, and Acquisition of Medical Countermeasures for CBRN Incidents." *Toxicology Mechanisms and Methods*, vol. 31, no. 4, 2021, pp. 308–321. doi:10.1080/15376516.2020.1853294
Novack, Gary D. "Expedited Regulatory Product Approval in the Time of COVID-19." *The Ocular Surface*, vol. 26, 2022, pp. 345–348. doi:10.1016/j.jtos.2022.05.007
Polish Center for International Aid. *"Humanitarian and Development Aid."* https://pcpm.org.pl/en/humanitarian-and-development-aid. Accessed 3 December, 2023.
Salehinejad, S., N. Jannati, R. Ershad Sarabi, and K. Bahaadinbeigy, 2020. Use of Telemedicine and e-Health in Disasters: A Systematic Review. *Journal of Emergency Practice and Trauma*, vol. 7, pp. 56–62. https://doi.org/10.34172/jept.2020.34
Sauveterus Sans Frontiers. "*Our Medical Missions."* https://www.sauveteurs.org/our-missions/our-medical-missions. Accessed 3 December, 2023.
Save the Children. *"Child Health."* https://www.savethechildren.org/us/what-we-do/health/child-health. Accessed 3 December, 2023.
Sever, Mehmet S, Ekrem Erek, Raymond Vanholder, Birsen Yurugen, Gulcin Kantarci, Mahmut Yavus, Hulya Ergin, Semra Bozlakioglu, Serran Dalmak, M. Yasar Tulbek, Ualuk Kiper, and Norbert Lamiere. "Renal Replacement Therapies in the Aftermath of the Catastrophic Marmara Earthquake." *Kidney International* vol. 62,6 (2002): 2264–71. doi:10.1046/j.1523-1755.2002.00669.x
Tavakoli, Nahid, Mohammad H Yarmohammadian, Reza Safdari, and Mahmour Keyvanara. "Designing a Model of Patient Tracking System for Natural Disaster in Iran." *Journal of Education and Health Promotion*, vol. 6, 9 Aug. 2017, p. 77, doi:10.4103/jehp.jehp_135_16
Team Rubicon. *Annual Report*. 3-67. 2022. https://teamrubiconusa.org/app/uploads/2023/09/2022AnnualReport.pdf
TRACIE. "General Overview of Healthcare Coalitions." https://files.asprtracie.hhs.gov/documents/aspr-tracie-general-overview-hccs.pdf Accessed 4 December, 2023.
UK-MED. *"What We Do."* Accessed 3 December, 2023. https://www.uk-med.org/
UNOCHA (United Nations, Office for Coordinated Affairs). "*We Coordinate*." https://www.unocha.org/we-coordinate. (n.d.) Accessed 4 December, 2023.
US Department of Defense (DOD), US Special Operations Command. *Joint Special Operations Command.* https://www.socom.mil/pages/jsoc.aspx. Accessed 3 December, 2023.
US Department of Health and Human Services (HHS), Administration for Strategic Preparedness and Response (ASPR), Administration for Strategic Preparedness and Response. ASPR-TRACIE Fact Sheet. Accessed 3 December, 2023. https://files.asprtracie.hhs.gov/documents/aspr-tracie-fact-sheet.pdf
US Department of Health and Human Services (HHS), Administration for Strategic Preparedness and Response (ASPR). *"CHEMPAK."* https://aspr.hhs.gov/SNS/Pages/CHEMPACK.aspx. Accessed 4 December, 2023.
US Department of Health and Human Services (HHS), Administration for Strategic Preparedness and Response (ASPR), Administration for Strategic Preparedness and Response. "Disaster Medical Assistance Teams." Accessed 4 December, 2023.
US Department of Health and Human Services (HHS), Administration for Strategic Preparedness and Response (ASPR). "*The Emergency Prescription Assistance Program.*" https://aspr.hhs.gov/EPAP/Pages/about-epap.aspx.

US Department of Health and Human Services (HHS), Administration for Strategic Preparedness and Response (ASPR). "Health Care Coalitions (HCCs)." https://aspr.hhs.gov/HealthCareReadiness/HPP/Documents/Health%20Care%20Coalitions%20(HCCs)%20One-Pager.pdf. Accessed 4 December, 2023.
US Department of Health and Human Services (HHS), Administration for Strategic Preparedness and Response (ASPR). "Hospital Preparedness Program (HPP)." https://aspr.hhs.gov/HealthCareReadiness/HPP/Pages/default.aspx. Accessed 4 December, 2023.
US Department of Health and Human Services (HHS), Administration for Strategic Preparedness and Response (ASPR). "Organization Chart." https://www.hhs.gov/about/agencies/orgchart/aspr/index.html. Accessed 4 December, 2023.
US Department of Health and Human Services (HHS), Administration for Strategic Preparedness and Response (ASPR). *"Regional Disaster Health Response System: An Overview."* https://aspr.hhs.gov/RDHRS/Pages/rdhrs-overview.aspx. Accessed 4 December, 2023.
US Department of Health and Human Services (HHS), Administration for Strategic Preparedness and Response (ASPR). *"Trauma and Critical Care Teams."* https://aspr.hhs.gov/NDMS/Pages/tcct.aspx. Accessed 4 December, 2023.
US Department of Health and Human Services (HHS), Administration for Strategic Preparedness and Response (ASPR), Food and Drug Administration. *Emergency Use Authorization of Medical Products and Related Authorities*. 2017. US Department of Health and Human Services Food and Drug Administration. https://www.fda.gov/media/97321/download.
US Department of Homeland Security, Federal Emergency Management Agency. *Emergency Support Function #8 - Public Health and Medical Services Annex*. Federal Emergency Management Agency. 2008.
US Department of Homeland Security, Federal Emergency Management Agency. *A Whole Community Approach to Emergency Management: Principles, Themes, and Pathways for Action*. FDOC 104-008-2, Federal Emergency Management Agency. December 2011.
Wolinsky, Philip R, Nirmal Tejwani, Noel n. Testa, and Joseph Zuckerman. Lessons Learned from the Activation of a Disaster Plan: 9/11. *The Journal of Bone & Joint Surgery*, vol. 85, no. 9, September, 2003, pp. 1844–1846.
World Health Organization (WHO). "Mental Health in Emergencies." https://www.who.int/news-room/fact-sheets/detail/mental-health-in-emergencies. Accessed 4 December, 2023.
World Health Organization (WHO). *"National IHR Focal Points."* WHO. https://www.who.int/teams/ihr/national-focal-points. Accessed 1 December 2023.
World Health Organization (WHO). "*Global Classified Teams*." (nd) https://www.who.int/emergencies/partners/emergency-medical-teams/emt-global-classified-teams. Accessed 4 December, 2023.
World Health Organization (WHO). *Emergency Response Framework* - 2nd ed. Geneva, World Health Organization, 2017a. License: CC BY-NC-SA 3.0 IGO.
World Health Organization (WHO). "*World Bank and WHO: Half the World Lacks Access to Essential Health Care Services, 100 Million Pushed Into Extreme Poverty Because of Health Expenses*." WHO, 13 December 2017b, https://www.who.int/news/item/13-12-2017-world-bank-and-who-half-the-world-lacks-access-to-essential-health-services-100-million-still-pushed-into-extreme-poverty-because-of-health-expenses. Accessed 4 October, 2023.

4

DEBRIS MANAGEMENT AND REMOVAL

S. D. Burris

Introduction

Through their physical impacts on the natural and built environment, disasters produce varying amounts of debris. Disaster debris is generally defined as waste materials created by or in the aftermath of a natural or human-made disaster (Luther, 2017). Disaster debris can include materials from damaged buildings, vegetative debris (e.g., downed trees, limbs, and soil), damaged vehicles or household appliances, hazardous materials, and even putrid materials like rotting food or dead animals. Debris of various types can affect human health and safety, restrict transportation of goods and services, or prevent rapid emergency response. Debris removal is largely a responsibility of the local government with cost-sharing opportunities through the federal government. Prompt and well-orchestrated debris removal can help communities cope with disaster recovery and clear the area so restoration and rebuilding can begin.

Debris management refers to an all-encompassing strategy that includes not only the removal of debris but also its reduction, recycling, and disposal. It aims to optimize the recovery of priceless resources while minimizing the environmental impact of debris removal. Debris removal is the methodical and coordinated efforts to gather, transport, and dispose of debris associated with disasters in a way that is safe, effective, and environmentally conscious. Debris management activities include planning for debris operations before a disaster, ensuring debris collection sites are properly identified and prepared, disposing of the debris, and verifying different types of debris get to the right place. Jurisdictions document their debris operations through debris management plans.

Disaster debris removal involves a variety of stakeholders, including the impacted community, non-governmental organizations (NGOs), local, state, and federal government agencies, and private contractors. The success of debris clearance depends on these stakeholders working effectively together and coordinating their efforts. Debris removal also requires adherence to laws and regulations to protect lives and property, as well as maximize opportunities for financial cost-sharing. This chapter guides emergency managers and government officials through the pre-disaster planning process to build capabilities and capacities to plan for and manage disaster debris removal.

DOI: 10.4324/9781003414834-7

Types of Debris

Disaster debris is a unique combination of our natural and built environments. Disasters produce and disperse debris indiscriminately across geographies. Depending on the kind and extent of the disaster, debris might differ greatly in type, size, and amount. For example, earthquake and tornado debris commonly includes building materials, and flooding carries debris away from its original location. Debris can range from household objects, vehicles, and hazardous materials to fallen trees and destroyed buildings. As part of debris management, debris is categorized by type. Types of debris include the following:

- *Vegetative*: Whole trees, tree stumps, tree branches, tree trunks, and other leafy materials. This type may also include soil, mud, and sand deposited by floods, landslides, and storm surge.
- *Construction and Demolition (C&D)*: Damaged building materials and contents, utility equipment (e.g., utility poles and wiring), and metal debris from roofing, mobile homes, or vehicles.
- *Household Goods*: Household furnishings and personal effects; white goods (e.g., refrigerators, washing machines, and stoves); common household hazardous waste (HHW) (e.g., paint, cleaning supplies, and propane); and electronic waste (e-waste) (e.g., televisions, computers, and batteries).
- *Hazardous Waste*: Substances that are potentially harmful to human health and the environment, including ignitable, corrosive, reactive, toxic, and infectious substances. This type may also include putrescent debris such as carcasses of agricultural, domestic, and wild animals.

Importance of Debris Removal

Local jurisdictions are charged with performing debris removal, generally in accordance with state laws and federal requirements. Effective debris management planning also helps communities comply with federal regulations and reduce the risk of negative environmental impacts. Most importantly, debris removal after a disaster is critical for community recovery and to protect public health and safety. Debris not picked up in a timely manner can pose safety hazards, health risks (e.g., attracting vermin and mold), or create a nuisance (e.g., odors, physical barriers to homes and businesses). Debris not removed physically prevents or delays response activities, especially debris on roadways preventing transportation access in the area. Debris removal also plays an important role in psychological wellbeing for disaster survivors and affected communities to help restore a sense of normalcy.

Debris Removal Contracting and Public Assistance

All disasters are local, and disaster debris is the responsibility of local jurisdictions to work with contractors, private entities, and NGOs to manage debris in their communities using the Incident Command System (ICS) and National Incident Management System (NIMS). Jurisdictions commonly procure contracts to perform debris removal and monitoring operations. Jurisdictions should undertake a competitive procurement process for debris removal services that are compliant with state and federal laws. Often, jurisdictions choose more than one debris contractor to be available for support after a disaster.

The Federal Emergency Management Agency (FEMA) Emergency Support Function (ESF) #3 – Public Works and Engineering – aids by coordinating and organizing federal government capabilities and resources to facilitate the delivery of services, technical assistance, engineering expertise, construction management, and other supporting activities in federally declared disasters. States frequently offer assistance to their jurisdictions. While direct federal aid is not the norm in most disaster debris removal operations, grants and reimbursement for the costs of debris removal are included in the Robert T. Stafford Act for federally declared disasters. Under the Stafford Act, the president can declare a major disaster for natural events (e.g., hurricanes, tornadoes, storms, flooding, tsunamis, earthquakes, volcanic eruptions, landslides, and fires) where the damage is so severe that local and state governments are unable to effectively respond. This declaration provides a wide range of federal financial assistance for governments and individuals affected by disaster.

On average, debris management costs account for 27% of a disaster's costs (Federal Emergency Management Agency (FEMA), 2007). Disaster debris can amount to the equivalent of years or even decades of normal amount of solid waste production in the community, resulting in a surge of solid waste requiring immediate attention. Communities may not have the resources to pick up the debris and often nowhere safe to store or dispose of it. Communities without available resources to manage with this surge will leverage contracts to perform the debris removal. These contracts are often a reimbursable expense by federal assistance if certain eligibility criteria are met. To be eligible for federal disaster assistance for emergency debris removal, the debris must have been generated by a presidentially declared disaster under the Stafford Act, within the declared disaster area, and be the legal obligation of the jurisdiction to remove, and pose an immediate threat to life, improved property, or public health and safety.

Eligibility requirements and reimbursement procedural requirements are outlined in FEMA's Public Assistance Program and Policy Guide (PAPPG). Emergency debris removal is covered in Category A: Debris Removal. Federal assistance shares in 75% of the eligible cost, with the remaining 25% split between the state and local government. This program only reimburses for eligible expenses, including overtime for force account labor, costs for force account equipment, and the cost for contract labor and equipment. Force account labor describes the work performed by the jurisdiction's permanent, full-time, or temporary employees and volunteers. Jurisdictions should document these hours worked and the equipment used to complete the eligible debris removal work. Force account labor is critical for the response phase of the disaster because local employees and volunteers are on the ground right away to address debris removal priorities.

Government agencies and emergency management officials should note if the disaster is not federally declared through the Stafford Act, federal disaster debris removal assistance is not generally available. The cost of debris removal is the responsibility of local jurisdictions, though states may offer financial assistance. Corporations may be held financially responsible for debris resulting from industrial accidents, such as train derailments or oil spills under the Oil Pollution Act of 1990. Federal or local agencies may control the cleanup activities while ordering the responsible party to pay for the costs.

Debris Management Plans

Debris management plans are documents to help communities expedite recovery efforts, promote public health and welfare, and facilitate a coordinated removal and disposal of debris after a disaster. Government agencies develop and maintain debris management plans as a standalone document or as an annex or appendix to their comprehensive emergency management plans (CEMPs). Debris management plans are flexible, living documents that are adaptable to jurisdictions' specific needs, resources, and disaster types. These plans should integrate with the jurisdiction's other plans for mutual aid, public information, resource management, and recovery.

Plan templates are available from FEMA, the U.S. Army Corps of Engineers, and the Environmental Protection Agency (EPA). While plan review was required in the past, FEMA no longer reviews or approves debris management plans. While FEMA does not require debris management plans to receive financial reimbursement for debris removal operations, these documents are recommended to ensure communities can effectively perform debris removal and facilitate recovery from the disaster.

The contents of these plans will vary by a jurisdiction's needs but generally include how the jurisdiction will prepare for managing the debris, including procurement, contracted resources, identifying the location of temporary staging and permanent disposal sites, health and safety requirements, environmental considerations, regulatory requirements, and public information. Ultimately, these plans provide assurance that communities understand and can meet post-disaster debris removal requirements before a disaster strikes. Waiting until after a disaster strikes to develop debris removal plans could result in delayed recovery and negative impacts to public health and safety.

Planning Process

Debris management plans are written documents that describe procedures and guidelines for managing disaster debris to ensure the activities coordinated across stakeholders, environmentally responsible, and as cost-effective as possible. These plans can also help state and local governments comply with federal regulations and demonstrate eligibility for federal disaster aid. Debris management plans delineate roles and responsibilities of agencies and individuals, describe how resources are procured and distributed, and provide estimates or forecasts of the potential amount, types, and location of disaster debris based on likely and severe hazard scenarios. Disaster management plans enable communities to become better prepared to restore public services and ensure the public health and safety after a disaster and can help communities become better positioned to receive the highest levels of federal disaster assistance.

Debris management plans describe in detail the roles and responsibilities of each agency involved in operations. Individuals within each agency are assigned responsibilities to perform the various steps in the plan. By outlining the roles and responsibilities in advance, the plan enables coordination and reduces delays of identifying responsible parties after the disaster. Often, just-in-time debris contracting is not available. Vendor contracts should be in place well before a disaster strikes. The plan should additionally include regularly updated organizational charts, rosters, and contact information.

Planning can help jurisdictions understand the expected volume of debris created by various disaster scenarios, so they can find or develop storage and disposal capacities, staff, and fund the debris removal process. Debris management planning is a cyclical process of pre-planning, planning, testing or executing the plan, and improving the plan. Pre-planning includes integrating debris management efforts with other jurisdiction plans, such as local emergency response plans, CEMPs, and environmental management plans. Working across agencies and with the public can help reduce the potential volume of disaster debris through mitigation activities. Pre-planning should identify applicable laws and authorities that must be addressed by the plan. Some states may require local plan approvals. Debris management plans should be reviewed and revised on a regular schedule and revised after any event, exercise, or training. Pre-planning also identifies the available resources to be leveraged, such as funding, personnel, equipment, facilities, and agreements with other agencies. Finally, communities may involve the public to participate in the planning process, and to review and provide comments on the plans.

Pre-Incident Activities

Disasters commonly overwhelm local jurisdictions' capacity for debris management. Communities often rely on pre-qualified disaster contractors to secure the labor and equipment necessary to support debris removal. Major debris contracting companies are familiar with federal regulations and FEMA requirements. Communities develop debris removal plans to identify the contractors and their responsibilities, and to outline how the jurisdiction will prepare for and manage debris removal.

To develop an effective debris management plan, jurisdictions must be able to describe and predict the amounts and type of disaster debris they could expect. The plan can leverage a hazard analysis from other emergency management planning documents, such as the Threat and Hazard Identification and Risk Assessment (THIRA), Hazard Mitigation Plan (HMP), CEMP to identify the types and the severity of hazards that could result in disaster debris. The THIRA is a tool to help communities understand its threats and hazards and when and how they could be impacted. Once communities identify possible hazards, they can begin to estimate what types and how much debris could be generated. Forecasting tools and methods can further help jurisdictions predict and anticipate debris quantities before a disaster.

Debris management plans should identify critical facilities within that may be impacted by or inaccessible due to disaster debris, including hospitals, shelters, critical transportation routes, and utility providers that should be prioritized for debris removal. Planning can also identify areas of potentially high debris volume, such as population-dense areas. FEMA offers a free software, Hazus, used for predicting the number of debris from earthquakes, floods, and hurricanes. Hazus is a GIS-based desktop software application with a growing collection of simplified open-source tools. Another forecasting model, the U.S. Army Corps of Engineers' SimSuite, is often used to forecast hurricane debris within three days of landfall.

These plans should also identify and permit appropriate staging and storage sites aligned with regulatory requirements for environmental protection. Debris management sites (DMSs) serve as staging areas to sort debris so that it is properly disposed of. DMSs are generally located on public land (i.e., jurisdiction owned) and include equipment such as incinerators, grinders, and HHW storage containers. DMSs enable debris to be quickly removed off the roadways. The site also begins the recycling procedures that can help the community save landfill space.

Post-Disaster Activities

There are two phases of debris removal after a disaster: the response or debris clearance phase, and the recovery phase that includes debris collection and management. The response phase focuses on reopening emergency access routes and main roadways, then moving on to debris removal for critical infrastructure (e.g., electricity and communications), emergency operations, and supply distribution centers, followed by access to water and wastewater infrastructure. The recovery phase includes collecting the remaining debris, and reducing, recycling, and final disposal.

When the situation has stabilized so emergency response personnel can safely access the impact area, officials will assess the amount and type of debris as part of a preliminary damage assessment (PDA) or a debris-specific assessment. PDAs are conducted to enable local jurisdictions to report the magnitude of damage to state and federal officials, as part of the request for federal disaster assistance. Jurisdictions may be required to provide a debris quantity calculation sheet and annotated maps or Global Positioning System (GPS)-tagged debris locations within the impacted area. In some cases, jurisdictions may not be able to enter an area to visually confirm debris, such as flood waters, landslides, or mudslides blocking access or destroyed roads and bridges blocking access. Jurisdictions seeking federal assistance for debris removal should also supply statements describing the amount of debris removal already completed at the time of the assessment, photographs, calculation sheets, details on the costs and funding sources for associated labor and equipment, and information on the public health, community, environmental, and historical impacts of the debris and removal (Federal Emergency Management Agency (FEMA), 2021).

Debris removal begins with removing debris from transportation networks to allow emergency personnel to safely access the impacted area, to help residents to evacuate, and to enable emergency assistance to reach the area. Clearing roadways requires coordination of jurisdictions at the local, state, and federal levels where the ownership of these roadways overlaps. State and local departments of transportation (DOTs) must invest in training, resources, and debris management resources, training, and plans; and understand the timeline, roles, and responsibilities of each agency in debris removal on roadways. These roadways should be identified and prioritized in advance. Emergency officials should plan to reduce and remove debris blocking access to critical infrastructure, such as hospitals, schools, telecommunications, power generation, and sanitation facilities (e.g., drinking water, sewage, and storm water treatment).

Emergency officials should communicate with the public regarding preparing for debris removal and their important role. Jurisdictions should provide community education, so residents and businesses know what to expect and the activities required of them. Homeowners may be asked to move their debris to the right-of-way, or the curb, for collection or take their debris to a collection site. Collection sites may be more convenient for centralized collection, while curbside pickup provides residents with the highest level of service. Collection sites are better suited for rural, sparsely populated areas or where roads/geography are difficult for debris removal trucks and equipment to access. Communities may also opt to host HHW drop-off events to encourage residents to safety and legally dispose of HHW to avoid comingling the waste with other debris.

When debris is moved to the curb, residents are asked to separate their debris by type for collection. How quickly debris removal can be performed depends on how quickly residents can move debris to the right-of-way. This process can begin several days after the disaster and last for weeks or months, depending on the volume and available resources. Severe impacts from disasters, where the damage is extensive, may delay the debris removal process. Additionally, clearing debris after a human-made disaster, industrial accident, or terrorist attack could be delayed due to criminal evidence collection (Federal Emergency Management Agency (FEMA), 2002). Generally, emergency protective measures, including debris removal, should be completed within six months. However, catastrophic disasters may result in debris operations lasting several years. Initial cleanup from Hurricane Katrina in 2005 lasted more than three years (Luther, 2008). Disasters affecting hazardous materials manufacturing or storage sites, such as oil and gas refineries, may also experience cleanup delays while the jurisdiction prepares plans to address safety risks.

Each state and local community has laws governing rights of private property owners, and debris removal from private property must be addressed in the plan. Most jurisdictions do not remove disaster debris from businesses and private property. Business owners and gated communities may be excluded from disaster debris removal on private property. Debris removal from private property is generally ineligible for federal assistance because it is considered a property owner's responsibility that may be covered by private insurance. When debris poses a public health or safety risk, blocks first responders' access, or promotes economic recovery, local officials may act to remove debris from private property and roads (e.g., gated communities and homeowner associations).

Local government officials should also provide regular updates on the progress of debris removal, when neighborhoods can expect to have their debris picked up and address any environmental concerns the public may have. Information must be distributed quickly to reduce the risk of rumors and misinformation that could erode confidence in the management of the recovery. Because the public plays a large role in debris separation, emergency officials should provide easy-to-understand instructions and support. Social media was effectively used to help residents understand their roles and expedite debris separation and placement at curbside following Hurricane Harvey in Texas in 2017 (Page-Tan, 2021).

Separation and Collection Sites

As local landfill space is limited, communities generally attempt to divert as much waste from landfills by using the process of separation, recycling, or even composting. Mixed debris is segregated according to debris type before disposal to promote recycling and achieve operational and cost-effectiveness. Debris should be separated into vegetative, white goods (e.g., large appliances), C&D, hazardous waste, and electronics. Household waste is generally excluded from disaster debris pickup. Additionally, disaster debris must be separated so that it can be properly disposed of to meet federal regulations and eligibility for federal assistance. Types of debris can be separated before pickup. Residents are commonly asked to sort their debris on the curb to assist the debris separation process. Separation is often time consuming, difficult, and sometimes hazardous. The process requires residents to perform the separation, where debris can be mixed to the point that separation is not possible. Further, residents may have evacuated the area due to the disaster and not be able to return to perform the debris separation (Luther, 2008). Separating debris is necessary to comply with federal waste management laws. The Resource Conservation and Recovery Act (RCRA) governs the management of hazardous waste, which may be inextricably mixed with or contaminating disaster debris. The United States Environmental Protection Agency (EPA) issues regulations and guidelines to oversee the proper handling of these waste types and provides technical assistance to state and local governments for managing debris.

Hazardous and nonhazardous solid waste must be managed in compliance with relevant federal, state, local, and tribal regulations. States have the primary responsibility for enforcing hazardous waste regulations and implementing nonhazardous waste programs and generally require disaster debris removal as part of regulatory requirements for health, safety, and sanitation. Disaster debris that includes hazardous materials is subject to the U.S. Department of Transportation's Hazardous Materials Regulations (49 CFR parts 171–180), which requires hazardous materials classification, hazard communication, packaging, and training. Additional federal and state laws cover requirements for special types of disaster debris, such as hazardous materials and marine debris.

In addition to pick up, debris can be brought to a collection center or DMS. Collection centers can be vacant lots, fairgrounds, parks, parking lots, or warehouse facilities. When using collection sites, residents are asked to bring their debris to the predetermined locations. Short- and long-term temporary sites can be used as staging areas when debris must be transferred from one collection method to another for further transport or for processing (i.e., sorting) before being transferred to a disposal site.

Monitoring

Debris monitoring describes activities between the jurisdiction, as the applicant for Public Assistance, the state, and FEMA to ensure debris removal is efficient and effective, and confirm the debris is eligible for federal financial assistance. Debris monitors observe and document debris removal from the point of collection to the final disposal point to confirm the work meets federal assistance guidelines and federal, state, and local regulations. Failure to properly monitor debris removal operations may jeopardize federal financial assistance.

Monitors ensure trucks are fully loaded, as the exact amount of eligible debris must be properly documented to receive Public Assistance funds. Monitors are stationed at towers at the DMSs to track the trucks entering and leaving. Debris monitoring is often performed by force account labor. However, large-scale disasters can quickly overwhelm a jurisdiction's workforce. Debris monitoring contracting should be addressed in the plan, and when the jurisdiction will use monitoring contracts. Whether force account labor or contractors, the debris monitor should not have a financial relationship with the debris management contractor to avoid conflicts of interest.

Disposal and Recycling

A significant proportion of disaster debris is vegetative. When clean and not otherwise contaminated by hazardous waste, vegetative debris can be reduced and recycled by being chipped or ground up for re-use. Because of the large volume of vegetative debris, jurisdictions may seek to burn it rather than place it in a landfill. Both options create logistical and environmental impacts that must be weighed for the community's benefit.

Some debris can be recycled, including asphalt, wood, and metals. The debris management plan should determine if recycling is feasible and describe how the jurisdiction will identify, separate, transport, and recycle the materials. Counties that developed debris management plans recycled an average of 46% of their total disaster debris, while counties without these plans only recycled about 26% (Crowley, 2017).

Another sizable proportion of disaster debris is C&D waste. Since C&D is not considered hazardous or municipal waste, it can be disposed of in a landfill not meeting stricter federal design and operation standards. However, since C&D is often commingled, jurisdictions should dispose of it in appropriate landfills meeting the stricter requirements.

Planners should consider the location of debris disposal sites based on several factors. Final disposal sites may be in close proximity to the jurisdiction to minimize hauling distances. However, this distance should be located far from potable water sites and other bodies of water to reduce environmental impacts. Both temporary staging and final disposal sites should reduce the likelihood of becoming a public nuisance or health hazard by creating dust, odors, or traffic. Debris sites should reduce impacts to residential neighborhoods, schools, hospitals, historic, and environmental sites.

Debris disposal carries costs and potential liability risks, leading many jurisdictions to opt for permanent debris disposal sites on property they own. However, arrangements can be made to use another jurisdiction's landfill or other disposal sites. Hurricane Katrina in 2005 generated over 100 million cubic yards of disaster debris (Luther, 2008). Every million cubic yards of debris shortens the life of a typical landfill by five years (Environmental Protection Agency (EPA), 2008). Local planners should be prepared to address the demand on their local landfills or identify alternatives that align with costs and environmental requirements.

Post-Incident and Plan Maintenance

After a disaster and the completion of the debris removal process, emergency officials should perform an after-action review to identify how effectively the plan worked to meet the community's needs. This review identifies lessons learned that should be incorporated into the plan to improve its effectiveness. The debris management plan should be reviewed and revised annually with the most current regulations, community information, and relevant lessons from debris removal operations in other jurisdictions. These activities can also be practiced and tested as part of an emergency management exercise program.

Special Considerations for Debris Removal

Special considerations or alternative debris removal laws and processes apply to unique situations such as commercial airline accidents, debris involving military operations or property, or biohazard facilities. Emergency planners should identify geographic locations of higher risk facilities and understand the types of materials at each location, to plan for the types of debris that could be expected after a disaster. Emergency planners may opt for site visits or facility operators to submit disaster plans that demonstrate how the site owner will protect its property and surrounding area from potentially harmful disaster debris.

Disaster debris often contains the remnants of people's homes and belongings and may carry personal significance. Disasters can destroy culturally significant structures and lands that are meaningful to a community. Emergency planners should recognize how removing these destroyed buildings and other debris will affect individuals and groups and outline how personal property and valuables are handled and processed during debris operations. Debris removal from disasters can pose serious public health threats, as in the case of survivors and responders to the 9/11 terrorist attacks on the World Trade Center in New York. Responders and nearby residents reported a range of illnesses soon after working at the site (Centers for Disease Control and Prevention (CDC), 2023).

Destroyed sites, such as a collapsed building after an earthquake, may require clearance from search and rescue personnel to ensure human remains are properly removed before removing large amounts of debris. Debris operations personnel should notify authorities if human remains are found. Disasters, especially floods, can disinter coffins and carry them far from their original site. Technological disasters such as accidental explosions, airplane crashes, and human-made disasters like terrorist attacks can also result in significant disaster debris. Debris considered crime scene evidence must first be cleared by officials before debris removal can begin.

Marine debris, or disaster debris that has entered waterways, requires unique planning to consider environmental and regulatory impacts. Marine debris can include boats, ships, and fishing gear, but also debris from land that has been pushed into the water. Marine debris can pose a threat to navigation that could create difficulties reaching people in need. Further, marine debris can threaten natural habitats, marine animals, aquaculture, shipping, transportation, and sources of drinking water.

FEMA reviews each Public Assistance project to ensure that the debris management operations comply with applicable federal historic preservation laws. Some of these laws include the National Environmental Policy Act, the National Historic Preservation Act, the Endangered Species Act, the Clean Water Act, the Clean Air Act, the Coastal Barrier Resources Act, the Resource Conservation and Recovery Act, the Coastal Zone Management Act, the Farmland Protection Policy Act, the Fish and Wildlife Coordination Act, and the Wild and Scenic Rivers Act. Local planners should identify how debris is managed on potentially historic or archaeologically significant sites in compliance with historic preservation laws. Debris removal and disposal must also consider environmentally sensitive area such as wetlands and critical animal and plant habitats. For example, the BP Deepwater Horizon drilling disaster and oil spill cleanup on Florida beaches required planning for protected species of beach mice and shore birds (Department of the Interior (DOI), 2010).

Conclusion

Disaster debris removal is critical for local jurisdictions to enable emergency response, protect public safety, and facilitate community recovery after a disaster. Debris management is closely tied to the United States' emergency management system and federal disaster relief policies, as federal financial assistance is largely used to pay for debris removal. This complex process includes coordinated participation from government stakeholders, contractors, NGOs, and the public. Debris management plans help communities prepare for the overwhelming amount of debris and resources required to remove it. These plans can help communities expedite debris removal by forecasting the type, location, and amount of debris expected, and assigning roles and responsibilities to the various stakeholders. Emergency planners should stay abreast of federal policy changes, environmental regulations, and special types of debris requiring additional resources and management expertise within their communities.

References

Centers for Disease Control and Prevention (CDC). "Toxins and Health Impacts." 2023, https://www.cdc.gov/wtc/exhibition/toxins-and-health-impacts.html

Crowley, Julia. "A Measurement of the Effectiveness and Efficiency of Pre-Disaster Debris Management Plans." *Waste Management*, vol. 62, 2017, pp. 262–273.

Department of the Interior (DOI). "Assessing Damages from the BP Deepwater Disaster." 2010, https://www.doi.gov/ocl/hearings/111/DamagesFromBPDisaster_072710

Environmental Protection Agency (EPA). "Planning for Natural Disaster Debris." 2008, https://www.epa.gov/sites/default/files/2015-08/documents/pndd.pdf.

Federal Emergency Management Agency (FEMA). "Managing the Emergency Consequences of Terrorist Incidents." 2002, https://www.fema.gov/pdf/plan/managingemerconseq.pdf

Federal Emergency Management Agency (FEMA). "Public Assistance Debris Management Guide." 2007, https://www.fema.gov/sites/default/files/2020-07/fema_325_public-assistance-debris-mgmt-plan_Guide_6-1-2007.pdf.

Federal Emergency Management Agency (FEMA). "2021 Preliminary Damage Assessment Guide." 2021, https://www.fema.gov/disaster/how-declared/preliminary-damage-assessments/guide.

Luther, Linda. *Disaster Debris Removal After Hurricane Katrina: Status and Associated Issues*. Congressional Research Service, the Library of Congress, 2008. http://www.opencrs.com/document/RL33477/. Accessed 25 Nov. 2023.

Luther, Linda. *Disaster Debris Management: Requirements, Challenges, and Federal Agency Roles*. Congressional Research Service, 2017.

Page-Tan, Courtney. "An Analysis of Social media Use and Neighbor-Assisted Debris Removal in Houston Following Hurricane Harvey." *International Journal of Disaster Risk Reduction*, vol. 63, 2021. https://doi.org/10.1016/j.ijdrr.2021.102450

Section II

Channeling Disaster Response

5

FLASH APPEAL AS A FRAMEWORK FOR DISASTER RESPONSE

Subarna Chatterjee

Introduction

A disaster is 'a serious disruption of the functioning of a community or society, which involves widespread human, material, economic or environmental impacts that exceed the ability of the affected community or society to cope using its own resources' (UN General Assembly, 2016, 13). Disaster management aims to minimize the impacts of disasters through mitigation, preparedness, response, and recovery (Koob, 2000). While prevention and preparedness phases in disaster management are anticipatory (i.e., they are carried out before the disaster strikes), response and recovery measures are undertaken post-disaster and include emergency relief, early recovery (ER), mid- to long-term recovery, and development phases (Ingram et al., 2006; Vallance, 2011; Thomalla et al., 2017).

The emergency relief measures include provision of lifesaving needs, such as food, water, medical care, and other necessities, to keep people alive. With time and active efforts from humanitarian agencies, this phase slowly transitions into 'recovery,' as the affected population now mobilizes resources toward rebuilding their communities and increasing resilience (Winkworth, 2007). While it is the responsibility of the government(s) of the affected nation(s) to fund disaster response and recovery, sometimes the scale of the disaster exceeds the government's capacities. The United Nations issues a flash appeal (FA) to secure immediate international humanitarian assistance to support a government's response to an emergency and some ER projects (IASC, 2009a; Kusumasari et al., 2010).

An FA is an internationally recognized tool for structuring a coordinated, interagency humanitarian response for the first three to six months after an emergency (IASC, 2006a). It is generally issued by the United Nations Office for the Coordination of Humanitarian Affairs (UNOCHA) within about three to seven days following an emergency. Sometimes, minor appeals are issued by Red Cross Societies and international nongovernmental organizations (NGOs) (Kukkonen, 2018). There is no specific list of emergencies or types of disasters that qualify for an FA, nor is there a threshold of loss and damage that, when crossed, may trigger an FA. The only criterion to qualify for an FA is when the funds required for an emergency exceeds the government capacity to respond (IASC, 2009a). An appeal can be issued for both rapid-onset disasters such as an earthquake (e.g., 2023 Turkey) (UNOCHA, 2023a)

DOI: 10.4324/9781003414834-9

or a flood (e.g., 2023 Libya) (UNOCHA, 2023b) and slow-onset disasters/hazards such as an ongoing drought (e.g., Kenya, Oct. 2021 to Oct. 2022) (UNOCHA, 2022). FAs are also issued in response to wars and political crises (e.g., Occupied Palestinian Territory, Oct. 2023) (UNOCHA, 2023c). Issuing an FA as a response to slow-onset disasters is more challenging. For slow-onset hazards, the requirements may be more recovery-based than humanitarian or may require funds and other resources to prepare for the worsening nature of the crisis. Slow-onset hazards are also prolonged, which makes it difficult to identify the correct time to issue the appeal. FAs for regional crises, which involve two or more countries, are less common in number. This is because it is challenging to bring together all relevant actors, mobilize the necessary information from different nations, and still be able to issue the appeal within three to five days of the crisis (IASC, 2009a).

Issuing a Flash Appeal

The UN Resident Coordinator (RC) or the Humanitarian Coordinator (HC) with support from UNOCHA initiates the FA. The office of the RC/HC decides the scope of the appeal. It also specifies certain parameters of international response to help structure the duties of the involved agencies/clusters and to avoid confusion and delay in response. The humanitarian cluster leads then conduct needs assessments and outline the response plans for their respective sectors within the parameters set by the RC/HC (IASC, 2009a). Clusters are groups of humanitarian organizations that manage different sectors of humanitarian actions (IASC, 2006b). They are food security, health, logistics, nutrition, protection, shelter, water, sanitation and hygiene (WASH), camp coordination and camp management, ER, education, and emergency telecommunications. The sectors are designated by the Inter Agency Standing Committee (IASC) and have their own responsibilities while responding to an emergency (IASC, 2000). The role of the clusters is to identify the sector/cluster specific losses, suggest priority actions, and determine the required funding. For instance, after the 2015 earthquake in Nepal, the food security cluster required $128 million in funding. The cluster identified that about 3.5 million people might need food assistance, and one of their priority actions was emergency general food distribution and provision of 'high energy biscuits' to the affected (UNOCHA, 2015). The cluster leads are not required to be part of an UN agency. They may be parts of other partner agencies with abilities to efficiently lead the sectors. The cluster approach allows a comprehensive action plan that covers the essential sectors in a relief operation.

An FA is prepared in consultation with, and not limited to, other UN agencies, donor organizations, NGOs, the International Committee of the Red Cross (ICRC), the International Federation of Red Cross and Red Crescent Societies (IFRC), the national Red Crescent Society, and government bodies. An FA can include projects from UN agencies and NGOs. It may also partner with the national Red Cross and Red Crescent Societies of the respective countries. The government cannot directly ask for funds from the FA but can partner with the UN and other NGOs in their relief and rehabilitation projects (IASC, 2009a). Funding applications can be made to the Central Emergency Response Fund (CERF) simultaneously with the FA. CERF is a standby fund established by the United Nations General Assembly (UNGA) for humanitarian responses to emergencies. It mobilizes funds from donors and provides seed funds to humanitarian operations after the first appeal following an emergency and can sanction more funds afterward if needed. FAs have lists of projects to be funded by CERF. After an emergency, both FA and CERF applications must be developed in parallel for the CERF to sanction funds based on the appeal. If CERF has already sanctioned funds before the publication of the FA, it must be mentioned in the appeal (IASC,

2009a). For instance, the FA following the 2015 Nepal earthquake allocated US$15 million from CERF to support critical needs for survivors by the Under-Secretary-General and Emergency Relief Coordinator (ERC) on April 28, 2015 (UNOCHA, 2015).

If funding is likely to be insufficient, the RC/HC appeals to the ERC for initial funding based on available information about the scale of the emergency and intensity of loss and damage. CERF has funding priorities, and the HC/RC requests the clusters/sectors to identify projects that are of the highest priorities within their respective clusters/sectors. All selected proposals are sent to the ERC and CERF Secretariates.

Permission of the affected government is not required to issue an FA. It is, however, recommended by the UNGA that clusters should work in consultation with the concerned government(s) both ahead of the emergency as a part of the contingency planning process and during the post-emergency relief operations. This includes involving the government(s) in disaster preparedness, post-disaster response planning, and implementation of projects. Such actions may establish better accountability between the government and humanitarian partners. Moreover, identifying the potential requirements in each sector/cluster for the concerned country ahead of the emergency enhances a more coordinated, effective, and in-time response during and after an emergency. An FA should specify the areas being funded by the government and identify additional requirements that are to be covered by international assistance. Sometimes governments might be opposed to seeking international assistance, leading to delays in the implementation of the response plans. Such reluctance may stem from perceptions that, by accepting international aid, the government might lose sovereignty and/or control over the country, 'humiliation' that the country must rely on international aid, or stances toward certain donor organizations or countries. In such cases, IASC suggests that understanding and addressing the reasons for reluctance may help with the process (IASC, 2009a).

Timeline for a Flash Appeal

Apart from some extreme events (Table 5.1), FAs must be issued within three to seven days of the disaster's onset based on available information, best guesses, and inferences rather than waiting for more detailed information and assessments (IASC, 2009a). FAs within a week of the emergency help raise funds swiftly as the crisis is still in the donors' fresh memories. FAs act as the quick guides used by donors to estimate the scale and nature of damage. Some donors may be able to access funds only after the issuance of an FA and based on the urgency of the crisis. Moreover, the credibility of humanitarian actors to respond to the crisis may be questioned, and

Table 5.1 Flash Appeal Information for Selected Disasters

Year/country/event	*Disaster occurred/ start date*	*FA issued*	*FA length (no. of pages*)*
2010 Haiti Earthquake	12 January	16 January	87
2013 Libya Flood	10 September	14 September	14
2015 Nepal Earthquake	25 April	29 April	17
2023 Syria Earthquake	6 February	13 February	38
2023 Turkey Earthquake	6 February	16 February	26

Notes: *Excludes cover page
Sources: UNOCHA (2010, 2015, 2023a, b, d).

hence donations may be delayed if they fail to put together an appeal within a few days of the crisis. In case an FA takes more time to be issued, different agencies may issue individual appeals, creating confusion among donors and stakeholders who may seek a more trustworthy and coordinated response from humanitarian actors.

FAs have a fixed duration for the response/projects for which funds are collected, usually three to six months from an emergency. This time frame, however, is flexible and can be modified (extended or shortened) during the revision process of the initial appeal, depending on more available information and project accomplishments. The usual duration of the appeal is three months. In case projects need to be extended beyond six months, the appeal can be converted into a longer term or a Consolidated Appeal. Given that the appeals take time to be funded, it is common for projects to run beyond the specified time frame (IASC, 2009a).

On the first day after an emergency, an FA is initiated by the RC/HC in consultation with the humanitarian country team (HCT). The RC/HC appoints one organization to lead the response in each priority sector/cluster. If there is no UNOCHA in the affected country, the RC/HC assigns an appeal focal point to coordinate response and input among the various agencies working in the field. The RC/HC consults the affected government. This is followed by rapid needs appraisal on day 2 after the emergency, and the RC/HC communicates the scope of the appeal and projects to the HCT and cluster leads. The cluster leads meet on a national level to determine requirements, discuss the respective cluster capacities, and assign responsibilities within the clusters. On day 3, the RC/HC is required to draft sections of the FA. The RC/HC decides the duration of the FA and communicates with the HCT.

All organizations in the relevant clusters/sectors meet to decide upon their respective response strategies, select team members, and propose projects. At this stage, the cluster members also review baseline data and other information derived from their work done ahead of the emergency and decide upon the parameters to use the information. The cluster leads then draft a response plan and finalize the inclusion of selected projects within the response plan and share it with the OCHA or the assigned focal point. OCHA/focal point then combines each draft and finalizes the FA. The RC/HC then approves the final draft of the FA and sends it to UNOCHA Consolidated Appeals Process (CAP) Section in Geneva, Switzerland; no further changes can be made to the FA once it is sent to Geneva. The OCHA CAP Section headquarters may contact the RC/HC in case a major change is required in the FA (IASC, 2009a). Otherwise, the appeal is officially through a donor meeting or through press release. This timeline may vary according to the situation in the affected country (IASC, 2006a).

Even though FAs are meant for immediate humanitarian assistance, they may request funding for ER projects. When ER is systematically planned, it allows humanitarian funds to be used in more outstanding time-critical emergencies. For the first edition of the FA, ER projects may not be appealing to donor organizations because they may prefer funding immediate lifesaving humanitarian needs. ER projects may still find a place in the first edition of the appeal only if the requirements are such that the projects have an advantage to be started at the beginning, the projects address time-critical needs, and the projects can be completed within the regular six months' time frame of the FA. It must also justify why the ER must be started earlier. A startup cost or ER projects can be included in the FA. More assessment-based ER needs, such as post-disaster needs assessments, can be included in the FA revision. The IASC suggests that including ER projects in the FA may be essential to the affected people as they may provide a more 'durable' solution to emergencies. In case the ER projects need more time to be implemented, the six-month time frame can be revised. ER projects may also address lifesaving activities that are compatible with the scope of the FA (IASC, 2009a; UNDP, 2015).

Scope and Contents of a Flash Appeal

FAs generally provide information about the overall and cluster-specific losses and damages due to the disaster, outline the immediate relief activities and ER projects, and specify the funds required for their implementation. More specifically, FAs contain the following sections.

Overview of the Crisis

The first section of any appeal provides a brief overview of the crisis. It tells us about the nature and scale of the disaster. For instance, in the case of an earthquake, it provides information about its intensity, the number and intensities of aftershocks, location of the epicenter, and maps of the affected areas. It also provides an overview of the reported numbers of fatalities, injuries, losses, and damages to properties in USD and essential services damaged and/or destroyed (e.g., hospitals, medical facilities, maternity facilities, and schools) (UNOCHA, 2023a). The impacted population is then disaggregated by socio-economic and demographic characteristics, regions, and the extent of sustained damage. These numbers are also represented by relevant figures and tables for ease of understanding by the donors. If the country was already experiencing a socio-political crisis or war, the extent of additional damage caused by the natural disaster to the affected population and on the refugee population (if any) is also mentioned. At the end, this section contains a justification for international funding requirements (UNOCHA, 2015, 2023a, b).

Emergency Needs

Assessments of emergency needs are conducted based on existing information and preliminary field observations (IASC, 2009a). This section lists the sector-wise requirements of immediate humanitarian needs, such as the requirements for: (a) emergency health services (e.g., medicines, first aid kits, body bags, surgical kits, other hospital equipment); (b) emergency food and nutrition; (c) emergency shelters and safe spaces for women, children, and vulnerable populations; (d) heavy machineries and other tools to remove debris (if needed); (e) specialists for the management of mental health and trauma from experiencing death and destruction; and (f) clean water, cleaning of water bodies, and removal of garbage and solid waste to mitigate waterborne diseases (UNOCHA, 2015, 2023b).

Response and International Solidarity

This section provides the overall approach of the affected government and humanitarian partners toward the emergency. It lists the amount of funds and other necessities already allocated by the government. For example, the FA issued after the 6 February 2023 earthquake in Turkey stated that nearly 249,000 search and rescue (SAR) workers and more than 12,200 vehicles, including excavators, tractors, cranes, and bulldozers, were deployed in affected areas by the Turkish government. Twenty-six ships were employed in the relief efforts, and more than 4,200 sorties had been undertaken by 170 helicopters and 76 aircraft by 14 February (UNOCHA, 2023a).

This section also lists the involvement of humanitarian partners in SAR operations, food distribution, distribution of clean water and hygiene kits, emergency cash distribution, and health care. This includes the provision for psychological care (mostly for vulnerable groups), arranging temporary learning spaces for education, supporting groups that work toward preventing violence toward women and the elderly, assessing water and hygiene services to determine the needs, and providing temporary shelter, clothing, blankets, mattresses, etc. (UNOCHA, 2015, 2023a).

Strategic Objectives

The Strategic Objectives section lists the goals of the UN and their humanitarian partners in the emergency response support to the affected government(s). This may include and is not limited to: the provision of lifesaving and timely assistance, including education, food, clean water, and protection; timely delivery and recovery of essential services such as schools, health centers, water facilities, and mental health support (UNOCHA, 2023a); provision of livelihood in the affected areas by generating temporary employment in the relief operations; strategies to ensure the protection of vulnerable populations through protection monitoring, child protection plans, and mental health support; and establishment of proper communication within the affected communities to circulate information about the range of services people can avail (UNOCHA, 2023d).

Response Strategy

Keeping the welfare and protection of people as the main criteria of the Response Strategy. This section ensures the best use of the local actors and national agencies, including the national Red Cross and Red Crescent Societies, in the disaster response. However, given the magnitude of the event and limited governmental capacity, unified international support from various organizations may be ensured under the FA. The response strategy also plans to provide cash assistance to people wherever feasible. Cash assistance is more 'cost-effective' than in-kind assistance and allows people to prioritize their needs accordingly. Otherwise, material convergence may emerge, which is defined as the movement of solicited and unsolicited donated relief goods toward disaster sites (Paul, 2019). However, in the case that cash assistance is not feasible, humanitarian partners ensure help through other forms of assistance (e.g., in-kind) (UNOCHA, 2023a).

The Phases of Response

FAs outline the various phases of planned response operations. This includes SAR operations focused on rescuing trapped people from debris or rescues via boat or aircraft, for instance. This phase is the most critical, as it involves people who are at direct risk of losing their lives. This is followed by the second phase of immediate lifesaving response, which begins with the assessments of lifesaving requirements. At this phase, humanitarian actors provide lifesaving resources, such as food, water, shelter, clothing, blankets, emergency medical facilities, and sanitation facilities. This phase also involves identifying and planning protection mechanisms for vulnerable groups and acting toward reunifying families. The third phase involves building resilience through community-based recovery projects, reestablishing livelihoods, rebuilding infrastructure, and providing basic social services. This phase involves a transition toward longer-term recovery efforts.

Sectoral Response

To enable a more comprehensive action plan, FAs outline the responsibilities of different sectors/clusters, namely: education, emergency shelter and non-food items, food security and livelihoods, health and nutrition, protection, temporary settlement support, WASH, multipurpose cash and social protection, ER, and debris removal. The FAs specify losses and damages, lay out funding requirements, and outline a response strategy for each sector. Sectoral responses may differ across emergencies due to differences in the nature of the emergencies, their impacts, and the geography of the affected place. However, the sectoral response activities have similar goals to a certain degree (UNOCHA, 2023a, b, d, e).

Figure 5.1 Temporary Grade School Opened after the 2015 Earthquake in Chautara, Nepal
Source: Image from author.

The education sector provides a safe learning environment and temporary learning spaces to children (Figure 5.1), provides emergency education kits, supports renovations to damaged schools, trains teachers to identify children's psychological issues post-disaster, and provides general support to educators and caregivers (UNOCHA, 2015, 2023a). FAs issued for slow-onset hazards, such as droughts, have additional longer-term priorities. For instance, for the FA issued for Kenya drought in 2021–2022, priorities in the education sector were increasing the numbers of school meals, providing storage facilities for food commodities, transferring money to schools to provide food to vulnerable households, providing safe water, water harvesting in schools, attempts to bring existing boreholes to life, encouraging students to attend school, and encouraging hygiene among students (UNOCHA, 2022).

The emergency shelter and non-food item sector prioritizes the provision of emergency shelters to the displaced and critical relief items, such as tents, blankets, sleeping bags, and warm clothing. It also prioritizes identifying the at-risk populations; ensuring safe spaces for children, women, and the elderly; providing emergency cash assistance for short-term rentals; and providing construction tool kits and technical support toward renovation of damaged houses (UNOCHA, 2015, 2023a, e).

The food security and livelihood sector focuses on delivering cooked meals, food packets, and cash vouchers. In agricultural areas, it prioritizes the distribution of relevant agricultural equipment or cash assistance to farmers and aids livestock farmers through animal health monitoring and vaccinations (UNOCHA, 2015).

The health and nutrition sector prioritizes coordination of medical teams throughout the emergency period, deployment of mobile medical units, provision of lifesaving medical equipment, provision of nutrition supplies, the training of healthcare workers, and measures to control waterborne diseases. It also prioritizes the provision of essential support to the primary healthcare sector and promotes mental health awareness and access to mental health facilities and trauma response centers. In addition, this sector establishes temporary diagnostic centers; rebuilds medical infrastructure; and promotes access to reproductive and sexual health services, immunization services, and services to amputees (UNOCHA, 2015, 2023a, b, d). For slow-onset hazards, the additional sectoral response strategy includes identifying severely malnourished children and pregnant and lactating women through mass screenings. It also prioritizes the provision of therapeutic and supplementary food (UNOCHA, 2022).

The protection sector prioritizes providing security to affected people and refugees (if any), supporting the evacuation of children, providing recreational supplies to children, establishing survivor centers, using community-based identification systems to identify at-risk women and ensure their protection, distributing maternity and menstrual hygiene kits, and ensuring protection and health to front-line workers (UNOCHA, 2023a).

The water, sanitation, and hygiene (WASH) sector installs temporary sanitation facilities, mostly in the areas where displaced people have gathered. The WASH sector also assists people in hospitals, other healthcare units, and temporary shelters by providing safe drinking water, toilets, and bath spaces. In temporary shelters housing people displaced due to the emergency, the sector distributes hygiene kits, promotes hygiene to reduce outbreaks of waterborne and other infectious diseases, and provides limited facilities for the collection of solid waste (UNOCHA, 2015, 2023a).

The multipurpose cash and social protection sector focuses on providing cash assistance for social protection for three months following the emergency (UNOCHA, 2023a).

Debris removal is one of the important areas to address in an FA issued after an earthquake. This addresses removal of debris, assistance in the SAR operations, restoration of access roads, and service sectors to help mobilize resources and strengthen the overall restoration of infrastructure (UNOCHA, 2015).

Early recovery (ER) priorities include restoration of livelihood infrastructure, generation of income by debris removal and recycling, provision of small-scale financing to small businesses toward their recovery and to boost employment, and aid in the primary and secondary economic sectors for their revival. It also includes opening community assistance centers for the protection of vulnerable populations, assistance in accessing social security, and adopting action plans to restore and preserve the cultural heritages in the affected areas (UNOCHA, 2015, 2023a).

The emergency telecommunications sector aims to enhance communication facilities in the affected areas. It prioritizes maintaining information sharing platforms and security communication services to support rescue operations and using radio communication services to communicate with people in remote areas (UNOCHA, 2015). It also prioritizes training humanitarian actors on the proper use of telecommunication services in emergency relief and encourages the use of a common platform for humanitarian partners to communicate with the affected populations throughout the emergency areas (UNOCHA, 2023a). Additionally, this sector seeks to provide internet facilities to the affected people in select areas, help desk services, and charging stations.

Finally, the *logistics sector* aims to enhance the flow of humanitarian resources into the affected areas, support better coordination at entry points for the effective flow of resources, and encourage the sharing of logistics and logistical constraints across the various activity sectors (UNOCHA, 2023a).

Preparedness for a Flash Appeal and Contingency Planning

According to IASC (2009a), planning response strategies prior to the disaster is important. Effective preparedness includes prior identification of the areas of risks and vulnerabilities, identification of baseline data, and simulation exercises, all of which may help HCTs with the response mechanisms. Additionally, working with government agencies on disaster preparedness may help coordinate the government's response post-emergency. Prior formation of clusters and clear assignments of their roles and responsibilities are also important. Drafting generic projects based on prior risk assessments may also be helpful.

Interagency contingency planning is also a key component in disaster preparedness. It is a tool to assess the impacts of potential hazards so that adequate response strategies can be formulated ahead of emergency. This is important when the scale of the potential emergency may require coordinated action from multiple national and international agencies. Interagency contingency planning must operate on regional, national, and local levels. It should identify the potential emergencies, analyze the potential impacts of the emergencies on the population, identify strategies to respond to the emergencies in a timely manner, and analyze the response strategies and policies, including critical actions to be undertaken during and after the emergencies. The interagency contingency planning should involve the governments of different countries as it is the primary responsibility of the governments to provide post-disaster relief. The planning should consider the capacity of the system of national and local governments. Proper contingency planning helps in the timely issue of an FA, especially in places with a dearth of humanitarian response resources or with no OCHA. A pre-prepared FA may also be a part of the contingency planning based on the areas of potential risk, baseline data on demographics and vulnerability, and response strategy of international humanitarian organizations (IASC, 2009a).

Flash Appeal Revision

The first version of the FA that is issued within three to ten days of the emergency is based on pre-emergency baseline data/information along with best guesses and inferences. With time, more available information on the disaster-affected areas, and ongoing donor support, it becomes necessary to revise the original FA. The cluster/sector leads, in consultation with the RC/HC, maintain a record of the ongoing projects (including donor support and expenditures) and add potential projects and funding requirements in Financial Tracking Service (FTS), an online platform managed by OCHA. Within one month from the date of issue of the original appeal, the HCT is required to publish a revised version of the appeal specifying the new projects, funding requirements, and updated response strategies. Considering the funding received and project achievements, the HCT may change the response strategies of the existing projects, add new projects, and scrap older projects if needed. While assessing the funding situation, the HCT must consider donations received outside the appeal. The HCT may also decide whether the updated strategy under the revised FA still prioritizes lifesaving humanitarian needs or a wider range of ER projects with longer-term impacts (IASC, 2009a).

The revised version of the FA presents an updated report with international and national responses to date. Based on newly available information, it provides the figures of the affected population disaggregated by demographic, socio-economic, and geographical variations. It also provides the most updated assessments of the respective cluster/sectoral responses and the funding received by each of them. It identifies gaps in the response operations, tells whether the project strategies produced their anticipated effect, proposes new/updated strategic plans, and incorporates new projects with proper justifications. The revised appeal now includes the organizations that were not a part of the original FA. If the project does not need any additional funding, then a simple revision is made to it. The revised appeal should also mention whether the projects will be more humanitarian or longer-term ER (IASC, 2009a).

The RC/HC is responsible for the revision of the project under the guidance of the HCT. The cluster/sector leads now have the responsibility to review and confirm the status of the projects and identify the priority areas. This must be done in consultation with partner agencies and all donor organizations who were involved in the humanitarian response to ensure that there is an agreement among the RC/HC and donors. The revised appeal prioritizes the projects that are most urgent. Project prioritizations are difficult for the original appeal but are crucial to have in the revised edition and are essential for understanding whether the most urgent needs are addressed. According to the IASC guidelines, lifesaving priorities top the list, but not all priorities are lifesaving. Priorities should be in accordance with the nature of the emergency and the respective requirements. The project's priorities may be more humanitarian or ER, and the most significant part of the specific crisis that needs to be prioritized may be the operational part of the project and may focus on a particular region, groups of people, or its time sensitivity. Sometimes ER projects may be prioritized, such as the provision of seeds and other agricultural equipment, to reduce future dependence on aid (IASC, 2009a).

Financial Tracking Service (FTS)

The FTS is a database managed by OCHA that records all reported humanitarian aid from donor organizations (IASC, 2009b). Contributions to organizations and projects that are included in the appeal and outside the FA are reported in the FTS database. According to IASC (2009b), most donors express interest in counting their donations outside the FA. The FTS specially focuses on FAs, as they cover the major emergencies requiring humanitarian aid. FAs also provide a concise list of project requirements, which allows the FTS to analyze the funding gaps. The FTS is used by HCTs for preparing and revising FAs. In the preparation of the first edition of the appeal, the HCT analyzes the reported funding that has already been given to the emergency. During revisions, it allows HCTs to analyze the received funds by time, location, priority, and type/sector. The FTS also allows HCTs to compare project requirements of various types of emergencies, help review the original projects, and analyze future project requirements and activities needing strategic planning (IASC, 2009b).

Limitations and Conclusion

FAs initiate quick humanitarian actions to support government responses to emergencies. They are effective in mobilizing resources to address the immediate needs of affected populations and initiate ER responses. FAs systematically cover the sectoral/cluster requirements and provide a holistic response framework and strategy. While immensely important, this disaster management tool does have certain limitations. For instance, FAs must be issued within a very short time span after a rapid-onset disaster based on information that may not be accurate and therefore results in poor planning. Dependence on donors for the fulfillment of the proposed projects may also compromise the efficacy of the responses. According to IASC (2009a), projects in the FAs are often delayed due to delays in funding. Moreover, required funding is estimated based on best guesses and inferences on the nature and scale of the disaster (IASC, 2009a). Without proper contingency planning, there may be gaps in the estimated and actual requirements. Additionally, the proposed projects in FAs do not readily consider the geographical differences among places and people or local knowledge, which may be important for robust ER projects. Nevertheless, FAs are internationally accepted and one of the most potent tools for disaster response. It is important that we acknowledge and address the limitations to produce an even more effective humanitarian response.

References

IASC (Inter-Agency Standing Committee). 2006a. *Guidelines for Flash Appeal*. New York: The United Nations. https://www.humanitarianlibrary.org/sites/default/files/2014/02/Guidelines_for_Flash_appeals.pdf. Accessed 30 Nov. 2023.

IASC (Inter-Agency Standing Committee). 2006b. *Guidance Note on Using the Cluster Approach to Strengthen Humanitarian Response*. New York: The United Nations. https://interagencystandingcommittee.org/sites/default/files/migrated/2021-03/Guidance%20Note%20on%20Using%20the%20Cluster%20Approach%20to%20Strengthen%20Humanitarian%20Response.pdf. Accessed 30 Nov. 2023.

IASC (Inter-Agency Standing Committee). 2009a. *Humanitarian Coordination Competencies, 2009*. New York: The United Nations. https://interagencystandingcommittee.org/iasc-sub-working-group-leadership-and-humanitarian-coordination/humanitarian-coordination-competencies-2009. Accessed 1 Dec. 2023.

IASC (Inter-Agency Standing Committee). 2009b. *Revised Guidelines for Flash Appeals*. New York: The United Nations. https://interagencystandingcommittee.org/sites/default/files/migrated/2021-03/IASC%20Guidelines%20for%20Flash%20Appeals.pdf. Accessed 30 Nov. 2023.

Ingram, J.C., G. Franco, C. Rumbaitis-del Rio, and B. Khazai. 2006. Post-Disaster Recovery Dilemmas: Challenges in Balancing Short-Term and Long-Term Needs for Vulnerability Reduction. *Environmental Science and Policy* 9(7–8): 607–613. https://doi.org/10.1016/j.envsci.2006.07.006. Accessed 1 Dec. 2023.

Koob, P., ed. 2000. *Emergency Risk Management: Applications Guide*. Canberra: Australia Emergency Management Australia.

Kukkonen, E. 2018. Inclusion of Evidence-Based Approach to Humanitarian Needs Assessment in Flash Appeals. PhD Thesis, Lund University, Lund, Sweden. https://lup.lub.lu.se/luur/download?func=downloadFile&recordOId=8949003&fileOId=8949007. Accessed 2 Dec. 2023.

Kusumasari, B., Q. Alam, and K. Siddiqui. 2010. Resource Capability for Local Government in Managing Disaster. *Disaster Prevention and Management: An International Journal* 19(4): 438–451. https://doi.org/10.1108/09653561011070367. Accessed 30 Nov. 2023.

Paul, B.K. 2019. *Disaster Relief Aid: Changes & Challenges*. Gewerbestrasse: Palgrave Macmillan. https://www.amazon.com/Disaster-Relief-Aid-Changes-Challenges/dp/3319772813. Accessed 30 Nov. 2023.

Thomalla, F., L. Lebel, M. Boyland, D. Marks, H. Kimkong, S.B. Tan, and A. Nugrobo. 2017. Long-Term Recovery Narratives Following Major Disasters in Southeast Asia. *Regional Environmental Change* 18: 1211–1222. http://doi.org/10.1007/s10113-017-1260-z. Accessed 30 Nov. 2023.

UN General Assembly (UNGA). 2016. *Report of the Open-Ended Intergovernmental Expert Working Group on Indicators and Terminology Relating to Disaster Risk Reduction*. New York: The United Nations.

UNDP (United Nations Development Programme). 2015. *Global Clusters for Early Recovery Strategic Plan*. New York: The United Nations. https://www.undp.org/geneva/global-cluster-early-recovery-gcer. Accessed 30 Nov. 2023.

UNOHCA (United Nations Office for the Coordination of Humanitarian Affairs). 2010. *Flash Appeal Haiti Earthquake*. New York: The United Nations. https://reliefweb.int/report/haiti/consolidated-appeals-process-cap-haiti-earthquake-flash-appeal-2010. Accessed 27 Nov. 2023.

UNOCHA (United Nations Office for the Coordination of Humanitarian Affairs). 2015. *Nepal Flash Appeal for the Response to the Nepal Earthquake April*. New York: The United Nations. https://reliefweb.int/report/nepal/nepal-flash-appeal-response-nepal-earthquake-april-july-2015. Accessed 30 Nov. 2023.

UNOCHA (United Nations Office for the Coordination of Humanitarian Affairs). 2022. *Kenya Drought Flash Appeal*. New York: The United Nations. https://reliefweb.int/report/kenya/kenya-drought-flash-appeal-october-2021-october-2022-revised-may-2022. Accessed 30 Nov. 2023.

UNOCHA (United Nations Office for the Coordination of Humanitarian Affairs). 2023a. *Turkiye Earthquakes Flash Appeal 2023*. New York: The United Nations. https://reliefweb.int/report/turkiye/flash-appeal-turkiye-earthquake-february-may-2023-entr. Accessed 30 Nov. 2013.

UNOCHA (United Nations Office for the Coordination of Humanitarian Affairs). 2023b. *Libya Flood Response Flash Appeal*. New York: The United Nations. https://reliefweb.int/report/libya/libya-flood-response-flash-appeal-sept-2023-dec-2023-issued-september-2023. Accessed 30 Nov. 2023.

UNOCHA (United Nations Office for the Coordination of Humanitarian Affairs). 2023c. *Hostilities in Gaza and Israel – Flash Appeal for the Occupied Palestinian Territory*. New York: The United Nations. https://reliefweb.int/report/occupied-palestinian-territory/hostilities-gaza-and-israel-flash-appeal-occupied-palestinian-territory-version-1-12-october-2023. Accessed 30 Nov. 2023.

UNOCHA (United Nations Office for the Coordination of Humanitarian Affairs). 2023d. *Syrian Arab Republic Appeal for the Response to the Syrian Arab Republic Earthquake*. New York: The United Nations. https://reliefweb.int/report/syrian-arab-republic/flash-appeal-syrian-arab-republic-earthquake-february-may-2023-. Accessed 30 Nov. 2023.
Vallance, S. 2011. Early Disaster Recovery: A Guide for Communities. *Australasian Journal of Disaster and Trauma Studies* 2011(2): 19–25.
Winkworth, G. 2007. *Disaster Recovery: A Review of Literature*. Canberra: Australian Catholic University.

6

POST-DISASTER NEEDS ASSESSMENTS (PDNAs)

Madison Kueny, Grace Blazek, and Bimal Kanti Paul

Introduction

A Post-Disaster Needs Assessment (PDNA) is an internationally accepted methodology to determine damage and losses caused by major natural disasters and to promote effective recovery in lesser-developed countries that have limited resources and finances (Jeggle & Boggero, 2018; Paul, 2019). According to Coffey et al. (2017, 6), a PDNA is a report with 'information on the physical impacts of a disaster, the economic value of the damages and losses, the human impacts as experienced by the affected population, and the resulting early and long-term recovery needs and priorities.' Thus, a PDNA provides detailed data on the physical and human impacts of an extreme event, the monetary value of damages and losses, and the corresponding short- and long-term recovery needs and priorities (GFDRR, 2013).

Immediately following a major natural disaster, the government of the affected developing country asks the international aid community and development partners to prepare a PDNA report; a PDNA cannot occur without a direct request from the affected country (Hinzpeter & Sandholz, 2018). Sometimes, the affected country's government provides input in preparing the consolidated PDNA report (GoN, 2015).

PDNAs require the collaborations with and among the United Nations Development Group (UNDG), World Bank (WB), European Union (EU), and other international agencies such as the Asian Development Bank (ADB) and the International Federation of Red Cross and Red Crescent Societies (IFRC) (Jeggle & Boggero, 2018; Paul, 2019). WB participation in a PDNA is conditioned by an official request, preferably in writing. In addition, the International Recovery Platform (IRP) is a global entity organized by the United Nations Office for Disaster Risk Reduction (UNDRR) to promote resilient and 'build back better' recovery, rehabilitation, and reconstruction using PDNAs. Participation from these agencies improves the credibility of the PDNA, and government involvement at the start of the PDNA process is critical (Jones, 2010). Ideally, the PDNA should commence within one week of a disaster event (Paul, 2019).

The first PDNA was published in April 2008 after Tropical Cyclone Sidr impacted Bangladesh in November 2007 (GoB, 2008). In 2008, seven Joint Damage, Loss, and Need Assessment (JDLNA) reports—as they were referred to then—were published for seven major disasters that occurred in Bangladesh, Bolivia, Haiti, India, Madagascar, Myanmar, and Yemen (Paul, 2019).

DOI: 10.4324/9781003414834-10

In 2009, the reports were officially classified as PDNAs. Approximately 65 PDNAs have been published since 2008, with the highest number (10) published in 2009 (GFDRR & WB, n.d.; IRP, n.d.). Recently, some PDNAs have also required a Rapid Damage and Needs Assessment (RDNA), such as for the 2023 earthquake in Syria (WBG & GFDRR, 2023).

To date, PDNA reports have primarily been conducted in Asia followed by Africa, island states in the South Pacific, Caribbean nations, and then Europe, Central America, and South America. The underlying disasters in descending order are tropical cyclones (including typhoons, hurricanes, and tropical storms) and floods (including landslides), earthquakes, and volcanic eruptions. Only one drought and one tsunami have led to a PDNA. Most PDNAs are published in English, while a few have been published in French and Portuguese and one in Vietnamese (GFDRR & WB, n.d.; IRP, n.d.).

Purpose of a PDNA

The primary purpose of a PDNA is to assess the impact of the disaster, identify priority support needed by the affected government in short-term response and long-term recovery, and recommend interventions for a sustainable and holistic recovery strategy, including a timeframe, costs, objectives, interventions, and expected results (Jones, 2010; Hinzpeter & Sandholz, 2018; Paul, 2019). PDNAs support governments by improving their disaster recovery practices and providing a coordinated approach for identifying and prioritizing recovery activities (EU et al., 2013; Jeggle & Boggero, 2018). PDNAs typically encompass two different perspectives: the quantification of physical damage and economic loss, and the identification of socioeconomic recovery needs in the affected country.

Assessment of the immediate impact on infrastructure, such as houses, roads, and utilities (e.g., electrical and water supply systems), is necessary to quantify the physical damage and economic loss. ‘Damage’ is defined as the total value of destroyed and partially damaged physical assets, capital stock, and material goods in the affected area, while ‘loss’ refers to negative impacts of natural disasters that cannot be repaired or rebuilt (Paul, 2019). Results of loss lead to changes in economic flows as measured from the medium-term effect of the disaster until full recovery of the affected country is achieved, which can take five to ten years depending on the magnitude and extent of affected area (Paul, 2022).

The second perspective, the identification of socioeconomic recovery needs, utilizes information collected from people in the disaster-affected areas (Coffey et al., 2017). The effects of disasters are spread unevenly between the wealthy and poor, male and female, rural and urban areas, as well as the public and private sectors. Combining the two perspectives, the PDNA is an efficient tool for joint assessment, medium-term funding, and long-term recovery planning.

PDNA Structure

PDNAs are structured into four sectors—social, productive, infrastructure, and cross-cutting issues—in order to categorize and aid recovery initiatives (Hinzpeter & Sandholz, 2018). Although the sectors remain the same, the number of sub-sectors changes over time and varies due to the physical and economic characteristics of disaster-affected countries (Table 6.1). For example, the PDNA prepared for the April 2015 Gorkha earthquake in Nepal included irrigation in a new sub-sector (finance) under the productive sector (GoN, 2015). See Table 6.1 for examples of sub-sectors that have been included in PDNAs. PDNAs often comprise volumes A and B. Volume A facilitates planning, organization, and the assessment approach, and outlines the PDNA process and methodology

Table 6.1 Comparison of Sub-Sectors Addressed across PDNAs in Bangladesh, Nepal, and Afghanistan

Sector	*Sub-sectors addressed in PDNA*		
	Cyclone Sidr (2007) in Bangladesh	*Earthquake (2015) in Nepal*	*Earthquake (2023) in Afghanistan*
Social	Health and nutrition	Housing and human settlement	Housing
	Education	Human and health	Health and nutrition
		Education	Education
		Cultural heritage	Cultural heritage
Productive	Agriculture	Agriculture	Agriculture, livestock, and minor irrigation
	Industry	Irrigation	Commerce, private sector, and industry
	Commerce	Commerce	
	Tourism	Industry	
		Tourism	
		Finance	
Infrastructure	Housing	Electricity	Water and sanitation
	Transport	Communications	Energy
	Electricity	Community infrastructure	Transport and telecommunication
	Water and sanitation	Transport	
	Urban and municipal	Water and sanitation	
	Water resource control		
Cross-Cutting Issues	Environment	Governance	Mine action
		Disaster risk reduction	Disaster risk reduction
		Environment and forestry	

Sources: Compiled from GoB (Government of Bangladesh) (2008); GoN (Government of Nepal). (2015); and the UN (the United Nations) et al. (2024). Table created by authors.

or key findings. Meanwhile, volume B provides sector-specific guidance for each of the four sectors (EU et al., 2013). Overlaps across the sectors are common.

The efforts to carry out such a study can be massive. For example, the PDNA for the April 2015 Nepal earthquake utilized more than 250 national and international experts assembled into 23 teams to conduct one-month assessments across 31 districts (Hinzpeter & Sandholz, 2018). Each sub-sector provided responsibility to a group of experts ranging from 2 to 15 persons. Sometimes, depending on the level of development and/or stability of the disaster-affected country, all experts may belong to the international agencies. For example, all 126 experts were from international agencies in conducting the PDNA for the 2023 Herat earthquake in Afghanistan (UN et al., 2024). No clear regulations govern the number of required experts or the timeframe for conducting a PDNA; thus, training on PDNA study implementation would be beneficial to government-appointed experts.

PDNA reports typically begin within one week and are generally prepared within one to four months following a disaster (Paul, 2019). Jones (2010) recommends that an approximate timeline (week by week) with specified activities (stages) should be followed during the preparation of a PDNA report (Table 6.2). The timeline of the PDNA can and should be adjusted to meet the needs of the country and the scale of the disaster being assessed. However, it should also be kept in mind that the timeliness of preparing a PDNA may sometimes be of more importance than ensuring

Table 6.2 Stage, Approximate Timeline, and Activities of the PDNA Preparation Process

Stage	*Week*	*Activity*
Planning	1	Baseline data collection Agree on terms of reference (TOR) Team composition for methodology Training on damage and loss assessment (DaLa)
Assessment	2	Field study Sector-by-sector assessment Calculate DaLa Human needs assessment
Analysis	3	Macroanalysis Report drafting Estimate reconstruction and recovery needs
Dissemination	4	Publish and launch of the PDNA

Source: Modified by authors from Jones (2010). Table created by authors.

100% accuracy of data. Data accuracy is of critical importance, but the prompt completion of a PDNA facilitates the timely flow of much-needed resources and finances to affected communities (and minor flaws can be adjusted during the recovery process). Overall, acquisition of assessment results as quickly as possible is a key element in recovery and the local government's decision making (Jones, 2010; Perwaiz & Sinha, 2018).

A PDNA comprises four deliverables: (1) a merged assessment report with damages and losses, impacts, and effects in each sector and sub-sector and the respective recovery needs; (2) a recovery strategy with corresponding timeframe, costs, objectives, interventions, and expected results; (3) a written basis for the mobilization of recovery resources; and (4) a design for country-led implementation of recovery activities (Hinzpeter & Sandholz, 2018). These deliverables are contained in a damage and loss assessment (DaLa), Human Recovery Needs Assessment (HRNA), and a recovery framework (RF). Each of these documents is discussed in later sections of the chapter.

PDNA Preparation

In terms of preparation, it should first be noted that PDNA preparation costs can exceed $100 million USD—a cost that is typically incurred by international agencies, not the disaster-affected country. Furthermore, in addition to the official PDNA, local and international agencies often publish other reports to evaluate disaster damage for their own purposes. Due to the scale of the undertaking and wide range of full-time, part-time, and expert staff demanded, clear roles of all coordinating agencies and team members are required throughout the PDNA preparation process, and all individuals involved in PDNA production must be identified. Coordinating agencies should also remain uncompetitive during the assessment. This is particularly critical in current- and post-conflict states as well as when cronyism, nepotism, state-favored contractors, corruption, and other cases have the potential to arise. Therefore, clear roles should be delineated to avoid unnecessary competition, data duplication, assessment gaps, and issues of (perceived) dishonesty, misrepresentation, or favoritism.

Jones (2010) has created a resource chart to outline roles and priorities of those involved in PDNA preparation. Recommendations include promoting inclusive ownership of the PDNA process and conducting joint (not competing) assessments. Jones (2010) warns against beginning the PDNA when it may impede the delivery of humanitarian aid and also states that logistics such as interpreters and transportation should not be neglected. Multiple, competing assessments should not be created and no sectors should be ignored due to favoritism or development priorities. Furthermore, effective assessment teams should comprise both technical experts and policy makers, and an appropriate balance of speed and accuracy is required when assessing damages and in production of the final report. As mentioned, producing data in a timely manner is essential; balancing time and data credibility facilitates the prompt production of figures to enable the acquisition of funds to aid recovery needs. Additionally, providing early access to the data and figures allows the mass media to utilize initial numbers to influence recovery through increased aid donations (Jones, 2010). Local media can also influence recovery by reporting estimated damage and loss numbers, thereby decreasing the amount of misinformation, building transparency with the public, and reducing speculation before the PDNA report is released.

Damage and Loss Assessment (DaLa)

As mentioned, a PDNA utilizes a DaLa, an HRNA, and an RF as a streamlined method to assess the severity and range of disaster impacts and how to best allocate resources for recovery. A key objective of the PDNA is to critically analyze the damages, losses, and prioritize recovery needs. Consequently, a variety of stakeholders including governmental organizations (GOs), non-governmental organizations (NGOs), international agencies, and corporations must research, analyze, and prepare the PDNA. If the relevant agencies can coordinate, and the scale of the disaster necessitates it, then a PDNA is more likely to be completed.

The DaLa methodology was first introduced by the UN Economic Commission for Latin America and the Caribbean (ECLAC) in 1972. Partners such as the World Health Organization (WHO), Pan American Health Organization (PAHO), WB, Inter-American Development Bank (IDB), United Nations Educational, Scientific and Cultural Organization (UNESCO), and International Labor Organization (ILO) have improved the methodology over time. The DaLa approach can also be adapted to mesh with a diverse range of governmental structures and standards, potentially aligning with the governmental structure of the disaster-affected country (CRED, 2011; Paul, 2019). The methodology is rigorous and vital to decision makers at all governmental levels.

A DaLa, the first element of a PDNA, should include the following figures: valuation of partially and completely damaged property and other tangible assets; economic losses associated with the interruption of economic transactions and behaviors; expected impact to macroeconomic performance (especially impact to economic growth and gross domestic product (GDP)); and the balance of payments and finances of the affected country's government (GFDRR & World Bank, 2022). The DaLa utilizes quantitative data and analyses to understand the economic impacts of a disaster, which last significantly longer than the duration of the disaster event itself. Economic and livelihood loss should also be associated with each stage of a disaster, including the pre-disaster stage. Livelihood loss often persists well after the disaster, extending beyond the migratory or evacuation period. For example, economic and social growth and performance are still affected years after Hurricane Katrina devastated New Orleans in 2005, although the city has substantially recovered its population and economic base.

Human Recovery Needs Assessment (HRNA)

The HRNA, the second key element of a PDNA, assesses population's coping capacity to recover from the disaster. The objective of the HRNA is to identify the requirements for a resilient recovery of human development of the affected populations, including the restoration of governance systems (UNDP, 2010). Resilient recovery includes the sustainable restructuring of previous systems to increase preparedness for future disasters. It should also mediate levels of inequality since inequalities intensify the effects of disasters upon vulnerable and marginalized groups.

Although the DaLa and HRNA are similar in practice, neither is meant to replace the other; economic and human development are intertwined and expected to overlap. Thus, these two documents are meant to act as complementary streams of analyses to aid in PDNA efficiency and usability (UNDP, 2010). While the DaLa reveals the loss of human livelihood and social impacts, most of the assessment (and others) focuses solely on national accounts and economy. Therefore, the HRNA is necessary to remedy deficiencies in the DaLa and to provide further data to foster the recovery of human development. Like the DaLa, the HRNA is government-led and often supported by a specialized UN agency per official mandates and global conventions (UNDP, 2010). An HRNA must be informed by humanitarian institutions, processes, and principles to ensure that human life is restored appropriately and ethically in a disaster-stricken region.

The UN created HRNA guidelines to streamline the process and ease interactions between and among countries. The EU et al. (2013) identified seven tasks to prepare a human impact assessment: (1) determine the government ministry that will lead the assessment; (2) identify the assessment team members who will join in the report; (3) evaluate HRNA guidelines, adjust them to the PDNA, and then select relevant indicators and sub-indicators (discussed below); (4) collaborate with sector leads to collect relevant pre-disaster baseline and post-disaster data; (5) train all team members on how to prepare a PDNA; (6) arrange and coordinate field visits; and (7) organize and plan logistics, budgets, and implementation if household and community survey data are collected.

Five indicators, including sub-indicators, can be used to evaluate the human impact of disasters. The first indicator that fully satisfies the requirements and scope of the HRNA is living conditions, health, and education using the Multidimensional Poverty Index (MPI) to evaluate impacts on water, sanitation, electricity, cooking fuel, housing, basic household assets, health, and education. The second indicator is livelihoods, which is measured in terms of access to occupations, income, productive assets, and resources. The third indicator, food security, focuses on the three pillars of food security (availability of food, adequate income to acquire food, and increasing productivity in agriculture) as well as household coping strategies. The last two indicators, gender inequality and social inclusion, are closely aligned. Together, they highlight disproportional impacts on various genders and unequal access and denial of opportunities to vulnerable populations (EU et al., 2013).

After the human impact analysis fieldwork of the PDNA is completed, data compilation is necessary to fully understand the breadth of a disaster's impact. PDNA teams must work cooperatively to analyze and interpret the collected data. During these final stages, data and information must be cross-checked and triangulated with the sector teams to identify inconsistencies and confirm the human impact findings (EU et al., 2013). The final presentation of this data can advise government officials on where to allocate resources and how recovery should be approached so that sustainability and resiliency are embedded into their systems. A recovery strategy and timeline should be developed that utilizes and creates programs, policies, and projects that both recognize disaster risk reduction priorities and lead to appropriate human development outcomes. Ideally, an HRNA should be complied before a disaster occurs in order to streamline the recovery process and so data are already available and can be quickly presented.

Challenges, however, may complicate the efficacy of an HRNA. For example, although an HRNA is flexible for many governments, it may be difficult for institutions to use if they have different expectations and perspectives. Additionally, more resources may be allocated to conduct the DaLa, so sometimes the HRNA does not receive the resources required to provide a complete analysis of the disaster's human impact and associated recovery needs. Furthermore, personal experience from individuals in disaster-affected areas must be included in the human impact analysis; otherwise the human recovery strategy will not be fully informed and will fail to meet recovery standards. Finally, timing is a challenge during humanitarian situations, and creation of the HRNA may be sidelined for the DaLa—as occurred in recent disasters such as the 2022 flood in Pakistan (GoP et al., 2022)—emphasizing the imperative that an HRNA framework be created before a disaster strikes. The PDNA report for the 2010 Haiti earthquakes contains both the DaLa and the HRNA. UN guidelines were used to evaluate impacts of the earthquakes and create a detailed analysis of community needs. However, the PDNA analysis focused on short-term community needs rather than confronting and remedying potential inequalities that existed before the disaster occurred.

PDNAs prioritize governance, regional development, risk reduction, environmental management, social sectors, the production sector, infrastructure, cross-cutting themes, and utilizing a macroeconomic framework (UNEP, 2010). Thus, PDNA and DaLa assessments focus on macroeconomic impacts, while the HRNA emphasizes microeconomic scales of impact. Despite their differences, including an HRNA in the PDNA complements and significantly aids reconstruction and recovery processes.

Recovery Framework (RF)

The RF, or disaster RF (DRF), is the final component of the PDNA and is meant to espouse the recovery methods prescribed in the DaLa and HRNA. An RF promotes effective recovery activities for disaster-impacted countries by outlining consensual, coherent, and concrete actions to address recovery needs over a period of 18 months (UNEP, 2010). The RF, which includes early and long-term recovery needs and priorities for each sector, is a concise report of strategies and objectives to be implemented over time to ameliorate the impacts of the disaster (Paul, 2019).

As mentioned, a PDNA should commence as soon as possible to facilitate an efficient, rapid recovery. A primary objective of the PDNA is to support application of the RF in time to revise the humanitarian flash appeal (an initial inter-agency humanitarian response strategy), which is typically five to six weeks following a disaster (ReliefWeb, 2008). This results in more in-depth assessments, a more informed recovery process, and a more seamless transition to supporting long-term development. Although the RF is a key element of the PDNA, very little literature is available on the topic, meaning further study is necessary for it to be fully incorporated into the PDNA process.

Criticisms of PDNAs

Despite the advantages, PDNAs have noted disadvantages. One of the main concerns revolves around the cost-effectiveness of PDNAs, which can exceed $100 million USD. Further, the preparation of a PDNA coincides with the early stages of disaster response and recovery when disaster survivors are most at risk and most in need of urgent care. Thus, PDNA preparation demands a narrow timeframe that can conflict with the humanitarian response timeline. Additionally, the focus on expediency can limit data collection, meaning that speed can be deemed more important than data accuracy (Jones, 2010). This creates a barrier for PDNA team members who must propose realistic recovery measures, ultimately leading to inefficient resource utilization and less impactful recovery outcomes for disaster-affected communities (Hinzpeter & Sandholz, 2018).

Another criticism of PDNAs is that they duplicate other post-disaster reports, such as flash appeals and the cluster approach, that are often authored by the same agencies. These agencies (e.g., the UN) and PDNA experts collect nearly identical data and produce similar findings, which generates duplication, wasted human/labor resources, and overuse of scarce physical resources and assets. Duplication—which can create confusion among various agencies, actors, and team members—often occurs due to lack of coordination, lack of communication, and lack of information sharing (EU & UNDP, 2017). In addition to duplication of efforts, the use of several languages among participating organizations can lead to confusion in conceptualizing, writing, and communicating key terms and findings.

A PDNA also does not provide detailed recovery projects or in-depth sectoral analysis (GFDRR, 2016). Further, overlaps exist across the four sectors, with some experts claiming that the cross-cutting issues sector is especially problematic because its environment sub-sector overlaps with other sectors and sub-sectors (Hinzpeter & Sandholz, 2018). Jeggle and Boggero (2018) also found a series of PDNA limitations and challenges, including the limited use of PDNAs due to their inability to both provide in-depth disaster analysis and tailor recommendations to actual government needs. Therefore, an alternate assessment procedure may be more appropriate for small countries, minor disasters, or atypical recovery requirements. In addition, minimal dialogue has occurred with governments to identify country-specific needs, the suitability of alternative methodologies, priority objectives, or timing and duration of the PDNA process. The time required to conduct PDNAs and disseminate outputs can also limit the beneficial impacts of the exercise, and accountability to affected communities and how PDNAs serve their interests also remain unclear. Finally, the current PDNA process has struggled to increase inclusion of the most vulnerable, socially marginalized groups, and persons with special needs.

Conclusion

This chapter provided a detailed summary of PDNAs. This chapter identified the purpose of PDNAs, explored the process of preparing PDNAs, and detailed how lesser-developed countries affected by a natural disaster can request a PDNA report. This chapter also discussed key roles in PDNA production and highlighted the importance of conducting PDNAs efficiently and cost-effectively with accurate data and figures. The documents that comprise a full PDNA report were described, including a DaLa, HRNA, and RF. Finally, critiques of the PDNA process were offered.

To prevent the criticisms and limitations from affecting the success of a PDNA, all components must be fully met and conducted completely. Because the PDNA and its elements are used frequently, the potential pitfalls and limitations may be ameliorated by formulating a more refined and practical PDNA methodology and overall framework. For example, the PDNA methodology could be improved by increasing the training and coordination of stakeholders. Additionally, Hinzpeter and Sandholz (2018) suggest that PDNAs could be enhanced by implementing shorter timeframes, increasing data sharing, and collecting more primary data and case studies to support decision making. Professional leadership and/or leadership training, impartiality assessments, training on biases, and more local inputs would also likely benefit future PDNAs. Ultimately, more research is needed to increase the accessibility and practicality of PDNAs in order to promote disaster resilience, the recovery of affected communities, and long-term development.

References

Coffey, M., W. Yasmina, & S. Saut. 2017. *The Role of Post Disaster Needs Assessments in Addressing Vulnerability of Internally Displaced Persons in the Post Disaster Recovery Process*. Working Paper Series, No. 1. Bandung: Resilience Development Initiative.

CRED (Center for Research on the Epidemiology of Disasters). 2011. *Disaster Loss Characterization: Review of Human and Economic Impact Indicator Definitions*. Brussels: Catholic University of Louvain.

EU (European Union) & UNDP (UN Development Program). 2017. *Progress Report: Strengthening Capacities for Post Disaster Needs Assessment and Recovery Preparedness*. New York: UNDP. https://info.undp.org/docs/pdc/Documents/H71/PDNA%20Rollout%20II%20Progress%20Resport%201st%20year%202017.pdf.
EU (European Union), UN (the United Nations), & WB (the World Bank). 2013. *Guidelines for Assessing the Human Impact of Disasters*. New York: United Nations Development Program.
GFDRR (Global Facility for Disaster Reduction and Recovery). 2013. *Annual Report '13: Inroads to Resilience*. New York: United Nations Development Program. New York: The UN. https://www.gfdrr.org/en/publication/annual-report-2013.
GFDRR (Global Facility for Disaster Reduction and Recovery). 2016. *Annual Report 2015: Bringing Resilience to Scale*. New York: The UN. https://www.preventionweb.net/publication/gfdrr-annual-report-2015-bringing-resilience-scale.
GFDRR (Global Facility for Disaster Reduction and Recovery) & World Bank. n.d. *Post Disaster Needs Assessments*. New York: The UN. https://www.gfdrr.org/en/post-disaster-needs-assessments.
GFDRR (Global Facility for Disaster Reduction and Recovery) & World Bank. 2022. *Damage, Loss, and Needs Assessment – Tools and Methodology*. New York: The UN. https//www.gfdrr.org/en/damage-loss-and-needs-assessment-tools-and-methology.
GoB (Government of Bangladesh). 2008. *Cyclone Sidr in Bangladesh: Damage, Loss and Assessment for Disaster Recovery and Reconstruct*. Dhaka: GoB.
GoN (Government of Nepal). 2015. *Nepal Earthquake 2015: Post Disaster Needs Assessment, Vol. A: Key Findings*. Kathmandu: National Planning Commission.
GoP (Government of Pakistan), ADB (Asian Development Bank, EU (European Union), UNDP (United Nations Development Programs), & WB (the World Bank). 2022. *Pakistan Floods 2022: Post-Disaster Needs Assessment (PDNA) – Main Report*. Islamabad: GoP. https://thedocs.worldbank.org/en/doc/4a0114eb7d1cecbbbf2f65c5ce0789db-0310012022/original/Pakistan-Floods-2022-PDNA-Main-Report.pdf.
Hinzpeter, K. & S. Sandholz. 2018. Squaring the Circle? Integrating Environment, Infrastructure and Risk Reduction in Post Disaster Needs Assessments. *International Journal of Disaster Risk Reduction* 32: 113–124. https://doi.org/10.1016/j.ijdrr.2018.05.016.
IRP (International Recovery Platform). n.d. PDNA: Post-Disaster Needs Assessments. https://recovery.preventionweb.net/build-back-better/post-disaster-needs-assessments.
Jeggle, T. & M. Boggero (eds.). 2018. *Post-Disaster Needs Assessment: Lessons from a Decade of Experience*. Washington, DC: The World Bank.
Jones, B. 2010. *Managing Post-Disaster Needs Assessments (PDNA)*. EAP DRM Knowledge Notes; No. 19. Washington, DC: World Bank. http:/hdl.handle.net/10986/1032.
Paul, B.K. 2019. *Disaster Relief Aid: Changes & Challenges*. Gewerbestrasse: Palgrave Macmillan. https://doi.org/10.1007/978-3-319-77282-0.
Paul, B.K. 2022. Recovery. In *Routledge Handbook of Environmental Hazards and Society*, edited by T. McGee & E. Penning-Rowsell, pp. 425–441. London: Routledge.
Perwaiz, A., & S. Sinha. 2018. Post Disaster Needs Assessment for Resilient Recovery Using Space Applications and Innovative Technologies. In *Science and Technology in Disaster Risk Reduction in Asia*, edited by R. Shaw, K. Shiwaku, & I. Takako, pp. 343–365. Cambridge, MA: Academic Press. https://doi.org/10.1016/B978-0-12-812711-7.00020-1.
ReliefWeb. 2008. *Glossary of Humanitarian Term*. New York.
UN (the United Nations), WB (the World Bank), EU (European Union), & APD (the Asian Development Bank). 2024. *Herat Earthquakes 2023. Post-Disaster Needs Assessment*. Kabul: Afghanistan.
UNDP (The United Nations Development Program. 2010. *2010 Annual Report*. New York: Bureau for Crisis Prevention and Recovery, the United Nations.
UNEP (UN Environment Program). 2010. *Haiti Earthquake PDNA: Assessment of Damages, Losses, General and Sectoral Needs*. New York. https://www.unep.org/resources/report/haiti-earthquake-pdna-assessment-damage-losses-general-and-sectoral-needs. https://wedocs.unep.org/bitstream/handle/20.500.11822/8868/Haiti_earthquake_PDNA.pdf?sequence=3&amp%3BisAllowed=
WBG (World Bank Group) & GFDRR (Global Facility for Disaster Reduction and Recovery). 2023. *Syria Earthquake 2023: Rapid Damage and Needs Assessment (RDNA)*. Washington, DC: International Bank for Reconstruction and Development/The World Bank.

7

EMERGENCY RELIEF FLOWS AND DETERMINANTS OF DISASTER AID

Munshi Khaledur Rahman, Bimal Kanti Paul, Nafeeul Alam Walee, and Md Sariful Islam

Introduction

Disasters are events that inflict great damage, destruction, and human suffering. Disasters can be natural, human-made, technological hazards, or originate from other factors that influence the exposure and vulnerability of a community (IFRC, 2022). Natural disasters may include events driven by earthquakes, volcanic activity, blizzards, landslides, drought, wildfires, hurricanes, and flooding that are associated with geophysical, meteorological, and climatic events and processes. In 2022, a total of 387 disasters occurred globally, more than 1 every day (NDD, 2022). In the case of any cataclysmic event, the results can be injury, loss of human life, damage to the natural environment, and public and private property loss. Over the years, natural disasters around the world have damaged assets totaling hundreds of billions of dollars. According to the Centre for Research on the Epidemiology of Disasters' Emergency Events Database (EM-DAT), an estimated 387 natural disasters caused the loss of 30,704 lives and affected 185 million individuals with economic losses of over $223 billion USD in 2022 alone (EM-DAT, 2023). The most devastating disasters recorded in 2022 alone were floods in Pakistan, droughts in Africa (e.g., Democratic Republic of the Congo, Ethiopia, Nigeria, Sudan, China, Burkina Faso, Malawi, Cameroon, Central African Republic, Chad, Niger, and Mali), typhoons in the Philippines (e.g., Megi in April and Nalgae in October), and hurricanes (e.g., Hurricane Ian in Cuba and Hurricane Fiona in Dominican Republic) (NDD, 2022). It is reported that in 2022 a total of 308 climate-related disasters were recorded worldwide, which indicates an upward trend over the last 30 years (1992–2021) accounting for an annual average of 16 droughts, 147 floods, and 100 storms (NDD, 2022). The *Humanitarian and Disaster Risk 2021* report revealed that the total dollar value of philanthropic aid was $33.4 billion USD and bi/multilateral aid totaled $191.3 billion in year 2023 (Candid, 2023).

Given this backdrop, the aims of this chapter are two-fold. The first section presents relevant issues on the flow of relief aid and pledged amounts. Second, a closer look at the determinants of foreign aid in the aftermath of major natural disasters is explored. Sources for this chapter are based on a literature review and our experiences with several natural disasters.

DOI: 10.4324/9781003414834-11

Flow of Relief Aid and Disaster Pledges

A country affected by a natural disaster can formally request that other countries help with emergency aid through a 'flash appeal.' Such a request indicates that the government of the affected entity cannot cope with the disaster independently. In most cases, recipient countries are willing to accept emergency aid because without it they cannot efficiently respond to the extreme event with their own resources (Bradshaw, 2013). Note that some recipient government may refuse offers of assistance from foreign countries for political and/or other reasons (Paul, 2019). However, generally, governments utilize three channels to provide and receive disaster aid: (1) bilateral assistance, (2) multilateral assistance, and (3) indirect assistance through non-governmental organizations (NGOs).

Bilateral channels tend to be more efficient and expedient as aid is distributed directly from a donor country to a recipient country and thus incurs lower administrative costs (Biscays, n.d.). However, this channel is more politicized than the multilateral channel. Through the use of multilateral channels (e.g., various United Nations agencies, the World Bank, and the Asian Development Bank (ADB)), a donor country avoids many of the political implications of bilateral assistance. Thus, this channel tends to be more apolitical and represents a significant advantage of aid dispersal. However, multilateral channels often suffer from a lack of clarity, overlap, inter-agency rivalry, and lack of coordination (Paul, 2019). Foreign private donations are primarily channeled through NGOs.

Examples of Accepting and Rejecting Aid

Generally, emergency assistance flows from relatively higher income countries to relatively lower income countries. However, almost all countries have provided some form of disaster assistance to other countries, often to regional neighbors and/or political, economic, and military allies. Some disaster-affected countries have refused to accept disaster aid based on ideological, economic, religious, or political reasons (Bradshaw, 2013). In some cases, the decision to reject international aid is exercised as an indication that the country has achieved an improved standard of living. Refusing disaster aid thus becomes a method for countries to demonstrate their self-sufficiency and ability to manage disasters with their own resources and expertise. In other words, aid rejection decisions are primarily inspired by concerns for a country's international reputation (Carnegie & Dolan, 2015).

After the Indian Ocean Tsunami (IOT) in 2004, India and Thailand initially refused to accept foreign aid for tsunami survivors. In doing this, both countries sought to portray that they are economically and politically better off compared to the other tsunami-affected countries (Paul, 2007; Carnegie & Dolan, 2015). Ultimately, both countries were forced to accept aid from donor countries. For example, as of 31 December 2016, India received $151 million USD from donor countries to help mitigate the effects of the tsunami that killed more than 12,000 and displaced over 600,000 people in the country (Paul, 2019). India and Thailand also participated in emergency efforts to assist other tsunami-affected countries. In fact, India not only dispatched immediate assistance to Indonesia, Sri Lanka, and the Maldives but also participated in relief operation. Furthermore, India donated $25 million in emergency aid to Sri Lanka.

Indeed, India turned down emergency assistance for each subsequent disaster after the 2004 IOT. In 2018 the Indian state of Kerala experienced a severe flood. It is noteworthy that Kerala has sent many guest workers to countries throughout the Middle East, particularly the United Arab Emirates (UAE). It is estimated that 80% of Indian workers in the UAE are from Kerala, and there is no disputing the fact that the Kerala guest workers made significant contributions to the economy of this region. The UAE and Qatar offered $100.5 million USD for flood relief aid, yet the Indian government declined the assistance. Interestingly, the Kerala state government was keen to accept aid from such sources. However, as a matter of policy, since 2005 the Indian government has not requested or accepted foreign assistance for disaster relief (Ramekar, 2021). It is important to mention that unless an appeal is issued, no international actors can formally respond to a disaster. Such a request indicates that the affected government cannot cope with the extreme event.

Cyclone Nargis struck Myanmar on 14 May 2008. The event caused the deaths of more than 100,000 and injured over 19,000 (Paul, 2021). Unfortunately, since the 1960s, the country had endorsed a self-reliance policy and thus refused to accept aid from international sources. This policy was a key factor behind the initial refusal of relief aid. Ultimately, leaders of the Association of Southeast Asian Nations (ASEAN) convinced the Myanmar government to allow the delivery of international aid for the devastating cyclone. Mounting international pressure surrounding human rights along with Myanmar being a member of ASEAN finally led to the government granting access to foreign aid workers and emergency aid (Paul, 2021).

Japan, a relatively developed and high-income country, regularly experiences major disasters such as earthquakes, tsunamis, floods, typhoons, and volcanic eruptions. The country was hit by a magnitude 9.0 earthquake on 11 March 2011. The earthquake originated in the sea, not far off Japan's east coast, which triggered a massive tsunami that subsequently caused extensive damage to the Fukushima Nuclear Power Plant. The horrendous triple disaster resulted in 15,780 deaths and 4,122 missing persons (Karan, 2016). Being a wealthy country, Japan initially refused international aid, yet ultimately did accept such assistance selectively. The U.S. offered technical assistance if the reactors overheated, but the Japanese government and the Tokyo Electric Power Company (TEPCO) insisted that they could handle the situation themselves. A follow-up U.S. offer was also rejected, and the refusal of these international offers were heavily criticized by researchers (see Choate, 2011; Karan, 2016).

Except for two cases, the U.S. has refused to accept emergency aid in the aftermath of major disasters. After the 2005 Hurricane Katrina disaster, the U.S. government initially refused all foreign aid. In the end, however, the government reversed its policy and welcomed relief from 120 nations that pledged more than $700 million in assistance to Katrina survivors (Paul, 2019). Aid arrived from America's allies (e.g., the U.K., Canada, and Germany) as well as their foes (e.g., Cuba, Iran, and Venezuela). Kuwait made the largest donation of $500 million USD. Other large contributions were made by Qatar ($100 million), India ($5 million), China ($5.1 million), and Bangladesh ($1 million) (Paul, 2019). Some of these countries also sent donations after the great Midwestern floods of 1993 (Tobin & Montz, 1997).

Gaps in Pledged Amounts and Competition

International donor countries typically make an assistance pledge following a natural disaster. The affected country makes the request for assistance through a flash appeal. A pledge is a formal promise to give money to the government of a disaster-affected country for rapid response and long-term recovery from an extreme event. Often, the 'total pledge amounts exceed the estimated losses and damage. For example, the pledge of $10 billion for the 2004 IOT exceeded by at least one-third' (Paul, 2019, 54). Additionally, pledges do not always take place or get distributed in the immediate aftermath of a disaster. For example, 52% of all pledged amounts for the IOT occurred in early 2005, just shortly after the disaster. By 31 December 2016, over a decade later, the total amount pledged and released had increased to 92% of the total, with the remaining 8% to occur later. It is interesting to note that 62% of the funds came from private sources, with the rest originating from official government sources (Paul, 2019).

However, for some countries, there has been an inconsistency between the pledged and released amounts. For example, Iran, Russia, and Saudi Arabia each pledged more money for IOT relief than they actually released; in fact, only 13% of their total pledge amount was actually distributed (United Nations Financial Tracking Service [UN-FTS], 2017). Meanwhile, Brunei, Cameroon, Georgia, Jamaica, Macedonia, and Niger made pledges but did not contribute anything after the 2004 IOT disaster. In contrast, 64 countries, including Canada, Denmark, Greece, Germany, Japan, the Netherlands, and Switzerland, provided 100% of their pledge amount by 2016 while the U.S. contributed only 38% of its pledge amount (UN-FTS, 2017).

Donor countries often need to increase their original pledge amount several times. For example, the U.S. initially pledged only $15 million USD for the 2004 IOT, but in response to criticism from the United Nations, the U.S. added another $20 million to its original pledge (Margesson, 2005). As the death toll rose, the amount increased to $350 million. Finally, when contributions from Europe soared, the U.S. again increased its pledge to $857 million (Wehrfritz & Cochrane, 2005).

Donor countries promised $1.1 billion USD after the 2003 Bam earthquake of 6.5 magnitude in Iran, which caused approximately 3,000 deaths and rendered 75,000 homeless (Chatterjee, 2021). However, the donor nations delivered only $17.5 million, which translated to 1.5% of all pledged amounts. Three reasons help to explain the gap between the pledged amount and the amount released. First, donor countries often feel hidden peer pressure to pledge more funds than they can actually afford. Telford and Cosgrave (2007) refer to this as a donor 'beauty contest.' A second reason is due to double counting of funds, particularly when a donor nation promises funding for an initial disaster but the resources are diverted to other disasters. Finally, diplomatic competition among donor countries may increase the pledged amounts, which happened between China and Japan after the 2004 IOT. Japan desired to provide more funding to tsunami survivors than China to demonstrate that Japan is more economically powerful than its traditional rival (Tammoto, 2005). Japan also competed with the U.S. since the disaster occurred in its backyard in Asia, with Japan contributing the most among all donor countries (Paul, 2019).

Determinants of Disaster Aid

Generally, disasters that create humanitarian crises attract attention through electronic, print, and social media coverage. The damage leads many philanthropic organizations, including but not limited to governments, multilateral organizations, NGOs, private charities, trusts, and philanthropists, to provide aid to affected populations. In the absence of aid, both from within and beyond the national boundary, survivors exhibit greater difficulty coping with the impacts of extreme events, particularly in lesser developed countries. Thus, the availability of various sources of support in the form of monetary funds, technology, expert assistance, resources, heavy machinery, volunteerism, and other assistance is crucial to disaster survivors. This chapter progresses in two parts. The first part discusses the general factors that are assumed to be associated with disaster aid. The next part examines the determinants of the level of emergency assistance based on a review of the literature and empirical studies.

Factors Associated with Disaster Aid: General Discussion

Disaster assistance to foreign countries largely depends on which country experiences the extreme event. If the disaster-affected country or countries are located in the same region, donor governments may distribute higher amounts of emergency aid. As already noted, after the IOT, Japan—a fellow Asian country—claimed that it felt compelled to contribute more aid than the U.S. due to its proximal location (Tammota, 2005). Thus, geography plays a significant role in determining the provision of disaster aid. In addition, regions prone to recurrent disasters may receive more aid due to their heightened vulnerability. Moreover, the severity and scale of the crisis impacts the allocation of aid, with larger and more devastating events often attracting greater international attention and hence support (Enia, 2020; Buszta, 2023). This implies that the magnitude and impact of natural disasters tends to determine the amount of aid that is received. Having accurate information about the scale and severity (usually measured in terms of the number of deaths, injuries, or displaced persons) of a natural disaster influences the amount of direct foreign aid distributed, as do the absolute number and relative frequency of disasters (Enia, 2020; Buszta, 2023).

The affected country's socioeconomic level of development and capacity to cope with the disaster on its own are important factors that determine whether foreign governments will allocate disaster aid. Countries with limited resources and infrastructure may attract more external disaster aid from both the local/regional and global disaster aid community. Additionally, the political landscape, both nationally and internationally, is a key determinant of disaster aid allocation. Donor countries may prioritize aid to strategically important regions or those aligned with their foreign policy objectives for politico-economic, geopolitical, or diplomatic gains. Similarly, recipient countries' political relationships with donors also influence the amount and type of aid that is received. Thus, political and historical alliances—along with their power dynamics—can shape the distribution of aid, reflecting broader considerations and motivations. For example, donor governments often display disaster aid biases in favor of their former colonies (Becker,2020; Cheng et al, 2023).

Disasters, through the resultant exchange of aid and other humanitarian assistance, can shape formal relationships both between and among countries (Enia, 2020). Strong alliances and diplomatic ties, such as ASEAN and the South Asian Association for Regional Cooperation (SAARC), may lead to more substantial support for disaster aid to a particular nation or country. The amount of aid is also influenced by international dynamics such as deliberate decisions made by multilateral and supranational organizations and implicit and explicit pressure exerted by major donors that compel other countries to contribute (or increase their contribution).

The cultural context is a determinant of disaster aid. Culture plays a role in how people experience disasters, develop coping strategies, and respond to disasters. Moreover, culture shapes how individuals and communities perceive the acceptability of receiving external aid and support. Language barriers and cultural differences can impact the effective and equitable distribution of disaster aid. Among cultural elements, religion (e.g., religious affinity) is one of the biggest determinants of disaster aid. For example, after a 7.6 magnitude earthquake struck the Pakistan-administered portion of Kashmir in 2005, the Islamic world was extremely generous and quick to provide aid to fellow Muslim survivors (Ozerdem, 2006). The Islamic world raised large sums of aid, including from expatriates and Muslims in non-Muslim majority countries (particularly in Britain and the U.S.). Further, the earthquake occurred during the holy month of Ramadan, during which Muslims fast. Thus, individuals were especially generous and donated more willingly to charities. According to Rashid (2005), even Pakistanis themselves were surprisingly generous.

In contrast to this event, a massive earthquake measuring 7.8 magnitude struck Nepal on 25 April 2015. Neighboring India and China competed against each other to see which would send more aid after the earthquake. However, the majority religion in both India and Nepal is Hinduism. Not surprisingly, the amount of money pledged by India was two times greater than that of China. Interestingly, from September 2015 through September 2016, Nepal experienced an unofficial Indian trade embargo that was known as a border blockade. It resulted in economic turmoil throughout Nepal and the population suffered tremendously (Ruszczyk, 2018). India has an open border with Nepal and millions of Nepalis work there; in contrast, Nepal's interactions with China are less frequent. However, the Nepalese people have a somewhat rosier perception of China due to the blockade that came on the heels of the earthquake (Shakya & Gurung, 2015).

Finally, at the core of disaster aid is the humanitarian imperative to address acute needs and alleviate suffering. When addressing humans' basic needs, aid organizations strive to prioritize aid based on the scale of need while simultaneously aiming to ensure equitable distribution and accessibility to affected populations. However, challenges obviously arise in accurately assessing and addressing needs, particularly in complex emergencies and protracted crises.

Factors Associated with Disaster Aid: Empirical Studies

Economists, geographers, sociologists, and scholars from numerous other disciplines have studied and identified the determinants of disaster aid. This chapter will attempt to crystallize a selection of these studies from 2000 to 2022 based on the inclusion of various temporal, geographic, and cross-sectional dimensions. Selected studies have examined a single year after an extreme event, several years after an extreme event, multiple countries for a single year after an extreme event, and multiple countries for multiple years after extreme events. The studies considered multiple donor types, and data primarily came from three secondary sources: EM-DAT, Organization for Economic Cooperation and Development Assistance Committee (OCECD-DAC), and United Nations Financial Tracking Service (UN-FTS).

The first data source maintained by the Center for Research on Epidemiology of Disasters (CRED) at the Catholic University of Louvain in Brussels, Belgium. This publicly available disaster database is known as EM-DAT. For a disaster to be entered into EM-DAT, at least one of the following four criteria must be met: (1) 10 or more people reported killed; (2) at least 100 people affected; (3) a call for international assistance; or (4) a state of emergency is declared. Detailed data on any aid flows are available from OECD-DAC, which is a unique forum where the democratic, market-based governments of 37 countries collaborate to develop policy standards to promote sustainable economic growth. Finally, the United Nations Office for the Coordination of Humanitarian Affairs (UNOCHA) publishes the UN-FTS, which processes all funds transferred from donors to disaster-affected countries. All of these data sources have their advantages and disadvantages, which is beyond the purview of this chapter. Following is a discussion of some of the major studies and their findings on determinants of disaster aid.

Raschky and Schwindt (2012) contend that donors primarily consider two criteria when allocating disaster assistance: the number of fatalities and the number of people affected. Simon (1997) argues that the fatality rate reported by media outlets influences how much donors allocate to a particular disaster. Becerra et al. (2014) concur and found that disaster aid pledges are at their highest when media coverage is at its peak. Strömberg (2007) further found that the amount of aid is influenced by news coverage as well as having a previous connection with the affected country. Thus, disaster media coverage, particularly mass televised news, is a major determinant of disaster assistance and has been termed the 'CNN effect' (Tornau, 2006).

Several studies have reported that disaster aid surges as a function of the severity of the disaster, the country's per capita gross domestic product (GDP), and the amount of foreign exchange reserves (Becerra et al., 2010, 2014; Raschky & Schwindt, 2012). Fink and Redaelli (2011) claim that donors favor small nations that are geographically closer when considering disaster aid allocation. They further argue that donors might provide more assistance if the recipient country has fewer political alliances compared to other countries.

Van Wassenhove (2006) specifies that rapid onset disasters, such as earthquakes and tropical cyclones, receive more media attention compared to slow onset events and thus tend to garner more assistance. Geostrategic interests can also be a determining factor for foreign aid volume, and demand-factors—in particular the recipient country's readiness to absorb new flows through NGOs—are important determinants of aid flows (Olsen et al., 2003). The literature argues that geostrategic and political interests play an especially large role in determining American aid allocations across time and geography (Drury et al., 2005; Fleck & Kilby, 2010). Demirel-Pegg and Moskowitz (2009) argue that humanitarian motivation is an important factor of disaster aid distribution, while Easterly and Pfutze (2008) contend that the presence of efficient institutional capacity to implement aid is an important factor.

Yang (2008) reported that foreign aid (as a whole) increases significantly after disasters, with disaster aid often comprising the majority of foreign aid received by a country over a given period of time—especially among lesser developed countries. A study conducted by Becerra et al. (2010) documented the determinants of disaster aid flows by using regression analysis. Their results suggest that the median increase in realized post-disaster aid is approximately 18% compared to pre-disaster flows. In terms of the determinants, they identified the intensity of the event, the affected country's level of development, and country's size or population as positively associated with foreign disaster aid allocation (Becerra et al., 2010). They did not find that post-disaster aid was positively associated with political affinity and the geostrategic interests of donor countries. This is supported by Fink and Redaelli (2011), who argue that humanitarian need appears to best determine whether a country will provide disaster aid. A study of 22 donor countries from 1997 to 2008 by Nelson (2012) reported that aid activity is determined by humanitarian needs along with political variables (e.g., trading partners, former colonies, and military allies).

Individual and household levels have also considered. Several studies (see Tiehen, 2001; Auten et al., 2002; Gittell & Tebaldi, 2006; Bakija & Heim, 2008) focusing on household level disaster aid found positive relationships between disaster donations and household income, level of education, religious affiliation, and tax breaks (Paul, 2019). A study conducted by Ishida (2017) after the 1995 Kobe earthquake and 2011 Tohoku earthquake in Japan revealed that past experience with disasters, household income, savings, and age were positively associated with donations to survivors of the two earthquakes. Furthermore, donations were a function of geographical distance from the disaster sites and the location of individual donors. As a summation, Table 7.1 outlines the major determinants of disaster aid based on the literature available through the Web of Science (WoS).

Conclusion

This chapter discussed the flow and determinants of disaster relief aid. Understanding the interplay between disasters and the allocation of aid by donors—including governments, NGOs, supranational agencies, and private entities and individuals—is essential for helping disaster-affected communities. There exist diverse rationales behind the provision of disaster aid, and there often exist differences between the pledged amount and actual amount provided. Competition among donor countries and the refusal of aid by disaster-affected countries have also occurred. The determinants of disaster aid can vary based on the nature of the disaster, the affected area's geography, the donor country's geopolitical interests, and the response mechanisms in place. Additionally, the magnitude and impact of the disaster, media coverage, and affected country's economic status influence the flows of disaster aid. Cultural and social factors as well as widely held humanitarian principles such as neutrality, impartiality, and independence also motivate countries to provide disaster aid. It is important to understand these determinants to facilitate more effective response strategies and address humanitarian needs in a timely and efficient manner.

Table 7.1 Identified Determinants of Disaster Aid Flows from Donor Countries to Affected Countries

Determinants of disaster aid	*Source*
Population density, number affected, number of fatalities, GDP per capita, trade openness, geographic distance, affinity, former colony, 'Oil dummy'	Fink and Silvia (2011)
Geographic distance, oil production, colonial history, 'Coalition of the Willing,' alliances, regime type	Nelson (2012)
GDP per capita, number affected, number of fatalities, commercial interests, geographic proximity, media coverage, shared colonial history	Robinson et al. (2017)
Prior connection with affected country, media coverage, development level, country size, stock of availability of foreign reserves, economic damage, number of fatalities, initial aid level, real GDP, population size, political affinity	Becerra et al. (2010)
Economic damage, number of fatalities, media coverage, real GDP, international reserves over GDP, population size, political affinity, openness to international trade, existence of armed conflict, small island state, former colony, official development assistance (ODA) total net disbursement	Becerra et al. (2014)
Geostrategic interest	Olsen Carstensen, and Høyen (2003)
Duration, number of fatalities, number affected, total estimated damage	Lenard et al. (2023)
Number affected, number of fatalities, GDP per capita, population density, control of corruption, United Nations Security Council (UNSC) member, mineral and energy depletion, distance, geographic neighbor, colonial history, official language, major religion, migration in/out, exports	Fuchs and Nils-Hendrik (2013)
Hazard severity, number affected, state fragility, former colony, oil endowment, voting with the U.S. on the United Nations General Assembly (UNGA)	Dellmuth et al. (2021)
U.S. military assistance, donor-recipient arms transfer, donor exports, donor imports, UNSC membership, UN vote distance, former colony, population size, income, geographic distance, migration	Bermeo (2017)

Source: Compiled from different sources, including Web of Science.

Understanding the nature of flows and determinants of disaster aid is essential for improving the impact of aid, disaster diplomacy, and equity in global aid distribution. Recognizing the complexities and connectivity of factors can inform decisions to optimize the allocation of scarce resources and ensure that aid promptly reaches those most in need. Due to the impacts associated with anthropogenic climate change, our lives, livelihoods, and infrastructure/assets will increasingly be threatened by extreme weather events. In addition to ongoing disaster mitigation and risk reduction, the formulation of more effective disaster relief mechanisms is needed.

References

Auten, Gerald E, Holger Sieg, and Charles T. Clotfelter. "Charitable Giving, Income, and Taxes: An Analysis of Panel Data." *American Economic Review* 92, no. 1 (2002): 371–82.

Bakija, Jon, and Bradley Heim. *How Does Charitable Giving Respond to Incentives and Income? Dynamic Panel Estimates Accounting for Predictable Changes in Taxation.* Cambridge, MA: National Bureau of Economic Research, 2008.

Becerra, Oscar, Eduardo Cavallo, and Ilan Noy. "Foreign Aid in the Aftermath of Large Natural Disasters." *Review of Development Economics* 18, no. 3 (2014/08/01 2014): 445–60. https://doi.org/https://doi.org/10.1111/rode.12095. https://doi.org/10.1111/rode.12095.

Becerra, Oscar, Eduardo Cavallo, and Ilan Noy. *In the Aftermath of Large Natural Disasters, What Happens to Foreign Aid?* University of Hawaii, Manoa. Working Paper No. 10-18, (2010).

Becker, Bastian. "Colonial Legacies in International Aid: Policy Priorities and Actor Constellations." *From Colonialism to International Aid: External Actors and Social Protection in the Global South* (2020): 161–85.

Bermeo, Sarah Blodgett. "Aid Allocation and Targeted Development in an Increasingly Connected World." *International Organization* 71, no. 4 (2017): 735–66.

Biscays, Pierre. "Donors' Dilemma: Bilateral versus Multilateral aid Channel (n.d.). https://cdn-odi-production.s3.amazonaws.com/media/documents/10492.pdf.

Bradshaw, Sarah. *Gender, Development and Disasters*. Northampton, MA: Edward Elgar, 2013.

Buszta, Julia, Wójcik, Katarzyna, Guimarães Santos, Celso Augusto, Kozioł, Krystian, and Maciuk, Kamil. "Historical Analysis and Prediction of the Magnitude and Scale of Natural Disasters Globally." *Resources* 12, no. 9 (2023): 106. https://doi.org/10.3390/resources12090106.

Buyukkidik, Serap. "A Bibliometric Analysis: A Tutorial for the Bibliometrix Package in R Using IRT Literature." *Journal of Measurement and Evaluation in Education and Psychology* 13, no. 3 (2022): 164–93.

Candid, "More Than Just Data. Insights That Drive Impact." Foundation Maps, 2023, 18 December 2023, https://maps.foundationcenter.org/home.php.

Carnegie, Allison, and Dolan, Lindsay. *The Effects of Aid on Recipients' Reputations: Evidence from Natural Disaster Responses*. New York: Department of Political Science, Columbia University, 2015. https://docs.aiddata.org/ad4/files/carnegiedolan_2015_aid_legitimacy.pdf.

Chatterjee, Subarna, "Bam Earthquake, Iran." In *Natural Hazards and Disasters: From Avalanches and Climate to Water Spouts and Wildfires, Volume 2: Natural Disasters*, edited by Paul, Bimal K., 1–6. Santa Barbara, CA: ABC-CLIO, 2021.

Lenard Cheng, Attila J Hertelendy, Alexander Hart, Lawrence Siu-Chun Law, Ryan Hata, Georgina Nouaime, Fadi Issa, *et al*. "Factors Associated with International Humanitarian Aid Appeal for Disasters from 1995 to 2015: A Retrospective Database Study." *PLoS One* 18, no. 6 (2023): e0286472. doi: 10.1371/journal.pone.0286472.

Choate, Allen. *In Face of Disaster, Japanese Citizens and Government Pull from Lesson Learned*. San Francisco: The Asia Foundatation, 2011. https://asiafoundation.org/2011/03/16/in-face-of-disaster-japanese-citizens-and-government-pull-from-lessons-learned/.

Dellmuth, Lisa M., Frida A.-M. Bender, Aiden R. Jönsson, Elisabeth L. Rosvold, and Nina von Uexkull. "Humanitarian Need Drives Multilateral Disaster Aid." *Proceedings of the National Academy of Sciences* 118, no. 4 (2021): e2018293118.

Demirel-Pegg, Tijen, and James Moskowitz. "Us Aid Allocation: The Nexus of Human Rights, Democracy, and Development." *Journal of Peace Research* 46, no. 2 (2009): 181–98.

Drury, A. Cooper, Richard Stuart Olson, and Douglas A. Van Belle. "The Politics of Humanitarian Aid: Us Foreign Disaster Assistance, 1964 1995." *The Journal of Politics* 67, no. 2 (2005): 454–73.

Easterly, William, and Tobias Pfutze. "Best Practice for Foreign Aid: Who Knows Where the Money Goes?". *Journal of Economic Perspectives* 22, no. 2 (2008): 29–52.

EM-DAT. "Inventorying Hazards & Disasters Worldwide since 1988." Centre for Research on the Epidemiology of Disasters (CRED), 2023, 18 December 2023, https://www.emdat.be/.

Enia Jason. "Disaster Diplomacy", 2020, accessed 18 December 2023, https://www.oxfordbibliographies.com/display/document/obo-9780199743292/obo-9780199743292-0135.xml.

Fink, Günther, and Silvia Redaelli. "Determinants of International Emergency Aid—Humanitarian Need Only?" *World Development* 39, no. 5 (2011): 741–57.

Fleck, Robert K, and Christopher Kilby. "Changing Aid Regimes? Us Foreign Aid from the Cold War to the War on Terror." *Journal of Development Economics* 91, no. 2 (2010): 185–97.

Fuchs, Andreas, and Nils-Hendrik Klann. "Emergency Aid 2.0." *Beiträge zur Jahrestagung des Vereins für Socialpolitik* 1–53, (2013).

Gittell, Ross, and Edinaldo Tebaldi. "Charitable Giving: Factors Influencing Giving in Us States." *Nonprofit and Voluntary Sector Quarterly* 35, no. 4 (2006): 721–36.

IFRC (International Federation of Red Cross and Red Crescent Societies), "What Is a Disaster?" The International Federation of Red Cross and Red Crescent Societies, 2022, accessed 18 December 2023, https://www.ifrc.org/our-work/disasters-climate-and-crises/what-disaster.
Ishida, Ryo. "Determinants of Charitable Giving to Unexpected Natural Disasters: Evidences from Two Big Earthquakes in Japan." *Unpublished Paper.* 2017, accessed 11 January, 2024. *http://cis.ier.hit-u.ac.jp/Japanese/society/140110ws/ishidapaper.pdf.*
Karan, Pradyumna P. "After the Triple Disaster: Landscape of Devastation, Despair, Hope, and Resilience," In *Japan After 3/11: Global Perspectives on the Earthquake, Tsunami, and Fukushima Meltdown* edited by Karan, Pradyumna P. & Unryu, Suganuma, 1–42. Lexington, KY: University Press of Kentucky, 2016.
Margesson Rhoda. Indian Ocean Earthquake and Tsunami: Humanitarian Assistance and Relief Operations. CRS Report for Congress, 2005.
NDD. *Natural Disaster Databook 2022 an Analytical Overview*. Asian Disaster Reduction Center, 2022. https://reliefweb.int/attachments/f74f379b-0c50-4bce-9422-38db37d026e1/DataBook2022.pdf
Nelson, Travis. "Determinants of Disaster Aid: Donor Interest or Recipient Need?" *Global Change, Peace & Security* 24, no. 1 (2012): 109–26.
Olsen, Gorm Rye, Nils Carstensen, and Kristian Høyen. "Humanitarian Crises: What Determines the Level of Emergency Assistance? Media Coverage, Donor Interests and the Aid Business." *Disasters* 27, no. 2 (2003): 109–26. doi: 10.1111/1467-7717.00223.
Ozerdem, Alpaslan. The Mountain Tsunami: Afterthoughts on the Kashmir Earthquake. *Third World Quarterly* 27 (2006): 397–419.
Paul, Bimal Kanti. "2004 Tsunami relief efforts: An overview." *Asian Profile* 35, no. 5 (2007): 467–78.
Paul, Bimal Kanti. *Disaster Relief Aid: Changes and Challenges*. Gewerbestrasse: Palgrave Macmillan, 2019.
Paul, Bimal Kanti. "Cyclone Nargis, Myanmar, 2008." In *Natural Hazards and Disasters: From Avalanches and Climate Change to Water Spouts and Wildfires.* Volume 2: Natural Disasters, edited by Paul Bimal Kanti, 73–77. Santa Barbara: ABC-CLIO, 2021.
Ramekar, Avantika. "Kerala, floods, India, 2018." In *Natural Hazards and Disasters: From Avalanches and Climate Change to Water Spouts and Wildfire*. Volume 2: Natural Disasters, edited by Paul, Bimal Kanti, 225–230. Denver, CO: ABC-CLIO, 2021.
Raschky, Paul A., and Manijeh Schwindt. "On the Channel and Type of Aid: The Case of International Disaster Assistance." *European Journal of Political Economy* 28, no. 1 (2012/03/01/2012): 119–31. https://doi.org/https://doi.org/10.1016/j.ejpoleco.2011.07.001. https://www.sciencedirect.com/science/article/pii/S0176268011000747.
Rashid, Ahmed. "Post-Quake Giving United Pakistan. *The New York Times, October* 27 (2005).
Robinson, Tyler D., Thiago M. Oliveira, and Stephanie Kayden. "Factors Affecting the United Nations' Response to Natural Disasters: What Determines the Allocation of the Central Emergency Response Fund?" *Disasters* 41, no. 4 (2017): 631–48.
Ruszczyk, Hanna A. The Earthquake and Ideas Lying Around. In *Evolving Narratives of Hazard and Risk: The Gorkha Earthquake, Nepal 2015*, edited by Bracken Louise, Hanna A. Ruszczyk, and Tom Robinson, 125–39. Gewerbestrasse: The Palgrave Macmillan, 2018.
Shakya, Tsering, and Ashok, Gurung. Chinafile: Was Nepal a Soft Power Victory for China, 2015. https://foreignpolicy.com/2015/05/01/nepal-china-earthquake-aid-taiwan-power/.
Simon, Adam F. "Television News and International Earthquake Relief." *Journal of Communication* 47, no. 3 (1997): 82–93.
Strömberg, David. "Natural Disasters, Economic Development, and Humanitarian Aid." *Journal of Economic perspectives* 21, no. 3 (2007): 199–222.
Tamamoto, M. "After the Tsunami, How Japan Can Lead," *Far Eastern Economic Review* 168, no. 2 (2005): 15–19.
Telford, John, and John Cosgrave. "The International Humanitarian System and the 2004 Indian Ocean Earthquake and Tsunami." *Disasters* 31, no. 1 (2007): 1–28.
Tiehen, Laura. "Tax Policy and Charitable Contributions of Money." *National Tax Journal* 54, no. 4 (2001): 707–23.
Tobin, Graham A., and Montz, Burrell E. *Natural Hazards: Explanation and Integration.* New York: The Guilford Press, 1997.
Tornau, U. "CNN Effect: Power or Mean." A Study of Media Influence on Foreign Policy Decision—Making Comparative Analysis of Two Humanitarian Disasters: Ethiopia (1984–1985) and Somalia (1992). Master of Comparative History, Utrecht University, Utrecht, the Netherlands, 2006.

UN-FTS (United Nations Financial Tracking Service). Total Reported Funding 2017. New York: UNOCHA, 2017. https://fts.unocha.org/appeals/overview/2017.

Van Eck, Nees, and Ludo Waltman. "Software Survey: VOSviewer, a Computer Program for Bibliometric Mapping." *Scientometrics* 84, no. 2 (2010): 523–38.

Van Wassenhove, Luk N. (2006) "Humanitarian Aid Logistics: Supply Chain Management in High Gear." *Journal of Operational Research Society*, 57: 475–89. http://dx.doi.org/10.1057/palgrave.jors.2602125.

Wehrfritz, G., and J. Cocran. "Charity and Choas." *Newsweek*, January 17, 30–32, 2005.

Yang, Dean. "Coping with Disaster: The Impact of Hurricanes on International Financial Flows, 1970-2002." *The BE Journal of Economic Analysis & Policy* 8, no. 1 (2008): 1–43.

8

SOCIAL NETWORKS AND EMERGENCY RESPONSE

Shakil Bin Kashem and Hyung Jin Kim

Introduction

Social networks play a significant role in disaster response and recovery by facilitating information sharing, rescue operations, emotional support, and numerous other exchanges during and after a disaster event. People are influenced by family members, neighbors, and community leaders in their decision-making for emergency response, for example, whether to stay in place or relocate to a safer place (Casagrande et al., 2015). Immediate support for post-disaster recovery also typically comes from social network before formal institutional support is able to reach the community or affected households (Hurlbert et al., 2000). Social support networks and mechanisms can complement or even act as an alternative to institutional aid for individual and group recovery (Ibañez et al., 2004; Kaniasty, 2012). Social networks at the institutional level also contribute to the effectiveness of disaster response operations (Kapucu et al., 2010). Recent studies have examined the influence and implications of social networks on the disaster management cycle for flooding (Stewart et al., 2014; Casagrande et al., 2015; Malone & Kinnear, 2015; Haer et al., 2016), cyclones (Misra et al., 2017), earthquakes (Lu, 2017; Stevenson & Conradson, 2017), tsunamis (Joshi & Aoki, 2014), and other disaster events.

The role of social networks in disaster management is now a specialized area of research (Faas & Jones, 2017). To understand how social networks influence disaster response and recovery, we must investigate the structural characteristics of social networks in a specific community and explore how variations in network structures across communities contribute to differential outcomes. Studies have documented that there can be regional disparities in community collaboration and social networks themselves can evolve spatially and temporally before, during, and after a disaster event (Karunarathne & Lee, 2020). The structural pattern of relationships or the topology of a social network can significantly influence how different actors behave during a disaster (Bodin & Crona, 2009). Different methodological approaches for Social Network Analysis (SNA) are now being applied to study disaster response and recovery outcomes. In recent years, a range of scholarly works have utilized SNA methods to investigate the characteristics, behaviors, and impacts of social networks in disaster response and recovery (see Wasserman & Faust, 1994; Hurlbert et al., 2001; Stewart et al., 2014; Misra et al., 2017).

DOI: 10.4324/9781003414834-12

With greater access to the internet and smart phones, social media is now widely used to communicate hazard risk information (Feldman et al., 2016). Increasingly, more government institutions at all levels are using social media platforms to communicate with their citizenry (Mickoleit, 2014). The role of social media in the disaster warning and response process has been studied (Tyshchuk et al., 2012), and social media data has been used to analyze damage patterns post-disaster (Knox et al., 2013). There have been attempts to harness social media data to develop GIS-based real-time crisis maps to share with emergency agencies and non-governmental organizations (NGOs) (Middleton et al., 2014). Social media data has also been extracted to identify the needs of a disaster-affected community (Yin et al., 2012; Zhang et al., 2019). Thus, parallel to offline social networks, online social networks created through social media channels are now playing a significant role in disaster response and recovery (Rosen et al., 2010; Kane et al., 2014; Ellison & Vitak, 2015). As a result, there is a growing body of literature seeking to bridge online and offline social networks to enhance emergency response operations (Stephens & Poorthuis, 2015; Dunbar, 2016).

This chapter reviews the broad spectrum of research on social network and disaster management. The first section discusses how different aspects of social capital contribute to disaster response and recovery in different disaster contexts. The following section explores the connections and overlaps between formal and informal social networks and their roles in disaster response and recovery. Methodological approaches for SNA are briefly discussed utilizing case studies. Online social networks facilitated by social media and their roles in disaster management are examined, which is followed by a discussion on ways to link online and offline social networks. The concluding section posits future research directions on social networks and disaster management, and how mitigation planning can employ SNA approaches to plan for disaster response and recovery.

Social Capital for Disaster Response and Recovery

Social capital consists of some aspect of a social structure that facilitates certain actions from actors within the structure (Coleman, 1988). It can be measured by the strength of family, neighborhood, religious, and community ties. Stronger social capital has been found to support both physical health and subjective well-being (Helliwell & Putnam, 2004) and contribute positively to social mobility and information sharing (Adler & Kwon, 2002). Greater social capital also increases commitment to a community and the ability to mobilize collective actions (Ellison et al., 2007).

Many studies have identified the critical role of social capital in disaster response and recovery (Aldrich & Meyer, 2014). Nakagawa and Shaw (2004) studied the role of social capital in the post-earthquake rehabilitation and reconstruction programs in Kobe, Japan and Gujarat, India. They found that communities with strong social capital experienced successful and quick recovery, and that community leaders utilized social capital in the recovery process to facilitate collective decision-making. Nakagawa and Shaw (2004) identified three dimensions of social capital: bonding, bridging, and linking (Figure 8.1). Bonding social capital includes features such as: trust in the leader and among community members; a social norm of democratic decision-making; a high level of participation in community activities among community members; and having various community-based groups. Bridging social capital includes the interactions with various stakeholders such as planners, consultants, academicians, community groups, universities, NGOs, and other neighboring communities. Linking social capital includes formal interactions or collaborations with government officials through community development activities.

Several studies contend that disaster agencies and government decision-makers should seek to strengthen social infrastructure at the community level. For example, Aldrich (2011), in a study on

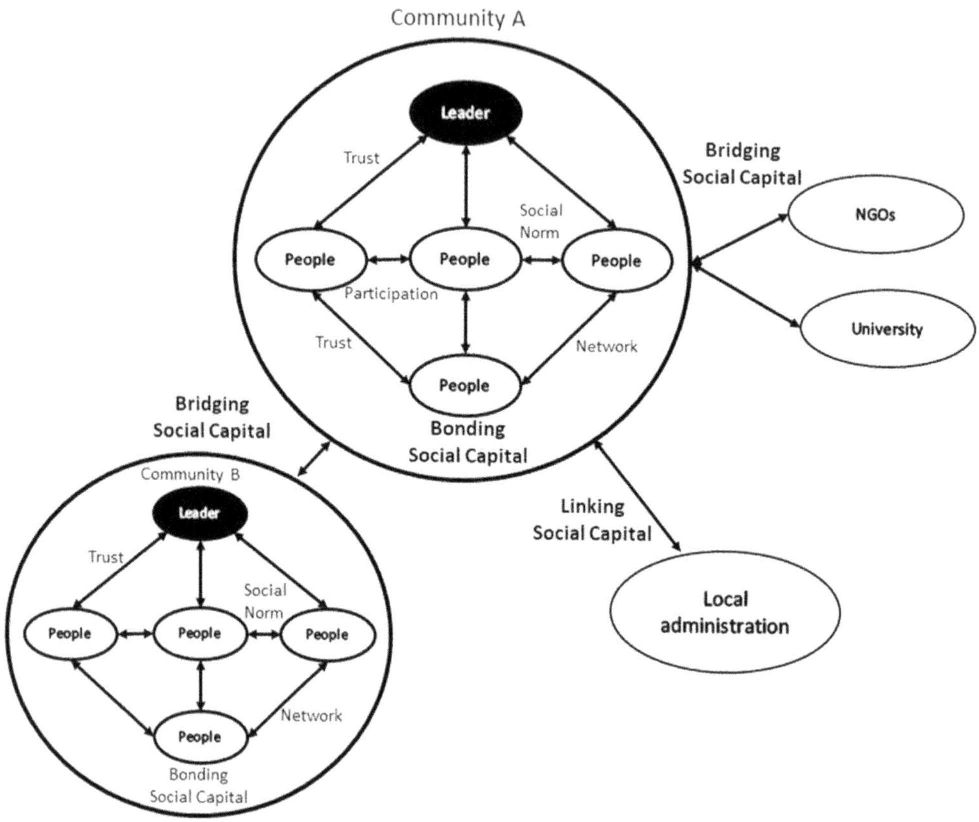

Figure 8.1 Conceptual Diagram Showing Different Aspects of Social Capital

Source: Figure Adapted from Nakagawa and Shaw (2004) and created by authors.

disaster recovery after the 1995 Kobe earthquake in Japan, found social capital to be the strongest and most robust predictor of post-disaster population recovery. One community with high social capital (measured by the number of NGOs per capita) was remarkably effective responding to fire events and providing care to one another, but another community with low social capital did not demonstrate the same degree of coordination. Joshi and Aoki (2014) also demonstrated the importance of social capital through their investigation of two districts affected by the tsunami in India. They found that social capital and the presence of strong community leadership were crucial for the implementation of a successful recovery program.

Social capital also contributes to a community's capacity for self-governance and the ability to deal with complex challenges. Grube and Storr (2014) studied how systems of self-governance prior to Hurricane Katrina supported post-disaster recovery in New Orleans. They found that community groups outperformed more centralized authorities, indicating how the relational nature or structure of social networks in a community can contribute to the advancement of social capital. Social networks, by forming the structural conditionality of social support, can either enable or prevent the emergence and diffusion of social capital (Vonneilich, 2022). The formation and sustenance of bonding, bridging, or linking social capitals—as depicted in Figure 8.1—relies on formal and informal social networks at different levels (Sabatini, 2009).

Formal and Informal Social Networks

As discussed in the previous section on different aspects of social capital, the formal, and informal processes through which social capital forms and operates necessitate further investigation. Bonding social capital at the community level is often operationalized as the degree of ties between individuals though informal associations (Faas & Jones, 2017). Bridging social capital with neighboring community or external agencies can be formed through both formal and informal interactions, while linking social capital is a more formal interaction between government agencies and community leaders. Studies on social networks have explored this broad spectrum of formal and informal linkages through studies at different scales, from individuals (Cassidy & Barnes, 2012) to the community (Varda et al., 2008) to the institutional level (Kapucu et al., 2010). These studies investigated the interactions among study units through the information, materials, and service flows among them, which occur through both formal and informal linkages. Karunarathne and Lee (2020) researched social networks at the household scale in Sri Lanka and how the networks change spatially and temporally for flood preparedness and recovery. Studying a cyclone-affected community in West Bengal, India, Misra et al. (2017) analyzed how social networks evolve during different phases of disaster. They identified unique networks at each phase while maintaining a few nodes or actors that remained central in more than one phase. General findings across the studies are that community and organizational networks usually evolve collectively in disaster situations, and the network itself can also evolve dynamically across the disaster cycle.

Formal and informal social networks, and the key actors within the network, can provide leadership and management support and augment social capacities by communicating, mobilizing, and collaborating with each other during a disaster. Rotberg (2010) demonstrated this though a study on a flood-prone village in Bangladesh, and Lu (2017) further established this in the recovery period after the 2008 Wenchuan earthquake in China. In the latter case, the emergence of new group and organizational actors took a primary role in overcoming the differences in institutional status and in developing a more resilient social structural environment.

Governance processes and outcomes can vary significantly due to differences in social network structures. Bodin and Crona (2009) showed this for the governance of natural resources, but it has also been identified in the case of disaster events (Hossain & Kuti, 2010). Stevenson and Conradson (2017) studied organizational support networks after the 2010–2011 earthquake in Canterbury, New Zealand, and identified a range of support types such as services, assistance, labor, emotional support, financial resources, and nonmonetary materials. They found that 'support tended to come from local supporters and from those with whom organizations had long-term relationships' (Stevenson & Conradson, 2017, 174). This case elucidates the implications of pre-disaster organizational networks and how characteristics of the networks contribute to providing support for disaster response and recovery.

Disaster management networks are actively used at all levels of government (e.g., local, state, and national) and can develop organically or by formal government mandate (Kapucu & Demiroz, 2017). Social networks at the community level usually form through informal reciprocal exchanges. Reciprocity can be viewed as an informal credit system in social networks in which a support provider can create an expectation of future support and the support recipient feels obliged to repay at some point (Coleman, 1988). In terms of reciprocity, support can be diverse during a disaster event. The reciprocity exchange can occur through information and resource mobilizations, evacuation and rescue, shelter, and financial and emotional support (Faas & Jones, 2017; Kapucu & Demiroz, 2017). However, studies have found that financial resources are usually perceived as less available for reciprocal exchanges through social networks compared to nonfinancial resources (Meyer, 2017). This indicates a shortcoming or gap in informal social networks that can be complemented by support from formal institutional networks. It also shows the importance of studying the structure of social networks and exploring how variations in network structure and its connection patterns contribute to disaster management. The next section discusses the methodological approach for analyzing social networks.

Social Network Analysis (SNA) Method

SNA is now widely applied in disaster response and recovery research (Faas & Jones, 2017). SNA can provide important insights on the process and mechanisms through which different social actors behave during and after a disaster event. SNA can help to identify the interactions within and between community networks, helping to build situational awareness, prepare more effective disaster management planning, and allocate resources more efficiently throughout the disaster cycle (Misra et al., 2017). A robust methodology utilizing SNA is emerging that offers the opportunity for cross-comparability research and policy analysis (Wilkin et al., 2019).

In this section we briefly review the methodological approach of SNA. The key aspect of SNA is the ability to determine whether and to what extent the network members or actors are connected among themselves through communication, resource sharing, and knowledge exchange (Bodin & Crona, 2009; Misra et al., 2017). Data for creating the network can be collected through focus group discussions, key informant interviews, or surveys of households and local institutions. After forming the network, different network indicators (see Table 8.1) are measured that can be related to network size and structural properties to indicate actors' position and the strength of their ties within the network.

As detailed in Table 8.1, network structures can be analyzed using attributes such as the density and size of network, degree of centrality and cohesiveness, and subgroup interconnectivity. *Network density* indicates the interlaced nature of community relations and a community's ability to undertake collective action. The level of network density can dictate how community members would get support during a disaster event. For example, through their study on Hurricane Andrew in the United States, Hurlbert et al. (2001) found that individuals with denser networks were more likely to receive informal social support while individuals with sparser social networks were more likely to receive support from institutions. *Average degree* or average number of ties per actor/node indicates the cohesiveness of actors in terms of collective actions in disaster preparedness and recovery. On the other hand, *degree centrality* can help to categorize actors/institutions in the network that play an influential role in mobilizing resources and disseminating information to the rest of network (Degenne & Forsé, 1999).

Closeness centrality of a network indicates how close an actor is to the rest of the actors in the network. A high degree of closeness centrality may support higher resource flow within the network, and actors with the shortest paths to other actors are expected to be more influential (Misra et al., 2017). Parallel to closeness centrality, *betweenness centrality* calculates the frequency or number of times an actor appears on the paths between two other actors (Wasserman & Faust, 1994). Actors with higher betweenness centrality can play a significant role in liaison between other actors and have potential control or sway over others (Freeman, 1978; Misra et al., 2017). Finally, a *structural hole* occurs whenever an actor has a relationship with someone who is connected to a separate cluster of actors yet has no direct or indirect connection with other actors in that cluster. Networks with more structural holes are likely to get exposed to more diverse information compared to networks with relatively fewer structural holes (Burt, 2004; Misra et al., 2017).

Table 8.1 Social Network Analysis Indicators and Their Relevance to Disaster Response and Recovery

Network property	*Description*	*Relevance for disaster response and recovery*
Network Density	Ratio between the existing ties and the possible ties of a particular network; when density is close to 1.0, the network is said to be dense/fully saturated, otherwise it is sparse	The interlaced nature of community relations and their ability to undertake collective action in disaster contexts
Average Degree	Average number of links per node/actor of a particular network (Total links/Total actors or nodes = Average degree)	Indicates the total number of helping ties connected to a particular household
Degree Centrality	Measure of number of ties that a node/actor has relative to the total number of ties existing in the network as a whole; higher degree centrality indicates actors with greater ties to other actors in the network	Identifies important persons or institutions in the networks at certain phases of a disaster; these actors are more important in mobilizing resources and disseminating information among the other actors across disaster support networks
Closeness Centrality	Measure of reciprocal of the geodesic distance (shortest path connecting two nodes) of a node to all other nodes in the network; a node is 'close' if it is a short distance from many other nodes in physical proximity	Important for disseminating service/information in a disaster; actors closer to the central actors may have advantages in disaster situations
Betweenness Centrality	Number of times an actor or node connects/occurs along the shortest path between two others	Actors/nodes with higher betweenness centrality can play the part of liaison, broker, or gatekeeper with potential control over other 'downstream' actors/nodes
Structural Hole	Occurs when a person (a) has a relationship with someone who is connected to a separate cluster of people, and (b) has no other direct or indirect connection with the people in that cluster	A network with more structural holes is likely receive more diverse information and opportunities compared to a network with fewer structural holes

Adapted from sources: (Misra et al., 2017; Karunarathne & Lee, 2020); Table created by Author.

Although most SNA studies have focused on offline social networks and their contributions to disaster management, studies of online social networks are gaining momentum given the increased use of social media for emergency response. For example, Kim and Hastak (2018) applied SNA to understand the characteristics of online social media networks during the 2016 flood in Louisiana. The next section discusses more on the use of online social networks for emergency management and the challenges of integrating online and offline social networks.

Online Social Networks for Emergency Management

With the advancement of internet technology, online social networks are now playing a critical role in disaster response and recovery. Social networking services, such as X (formerly Twitter) and Facebook, have become one of the most popular sources for accessing emergency information from individuals, disaster-related organizations, private and nonprofit entities, and state and local governments (Lindsay, 2011; Kim & Hastak, 2018). The use of social media in disasters can vary. However, emergency management agencies and officials regularly use it to facilitate communication with citizens, first responders, community organizations, and all levels of government by disseminating a wide range of emergency warnings and alerts, public safety and crisis information notifications, guidance on situational awareness, and requests for assistance (Lindsay, 2011). These communications often involve warnings for disasters, such as a cyclone, tsunami, or tornado, orders to evacuate or guidance to avoid certain areas, and requests to help bag sand for a flood or to be on the lookout for a specific individual, vehicle, or abnormal situation.

Disaster communication requires rapid dissemination and agile responses to situational information. From the perspective of emergency management agencies, online social networking via social media is one of the most effective ways to minimize communication gaps and establish vital two-way communications with the public (Fugate, 2011). Despite the effectiveness of social media in disaster communications among existing social connections, from the public's perspective, people tend to ignore or miss the official warnings delivered by state and local governments. Instead, they seek information from their friendship networks on social media and third parties or private companies such as Google (Wukich & Mergel, 2015). Also, online social media use in emergency management has encountered challenges that diminish its efficacy. In fact, emergency management use of social media can occasionally be counterproductive, such as through the propagation of inaccurate, outdated, and false information; malicious use of social media by actors during disasters; technical/logistical limitations (e.g., power outages that disrupt access to social media); and privacy concerns over the collection and retention of personal information by state and federal governments (Lindsay, 2011).

Additionally, with the lack of one-to-one and many-to-many communication strategies, emergency management agencies primarily rely on one-to-many strategies that resemble traditional communication strategies (e.g., television and radio) rather than collaborative and interactive social networks with citizens in online fora. Social media supports information flow and interpersonal communication differently compared to 'one-to-many' traditional mass media, such as via peer-to-peer communication, micro-blogging, content sharing, and (unregulated) discussion fora (Ellison & Vitak, 2015). While the functionality and reliability of some online social media networks have improved, this creates a challenge for trustworthy information sharing and communication in the disaster context.

Many attempts have been made to understand the impact of online communication on offline communications in terms of overall social interaction, especially since the advent of social media as one of the most popular online communication modes (Ellison & Vitak, 2015; Hage & Noseleit, 2015). Studies have highlighted the relationship between online and offline communications with a focus on social capital and social networks (DiMaggio et al., 2001; Rosen et al., 2010; Kane et al., 2014; Dunbar, 2016). These studies argue for an intersection between offline social networks and online social media networks to improve the outcomes of social interactions during disasters, but they mostly engage in theory (Boyd & Ellison, 2007; Kane et al., 2014). People use social media to communicate with others and to access a breadth of social relationships, but in reality there may be few connections between social media networks and offline interpersonal relationships, and the overall influence of each (let alone their joint impact) needs to be further studied (Kane et al., 2014).

Linking Online and Offline Social Networks

Although online social networks have gained popularity for spreading disaster-related information—particularly on social media platforms—it is not easy to determine the extent online and offline networks are related to each other and how much they overlap in disasters. There is growing attention being paid to the relationships between online and offline social networks, but a dearth of studies on the topic exists. Some studies suggest there are overlaps in certain parts of online-offline social networks, although the overlapping degree is low (Gaito et al., 2012). People do not build their relational networks unconsciously in online or offline settings. Rather, people knowingly manipulate the structures of their own social networks to prepare for, cope with, adapt to, resist, and recover from risks and disasters (Jones & Faas, 2016). In online networks, information diffusion patterns consist of influential nodes/actors and neighborhood nodes (general users) that are induced by the influential nodes; this is similar to how information transmits in offline social networks. Studies claim such influential nodes (e.g., respected friends and family members, trusted leaders and institutions, and celebrities and 'influencers') are critical to increasing information propagation and self-organized dissemination of information through online social networks in disasters (Yang et al., 2019; Fan, Jiang, et al., 2020; Fan et al., 2021).

A barrier to bridging the gap between online and offline social networks lies in the limitations or incongruencies of different scalability in spatio-temporal contexts. Although social networks are not a physical place, places are still important contexts to organize and facilitate social interactions among individuals' social networks (Piselli, 2007). Information dissemination through online social media can also be robust within a small geographical area, but its geographical reach extends to the regional, national, and even global scale given the absence of physical distance in online space, although network density obviously decreases over this distance (Stephens & Poorthuis, 2015). In disaster situations, local geographic actors/nodes can be effective in expanding their roles as hubs of situational information through social media sites (Fan et al., 2021). However, fundamental limitations remain in the unavailability of and mismatches and inequities in the spatial data of online social networks in ways that serve to reinforce offline networks in times of disaster. Thus, it is still difficult to collect and disseminate real-time situational information that is inclusive and tailored to local community needs. The amount of geotagged social media data is often too small (1–3%), and misrepresentation of these geotagged data further add to over- and under-estimation of the situational context (Morstatter et al., 2013; Fan, Esparza, et al., 2020;). Also, communities with larger population size gain more social media attention and thus tend to receive disaster-related information more quickly relative to those with smaller population size or sparsely populated, regardless of the level of damage or urgency of the situation (Fan, Esparza, et al., 2020).

Despite the challenges, linking online and offline social networks to better respond to disasters and emergencies is both valuable and inevitable considering ongoing societal trends in which individuals increasingly rely on online resources and services. We argue that spatio-temporal dynamics are still important factors even in the digital realm as disasters overlay upon particular places and particular communities with time and speed of communication being of the essence. The consideration of spatio-temporal dynamics can improve the accuracy, penetration, timeliness, and equity of disaster communication and disaster management writ large. Recognizing similarities and differences in the content and structure of online-offline social networks is also essential for

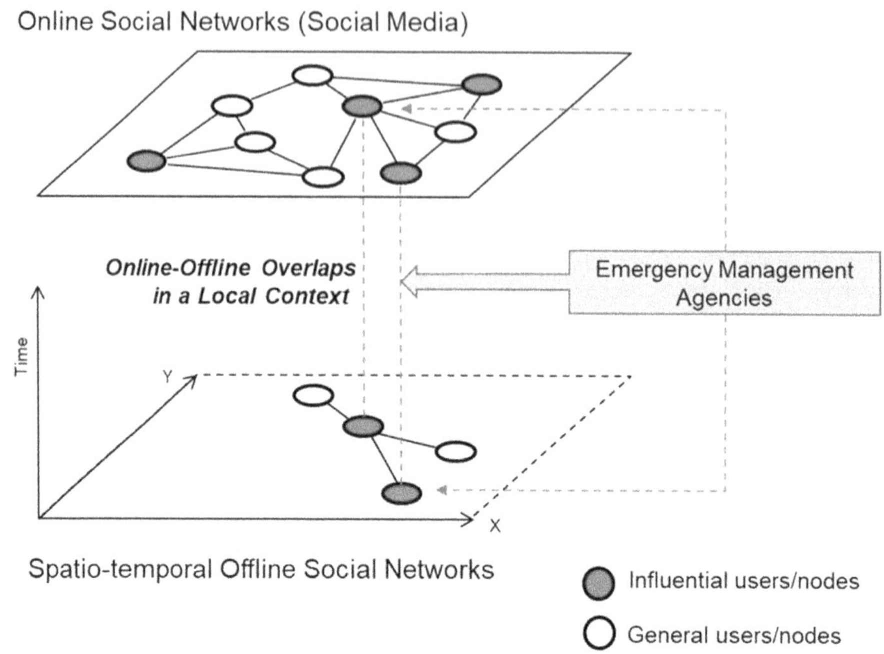

Figure 8.2 Conceptual Diagram Linking Online and Offline Social Networks for Disaster Communication
Source: Figure created by authors/original figure.

better understanding information diffusion at the local scale. Most importantly, and more practically, the role of local influencers and emergency management agencies on social media is also vital to creating a robust online-offline social network system. Figure 8.2 provides a topological conceptualization of linking online and offline social networks for disaster communication utilizing influential users/nodes.

Conclusion

Social networks, both online and offline, are forms of social capital that will continue to play a crucial role in disaster response and recovery. As contended in the previous section, considering the increased use of social media for emergency communication and in general, we must identify ways to better link online and offline social networks in the disaster sphere. How links between these networks and the network structures evolve spatially and temporally also needs to be studied. Problems of unequal access to social media and other smart engagement approaches, as has been documented in cases of public participation and policymaking (Conroy & Gordon, 2004; Kashem & Gallo, 2023), need to be recognized for more effective disaster management among vulnerable communities. Over-reliance on social media for emergency communication can itself be problematic considering the risk of electricity and internet disruptions during disaster events, which again buttresses the criticality of linking online and offline social networks.

In the case of offline social networks, studies have demonstrated that reciprocal support and exchanges mobilized by social networks are critical for disaster response and recovery. Reciprocity, mutuality, and egalitarianism represent fundamental elements in networked structures that help to foster responsible input and a sense of community from actors. However, unequal power relations can impinge the participation of low-income and underrepresented communities that may lack the resources or capacity to effectively respond to or recover from a disaster. While informal bonding social capital may allow such communities to respond to a disaster through reciprocal mobilization of resources (e.g., shelter, food, and emotional support), they may still need formal institutional support for long-term disaster recovery facilitated through bridging and linking social capitals (e.g., with government and aid agencies). SNA can be deployed in disaster planning to better understand, strengthen, and link social networks at the community/population and government/institutional levels. In any case, nurturing existing social networks pre-disaster can contribute to more effective disaster response and recovery.

References

Adler, Paul S., and Seok-Woo Kwon. "Social Capital: Prospects for a New Concept." *The Academy of Management Review*, vol. 27, no. 1, 2002, pp. 17–40. *JSTOR*, https://www.jstor.org/stable/4134367.

Aldrich, Daniel P. "The Power of People: Social Capital's Role in Recovery from the 1995 Kobe Earthquake." *Natural Hazards*, vol. 56, no. 3, Mar. 2011, pp. 595–611. *Springer Link*, https://doi.org/10.1007/s11069-010-9577-7.

Aldrich, Daniel P., and Michelle A. Meyer. "Social Capital and Community Resilience." *American Behavioral Scientist*, Oct. 2014. Sage CA: Los Angeles, CA, *journals.sagepub.com*, https://doi.org/10.1177/0002764214550299.

Bodin, Örjan, and Beatrice I. Crona. "The Role of Social Networks in Natural Resource Governance: What Relational Patterns Make a Difference?" *Global Environmental Change*, vol. 19, no. 3, Aug. 2009, pp. 366–74. *ScienceDirect*, https://doi.org/10.1016/j.gloenvcha.2009.05.002.

Boyd, Danah M., and Nicole B. Ellison. "Social Network Sites: Definition, History, and Scholarship." *Journal of Computer-Mediated Communication*, vol. 13, no. 1, 2007, pp. 210–30. *Wiley Online Library*, https://doi.org/10.1111/j.1083-6101.2007.00393.x.

Burt, Ronald S. "Structural Holes and Good Ideas." *American Journal of Sociology*, vol. 110, no. 2, Sept. 2004, pp. 349–99. *journals.uchicago.edu (Atypon)*. https://doi.org/10.1086/421787.

Casagrande, David G., Heather McIlvaine-Newsad, and Eric C. Jones. "Social Networks of Help-Seeking in Different Types of Disaster Responses to the 2008 Mississippi River Floods." *Human Organization*, vol. 74, no. 4, Jan. 2015, pp. 351–61. *Silverchair*, https://doi.org/10.17730/0018-7259-74.4.351.

Cassidy, Lin, and Grenville Barnes. "Understanding Household Connectivity and Resilience in Marginal Rural Communities through Social Network Analysis in the Village of Habu, Botswana." *Ecology and Society*, vol. 17, no. 4, Nov. 2012. www.ecologyandsociety.org, https://doi.org/10.5751/ES-04963-170411 https://doi.org/10.5751/ES-04963-170411

Coleman, James S. "Social Capital in the Creation of Human Capital." *American Journal of Sociology*, vol. 94, 1988, pp. S95–120.

Conroy, Maria Manta, and Steven I. Gordon. "Utility of Interactive Computer-Based Materials for Enhancing Public Participation." *Journal of Environmental Planning and Management*, vol. 47, no. 1, Jan. 2004, pp. 19–33. *Taylor and Francis+NEJM*, https://doi.org/10.1080/0964056042000189781.

Degenne, Alain, and Michel Forsé. *Introducing Social Networks*. SAGE, 1999.

DiMaggio, Paul, Eszter Hargittai, W. Russell Neuman, and John P. Robinson. "Social Implications of the Internet." *Annual Review of Sociology*, vol. 27, no. 1, Aug. 2001, pp. 307–36. *DOI.org (Crossref)*, https://doi.org/10.1146/annurev.soc.27.1.307.

Dunbar, R. I. M. "Do Online Social Media Cut through the Constraints That Limit the Size of Offline Social Networks?" *Royal Society Open Science*, vol. 3, no. 1, Jan. 2016, p. 150292. *royalsocietypublishing.org (Atypon)*, https://doi.org/10.1098/rsos.150292.

Ellison, Nicole B., Charles Steinfield, and Cliff Lampe. "The Benefits of Facebook 'Friends:' Social Capital and College Students' Use of Online Social Network Sites." *Journal of Computer-Mediated Communication*, vol. 12, no. 4, July 2007, pp. 1143–68. *academic.oup.com*, https://doi.org/10.1111/j.1083-6101.2007.00367.x.

Ellison, Nicole B., and Jessica Vitak. "Social Network Site Affordances and Their Relationship to Social Capital Processes." *The Handbook of the Psychology of Communication Technology*, edited by S. Shyam Sundar, 1st ed. Wiley, 2015, pp. 203–27. *DOI.org (Crossref)*, https://doi.org/10.1002/9781118426456.ch9.

Faas, Albert J., and Eric C. Jones. "Chapter 2 – Social Network Analysis Focused on Individuals Facing Hazards and Disasters." *Social Network Analysis of Disaster Response, Recovery, and Adaptation*, edited by Eric C. Jones and Albert J. Faas. Butterworth-Heinemann, 2017, pp. 11–23. *ScienceDirect*, https://doi.org/10.1016/B978-0-12-805196-2.00002-9.

Fan, Chao, Yucheng Jiang, Yang Yang, Cheng Zhang, and Ali Mostafavi. "Crowd or Hubs: Information Diffusion Patterns in Online Social Networks in Disasters." *International Journal of Disaster Risk Reduction*, vol. 46, June 2020, p. 101498. *ScienceDirect*, https://doi.org/10.1016/j.ijdrr.2020.101498.

Fan, Chao, Miguel Esparza, Jennifer Dargin, Fangsheng Wu, Bora Oztekin, and Ali Mostafavi. "Spatial Biases in Crowdsourced Data: Social Media Content Attention Concentrates on Populous Areas in Disasters." *Computers, Environment and Urban Systems*, vol. 83, Sept. 2020, p. 101514. *ScienceDirect*, https://doi.org/10.1016/j.compenvurbsys.2020.101514.

Fan, Chao, Yucheng Jiang, and Ali Mostafavi. "The Role of Local Influential Users in Spread of Situational Crisis Information." *Journal of Computer-Mediated Communication*, vol. 26, no. 2, Mar. 2021, pp. 108–27. *Silverchair*, https://doi.org/10.1093/jcmc/zmaa020.

Feldman, David, Santina Contreras, Beth Karlin, Victoria Basolo, Richard Matthew, Brett Sanders, Douglas Houston, Wing Cheung, Kristen Goodrich, Abigail Reyes, Kimberly Serrano, Jochen Schubert, and Adam Luke. "Communicating Flood Risk: Looking Back and Forward at Traditional and Social Media Outlets." *International Journal of Disaster Risk Reduction*, vol. 15, Mar. 2016, pp. 43–51. *ScienceDirect*, https://doi.org/10.1016/j.ijdrr.2015.12.004.

Freeman, Linton C. "Centrality in Social Networks Conceptual Clarification." *Social Networks*, vol. 1, no. 3, Jan. 1978, pp. 215–39. *ScienceDirect*, https://doi.org/10.1016/0378-8733(78)90021-7.

Fugate, Craig. *Understanding the Power of Social Media as a Communication Tool in the Aftermath of Disasters*. Federal Emergency Management Agency, 2011. *Google Scholar*, https://www.fema.gov/pdf/about/programs/legislative/testimony/2011/5_5_2011_use_of_social_media_in_disasters.pdf.

Gaito, Sabrina, Gian Paolo Rossi, and Matteo Zignani. "Facencounter: Bridging the Gap between Offline and Online Social Networks." *2012 Eighth International Conference on Signal Image Technology and Internet Based Systems*, IEEE, 2012, pp. 768–75. *DOI.org (Crossref)*, https://doi.org/10.1109/SITIS.2012.116.

Grube, Laura, and Virgil Henry Storr. "The Capacity for Self-Governance and Post-Disaster Resiliency." *The Review of Austrian Economics*, vol. 27, no. 3, Sept. 2014, pp. 301–24. *link.springer.com*, https://doi.org/10.1007/s11138-013-0210-3.

Haer, Toon, W.J. Wouter Botzen, and Jeroen C.J.H. Aerts. "The Effectiveness of Flood Risk Communication Strategies and the Influence of Social Networks—Insights from an Agent-Based Model." *Environmental Science & Policy*, vol. 60, June 2016, pp. 44–52. *ScienceDirect*, https://doi.org/10.1016/j.envsci.2016.03.006.

Hage, Eveline, and Florian Noseleit. *Changes and Variations in Online-and Offline Communication Patterns: Including Peer Effects*. 2015. *Google Scholar*, https://core.ac.uk/download/pdf/301366944.pdf.

Helliwell, John F., and Robert D. Putnam. "The Social Context of Well–Being." *Philosophical Transactions of the Royal Society of London. Series B: Biological Sciences*, vol. 359, no. 1449, Sept. 2004, pp. 1435–46. *royalsocietypublishing.org (Atypon)*, https://doi.org/10.1098/rstb.2004.1522.

Hossain, Liaquat, and Matthew Kuti. "Disaster Response Preparedness Coordination through Social Networks." *Disasters*, vol. 34, no. 3, 2010, pp. 755–86. *Wiley Online Library*, https://doi.org/10.1111/j.1467-7717.2010.01168.x.

Hurlbert, Jeanne S., Valerie A. Haines, and John J. Beggs. "Core Networks and Tie Activation: What Kinds of Routine Networks Allocate Resources in Nonroutine Situations?" *American Sociological Review*, vol. 65, no. 4, 2000, pp. 598–618. *JSTOR*, https://doi.org/10.2307/2657385.

Hurlbert, Jeanne S., Valerie A. Haines, and John J. Beggs. "Social Networks and Social Capital in Extreme Environments." *Social Capital*. Routledge, 2001.

Ibañez, Gladys E., Chad A. Buck, Nadya Khatchikian, and Fran H. Norris. "Qualitative Analysis of Coping Strategies among Mexican Disaster Survivors." *Anxiety, Stress, & Coping*, vol. 17, no. 1, Mar. 2004, pp. 69–85. *Taylor and Francis+NEJM*, https://doi.org/10.1080/10615800310001639628.

Jones, Eric C., and Albert J. Faas. *Social Network Analysis of Disaster Response, Recovery, and Adaptation*. Butterworth-Heinemann, 2016.

Joshi, Abhay, and Misa Aoki. "The Role of Social Capital and Public Policy in Disaster Recovery: A Case Study of Tamil Nadu State, India." *International Journal of Disaster Risk Reduction*, vol. 7, Mar. 2014, pp. 100–08. *ScienceDirect*, https://doi.org/10.1016/j.ijdrr.2013.09.004.

Kane, Gerald C., Maryam Alavi, Giuseppe Labianca, and Stephen P. Borgatti. "What's Different about Social Media Networks? A Framework and Research Agenda." *MIS Quarterly*, vol. 38, no. 1, 2014, pp. 275–304.

Kaniasty, Krzysztof. "Predicting Social Psychological Well-Being Following Trauma: The Role of Postdisaster Social Support." *Psychological Trauma: Theory, Research, Practice, and Policy*, vol. 4, no. 1, 2012, pp. 22–33. *APA PsycNet*, https://doi.org/10.1037/a0021412.

Kapucu, Naim, Thomas Bryer, Vener Garayev, and Tolga Arslan. "Interorganizational Network Coordination under Stress Caused by Repeated Threats of Disasters." *Journal of Homeland Security and Emergency Management*, vol. 7, no. 1, June 2010. *www.degruyter.com*, https://doi.org/10.2202/1547-7355.1629.

Kapucu, Naim, and Fatih Demiroz. "Chapter 3 – Interorganizational Networks in Disaster Management." *Social Network Analysis of Disaster Response, Recovery, and Adaptation*, edited by Eric C. Jones and Albert J. Faas. Butterworth-Heinemann, 2017, pp. 25–39. *ScienceDirect*, https://doi.org/10.1016/B978-0-12-805196-2.00003-0.

Karunarathne, Ananda Y., and Gunhak Lee. "The Geographies of the Dynamic Evolution of Social Networks for the Flood Disaster Response and Recovery." *Applied Geography*, vol. 125, Dec. 2020, p. 102274. *ScienceDirect*, https://doi.org/10.1016/j.apgeog.2020.102274.

Kashem, Shakil Bin, and Dora Gallo. "Smart Engagement in Small Cities: Exploring Minority Participation in Planning." *Urban Planning*, vol. 8, no. 2, Apr. 2023, pp. 44–56. www.cogitatiopress.com, https://doi.org/10.17645/up.v8i2.6607.

Kim, Jooho, and Makarand Hastak. "Social Network Analysis: Characteristics of Online Social Networks after a Disaster." *International Journal of Information Management*, vol. 38, no. 1, Feb. 2018, pp. 86–96. *ScienceDirect*, https://doi.org/10.1016/j.ijinfomgt.2017.08.003.

Knox, John A., Jared A. Rackley, Alan W. Black, Vittorio A. Gensini, Michael Butler, Corey Dunn, Taylor Gallo, Melyssa R. Hunter, Lauren Lindsey, Minh Phan, Robert Scroggs, and Synne Brustad. "Tornado Debris Characteristics and Trajectories Durinandhe 27 April 2011 Super Outbreak as Determined Using Social Media Data." *Bulletin of the American Meteorological Society*, vol. 94, no. 9, Sept. 2013, pp. 1371–80. *journals.ametsoc.org*, https://doi.org/10.1175/BAMS-D-12-00036.1.

Lindsay, Bruce R. *Social Media and Disasters: Current Uses, Future Options, and Policy Considerations*. Congressional Research Service, 6 Sept. 2011..

Lu, Jia. "Chapter 10 – Interorganizational Network Dynamics in the Wenchuan Earthquake Recovery." *Social Network Analysis of Disaster Response, Recovery, and Adaptation*, edited by Eric C. Jones and Albert J. Faas. Butterworth-Heinemann, 2017, pp. 143–59. *ScienceDirect*, https://doi.org/10.1016/B978-0-12-805196-2.00010-8.

Malone, Elizabeth L., and Susan Kinnear. "How and Why: Complementary Analyses of Social Network Structures and Cultural Values: Improving Flood Response Networks in Queensland, Australia." *Quality & Quantity*, vol. 49, no. 1, Jan. 2015, pp. 203–20. *Springer Link*, https://doi.org/10.1007/s11135-013-9982-6.

Meyer, Michelle. "Chapter 9 – The Family's Burden: Perceived Social Network Resources for Individual Disaster Assistance in Hazard-Prone Florida." *Social Network Analysis of Disaster Response, Recovery, and Adaptation*, edited by Eric C. Jones and Albert J. Faas, Butterworth-Heinemann, 2017, pp. 127–41. *ScienceDirect*, https://doi.org/10.1016/B978-0-12-805196-2.00009-1.

Mickoleit, Arthur. *Social Media Use by Governments: A Policy Primer to Discuss Trends, Identify Policy Opportunities and Guide Decision Makers*. OECD, 22 Dec. 2014. *OECD iLibrary*, https://doi.org/10.1787/5jxrcmghmk0s-en.

Middleton, Stuart E., Lee Middleton, and Stefano Modafferi. "Real-Time Crisis Mapping of Natural Disasters Using Social Media." *IEEE Intelligent Systems*, vol. 29, no. 2, Mar. 2014, pp. 9–17. *IEEE Xplore*, https://doi.org/10.1109/MIS.2013.126.

Misra, Sanchayeeta, Rupak Goswami, Tandra Mondal, and Rabindranath Jana. "Social Networks in the Context of Community Response to Disaster: Study of a Cyclone-Affected Community in Coastal West Bengal, India." *International Journal of Disaster Risk Reduction*, vol. 22, June 2017, pp. 281–96. *ScienceDirect*, https://doi.org/10.1016/j.ijdrr.2017.02.017.

Morstatter, Fred, Jürgen Pfeffer, Huan Liu, and Kathleen Carley. "Is the Sample Good Enough? Comparing Data from Twitter's Streaming API with Twitter's Firehose." *Proceedings of the International AAAI Conference on Web and Social Media*, vol. 7, no. 1, 2013, pp. 400–08. *ojs.aaai.org*, https://doi.org/10.1609/icwsm.v7i1.14401.

Nakagawa, Yuko, and Rajib Shaw. "Social Capital: A Missing Link to Disaster Recovery." *International Journal of Mass Emergencies & Disasters*, vol. 22, no. 1, Mar. 2004, pp. 5–34. *SAGE Journals*, https://doi.org/10.1177/028072700402200101.

Piselli, Fortunata. "Communities, Places, and Social Networks." *American Behavioral Scientist*, vol. 50, no. 7, Mar. 2007, pp. 867–78. *SAGE Journals*, https://doi.org/10.1177/0002764206298312.
Rosen, Devan, Michael A. Stefanone, and Derek Lackaff. "Online and Offline Social Networks: Investigating Culturally-Specific Behavior and Satisfaction." *2010 43rd Hawaii International Conference on System Sciences*, 2010, pp. 1–10. *IEEE Xplore*, https://doi.org/10.1109/HICSS.2010.292.
Rotberg, Fiona. "Social Networks and Adaptation in Rural Bangladesh." *Climate and Development*, Jan. 2010. world, *www.tandfonline.com*, https://doi.org/10.3763/cdev.2010.0031.
Sabatini, Fabio. "Social Capital as Social Networks: A New Framework for Measurement and an Empirical Analysis of Its Determinants and Consequences." *The Journal of Socio-Economics*, vol. 38, no. 3, June 2009, pp. 429–42. *ScienceDirect*, https://doi.org/10.1016/j.socec.2008.06.001.
Stephens, Monica, and Ate Poorthuis. "Follow Thy Neighbor: Connecting the Social and the Spatial Networks on Twitter." *Computers, Environment and Urban Systems*, vol. 53, Sept. 2015, pp. 87–95. *ScienceDirect*, https://doi.org/10.1016/j.compenvurbsys.2014.07.002.
Stevenson, Joanne R., and David Conradson. "Chapter 11 – Organizational Support Networks and Relational Resilience after the 2010/11 Earthquakes in Canterbury, New Zealand." *Social Network Analysis of Disaster Response, Recovery, and Adaptation*, edited by Eric C. Jones and Albert J. Faas. Butterworth-Heinemann, 2017, pp. 161–75. *ScienceDirect*, https://doi.org/10.1016/B978-0-12-805196-2.00011-X.
Stewart, Kathleen, Jennifer L. Glanville, and David A. Bennett. "Exploring Spatiotemporal and Social Network Factors in Community Response to a Major Flood Disaster." *The Professional Geographer*, July 2014. world, *www.tandfonline.com*, https://www.tandfonline.com/doi/full/10.1080/00330124.2013.799995.
Tyshchuk, Yulia, Cindy Hui, Martha Grabowski, and William A. Wallace. "Social Media and Warning Response Impacts in Extreme Events: Results from a Naturally Occurring Experiment." *2012 45th Hawaii International Conference on System Sciences*, 2012, pp. 818–27. *IEEE Xplore*, https://doi.org/10.1109/HICSS.2012.536.
Varda, Danielle M., Anita Chandra, Stefanie A. Stern, and Nicole Lurie. "Core Dimensions of Connectivity in Public Health Collaboratives." *Journal of Public Health Management and Practice*, vol. 14, no. 5, Oct. 2008, p. E1. *journals.lww.com*, https://doi.org/10.1097/01.PHH.0000333889.60517.46.
Vonneilich, Nico. "Social Relations, Social Capital, and Social Networks: A Conceptual Classification." *Social Networks and Health Inequalities: A New Perspective for Research*, edited by Andreas Klärner et al. Springer International Publishing, 2022, pp. 23–34. *Springer Link*, https://doi.org/10.1007/978-3-030-97722-1_2.
Wasserman, Stanley, and Katherine Faust. *Social Network Analysis: Methods and Applications*. Cambridge University Press, 1994.
Wilkin, Joanna, Eloise Biggs, and Andrew J. Tatem. "Measurement of Social Networks for Innovation within Community Disaster Resilience." *Sustainability*, vol. 11, no. 7, 7, Jan. 2019, p. 1943. *www.mdpi.com*, https://doi.org/10.3390/su11071943.
Wukich, Clayton, and Ines Mergel. "Closing the Citizen-Government Communication Gap: Content, Audience, and Network Analysis of Government Tweets." *Journal of Homeland Security and Emergency Management*, vol. 12, no. 3, Sept. 2015, pp. 707–35. www.degruyter.com, https://doi.org/10.1515/jhsem-2014-0074.
Yang, Yang, Cheng Zhang, Chao Fan, Wenlin Yao, Ruihong Huang, and Ali Mostafavi. "Exploring the Emergence of Influential Users on Social Media during Natural Disasters." *International Journal of Disaster Risk Reduction*, vol. 38, Aug. 2019, p. 101204. *ScienceDirect*, https://doi.org/10.1016/j.ijdrr.2019.101204.
Yin, Jie, Andrew Lampert, Mark Cameron, Bella Robinson, and Robert Power. "Using Social Media to Enhance Emergency Situation Awareness." *IEEE Intelligent Systems*, vol. 27, no. 6, Nov. 2012, pp. 52–59. *IEEE Xplore*, https://doi.org/10.1109/MIS.2012.6.
Zhang, Zhe, Hao Hu, Dandong Yin, Shakil Kashem, Ruopu Li, Heng Cai, Dylan Perkins, and Shaowen Wang. "A CyberGIS-Enabled Multi-Criteria Spatial Decision Support System: A Case Study on Flood Emergency Management." *International Journal of Digital Earth*, vol. 12, no. 11, Nov. 2019, pp. 1364–81. *IngentaConnect*, https://doi.org/10.1080/17538947.2018.1543363.

Section III

Disaster Relief and Aid Efforts

9

NON-GOVERNMENTAL ORGANIZATIONS (NGOs) AS PROVIDERS AND DISTRIBUTORS OF DISASTER ASSISTANCE

Julia Sherry and Kobel Dubique

Introduction

Non-governmental organizations (NGOs) play a pivotal role in disaster assistance and the distribution of humanitarian aid. Their significance in disaster assistance has grown in recent decades, and their effectiveness centers around their ability to deliver aid and assistance quickly and efficiently to affected populations. By definition, NGOs are autonomous local and international organizations that are not affiliated with a government (International Committee of the Red Cross, 1994; United Nations Office for Coordination of Humanitarian Affairs, 2018). While they vary significantly in size, scope, and area of focus, NGOs that operate in disaster settings share an overarching commitment to the humanitarian principles of humanity, impartiality, neutrality, and independence and share the goal of providing humanitarian assistance that saves lives and alleviates suffering among populations impacted by crises (International Committee of the Red Cross, 1994; Leaning, 2007; Sphere, 2018; United Nations Office for Coordination of Humanitarian Affairs, 2018). Some argue that while NGOs may share this commitment to humanitarian ideals in principle, in reality, NGOs may negatively impact the communities and nations they attempt to serve in disasters (Schuller, 2017).

NGOs are often at the forefront of providing and delivering aid in disasters, including but not limited to food, water, shelter, and health services. As aid distributors, they play a key role in coordinating disaster assistance efforts and participating in collaboration mechanisms alongside government and other organizations present in disaster settings. While the NGO community has long been involved in disasters, it is generally understood that their roles, objectives, and agendas have increased and evolved over time (United Nations Office for Coordination of Humanitarian Affairs, 2018; Sapat et al., 2019; Park & Yoon, 2022). Given their larger presence in disaster response activities in recent years, there has likewise been increasing interest among disaster scholars in the role of NGOs in the provision and delivery of disaster assistance, their advantages and efficiencies, and their challenges and criticisms (Farmer, 2011; Islam & Walkerden, 2015; Hsu & Schuller, 2020; Fitch-Fleischmann & Kresch, 2021; Arnaouti et al., 2022).

There is ongoing discussion over whether, in both disasters and long-term development contexts, the impact of NGO involvement is ultimately ‘good’ or ‘bad,’ or a foreign ‘invasion’ or a positive ‘infusion’ (Schuller, 2007). Regardless of the positive and negative impacts of NGO involvement, the functional reality is that humanitarianism involving individuals, states, agencies,

DOI: 10.4324/9781003414834-14

and NGOs "has become a potent force of our world," particularly during catastrophic events and disasters and thus must be acknowledged, studied, critiqued, and improved (Fassin, 2011, xi). In this vein, this chapter explores the dynamic and multifaceted roles played by NGOs involved in disaster assistance. Additionally, it offers insights on the opportunities and challenges related to NGO involvement in disaster response efforts with the ultimate goal of improving their operations and ultimately their impacts on disaster-affected communities.

Characteristics of NGOs Involved in Disaster Assistance

As key actors in the dynamic landscape of disaster assistance, each NGO is distinguished by unique characteristics that shape their response efforts. This section delves into NGOs' diverse traits, from their scope and scale to their varied areas of expertise. As Shah et al. (2022) note, this great "diversity in task, goal, and activities" among NGOs involved in disasters "strains any simple definition" of how NGOs operate in this setting (3). Knowledge of these characteristics is important for understanding the roles and capacities of NGOs in disasters.

Diversity in Scope and Scale

NGOs involved in disaster assistance vary significantly in terms of their scope and scale, ranging from community-based, grassroots organizations embedded in one community or country to massive international non-governmental organizations (INGOs) operating across the globe with thousands of staff and multi-million-dollar budgets (United Nations Office for Coordination of Humanitarian Affairs, 2018; Sapat et al., 2019).[1] The size and scope of NGOs impact their engagement strategies, operational capabilities, and timeline of involvement in each disaster (Park & Yoon, 2022). At each scope and scale, NGOs have relative advantages and disadvantages and unique roles and responsibilities throughout each stage of the disaster cycle (Shaw et al., 2003). Because collaboration is critical in disaster settings, these NGOs of diverse sizes and scales often work with other NGOs, United Nations (UN) agencies, the government, military, local agencies, volunteers, and various other partners to meet humanitarian needs (Islam & Walkerden, 2015; Park & Yoon, 2022).

Local and National NGOs

Local and national NGOs are generally classified as organizations working exclusively within their own community or country (United Nations Office for Coordination of Humanitarian Affairs, 2018). These grassroots organizations, community-based organizations (CBOs), and national NGOs often hold a wealth of knowledge of the local context and its sociocultural fabric (Perry, 2007; Islam & Walkerden, 2015). Given their contextual competency and on-the-ground presence prior to the disaster, such organizations are often well-positioned to meet immediate local needs and act as first responders through the ability to leverage their geographic proximity and established relationships with the community (Kidder, 2010; Fallah & Nejad, 2018; Fitch-Fleischmann & Kresch, 2021). The United Nations Office for the Coordination of Humanitarian Affairs (OCHA) notes that national NGOs are often numerous and frequently positioned as "essential partners in disaster response because they are known locally, and they know the area, the culture, the population, etc." (United Nations Office for Coordination of Humanitarian Affairs, 2018, 14).

However, local and national NGOs may be limited by scarce resources and difficulties in collaborating effectively with larger NGOs, government agencies, UN agencies, and international organizations involved in disaster response (Islam & Walkerden, 2015; Sapat et al., 2019; Shah et al., 2022). Furthermore, these smaller organizations may have challenges navigating the complex coordination mechanisms and communication channels utilized in global disaster response and funding management.

International NGOs (INGOs)

Larger INGOs, on the other hand, often have the advantage of higher operational capacity, more extensive resources, and greater experience in diverse disaster settings globally. They may be well-positioned to provide immediate, large-scale response efforts and have the specialized knowledge and expertise across the various sectors that need disaster assistance. OCHA (United Nations Office for Coordination of Humanitarian Affairs, 2018) notes that the "world's largest NGOs have budgets and resources exceeding that of many UN agencies" (14) and emphasizes the benefits of specialization across the disaster response sectors, skilled staff, capacity for rapid deployment, operational flexibility, and resources that INGOs offer in disasters.

Additionally, INGOs often have significant experience coordinating and networking in disaster settings. They may be able to coordinate quickly and efficiently with various stakeholders, including other (I)NGOs and government stakeholders (Shah et al., 2022). Many INGOs are familiar with widely used disaster management systems, including the UN Cluster Approach coordinated by OCHA, UN Financial Tracking Service (FTS), and working under the best practices and minimum standards in humanitarian response set by the organization Sphere, formerly known as the Sphere Project (Sphere, 2018; United Nations Office for Coordination of Humanitarian Affairs, 2018). INGOs also often possess technical expertise in sectors needed in disaster response, including healthcare, water and sanitation, and logistics.

On the other hand, INGOs may face challenges stemming from a lack of context-specific knowledge, dependence on external donors for sustained funding, poor coordination with local and national governments, and lack of accountability and transparency. In some cases, INGOs have been criticized for their lack of cultural sensitivity and poor alignment of interventions with the actual needs and preferences of affected communities (Paul, 2006; Perry, 2007; Schuller, 2017). Furthermore, INGOs may be more accountable to their donors rather than the state and populations they aim to serve. This dependency on private donors and lack of local accountability may limit the scope, duration, and impact of their disaster assistance efforts, leading to a short-term focus leaving gaps in long-term recovery efforts (Perry, 2007; Subbarao et al., 2010; Juran, 2015). As contended by Schuller (2017), "as private entities, NGOs do not have any structural accountability to the population they serve" (72), leading them to, in some cases, ignore the local and national government and undermine governmental disaster response, relief, and recovery efforts.

Varied Areas of Focus

Just as NGOs vary in scope and scale, they similarly differ in their focus areas, reflecting the diverse needs and challenges that disasters present. The focus area(s) vary based on the NGO's mission, sector of expertise, the unique context of any given disaster, and the needs of the affected communities. Areas of expertise range from general humanitarian aid to emergency medical care to distribution of food and non-food items (NFIs) to capacity building and beyond (Shaw, 2003). As OCHA notes, "NGOs tend to specialize in one or two fields, or to target their efforts towards one vulnerable population group" (United Nations Office for Coordination of Humanitarian Affairs, 2018, 14). Examples of specific focus areas include emergency medical services, water, sanitation, food distribution, shelter, logistical support, search and rescue (SAR), protection of vulnerable populations, and psychosocial support. These NGOs, along with their specializations and resources, are prepared to deploy in response to disasters. Examples include Médecins Sans Frontières (MSF), Mercy Corps, and CARE International. The specific specializations will be explored more in the subsequent section.

Not all NGOs involved in disaster response and recovery specifically focus on disaster settings as their primary operational context. Some NGOs, for example, focus on long-term development but participate in disaster response efforts when a disaster strikes areas where they are already present. One example of this is the NGO Partners in Health (PIH) and its Haitian branch, Zanmi Lasante (ZL), which participated in disaster response efforts following the 2010 earthquake in Haiti. ZL operates as a Haitian NGO and is staffed and led by Haitians rather than international workers. It was already operating in Haiti before the earthquake, working to build long-term capacity in the healthcare system and advocating for health equity. After the 2010 earthquake, PIH, through ZL, quickly mobilized to address emergency medical and healthcare needs (Kidder, 2010; Farmer, 2011). They focused on severely impacted areas in Port-au-Prince due to their presence prior to the disaster and the fact that the ZL staff "know their way around Haiti" geographically, politically, and culturally and could therefore provide practical assistance following the earthquake (Kidder, 2010, 770). The large variety of roles NGOs play illustrates the need for collaborative efforts and strong communication to navigate the complexities of disaster response while meeting both the short- and long-term needs of affected populations.

Role of NGOs in Disaster Response Efforts

While NGOs play myriad roles in disaster response, the following section outlines some of the most common and critical functions conducted by NGOs. In particular, NGOs play a prominent role in the distribution of aid, provision of humanitarian assistance, coordination and collaboration, advocacy and raising awareness, and long-term relief and recovery efforts.

Distribution of Emergency Aid and Provision of Humanitarian Assistance

NGOs are significant actors in distributing emergency aid and providing humanitarian assistance in the days, weeks, and months following a disaster. Assistance can include service delivery, technical support, distribution of goods, and general social and humanitarian assistance to affected populations. For example, after Cyclone Sidr in Bangladesh in 2007, NGOs distributed "immediate relief (food, water, medicine, household utensils), shelter (building materials, new houses), and livelihood assistance (microcredit, cropping seeds, livestock, fishing boats, and nets)" (Islam & Walkerden, 2015, 1707). In this case, "one hundred percent of respondent households received some kind of relief support from NGOs" (Islam & Walkerden, 2015, 1713), highlighting NGOs' extensive reach following disasters. This section will outline some specific areas of disaster aid and assistance that are routinely provided by NGOs.

Medical and Healthcare Services

Disasters often overwhelm the existing healthcare system and affect the local healthcare workforce (Kidder, 2010; Farmer, 2011). For this reason, NGOs often play a critical role in providing emergency medical and healthcare services following a disaster. They deploy trained healthcare professionals, set up field hospitals and emergency medical facilities to address urgent medical needs, treat injuries, conduct disease surveillance, distribute medicines and healthcare products, and address other public health concerns among affected populations (Jobe, 2011; Arnaouti et al., 2022). Notable examples of NGOs routinely involved in medical and healthcare services after disasters include MSF, PIH, and International Medical Corps (IMC).

Water, Sanitation, and Hygiene (WASH) Promotion

Following a disaster, one of the most urgent need is WASH. Access to safe water for drinking, cooking, and other uses, as well as access to adequate sanitation and the promotion of hygiene, are all vital for preventing the spread of disease and protecting the health of the community—and NGOs are often involved in these efforts (Yates et al., 2021; Hosseinpourtehrani et al., 2022). Examples of NGOs regularly involved in providing and promoting WASH services in the post-disaster context include Mercy Corps, World Vision, and Samaritan's Purse.

Food Security and Nutrition

NGOs participate in the promotion of food security and nutrition in numerous ways, including the provision and distribution of emergency food supplies, implementation of nutrition programs, and, through short- and long-term recovery support for local food and agriculture systems (Wentworth, 2020). Examples of NGOs involved in food security and nutrition include Oxfam, World Central Kitchen, and Action Against Hunger.

Emergency Shelter and Non-Food Items

After disasters, NGOs set up and manage temporary and transitional shelters, distribute shelter kits, and engage in the reconstruction of damaged and destroyed houses in impacted communities. Additionally, NGOs distribute NFIs such as blankets, clothing, hygiene kits, and other necessities required to meet basic human needs. Examples of NGOs that operate in this space include Habitat for Humanity, Catholic Relief Services, and Save the Children.

Psychosocial Support

NGOs are also involved in psychosocial support and emotional care following ters. Recognizing that disasters psychologically impact affected populations and may cause a great deal of trauma and stress, NGOs may be involved in counseling, individual mental health services, and community-based interventions (Jobe, 2011; Islam & Walkerden, 2015; Park & Yoon, 2022). Each disaster creates a unique set of circumstances, and each individual and community may react differently to the trauma and stress of a disaster or crisis. *The Sphere Handbook* (2018 notes that individuals who have been "forcibly displaced, separated from family members, survived violence or experienced previous mental health conditions" may be particularly distressed by disasters and thus argues that providing psychosocial services "in a socially and culturally appropriate way is essential to both prevent distress among affected populations and address discrimination" (15). In the case of Cyclone Sidr in Bangladesh, NGOs were involved in psychosocial support of the affected population, particularly centered on children through school-based recreational activities focused on post-traumatic stress disorder (PTSD) recovery (Islam & Walkerden, 2015). Examples of NGOs that specifically work in this area include the International Federation of Red Cross and Red Crescent Societies, Save the Children, and UNICEF.

Support for Vulnerable Populations

Some NGOs specifically work to address the needs of vulnerable populations, including children, pregnant women and new mothers, elderly populations, persons with disabilities, LGBTQ+ communities, and other marginalized and vulnerable groups. They tailor their assistance and may focus on protecting one or multiple vulnerable groups (Park & Yoon, 2022). Examples of NGOs explicitly involved in support for vulnerable populations include CARE International, Save the Children, and International Rescue Committee.

Involvement in Long-Term Recovery and Rehabilitation

NGOs play different roles at each phase of the disaster cycle, including involvement in long-term recovery and rehabilitation efforts, spanning months to years after the disaster (Shah et al., 2022). This work may include community resilience and risk reduction, disaster education and awareness, rebuilding or improving infrastructure, economic recovery and development, livelihood promotion and income generation, long-term advocacy for affected communities, and capacity building and training. Numerous NGOs are involved in long-term recovery and rehabilitation, and one of the greatest challenges of NGOs operating in the post-disaster context remains addressing short-term needs and suffering while simultaneously addressing long-term development that will serve to decrease future disaster vulnerability.

Coordination and Collaboration with Stakeholders

NGOs must collaborate with other organizations, local and national governments, and affected communities and individuals to adequately respond to disaster needs (Rouhi et al., 2019; Hosseinpourtehrani et al., 2022). There are many structures and approaches for fostering collaboration and coordination. Their relevance and use depend on the "affected Government's wishes, the particular needs of the situation, available resources and what is culturally, contextually and politically pertinent" (United Nations Office for Coordination of Humanitarian Affairs, 2018, 10). It is imperative to note that while the mechanisms for coordination may vary, the need for coordination does not. As Shaw (2003) states, "inter-agency collaboration serves as an integral component of disaster management" (9).

Sharing Information and Data Collection

NGOs are involved in all phases of the disaster cycle, from needs assessment to implementation to monitoring and evaluation (Shaw, 2003). Collecting data and sharing information with relevant stakeholders, including those impacted by and those responding to the disaster, is critical to enhancing management and situational awareness, ensuring the needs of all members of the affected population are being met, and eliminating duplication of support. NGOs may use activity reporting tools in addition to qualitative, quantitative, and mixed methods instruments to collect data and share knowledge with donors, NGO partners, and the scientific community. Collaborative information exchange among stakeholders is crucial in facilitating effective disaster response. *The Sphere Handbook* (2018) emphasizes the importance of collaboration, stating that "the sharing of information and knowledge between stakeholders, along with joint planning and integrated activities, can also ensure that organizations manage risk better and improve the outcomes of a response" (16).

Working with Governments

Ultimately, the government of the affected area has the legal responsibility for the welfare and safety of their people and should take the lead in coordinating disaster assistance and rehabilitation (Paul, 1998; Sphere, 2018; Shah et al., 2022). As recognized in *The Sphere Handbook* (2018), "the roles and responsibilities of humanitarian actors are, however, secondary to those of the state," thus "recognizing the primary role and responsibility of the host state" (16, 36). To this end, while generally seeking to remain politically neutral, NGOs are expected to work in alignment with the efforts and priorities of the national government and are essentially providing technical and financial support to the government (Shah et al., 2022).

It should be noted that while the state government is legally obligated to protect the welfare of their people and provide for the health of their citizens, the resources to accomplish these goals are limited, and, in many cases, the state is not able to fulfill this obligation both prior to and during disasters (Jobe, 2011; Shah et al., 2022). Thus, in some cases, "the role of humanitarian actors may be to encourage and persuade the authorities to fulfill their responsibilities" to protect the welfare of disasters-affected populations (Sphere, 2018, 36). Therefore, the relationship between NGOs and governments in the disaster arena can be complex (Hosseinpourtehrani et al., 2022; Park & Yoon, 2022). In some cases that are characterized by a weak national government (e.g., economically and in terms of institutional capacity), inadequate ability of the government to respond to disasters and provide relief, and increasing streams of aid being channeled through NGOs, it has been contended that NGOs are "in the process of taking over the traditional role of government in organized relief assistance for disaster victims" (Paul, 1998, 297). This is certainly not the role or goal of NGOs, and such an arrangement is not sustainable.

Collaboration and partnership between NGOs and agencies of the government at multiple levels (local, regional, national) is vital to ensure alignment of the response efforts. Memorandums of Understanding (MOUs) and other documents are often used to formalize agreements between NGOs and the government (Shah et al., 2022). Working in tandem with government agencies enables NGO and governmental stakeholders to better fill gaps, leverage scarce resources and expertise, and synergize activities to meet the joint priorities of alleviating suffering while planting seeds for long-term development (Juran, 2015; Park & Yoon, 2022).

Working within Coordination Mechanisms and the Cluster Approach

There are many mechanisms, strategies, and approaches used to foster coordination between agencies and NGOs (Rouhi et al., 2019). These channels and systems "are required to establish a clear division of labour and responsibility and to identify gaps in coverage and quality" as well as to "prevent the duplication of efforts and the waste of resources" (Sphere, 2018, 16). One of the most important strategies used to foster coordination, particularly among NGOs, is the UN Cluster Approach. The cluster approach, introduced in 2005 after the 2004 Indian Ocean Tsunami, aims to foster coordination within response sectors by setting up 'clusters' of "humanitarian organizations, both UN and non-UN, working in each of the main sectors of humanitarian action" in order to "strengthen system-wide preparedness and technical capacity to respond and to provide leadership and accountability" in disaster settings (United Nations Office for Coordination of Humanitarian Affairs, 2018, 10).

Use of the cluster approach has been found to contribute to enhanced collaboration, reduction of duplication of efforts, and overall improved outcomes (Rouhi et al., 2019; Shah et al.,

2022). Working within the cluster system also promotes adherence to established disaster management protocols, guidelines, and best practices. This helps NGOs to align their activities with sector-specific standards, including rules on minimum standards in humanitarian response set by *The Sphere Handbook* (2018), and adhere to standard guidelines, including the Code of Conduct published by the International Committee of the Red Cross (ICRC) (1994) that outlines principles and values that should be upheld by non-governmental stakeholders in disaster response efforts. However, as Rouhi et al. (2019) note, "the cluster approach is faced with some difficulties when it is implemented" in practice (6). For example, a lack of time and personnel for all organizations to participate, a lack of expertise in logistics and coordination, perceptions of coordination meetings as overly time-consuming, and a gap between theory and practice regarding the cluster system's effectiveness (Stumpenhorst et al., 2011).

Advocacy and Public Awareness

NGOs often play a critical advocacy role, along with the media, in showcasing the extent of the disaster to the outside world (Olsen et al., 2003). In this work NGOs serve as advocates for impacted communities, and through publicity and fundraising campaigns, they communicate the scale human suffering, raise awareness of the needs, advocate for international support, and lobby governments and individual donors for funding (Olsen et al., 2003; Paul, 2006). As a result, the volume of funding dollars for a disaster is impacted, in part, by the NGOs that are present in the country experiencing the crisis (Olsen et al., 2003).

Opportunities and Strengths of NGOs in Disaster Settings

Independence and Neutrality

While their approaches, focal areas, size, scope, and strategies vary, NGOs share an explicit commitment to the humanitarian principles of humanity, impartiality, neutrality, and independence along with the alleviation of suffering and preservation of lives among those afflicted by crisis (United Nations Office for Coordination of Humanitarian Affairs, 2018). Sphere, formerly known as The Sphere Project, articulates the central philosophy of NGOs involved in humanitarian response efforts as follows: "People affected by disaster or conflict have the right to life with dignity and, therefore, the right to assistance; and, all possible steps should be taken to alleviate human suffering arising out of disaster or conflict" (Sphere, 2018, 4). These principles unite the diverse NGOs involved in distributing and providing humanitarian aid in an overarching manner. That being said, it should be noted that the meaning and relevance of terms such as 'neutrality' and the reality of how such principles can be practically implemented are debated, with some asserting that they "do not believe that there is any such thing as a neutral humanitarian" (Leaning, 2007, 419).

As organizations that are formally unaffiliated with the government, NGOs can operate impartially based on human rights and humanitarian needs. This allows NGOs, at least in principle, to prioritize humanitarian need sans political, economic, or other biases. This also allows NGOs to establish transparency, build a trust ecology, and bolster cooperation among the myriad actors and affected communities that are part of the post-disaster sphere. In many (but not all) case studies, NGOs have proven more effective at allocating disaster aid impartially compared to governmental agencies (Paul, 2003; Paul, 2006).

Agility and Rapid Mobilization to Affected Populations

One distinct common characteristic of NGOs responding to disasters is their ability to deploy quickly and efficiently, providing relief where most needed (Paul, 2003). NGOs often have the reputation of providing "fast and flexible aid filling gaps in government relief actions" during disasters (Park & Yoon, 2022, 6). This agility allows for the rapid mobilization of supplies, trained staff and volunteers, and other resources immediately following a disaster. In the case of the 2010 earthquake in Haiti, for example, Farmer (2011) notes that "to move relief funds quickly, donors had relied heavily on NGOs and contractors" (157).

Some evidence that suggests NGOs, particularly those specializing in disaster response and relief, may be able to mobilize more quickly and nimbly compared to the government of the affected nation, especially in the distribution of aid to regions and households hardest hit by the disaster (Paul, 2006; Fitch-Fleischmann & Kresch, 2021). In the case of Hurricane Mitch in 1998, Fitch-Fleischmann & Kresch (2021) found "compelling evidence for a story in which a developing country government is slower to react in the immediate aftermath to a disaster than NGOs specializing in providing disaster relief" (8). Additionally, they noted that in the days immediately following the hurricane, NGOs were more effective than the local government at distributing aid targeted explicitly to those houses that were hardest hit by the storm, "while government aid allocations were not significantly higher in the hardest hit areas" (Fitch-Fleischmann & Kresch, 2021, 1–2). In some cases, it was found that NGOs may be "the more efficient channel for targeting affected populations in the immediate aftermath of the storm" (Fitch-Fleischmann & Kresch, 2021, 9).

Flexibility and Adaptability

NGOs working in disasters are often highly adaptable, ready to adjust their approach in response to evolving challenges to provide "suitable services at the right time" and "improve their activities according to the victim's suggestions" (Park & Yoon, 2022, 6). Because many NGOs routinely work in disaster settings, they can draw upon prior experience to refine their approach, tailor interventions for context-specific factors, and adapt to unique challenges and needs presented by each disaster and crisis.

In one poignant case of flexibility, a doctor working for the NGO ZL was at his home in Port-au-Prince when the 2010 Haiti Earthquake struck. While the doctor typically worked for ZL in a hospital setting, because of his prior experience and training, he was able to quickly "set up a clinic in his backyard and spent the night caring as best he could for about 50 injured people," highlighting the ability of NGOs to pivot quickly to meet context-specific needs posed by disasters (Kidder, 2010, 769).

Challenges and Criticisms

NGOs face numerous challenges navigating disaster contexts and many criticisms of NGO involvement in disasters have been posited (Farmer, 2011; Hsu & Schuller, 2020; Park & Yoon, 2022). These criticisms include, among others, NGO under-preparedness, insufficient communication and collaboration with other stakeholders, bypassing the local government, short-sighted relief efforts that may lead to the creation of aid dependency, limited accountability and transparency, and lack of cultural sensitivity (Schuller, 2007; Farmer, 2011; Jobe, 2011; Hsu & Schuller, 2020). As Islam and Walkerden (2015) poignantly claim, "NGOs make important yet flawed contributions to disaster recovery and resilience" (1724).

While the disaster arena is an inherently challenging environment characterized by unmet human need, limited resources, and urgent demands, NGO involvement, while well-intentioned, may also have unintended negative impacts on those impacted by disaster. In both disaster and long-term development contexts, beyond good intentions, it is argued that humanitarian aid by NGOs may do more harm than good, and that the flow of disaster aid and assistance can turn into a secondary disaster, or a 'disaster after the disaster' (Schuller, 2007; Schuller, 2008; Hogan, 2012; Kreamer, 2016; Juran, 2018). Thus, this section focuses on NGO critiques and challenges in disaster contexts. Addressing these weaknesses and challenges is critical for ensuring an effective humanitarian response to disasters and crises.

Lack of Coordination

NGOs face difficulties in coordinating and collaborating with other key stakeholders in the disaster for many reasons, leading to inefficient resource allocation and duplication and gaps in addressing the needs of affected population (Arnaouti et al., 2022). These coordination challenges can manifest for several reasons, such as: insufficient communication and unclear policies for communication; NGOs having poor knowledge of other NGOs; language barriers (including lack of ability to speak the language of the affected population and other stakeholders); lack of standardized indicators; and inefficient information sharing systems (Subbarao et al., 2010; Islam & Walkerden, 2015; Rouhi et al., 2019; Hsu & Schuller, 2020; Arnaouti et al., 2022).

In many cases, challenges with coordination can be attributed in part to the innate confusion of disaster settings, including competing and overwhelming demands on staff time (Hsu & Schuller, 2020; Arnaouti et al., 2022) and the "heat and 'fog' of the moment" that disasters often bring (Subbarao et al., 2010, 329). As Dr. Paul Farmer (2011) candidly noted when describing coordination challenges in the immediate days following the 2010 Haiti earthquake, "everyone wanted to help but no one knew exactly what to do. Each of the many tents erected by NGOs in the hospital courtyards became its own semi-autonomous world" (15).

In some cases, lack of coordination may result from NGOs competing to secure donations and funding to sustain their operations (Subbarao et al., 2010; Islam & Walkerden, 2015; Juran, 2015; Park & Yoon, 2022). This competition for donations can disincentivize information sharing and coordination among NGOs, each of which may "become so preoccupied with gaining visibility that they fail to share rapid needs assessments and surveillance data with other humanitarian organizations, let alone collaborate on funding proposals to donors" (Subbarao et al., 2010, 330). For instance, disaster practitioners involved in Cyclone Sidr relief efforts in Bangladesh stated that "an invisible competition exists among the local NGOs to get donor funding. The donor agents usually provide funding through an open bid. Therefore, local NGOs do not share information and coordinate with each other" (Islam & Walkerden, 2015, 1715). Ultimately, competition represents yet another avenue through which inefficiencies, duplication of efforts, and seeds of distrust can take root and undermine relief efforts (Subbarao et al., 2010; Islam & Walkerden, 2015).

Bypassing Local Government Agendas

NGOs have been critiqued for their lack of coordination and collaboration with governmental efforts and priorities. As a case in point, in a critique of NGO involvement in the 2010 Haiti earthquake, Schuller (2017) contends that "while accumulating power and resources, many NGOs ignored the Haitian government" (72). Yet, NGOs are expected to interface with ongoing governmental efforts and "respect the doctrine of national sovereignty" to empower "host nations so that they can take a leading role in relief efforts" (Arnaouti et al., 2022, 30). *The Sphere Handbook* (2018) maintains the same, noting that such organizations are expected to recognize the "primary role and responsibility of the host state" and should work to support the host government in their role of leading relief efforts (16).

However, countless case studies showcase challenges in this collaboration between NGOs and governments. For example, more than 100,000 NGOs were operating in Haiti prior to the 2010 earthquake (Kidder, 2010). As noted by Kidder (2010), with some exceptions, the "NGOs hadn't worked together before, and they haven't worked with Haitian authorities," arguing that the lack of government collaboration in essence led to the creation of a 'parallel government' comprised of private aid organizations (771). Schuller (2017) adds that "NGOs and other foreign entities took control of the humanitarian response to the 2010 earthquake, creating parallel structures that contributed to the very duplication and state weaknesses that NGOs were supposed to solve" (68).

While relief, recovery, and development efforts by NGOs should align with governmental goals and priorities, it has been noted that "NGOs and other foreign initiatives, accountable to their donors, bring their own priorities, which often diverge from government priorities and national strategies" (Kim et al., 2013, 205). However, "when funding dries up and NGOs pack their bags, the government remains" (Kim et al., 2013, 200). Ultimately, bypassing the government can lead to inefficiencies and stymie the achievement of long-term sustainable development.

Short-Term Focus and Creation of Aid Dependency

Numerous factors impact the timeline of an NGO's relief and response efforts, including availability of funding, donor priorities, staffing requirements, and other organizational priorities. For this reason, NGOs may operate only in the short-term and focus on the immediate needs of affected populations. Thus, NGOs may be criticized when their relief efforts do not align with longer-term development and recovery priorities or undercut local efforts and the local economy (Jobe, 2011).

Some have even argued that between the short timeline of NGO involvement and focus on relief over long-term recovery, "NGO practices often foster relief dependency" at the household level, leading to an overreliance on external resources (Islam & Walkerden, 2015, 1718). Islam & Walkerden (2015) add to this negative outcome by stating that most NGO programs are "short term, issue based, and donor funded. Therefore, despite the community needs of existing projects, [the projects of] NGOs cannot continue" (1718).

Limited Accountability and Transparency

In general, NGOs are predominantly accountable to their donors and ensuring their operational sustainability rather than to the national governments or the people they serve (Paul, 2006). As such, they sometimes face challenges developing and maintaining accountability and transparency, which are critical for building trust and rapport. Therefore, it is no surprise that NGOs have faced criticism over their management of funds (Paul, 2006; Hsu & Schuller, 2020).

The allocation of scarce resources can create ethical challenges for NGOs involved in relief and render it more difficult to ensure that aid reaches the intended beneficiaries. Some have noted that local NGOs with programs before the disaster may favor households that were involved in these prior programs when allocating disaster aid (Islam & Walkerden, 2015). This uneven distribution of resources, with preference given to certain members of the population (and even the appearance of favoritism), has served to damage the relationship between NGOs and the local community, create mistrust, and delay post-disaster recovery (Islam & Walkerden, 2015).

Cultural Sensitivity and Challenges in Community Engagement

In addition to the challenges inherent to disaster settings, NGOs often have the added difficulty of navigating cultural, linguistic, and contextual factors with which they may be unfamiliar. Since disasters overlay upon unique composites of communities, cultures, and people groups, foreign NGOs "may not have clear knowledge regarding the geography and culture" of the affected country and therefore face difficulty negotiating these complexities to the point of even causing harm (Paul, 2006, 214). In the case of the 2004 tsunami in Indonesia, for example, this lack of contextual knowledge by foreign NGOs led to the distribution of heavy winter coats, which have little use due to the warm climate throughout the year, as well as the "distribution of copies of the Bible, for example, among Muslim victims … by foreign Christian NGOs [which] created tension in the emergency relief effort" (Paul, 2006, 213–214). There are also cases in which NGOs have been criticized for intentionally excluding the local community and government in relief coordination, as in the case of the 2010 Haiti earthquake where the cluster meetings were held in English rather than Creole (Hsu & Schuller, 2020).

Conclusion

As disasters continue to increase in frequency and complexity, partly driven by anthropogenic climate change, NGO involvement in disaster settings must likewise continue to evolve. While NGOs possess numerous strengths such as rapid mobilization of resources, neutrality, and flexibility, they simultaneously face formidable challenges in their involvement in disaster settings. However, within these challenges lie opportunities for innovation, collaboration, and improvement.

Disasters unfold within complex social, cultural, religious, economic, and political systems, and adequate response necessitates strong collaboration between many partners, including NGOs. These collaborative partnerships with governments, local communities, and other NGOs may present the best opportunity for NGOs to leverage their involvement in disasters toward more equitable, collaborative, and sustainable response and recovery efforts (Dubique, 2014). To this end, NGOs must work toward greater accountability to national and local governments and the communities they serve as they collaborate to meet the needs of affected populations (International Committee of the Red Cross, 1994; Paul, 2006). As a positive step forward, NGOs must adopt participatory approaches that invite "community participation in the relief distribution process" as this will help facilitate collaboration, accountability, and sustainable disaster recovery (Islam & Walkerden, 2015, 1719).

Beyond merely adopting participatory approaches, NGOs must also implement models of collaboration that foster 'bidirectional' and 'side-by-side' relationships with other NGOs, the government, and the local community at all levels and every step of the disaster assistance and recovery process (Dubique, 2014; Kidder, 2010). In this type of collaborative partnership, termed 'accompaniment,' the goal is the transfer of functions as well as capacity to local counterparts (Farmer, 2011). This ensures that key functions and capacity can be sustained after NGOs leave. Additionally, NGOs must commit to working within existing governmental efforts and structures during disasters in order to avoid duplication of efforts and to work synergistically toward long-term development goals.

The future of NGO involvement in disaster response demands proactive and adaptive approaches that confront the challenges head-on while seizing opportunities to increase their impact. By fostering robust collaborative relationships, amplifying the voices of local communities (particularly vulnerable and marginalized groups), and embracing innovation and self-reflection, NGOs can establish themselves as primary actors in shaping a more effective and equitable response to disasters.

Acknowledgments

The authors thank Dr. Nadge Belizaire, Dr. Mary Clisbee, and collaborators at the Zanmi Lasante research department for their insights and input to this chapter.

Note

1 It should be noted that, in this chapter, the term 'NGO' is used as an umbrella term that encompasses local and national NGOs as well as INGOs.

References

Arnaouti, Matthew Keith Charalambos, Gabrielle Cahill, Michael David Baird, Laëlle Mangurat, Rachel Harris, Louidort Pierre Philippe Edme, Michelle Nyah Joseph, Tamara Worlton, and Sylvio Augustin Jr, and Haiti Disaster Response–Junior Research Collaborative (HDR-JRC). "Medical Disaster Response: A Critical Analysis of the 2010 Haiti Earthquake." *Frontiers in Public Health*, vol. 10, 2022, p. 1, doi:10.3389/fpubh.2022.995595.

Dubique, Kobel. Humanitarian Aid after the 2010 Haitian Earthquake: The Case of Accompaniment, *ProQuest Dissertations Publishing*, 2014.

Fallah, Saeed, and Jhila Hosseini Nejad. "The Role of Nongovernmental Organizations in Disaster Management: A Case Study of Bam Earthquake, Iran." *Journal of Disaster and Emergency Research*, vol. 1, no. 1, 2018, pp. 43–50.

Farmer, Paul. *Haiti after the Earthquake*. Public Affairs, 2011.

Fassin, Didier. *Humanitarian Reason: A Moral History of the Present*. University of California Press, 2011.

Fitch-Fleischmann, Benjamin, and Evan P. Kresch. "Story of the Hurricane: Government, NGOs, and the Difference in Disaster Relief Targeting." *Journal of Development Economics*, vol. 152, 2021, pp. 1–12, doi:10.1016/j.jdeveco.2021.102702.

Hogan, Tori. *Beyond Good Intentions: A Journey into the Realities of International Aid*. Seal Press, 2012.

Hosseinpourtehrani, Mina, Thayaparan Gajendran, Kim Maund, and Michael Sing. "Preconditions, Processes and Structures: Interorganisational Collaboration in the Provision of Post-Disaster Water, Sanitation and Hygiene (WASH) Services." *International Journal of Disaster Risk Reduction*, vol. 80, 2022, p. 103177, doi:10.1016/j.ijdrr.2022.103177.

Hsu, Kaiting J., and Mark Schuller. "Humanitarian Aid and Local Power Structures: Lessons from Haiti's 'Shadow Disaster'." *Disasters*, vol. 44, no. 4, 2020, pp. 641–665, doi:10.1111/disa.12380.

International Committee of the Red Cross. "Code of Conduct for the International Red Cross and Red Crescent Movement and Nongovernmental Organizations (NGOs) in Disaster Relief." International Committee of the Red Cross, 1994.

Islam, Rabiul, and Greg Walkerden. "How Do Links between Households and NGOs Promote Disaster Resilience and Recovery?: A Case Study of Linking Social Networks on the Bangladeshi Coast." *Natural Hazards (Dordrecht)*, vol. 78, no. 3, 2015, pp. 1707–1727, doi:10.1007/s11069-015-1797-4.

Jobe, Kathleen. "Disaster Relief in Post-Earthquake Haiti: Unintended Consequences of Humanitarian Volunteerism." *Travel Medicine and Infectious Disease*, vol. 9, no. 1, 2011, pp. 1–5, doi:10.1016/j.tmaid.2010.10.006.

Juran, Luke. "A Framework for Integrating Water Projects in Disaster Reconstruction Processes in the International Context." *Public Works Management & Policy*, vol. 20, no. 4, 2015, pp. 305–321.

Juran, Luke. "Disaster, Development, and Water: The Reconstruction and (Re)fabrication of Hazardous Waterscapes in Post-Tsunami India." *The Asian Tsunami and Post-Disaster Aid*, edited by S. Reddy. Springer Nature, 2018, pp. 79–99.

Kidder, Tracy. "Recovering from Disaster—Partners in Health and the Haitian Earthquake." *New England Journal of Medicine*, vol. 362, no. 9, 2010, pp. 769–772.

Kim, Jim Yong, Michael Porter, Joseph Rhatigan, Rebecca Weintraub, Matthew Basilico, Cassia Van Der Hoof Holstein, and Paul Farmer. "Scaling Up Effective Delivery Models Worldwide." *Reimagining Global Health: An Introduction*, edited by Paul Farmer et al. University of California Press, 2013, pp. 184–211.

Kreamer, David K. "Hydrophilanthropy Gone Awry: How People Promoting Clean Water Availability, Sanitation, and Hygiene (WASH) Can Actually Injure Communities." *Water Resources IMPACT*, vol. 18, no. 1, 2016, pp. 8–13. JSTOR, https://www.jstor.org/stable/27031929

Leaning, Jennifer. "The Dilemma of Neutrality." *Prehospital and Disaster Medicine*, vol. 22, no. 5, 2007, pp. 418–421. doi:10.1017/s1049023x00005148

Olsen, Gorm Rye, Nils Carstensen, and Kristian Høyen. "Humanitarian Crises: What Determines the Level of Emergency Assistance? Media Coverage, Donor Interests and the Aid Business." *Disasters*, vol. 27, no. 2, 2003, pp. 109–126, doi:10.1111/1467-7717.00223.

Park, Eun-Seon, and D. K. Yoon. "The Value of NGOs in Disaster Management and Governance in South Korea and Japan." *International Journal of Disaster Risk Reduction*, vol. 69, 2022, pp. 1–15.

Paul, Bimal K. "Coping with the 1996 Tornado in Tangail, Bangladesh: An Analysis of Field Data." *The Professional Geographer*, vol. 50, no. 3, 1998, pp. 287–301, doi:10.1111/0033-0124.00121.

Paul, Bimal K. "Relief Assistance to 1998 Flood Victims: A Comparison of the Performance of the Government and NGOs." *The Geographical Journal*, vol. 169, no. 1, 2003, pp. 75–89, doi:10.1111/1475-4959.04958.

Paul, Bimal K. "Disaster Relief Efforts: An Update." *Progress in Development Studies*, vol. 6, no. 3, 2006. SciTech Premium Collection. doi:10.1191/1464993406ps139oa.

Perry, Marcia. "Natural Disaster Management Planning: A Study of Logistics Managers Responding to the Tsunami." Haiti's 'Republic of NGOs.' *Current History* 37, 5, 2007, pp. 409–433.

Rouhi, Narges, Hasan A. Gorji, and Mohammadreza Maleki. "Nongovernmental Organizations Coordination Models in Natural Hazards: A Systematic Review." *Journal of Education and Health Promotion*, vol. 8, no. 1, 2019, pp. 44, doi:10.4103/jehp.jehp_201_18.

Sapat, Alka, Ann-Margaret Esnard, and Aleksey Kolpakov. "Understanding Collaboration in Disaster Assistance Networks: Organizational Homophily or Resource Dependency?" *American Review of Public Administration*, vol. 49, no. 8, 2019, pp. 957–972. doi:10.1177/0275074019861347.

Schuller, Mark. "Invasion or Infusion? Understanding the Role of NGOs in Contemporary Haiti." *Journal of Haitian Studies*, vol. 13, no. 2, 2007, pp. 96–119. JSTOR. http://www.jstor.org/stable/41715359.

Schuller, Mark. "Deconstructing the Disaster after the Disaster: Conceptualizing Disaster Capitalism." *Capitalizing on Catastrophe: Neoliberal Strategies in Disaster Reconstruction*, edited by Nandini Gunewardena and Mark Schuller. AltaMira Press, 2008, pp. 123–145.

Schuller, Mark. "Haiti's 'Republic of NGOs.'" *Current History*, vol. 116, no. 787, 2017, pp. 68–73. https://www.jstor.org/stable/48614231.

Shah, Ikram, Tahir Mahmood, Sajjad A. Khan, Noor Elahi, Muhammad Shahnawaz, Adnan A. Dogar, Fazli Subhan, and Khoula Begum. "Inter-Agency Collaboration and Disaster Management: A Case Study of the 2005 Earthquake Disaster in Pakistan." *Jamba*, vol. 14, no. 1, 2022, pp. 1088, doi:10.4102/jamba.v14i1.1088.

Shaw, Rajib. "The Role of Nongovernmental Organizations in Earthquake Disaster Management: An Asian Perspective." *Regional Development Dialogue*, vol. 24, no. 1; SEAS SPR, 2003, pp. 117–129.

Sphere. "*The Sphere Handbook: Humanitarian Charter and Minimum Standards in Humanitarian Response*, 2018 Edition. Sphere, 2018, pp. 1–406.

Stumpenhorst, Miriam, Rolf Stumpenhorst, and Oliver Razum. "The UN OCHA Cluster Approach: Gaps between Theory and Practice." *Journal of Public Health*, vol. 19, 2011, pp. 587–592.

Subbarao, Italo, Matthew K. Wynia, and Frederick M. Burkle Jr. "The Elephant in the Room: Collaboration and Competition among Relief Organizations during High-Profile Disasters." *The Journal of Clinical Ethics*, vol. 21, no. 4, 2010, pp. 328–334, doi:10.1086/JCE201021408.

United Nations Office for Coordination of Humanitarian Affairs. "*United Nations Disaster Assessment and Coordination: UNDAC Field Handbook*." United Nations Office for Coordination of Humanitarian Affairs, 2018, pp. 1–137.

Wentworth, Chelsea. "Unhealthy Aid: Food Security Programming and Disaster Responses to Cyclone Pam in Vanuatu." *Anthropological Forum*, vol. 30, no. 1–2, 2020, pp. 73–90, doi:10.1080/00664677.2019.1647830.

Yates, Travis, Hur Zannat, Nabila Khandaker, Dominique Porteaud, Franck Bouvet, and Daniele Lantagne. "Evidence Summary of Water, Sanitation, and Hygiene (WASH) Coordination in Humanitarian Response." *Disasters*, vol. 45, no. 4, 2021, pp. 913–938, doi:10.1111/disa.12463.

10

PARTICIPATION OF DOMESTIC AND FOREIGN MILITARY FORCES IN DISASTER RELIEF

Rachel Routley, Joshua Hallwright, and Matthew Clarke

Introduction

The participation of domestic and foreign military forces in disaster relief is a common component of humanitarian response. Where military forces have the capabilities, training, and clear rules of engagement, their presence can enhance and expand the impact of disaster relief both domestically and upon foreign soil, as this chapter will demonstrate. The advantages of military engagement in disaster response, whether domestic or foreign, include significant logistical capabilities and assets, medics ready to deploy for emergency response, and trained engineers and crisis leaders. The commanded military approach to the management of conflicts and crises (e.g., preparation, intelligence, logistics, and mass deployment) can be well suited to disaster response efforts that require rapid decision-making and the delivery of critical support. At the same time, to be as effective as possible, disaster response needs to be grounded in humanitarian principles of accountability, participation, human rights, and of operational principles around which the civilian international emergency response community centers its response efforts (IASC, 2023a). As will be established in this chapter—and as with any actor operating in the disaster context—the presence of military forces in disaster relief requires significant coordination, negotiation, and training.

Background

There is a robust policy framework for the participation of domestic and foreign military forces in disaster relief to ensure such support is appropriate, well-coordinated, and functions in support of the broader civilian response effort. The United Nations (UN) CMCoord field handbook (UN-CMC, 2015) and UN Office for the Coordination of Humanitarian Affairs (UN OCHA) guide for the military (UN OCHA, 2014) provide key principles and parameters for such actions. These documents are in line with the Oslo Guidelines (UN OCHA, 2007); the UN OCHA Guidelines on the Use of Military and Civil Defence Assets To Support United Nations Humanitarian Activities in Complex Emergencies (UN OCHA, 2013); the UN Inter-Agency Standing Committee (IASC) Reference Document on Civil-Military Relations in Complex Emergencies (IASC, 2004); and the UN IASC non-binding guidelines on the use of armed escorts for humanitarian convoys (IASC, 2013).

DOI: 10.4324/9781003414834-15

At the country level, there are often additional contextualized guidelines that provide a framework for military involvement in disaster operations. Furthermore, there exist civil-military co-operation (CIMIC) guidelines within each military that outline the roles and responsibilities in coordination and information sharing with civilian responders (ACMC, 2023). During situations in which international humanitarian law applies, additional frameworks exist to inform military actors on civilian engagement and advocacy (IASC, 2004; UN OCHA, 2013). Protection of civilians in disaster response setting is a critical convergence point of military forces and civilian disaster relief as both possess similar objectives to protect civilians and promote their wellbeing.

Critically, however, civilian and military definitions of protection differ significantly in some contexts, which can lead to confusion and frustration between actors (UN OCHA, 2012; UN-CMC, 2015). While military definitions differ, there is often a common focus on protection from physical violence. The international humanitarian community has a definition of protection that includes, inter alia, family separation, discrimination, denial of humanitarian access, denial of access to information, forced recruitment, gender-based violence, restrictions to freedom of movement, psychological abuse, as well as attacks on civilians and other unlawful killings (IASC, 2023). A robust set of protection guidance frames civilian engagement on protection in disaster response (Global Protection Cluster, 2024).

This chapter progresses by first considering the operational challenges of military involvement in humanitarian response. The next two sections provide illustrative examples drawing upon experiences of the Pakistani and Australian military in such actions. The next section suggests five key principles for the participation of domestic and international military within disaster relief settings before the chapter is concluded.

Operational Challenges

Prioritization of Funding and Clarity of Resourcing

The prioritization of humanitarian assistance and disaster relief (HADR) resourcing within the military is a challenge, as will be discussed further in the Australia example below. Due to the preeminent responsibility for security by military actors, resourcing can be difficult in terms of both human and financial resources—whether operating domestically or internationally. Military staff have been trained and are prepared to respond primarily to security threats and thus often have less familiarity with disaster response activities. Military funding and resourcing are focused on security/defense, so disbursement of funds and prioritization of assets for disaster relief remains challenging. Additionally, the provision of funding to militaries for disaster relief is prohibited from Official Development Assistance (ODA) (Tian & Lopes da Silva, 2020). This can create bureaucratic challenges to disaster relief when the responsible body at the national level is a military-led organisation, as will be evidenced by the example from Pakistan.

In accordance with UN Guidelines on Military and Civil Defence Assets (MCDA), such assets should be used only as a last resort in disaster response settings (UN OCHA, 2013). However, in some instances, civilian responders are heavily dependent on MCDA as the only way to reach populations in need (Lee, 2015). In such cases, robust processes need to be in place to facilitate the effective prioritization of MCDA resources across the different actors responding.

The prioritization of resources is a key element of effective civil-military coordination. When international actors are involved in a disaster response, there is often a robust civilian coordination mechanism already in place to direct identification of and response to the most acute needs based on humanitarian principles of humanity, impartiality, neutrality, and (in some instances) independence. Sufficient information exchange with military actors operating in the disaster response space is critical to enhance collaboration, make efficient use of specialized resources and abilities, and avoid duplication of efforts (UN-CMC, 2015).

Humanitarian Principles and Security

Military actors (domestic or foreign) cannot, by their nature, adhere to humanitarian principles since they are not operationally independent from government decision-making (UN OCHA, 2012). As such, military operational decision-making can be shaped by geopolitical goals, economic agendas, and diplomatic sensitivities. This explicit link with civilian governance structures, and hence perceptions of the military as a political entity, can impact the level of community acceptance of the military and ultimately the effectiveness of their response efforts (Al Jazeera, 2023).

There is a critical difference in approach to security between humanitarian responders and the military. While the military traditionally relies on physical protection to safeguard the security of the team and operations, the humanitarian approach to safety and security is through community acceptance. Far from necessitating access to materiel or security force capacity, this community acceptance approach relies on an operational (and perceived) distinction between humanitarian operations and all military or state actors (IASC, 2004). A neutral humanitarian will have a greater chance of community acceptance, and hence safety, than actors linked to a political or military agenda. This is particularly the case for disaster response and complex crisis situations. The implications of these two distinct approaches are that a strong understanding of each other's perspective is critical for coordination, collaboration, and best use of scarce resources and expertise. In Burkina Faso, for example, the food security of citizens is significantly impacted during disasters as the Burkinabe military refuses access to humanitarians on some routes without military escort. In this context, the use of military escorts increases the risk of harm to humanitarians due to their perceived non-neutrality (ACAPS, 2021).

There are also varying interpretations of the concept of protection of civilians (e.g., if they are perceived as a political and/or military foe). This can translate to risks to the wellbeing of civilian populations if humanitarian operations are not allowed to have full operational access to areas where populations are in need (ICRC, 2023). A more comprehensive view of protection of civilians is necessary to foster an efficient and impactful disaster response that goes beyond immediate life-saving to also ensure that no harm is inflicted (intentionally or unintentionally) upon civilian communities.

Community acceptance in conflict and post-conflict contexts is a significant challenge to military involvement in disaster response. During response to the 2014–2015 Ebola outbreak in Sierra Leone, military forces were deployed to enforce government health directives. However, the affected population had lived through the civil war of 2000–2002 and had very low levels of trust in government and military forces. This promulgated a less effective response than could have otherwise ensued as in many instances villagers fled from the approaching military forces (Parker et al., 2019). Likewise, the perspective of military forces that some segments of the population are the 'enemy' can represent a significant impediment to rapid and robust decision-making on civilian needs (The New Humanitarian, 2023).

Foreign Military Engagement

The provision of foreign military assistance in disasters can provide critical asset and technical support and contribute significantly to life-saving activities. In response to the 2015 Nepal earthquake, 17 countries sent military teams to assist in search and rescue (SAR). In light of the rugged mountainous terrain and damaged infrastructure of post-earthquake Nepal, foreign military air assets proved crucial in accessing remote communities in need of humanitarian assistance (Cook et al., 2016). To be effective, the foreign military response necessitated significant diplomatic engagement including the establishment of a well-supported Multi-National Military Coordination Centre from which to launch operations in support of the disaster response (UN OCHA, 2015). Note that challenges arose from day one of the response, with U.K. assets being delayed in India pending approval to land by the Nepal Government.

When there is more than one foreign military responding, the specific rules of engagement can differ from force to force. This is especially problematic during complex emergencies in which military operations are ongoing at the same time as disaster response. In such contexts, the clarification of rules of engagement and limits of the military force to act are extremely critical (Sjöberg, 2019). Structured, resourced coordination bodies capable of addressing questions of prioritization and rules of engagement are the most effective method for enabling smooth operations, protecting civilians, and improving overall disaster response. If such coordination exists between foreign and domestic forces, it can in turn contribute to enhanced communication and information dissemination to civilian responders to inform planning and response.

The Military and Disaster Response in Pakistan

The recent experience of Pakistan illuminates the operational challenges of integrating military forces in disaster operations, both in terms of the Pakistani military responding to disasters within the country as well as the involvement of foreign military in Pakistan for disaster relief.

Two major historical events continue to shape Pakistan today: its independence from British colonial rule in 1947 and the formation of the independent nation of Bangladesh (formerly East Pakistan) in 1971. The first event established a ‘political economy of defense’ rather than a ‘political economy of development’ that is more common among newly independent countries (Gregory, 2015). The latter event reinforced the effects of the former event, further deepening the power of the military in the internal politics and economy of Pakistan despite the transition to civilian rule in December 1971 (Haqqani, 2005; Siddiqa, 2017). This laid the foundation for a polity that prioritizes militarized responses to disasters and one that considers international interventions very carefully.

To a large extent, the two founding principles of Pakistan became pro-Islam and anti-India (Haqqani, 2005; Siddiqa, 2017). Pro-Islam in the sense of forming an independent country for the Muslims of Colonial India, as advocated for by the Muslim League in the Lahore Declaration of 1940 (Pardesi & Ganguly, 2019). It is anti-India in the sense of forming a separate country that stood apart from India and could defend itself against this vastly more populous neighbor (Jalal, 1985). These two sentiments are important threads throughout the history of Pakistan, providing explanatory insights into the behaviors of the political, economic, and military leadership—all of whom have influenced the resourcing of disaster response.

Although the military is no longer overtly in control of the government, it still dominates the Pakistani political economy (Siddiqa, 2019). The common position among academics and political commentators is that the military asserts its control due to a lack of competence and confidence in the political class. Furthermore, it is argued that the military intervenes in government affairs (covertly or overtly) when Pakistan is at risk from internal or external threats; its role is ‘guardian of the nation’ (Haqqani, 2005; Shah, 2014; Tudor, 2015). Thus, the military considers itself above the fray of party politics, able to arbitrate between conflicting leaders of the country. It has also strived to build an image as the only competent institution able to run the country without fear of corruption, nepotism, or petty political feuds (Haqqani, 2005; Shah, 2014).

Another viewpoint is that the military retains political control and influence due to its commercial and industrial investments, enabling it to become financially autonomous from the federal government (Siddiqa, 2017). These investments have nominally been made with the stated objective of providing welfare to former soldiers; however, the investments have become so substantial as to wield significant political influence (Yong, 1995; Siddiqa, 2017). In actuality, it is argued, 'welfare' provided to former members of the military in the form of money, land, and access to resources has been provided to reward loyalty and is doled out on a preferential basis to those with political capital or to buy allegiances (Yong, 1995, 2005; Long et al., 2015).

However, the military maintains its influence, it can be considered Pakistan's most powerful political entity, in many respects more successful in that role than in its primary role as a defensive and offensive military force (Behera, 2006; Siddiqa, 2017, 2019). The military continues to exert major influence on many parts of Pakistan's political economy, including its system for managing the risks and impacts of the many natural hazards that the country is exposed to. Its dominant role also affects the international resourcing of disaster response as it cannot receive ODA.

Acknowledging the impact and influence of the military in civilian governance, it is noteworthy that the Chairperson of the National Disaster Management Authority (NDMA) has always been a senior current or retired military officer, except the brief period in which Dr. Zafar Iqbal Qadir, a senior non-military civil bureaucrat, led the Authority. The disaster management system continues to rely heavily on the military, especially in response operations. For example, over 100,000 military personnel were deployed in response to the 2010 floods (NDMA, 2017). Yet, the Oslo Guidelines articulate the principle of 'last resort'—that military support for disaster response should only be used as a last resort (UN OCHA, 2007). This principle is regularly disregarded in Pakistan, which has led to tension with providers of international assistance and operational challenges in areas with an existing military presence (e.g., Baluchistan and Khyber Pakhtunkhwa) (Murtaza et al., 2011). That said, many still see the military support provided during disasters as beneficial and helping to get material goods to affected communities quickly and effectively. Interestingly, the commonly held notion of the military as the guardian of the nation—keeping Pakistan safe from internal and external threats—has not been investigated in the academic literature in relation to the military's involvement in disaster responses (in relation to the economy, see Siddiqa (2017), and in relation to civil administration, see Chengappa (1999) and Shah (2014)).

In addition to ODA, Pakistan has received high levels of support for its military, much of it provided by the U.S. as part of their Cold War strategy of containment to control the 'domino effect' (Nasir et al., 2012). Pakistan remains a major recipient of U.S. support to its military; following the War on Terror in 2019, 54% of all U.S. foreign assistance (US$684 million) provided to Pakistan was categorized as military support (U.S. Department of State, 2021). This financial support to one of the primary actors in Pakistan's political economy and its disaster management system has implications for the financial resourcing of disaster response actors and activities in Pakistan.

As discussed above, the military in Pakistan is involved in every aspect of the disaster management cycle—relief, response, and especially in the recovery and rehabilitation phases. Many recent disaster responses have involved direct military support during the relief and response phases (Madiwale & Virk, 2011; Polastro et al., 2011), and the military is also involved indirectly through its various foundations and companies, for example, through the Frontier Works Organisation and National Logistics Cell (Mochizuki et al., 2019). As noted above, the military receives substantial international finance, especially from the U.S., which adds complexity to disentangling how international finance supports and shapes disaster management in Pakistan.

There is a long-standing position that ODA cannot be granted or loaned to militaries nor are militaries included as principal actors in the international humanitarian (or development) system

(OECD-DAC, 2016). The humanitarian system has explicit guidance on when and how to work with national militaries in disaster management, which states that their involvement should always be considered a last resort (UN OCHA, 2007). This has fomented tension between the international humanitarian system and Pakistan's domestic disaster management system such that there is no financial relationship between the international system and the Pakistani military in times of disasters (Madiwale & Virk, 2011). Furthermore, this explains why the military is precluded from most discussions on the international financing of humanitarian action in Pakistan despite the military being a major actor in disaster management in the country. Overall, in the case of disaster management and the military in Pakistan, lines have been crossed and the roles are very blurry.

The Military and Disaster Response in Australia

How the Australian military supports disaster relief exhibits some parallels to the Pakistani case yet diverges in other important aspects. The Australian Defence Force (ADF) has responded relatively frequently to disasters in Australia. While this is similar to Pakistan, it differs in the use and perceptions of use of ADF resources. Further distinctions between the two cases are evident with the use of foreign militaries in support of disaster relief. In Pakistan, the country received significant foreign military support in domestic disasters. Meanwhile, in Australia, there is little to no foreign military assistance during domestic disasters; however, the ADF regularly deploys to respond to disasters in other countries, including in Pakistan.

The ADF comprises the three services of Army, Navy, and Air Force. Whilst under distinct command structures, the majority of Australia's military operations are joint in nature, including humanitarian response. The ADF has a long history of providing direct and indirect support to humanitarian emergencies across the Asia-Pacific region but more recently has provided more regular domestic support to emergencies in Australia. As described by a former Chief of the Defence Force, the need for the ADF to both 'actively work for peace, as well as prepare for war' (Hutchinson, 2009, 78) has been a cornerstone of defense planning for more than two decades. Australian military personnel have been at the forefront of relief and humanitarian response in many countries across the Asia-Pacific region (e.g., Vanuatu, Fiji, and Solomon Islands). Typically, this includes logistical support for the provision of relief supplies, emergency medical services, and trained workforce for rebuilding small-scale infrastructure. In Australia, similar responses are provided, but these operations were extended during the COVID-19 response to also include quasi-policing on internal borders and physical monitoring of home detention for those required to isolate (Department of Defence, 2023). ADF resources being drawn upon for these international and domestic efforts has given rise to debates on the sustainability of being adequately prepared to defend Australia and to deploy in order to defend national (and global) security (Royal Commission, 2020).

Illustrating examples of ADF international and domestic relief and response is useful to understand this tension between working for peace (i.e., disaster assistance) and maintaining preparations to defend national security (i.e., original and primary function of the ADF). Two ADF responses are described below: the first in response to the 2004 Indian Ocean Tsunami, and the second the domestic confluence of bushfires, floods, and COVID-19 two decades later.

In late 2004, a catastrophic earthquake occurred in the Indian Ocean. Measuring 9.0 on the Richter scale, it was the fourth largest earthquake recorded since 1900 and lasted 10 minutes. It resulted in a 1,200-km rupture at the junction of the Eurasian and Indo-Australian plates, shifting the ocean floor and leading to a massive tsunami that impacted 11 countries, caused 300,000 fatalities, and displaced another 1 million people. Within minutes, waves measuring up to 12 m in height destroyed both natural landscapes, human-made structures, and the physical geography of coastlines, deltas, and estuaries (Juran, 2020). In addition to housing, community assets such as drinking water supplies, drainage and wastewater infrastructure, refuse disposal sites, drainage, roads, schools, clinics, meeting halls, and places of worship were all damaged. Thus, the Australian Government committed the ADF to join the international response within two days of the tsunami's devastation.

The ADF response, named Operation Sumatra Assist, was a whole-of-government response that involved close collaboration and joint operations not only with Australia's Department of Foreign Affairs and Trade and AusAID (the government agency responsible for the Australian foreign assistance program) but also with Indonesian and other foreign military services (which was coordinated through Operation Unified Assistance). Defense personnel have unique skill-sets that are difficult to source in such time-constrained and surge capacity environments. Given the size and scope of the international response, there was enormous air transport activity in relatively under-resourced areas. The ADF was able to provide 15 air traffic controllers to safeguard air safety for a range of international carriers (fixed wing and rotary wing aircraft) at a regional airport. Sourcing such expertise would have been extremely difficult for civil society under such circumstances. Beyond this human capacity, the ADF also provided nearly 1,000 (unarmed) military personnel with medical and engineering training to meet the immediate needs of survivors (ABC, 2005). For example, within a short period, ADF engineers had ensured the supply of safe water for numerous affected communities.

In addition to this specialized response, ADF aircraft, helicopters, navy ships, and personnel:

- distributed 1,200 tons of humanitarian aid by air;
- transported 2,530 people by air evacuation;
- undertook 70 aero-medical evacuations;
- provided 3,700 medical treatments;
- produced 4.7 million liters of treated water;
- cleared 9,000 cubic meters of debris and 1,700 large drains; and
- salvaged six large fishing boats (The Cove, 2021).

The ADF response was sustained when additional natural disasters affected the region in the period immediately following the tsunami. Similar collaborations with local and international agencies—including military—resulted in further food distribution, medical assistance, engineering works, and the provision of clean water. One important aspect of the ADF's involvement in these international relief efforts was their ability to exit at an appropriate time and allow civil society and domestic agencies to assume responsibility for humanitarian and development work over the long term. Approximately three months after the tsunami, 'the ADF … transferred humanitarian tasks to specialist relief organisations and Indonesian authorities that [we]re better placed to provide the type of ongoing, long-term support that … [was] required' (ReliefWeb, 2005, 1). This planning and execution of transfer of operations was a critical component of the relief operations. While the ADF possesses specialized skills that were vital for immediate response, the tenure of their involvement was always time-limited as a consequence of their unsuitability to play a long-term role for reasons of national security (host countries as well as Australia), inability to provide resources for a sustained period, as well as the lack of appropriate skills to participate in reconstruction, recovery, and longer term involvement. By recognizing their strengths and limitations, the ADF was able to implement an intervention that was timely and suitable but only needed for a limited duration. Had the ADF stayed longer or not planned a handover and exit, the value of their contribution could have negatively impacted communities and governments both internationally and at home.

While the ADF has continued to provide regular support and assistance internationally in response to humanitarian events, its domestic presence in these environments has been more limited. However, recent floods, bushfires, and the COVID-19 pandemic have drawn heavily on the ADF domestically.

For a six-month period from late 2019 to early 2020, Australia's east cost was particularly ravaged by intense, uncontrolled bushfires (wildfires). Over a large expanse of the continent, uncontained bushfires destroyed natural ecologies and the built infrastructure, including farm equipment, housing, and commercial buildings. As a consequence, the ADF initiated a response named Operation Bushfire Assist. It involved 'logistics, bases for firefighting aircraft, catering, recovery efforts, engineering assistance, evacuation operations, search and rescue, and surveillance. The ADF also drew on personnel and assets from overseas forces' (Royal Commission, 2020, 189). Similar to their international operations, the ADF engaged widely with federal and state agencies and worked alongside civil society agencies.

The extent of tasks undertaken during the 2019–2020 bushfires was immense. Over 1,500 distinct programs were implemented, including, but not limited to, deploying over 8,000 military personnel (2,500 of whom were reservists); flying nearly 70 fixed and rotary-wing aircraft; clearing nearly 5,000 km of road; assisting with the distribution of 77,000 civilian meals; building or repairing over 1,200 km of fencing; and providing temporary housing accommodations to over 500 evacuees (Royal Commission, 2020). These programs occurred in multiple provinces (states), and indeed, it was jurisdictional demarcation that gave rise to issues of command and control. Within Australia, provinces hold responsibility for fire and emergency response with the commonwealth government having very limited resources and responsibilities in this area. As a result, province leaders were required to formally request the commonwealth for ADF support. In order to deliver the support requested, the ADF had to establish several joint task forces (JTFs) to facilitate delegated authority over ADF personnel and equipment. Four state-based JTFs were established during this period to provide for service delivery across multiple provinces.

Prior to these bushfires, the ADF had also worked closely with province authorities in Queensland in response to flooding. Again, this support, though smaller in scale to that of Operation Bushfire Assist, included equipment and personnel to supplement local resources and capabilities. Similar support by the ADF was provided to the same region in early 2022 when floods again impacted the region (Department of Defence, 2022). Then, during 2020 and 2021, the ADF was further called upon to provide a range of support in response to COVID-19. Their support included a range of activities beyond those previously seen in domestic response as the ADF provided community policing services in addition to numerous logistical and medical-based support services, particularly in support of commonwealth-funded elderly care facilities (Department of Defence, 2023).

This almost continuous use of the ADF to supplement domestic humanitarian responses did place significant pressure on the ADF. Despite public perceptions that the ADF does have a primary duty to assist domestic relief efforts, a Royal Commission into the response following the 2020 bushfires found that 'The ADF has finite capacity and capability. The capacity and capability of the ADF to respond to natural disasters can be affected by its commitment to its priorities, both domestic and international. The ADF should not be seen as a first responder for natural disasters, nor relied on as such' (Royal Commission, 2020, 189). This exhaustive use of the ADF for non-core purposes led former Chief of the Defence Force Admiral Chris Barrie, who stated in 1999 that the ADF should work for peace through relief operations, to acknowledge the following: 'That's why we need to think carefully about how we're going to head off the likelihood that these things [disasters] might happen, and then be prepared to deal with them. But not using ADF people' (Blenkin, 2022, 1).

There are real limits to the support the ADF can provide both internationally and domestically in response to humanitarian events. Whilst they have specific, in-demand resources and skills that are highly valued in this setting, the ADF's primary purpose must remain national security with humanitarian relief ancillary to this.

Key Principles for the Participation of Domestic and Foreign Military in Disaster Relief

As demonstrated above in the examples of Pakistan and Australia, domestic and foreign militaries are actively engaged in disaster response. However, the operational challenges and political implications of military involvement are significant. This suggests five key principles for the future participation of militaries in the disaster setting. Optimization of their involvement demands these principles, irrespective of whether the military forces are domestic or foreign.

1 **Distinction**: Establishing a clear distinction between military and civilian (particularly humanitarian) actors is critical for the safety and security of humanitarian responders and affected populations.
2 **Transparency**: Clarity of resourcing in disaster response is critical to minimize waste and duplication in what are often resource-poor environments and to maximize the reach and impact for people in need.
3 **Civil-Military Coordination**: A forum for information exchange and regular coordination prior to and during operational responses is vital to ensure an understanding of each other's rules of engagement, ways of working, priorities, guiding principles, and legal frameworks. Additionally, this demonstrates formal acknowledgement of domestic leadership (i.e., diplomacy and rapport building) and recognition of broader long-term development goals.
4 **Military-Military Coordination**: Foreign military coordination can take many forms, but a dedicated forum for coordination must be created in order to align strategic intent. The forum also serves to prioritize the use and allocation of scarce resources and skills, establish alignment with domestic military aims, and maximize impact while minimizing gaps in response.
5 **Familiarity**: Training of military personnel in disaster response and having established relationships with foreign militaries and governments prior to a disaster increases the efficacy of decision-making and negotiations during an operational response. To this end, joint preparedness exercises are highly effective. The training of military personnel is especially vital so that staff and agencies are equipped to respond to domestic disasters and can act in ways that are complementary to civilian-led responses and broader long-term development goals.

Conclusion

In addition to the clear policy frameworks that exist on the participation of foreign and domestic military in the disaster response setting, there are useful learnings from a range of operational contexts that provide lessons for future response activities. This is particularly true in terms of the operational challenges of including militaries in disaster response: the prioritization of funding and clarity of resourcing; application of humanitarian principles and implications for safety and security; and issues related to foreign military engagement and coordination.

Five key principles were identified to help guide the provision of military disaster response activities in the future. *Distinction* facilitates the safety and security of humanitarian actors and establishes clarity in engagement vis-à-vis affected communities. *Transparency* fosters the deployment of resources at both the right times and to the most appropriate places. *Civil-military coordination* ensures that there is a minimal understanding of respective actors' parameters of engagement, ways of working, and expectations of each other, thus leading to enhanced communication and information sharing. *Military-military coordination* maximizes resource use and supports civilian access to resources through the coordinated deployment of military assets. Finally, *familiarity* enables military disaster response actors to operate in an informed and coordinated way. Joint preparedness exercises represent an effective tool at the international scale, and at the domestic scale, familiarity prepares the military to respond to local disasters and coordinate with civilian actors and domestic governments. Collectively, these five principles provide functional methods to improve disaster response both at home and abroad in a manner that complements civilian-led efforts and longer term recovery and rehabilitation goals.

References

ABC (2005). ADF Field Hospital Arrives in Banda Aceh, https://www.abc.net.au/news/2005-01-08/adf-field-hospital-arrives-in-banda-aceh/615208, accessed 8 October, 2023.

ACAPS (2021). Briefing Note: Central Sahel, https://www.acaps.org/fileadmin/Data_Product/Main_media/20210121_acaps_briefing_note_central_sahel.pdf, accessed 5 October, 2023.

ACMC (2023). Same Space, Different Mandates, – Same Space – Different Mandates 2023 Edition, ACMC, accessed 2 January, 2024.

Al Jazeera (2023). Syria Longs for International Aid Amid Earthquake Devastation, Turkey-Syria Earthquake News, Al Jazeera, accessed 26 November, 2023.

Behera, N. C. (2006). *Demystifying Kashmir*. Brookings Institution Press.

Blenkin, M. (2022). Defence's Future Role in Humanitarian and Disaster Relief, https://www.themandarin.com.au/200186-defence-role-disaster-relief/, accessed 26 November, 2023.

Chengappa, B. M. (1999). Pakistan: Military role in civil administration. *Strategic Analysis*, *23*(2), 299–312. https://doi.org/10.1080/09700169908455048

Cook, A., Shrestha, M. and Htet, Z. (2016). International Response to 2015 Nepal Earthquake Lessons and Observations, NTS_Report_4_Nepal_final_04Oct2016.pdf (rsis.edu.sg), accessed 2 January, 2024.

Department of Defence (2022). ADF Support to South-East Queensland Flood Response Nears Completion, https://www.defence.gov.au/news-events/releases/2022-03-21/adf-support-south-east-queensland-flood-response-nears-completion, accessed 12 December, 2023.

Department of Defence (2023). Operation COVID-19 Assist, https://www.defence.gov.au/operations/covid19-assist, accessed 1 October 2023.

Global Protection Cluster (2024). Protection Analytical Framework, Global Protection cluster – Protection Analytical Framework, Global Protection Cluster, accessed 5 January, 2024.

Gregory, S. (2015). *Democratic Transition and Security in Pakistan*. Taylor & Francis Group. http://ebookcentral.proquest.com/lib/rmit/detail.action?docID=4045385

Haqqani, H. (2005). *Pakistan: Between Mosque and Military*. Brookings Institution Press. http://ebookcentral.proquest.com/lib/rmit/detail.action?docID=1076118

Hutchinson, S. (2009). Civil-military cooperation in complex emergencies. *Australian Army Journal*, *1*, 77–93.

IASC (2004). Reference Document on Civil-Military Relations in Complex Emergencies, IASC Reference Paper: Civil-Military Relationship in Complex Emergencies, 2004, IASC (interagencystandingcommittee.org), accessed 2 January, 2024.

IASC (2013). Non-Binding Guidelines on the Use of Armed Escorts for Humanitarian Convoys, https://disasterlaw.ifrc.org/sites/default/files/media/disaster_law/2021-03/Armed%20Escort%20Guidelines%20Cover%20Page.pdf, accessed 4 January, 2024.

IASC (2023a). Operational Policy and Advocacy Group, https://interagencystandingcommittee.org/operational-policy-and-advocacy-group#task-forces, accessed 2/1/2024.

ICRC (2023). 'What the Law Says, Humanitarian Access: What the Law Says, International Committee of the Red Cross (icrc.org), accessed 4 January, 2024.

Jalal, A. (1985). Inheriting the Raj: Jinnah and the Governor-Generalship Issue. *Modern Asian Studies*, *19*(1), 29–53.

Juran, L. (2020). Indian Ocean tsunami, 2004. In *Natural Hazards & Disasters: From Avalanches & Climate Change to Water Spouts & Wildfires* (pp. 197–202), B.K. Paul (ed.). ABC-CLIO.

Lee, A. (2015). Nepal Earhquake, https://theconversation.com/nepal-quake-the-challenges-facing-disaster-response-40869, accessed 4 January, 2023.

Long, R. D., Singh, G., Samad, Y., & Talbot, I. (2015). *State and Nation-Building in Pakistan: Beyond Islam and Security*. Taylor & Francis Group. http://ebookcentral.proquest.com/lib/rmit/detail.action?docID=4185728

Madiwale, A., & Virk, K. (2011). Civil–military relations in natural disasters: A case study of the 2010 Pakistan floods. *International Review of the Red Cross*, *93*(884), 1085–1105. https://doi.org/10.1017/S181638311200032X

Mochizuki, J., Hallwright, J., & Handmer, J. (2019). *Enabling Factors for Financing and Implementing Post-disaster Operations*. Asian Development Bank. https://doi.org/10.22617/WPS190479-2

Murtaza, D. N., Akhtar, D. S., Mohammed, N., Harrison, S., Bhatti, S., & Ferretti, D. S. (2011). The DEC Real-Time Evaluation Report, 36.

Nasir, M., Rehman, F. U., & Orakzai, M. (2012). Exploring the nexus: Foreign aid, war on terror, and conflict in Pakistan. *Economic Modelling*, *29*(4), 1137–1145. https://doi.org/10.1016/j.econmod.2012.04.004

NDMA. (2017). *Annual Report 2017*. National Disaster Management Authority (NDMA). http://web.ndma.gov.pk/publications/NDMA%20Annual%20Report%202017.pdf

OECD-DAC. (2016). *The Scope and Nature of 2016 HLM Decisions Regarding the ODA-Eligibility of Peace and Security-Related Expenditures* (p. 3). OECD-DAC. https://www.oecd.org/dac/HLM_ODAeligibilityPS.pdf

Pardesi, M. S., & Ganguly, S. (2019). Violent Punjab, quiescent Bengal, and the partition of India. *Commonwealth & Comparative Politics*, *57*(1), 1–30. https://doi.org/10.1080/14662043.2019.1554232

Parker, M., Hanson, T., Vandi, A., Sao Babawo, L., & Allen, T. (2019). Ebola and public authority: Saving loved ones in Sierra Leone. *Medical Anthropology*, *38*(5), 440–454, DOI: 10.1080/01459740.2019.1609472

Polastro, R., Nagrah, A., Steen, N., & Zafar, F. (2011). Inter-Agency Real Time Evaluation of the Humanitarian Response to Pakistan's 2010 Flood Crisis, 140.

ReliefWeb (2005). Operation Sumatra Assist – Phase II: Australian Defence Force Helping Earthquake Victims in Indonesia, https://reliefweb.int/report/indonesia/operation-sumatra-assist-phase-ii-australian-defence-force-helping-earthquake, accessed 8 October, 2023.

Royal Commission (2020). Royal Commission into National Natural Disaster Arrangements, https://www.royalcommission.gov.au/system/files/2020-12/Royal%20Commission%20into%20National%20Natural%20Disaster%20Arrangements%20-%20Report%20%20%5Baccessible%5D.pdf, accessed 5 October 2023.

Shah, A. (2014). Constraining consolidation: Military politics and democracy in Pakistan (2007–2013). *Democratization*, *21*(6), 1007–1033. https://doi.org/10.1080/13510347.2013.781586

Siddiqa, A. (2017). *Military Inc.: Inside Pakistan's Military Economy*. Penguin Random House India.

Siddiqa, A. (2019). Pakistan—From hybrid-democracy to hybrid-martial law. *Journal of South Asian and Middle Eastern Studies*, *42*(2), 52–72. https://doi.org/10.1353/jsa.2019.0008

Sjöberg, A. (2019). The protection of civilians in South Sudan: From UNMISS protection mandates to physical protection. *International Peacekeeping*, *26*(1), 83–108.

The Cove (2021). #KYR: Indonesia – Special Issue, https://cove.army.gov.au/article/kyr-indonesia-special-issue, accessed 1 October, 2023.

The New Humanitarian (2023). The New Humanitarian, 'They Just Shoot and Burn': Civilians Targeted in Nigeria's War on Boko Haram, accessed 4 January, 2024.

Tian, N. and Lopes da Silva, D. (2020). Military Spending and Official Development Assistance in Recipient States: Is There a Relationship? (sipri.org), No. 2020/8 August 2020, accessed 8 October, 2023.

Tudor, M. (2015). Pakistan's security state of mind. *Perspectives on Politics*, *13*(4), 1097–1102. https://doi.org/10.1017/S1537592715002376

U.S. Department of State. (2021). U.S. Foreign Assistance by Country, https://foreignassistance.gov/

UN OCHA (2007). Oslo Guidelines on the Use of Foreign Military and Civil Defence Assets in Disaster Relief– Oslo Guidelines: Guidelines on The Use of Foreign Military and Civil Defence Assets In Disaster Relief – Revision 1.1 November 2007, OCHA (unocha.org), accessed 4 January, 2024.

UN OCHA (2012). OCHA on Message: Humanitarian Principles [EN/AR] – World, ReliefWeb, accessed 5 October, 2023.
UN OCHA (2013). OCHA Guidelines on the Use of Military and Civil Defence Assets to Support United Nations Humanitarian Activities in Complex Emergencies – Guidelines on the Use of Military and Civil Defence Assets to Support UN Humanitarian Activities in Complex Emergencies, OCHA (unocha.org), accessed 2 January, 2024.
UN OCHA (2014). UN OCHA Guide for the Military – UN OCHA Guide for the Military v 1.0.pdf, accessed 4 January, 2024.
UN OCHA (2015). Nepal Earthquake, https://reliefweb.int/report/nepal/nepal-earthquake-foreign-military-deployed-assets-04-may-2015, accessed 8 October, 2023.
UN-CMCoord (2015). UN-CMCoord Field Handbook – UN-CMCoord Field Handbook (v1.0) [EN/RU], OCHA (unocha.org), access 2 January, 2024.
Yong, T. T. (1995). Punjab and the making of Pakistan: The roots of a civil-military state. *South Asia: Journal of South Asian Studies*, *18*(sup001), 177–192. https://doi.org/10.1080/00856409508723250
Yong, T. T. (2005). *The Garrison State: Military, Government and Society in Colonial Punjab, 1849-1947*. SAGE Publications India.

11

INTERNATIONAL AND DOMESTIC DIPLOMACY WITH DISASTER AID

Rob Grace

Introduction

This chapter examines the intersection between diplomacy and disaster aid. As this chapter describes, the 'diplomacy-disaster aid nexus' cuts in two directions. On the one hand, diplomacy is an essential tool in disaster response. Over the past couple decades or so, international humanitarian organizations have increasingly recognized the role that diplomacy and negotiation play in disaster relief. This acknowledgment is evident in the ever-growing volume of handbooks, workshops, trainings, case studies, and policy analyses of humanitarian diplomacy and negotiation. On the other hand, disaster response can be an essential tool of international diplomacy. Government leaders have leveraged disaster response as a means of building and strengthening alliances, pursuing public diplomacy to garner good will with a foreign populace (i.e., 'winning hearts and minds'), and forging peace. The same is true in domestic politics, wherein governments leverage disaster relief to secure buy-in and support from key domestic constituencies. Many non-state armed groups (NSAGs), too, have been motivated to engage in disaster relief for reasons that include cultivating international and domestic legitimacy. These two dimensions—the first being diplomacy as a tool of disaster aid and the second being disaster relief as a tool of international and/or domestic diplomacy—frame the focus of this chapter.

The chapter examines these issues in six sections. After this introductory section, 'Humanitarian Diplomacy as a Tool of Disaster Response' section offers an overview of the field of humanitarian diplomacy, probing the ways that humanitarians and government leaders engage in diplomacy to further humanitarian ends. 'Disaster Response as a Tool of International Relations' section examines how governments leverage disaster relief toward non-humanitarian ends (including geostrategic, security-related, and economic aims). 'The Domestic Politics of Disaster Response' section delves into these types of dynamics in the context of domestic politics. 'The Domestic Politics of Disaster Response' section focuses specifically on the international and local political dynamics at play when NSAGs engage in disaster relief as a means of pursuing non-humanitarian objectives, such as consolidating political control over territory or strengthening relationships with third-party governmental sponsors. 'Conclusion' section offers concluding reflections.

DOI: 10.4324/9781003414834-16

Humanitarian Diplomacy as a Tool of Disaster Response

Diplomacy has always been a tool of disaster response. Consider Henry Dunant, one of the founders of the organizational network that grew into the International Red Cross and Red Crescent Movement. As the well-known (at least in humanitarian circles) story goes, Dunant witnessed widespread horror in the aftermath of the Battle of Solferino in 1859 in the form of wounded soldiers lacking adequate access to medical relief. He engaged in frontline humanitarian negotiation by advocating with the French military to allow detained Austrian doctors to help treat the wounded (Lempereur, 2023). Afterward, the experience inspired him to propose the idea that states should erect standing civilian relief societies that could be quickly mobilized when wars erupt. Intensive diplomatic lobbying with European governments—bolstered by the publication and dissemination of Dunant's book about his experiences, *A Memory of Solferino*—resulted in states negotiating the First Geneva Convention of 1864, a foundational document that would lay the groundwork for the further development of international humanitarian law (IHL) over the next 150-plus years (Moorehead, 1998). In short, modern institutionalized humanitarianism owes its legacy to the practice of humanitarian diplomacy.

Nevertheless, humanitarianism was slow to embrace the central role that diplomacy and negotiation play in disaster response. It was not until the 1990s and early 2000s that 'humanitarian diplomacy' and 'humanitarian negotiation' emerged as terms of art for disaster responders (Régnier, 2011). These two terms have somewhat fuzzy definitional contours and are sometimes used interchangeably. One definition—which encompasses the key attributes typically considered relevant to these two terms—explains 'humanitarian negotiation' as 'the interaction between humanitarian organisations and their counterparts to: establish and maintain the presence of humanitarian organisations in crisis environments (e.g., conflicts, disasters, migration flows, and epidemics), ensure humanitarian access to people in need, and deliver humanitarian aid and implement protection activities' (Centre of Competence on Humanitarian Negotiation, n.d.). The term 'humanitarian negotiation' (per this definition) encompasses engagement with high-level state actors (including heads of state), mid-level government personnel (for example, line ministries), and more low-level actors (such as checkpoint guards). Some consider 'humanitarian diplomacy' to encompass only higher-level engagement at the diplomatic level, also conceivably including situations when government officials negotiate toward humanitarian ends, including on issues of humanitarian access, civilian protection, and adherence to IHL. Relevant too in this issue area is diplomacy and negotiation related to the development and adoption of IHL treaties (i.e., the Geneva Conventions and their Additional Protocols), as well as (for humanitarian organizations specifically) standards and organizational accountability mechanisms, such as the Sphere Standards.

One can observe the key role that humanitarian diplomacy and negotiation play in various operational contexts around the world during responses to disasters stemming from armed conflict, natural hazards, and largescale public health emergencies (Bogatyreva, 2022). Diplomacy and negotiation can be essential for gaining and maintaining humanitarian access, a crucial issue area given the myriad types of humanitarian access obstructions (from governments, NSAGs, and other actors) that humanitarian organizations perpetually confront. These issues include bureaucratic restrictions (e.g., delays on approval of visas or travel authorizations), insecurity (e.g., threats or attacks against aid workers or military hostilities that inadvertently cause harm to humanitarian responders), and donor restrictions (by which governmental donors restrict what humanitarian organizations can and cannot do with donor funds, including restrictions that stem from counterterrorism concerns) (Kurtzer, 2019). Moreover, widespread concern persists about the state of IHL compliance more broadly, pointing toward the divide between IHL's normative aspirations and the realities of how contemporary combatants wage warfare (Pejic, 2016). It is also important to note that the domain of humanitarian diplomacy and humanitarian negotiation also encompasses negotiation between humanitarian organizations, internally between colleagues working for the same organization, and between humanitarian personnel and donors, all of which can be as challenging (if not more so) than negotiations with external counterparts, including host governments and NSAGs (Grace, 2020a).

At stake in these negotiations is the humanitarian character of aid organizations, which seek to provide relief in an apolitical manner in accordance with humanitarian principles (in particular, humanity, impartiality, neutrality, and independence). Fierce inter-organizational tensions and debates persist about how far is too far to compromise during these processes, especially in contexts where governments and/or NSAGs seemingly wish to coopt humanitarians to further political, economic, or security agendas. The perceived challenge for humanitarian actors is to strike a balance between the aspirations of humanitarian principles and the practical realities of aid work, especially in highly politicized settings (Labbé & Daudin, 2016; Schenkenberg van Mierop, 2016; Colona, 2017).

The field of humanitarian diplomacy and humanitarian negotiation that emerged over the past couple decades has made strides in informing how humanitarians can structure their thinking and improve strategic planning, competency building, and the practice of humanitarian diplomacy and negotiation. Humanitarians now have access to various humanitarian negotiation handbooks, including publications produced by the Centre for Humanitarian Dialogue, the UN Office for the Coordination of Humanitarian Affairs, the Centre of Competence on Humanitarian Negotiation, and Mercy Corps (Mancini-Grifoli &Picot, 2004; Mc Hugh & Bessler, 2006; Mercy Corps, 2018; Centre of Competence on Humanitarian Negotiation, 2019). Research has been produced in the form of case studies, as well as comparative analysis that draws linkages to theoretical insights from the broader field of negotiation scholarship (Minear & Smith, 2007; Magoneet al., 2012; Clements, 2019; Grace, 2020b; Sutton & Rhoads, 2022).

As the field of humanitarian negotiation continues to mature, researchers have much more to explore in terms of producing practically oriented analysis that informs practice while drawing on the existing body of negotiation theory and scholarship spanning other issue areas (i.e., business, legal, and international negotiations). Humanitarian diplomacy and negotiation entails relational dynamics (establishing a basis of trust with an interlocutor while not becoming 'too close'), cognitive dimensions (analyzing a counterpart's interests), emotional elements (such as managing stress and anger during negotiation, as well as effectively reading a counterpart's emotions), and cultural aspects (including context-specific rules of decorum and understanding what norms will likely appeal to an interlocutor). A rich body of existing broader negotiation scholarship probes these dynamics—and more—and field of humanitarian diplomacy and humanitarian negotiation has much more to analytically mine from this body of research to inform future humanitarian practice (Grace & Lempereur, 2021).

Moreover, existing case studies of humanitarian diplomacy and negotiation focus largely on contemporary or recent cases, leaving a large body of historical humanitarian diplomacy practice underexamined. Scholars have begun to explore this route. One example is a case study that examines John Rabe, who negotiated the Nanking International Safety Zone during World War II (Lempereur, 2016). Additional examples worthy of further examination include Herbert Hoover, who negotiated food relief for Belgium during World War I; Fridtjof Nansen, who negotiated the development of refugee law in the interwar period during the early 20th century; and Marcel Junod, who engaged in negotiations during World War II on issues spanning food relief and the treatment of prisoners of war (Nash, 1988; Marrus, 2002; Lempereur, 2023). As the study of humanitarian diplomacy continues to expand, looking back to negotiations of the past will help grow our understanding of how humanitarian negotiations have played out in different types of response contexts, with different interlocutors, from different eras. Such efforts can help grow our understanding of the causal mechanisms at play, as well as the extent to which one can generalize findings across different contexts.

Disaster Response as a Tool of International Relations

This section examines the flip side of the 'diplomacy-disaster aid nexus.' Whereas the previous section discussed diplomacy as a tool for humanitarianism, this section probes humanitarian action as a tool of international diplomacy. In this sense, this section examines the drivers of many of the aforementioned challenges of humanitarian diplomacy and negotiation. Whereas humanitarians seek to carve out apolitical space for disaster response based on needs and prioritizing the most urgent cases (i.e., in accordance with the principle of impartiality), governments often seek to leverage humanitarian action and disaster response to further particular political, economic, or security aims.

One context that illustrates these dynamics in quite stark terms is the post-9/11 era for Western humanitarian organizations. In a well-known quote from October 2001, U.S. Secretary of State Colin Powell referred to non-governmental organizations as 'a force multiplier' for the U.S. government and 'such an important part of our combat team' (Powell, 2001). During counterinsurgency operations in Afghanistan in the 2000s, while pursuing a strategy of 'winning hearts and minds,' U.S. military officials asserted that 'humanitarian aid is a key factor in any populationcentric operation,' referring to humanitarians as 'the guys who are going to win it for us' (Williamson, 2011, 1042). As these comments indicate, in this context, the United States perceived humanitarianism to be a diplomatic tool to be leveraged in service of U.S. security aims.

The post-9/11 U.S. example is not an outlier, however, but rather emblematic of a more general trend of states seeking to enhance soft power via disaster response. As one scholar writes about China's engagement in disaster response in Southeast Asia, 'Under President Xi, disaster diplomacy has gained greater recognition in China's foreign policy as Beijing seeks to diversify its diplomatic toolkit' (Gong, 2021, 97). Another scholar, writing about India, notes that supporting humanitarian aid in Middle Eastern countries has been 'a way to increase its soft power and visibility in the region. Through humanitarian aid, India can gain the amity and backing of the countries in the region. This will help India strengthen its political and economic influence in the region. In the long run, humanitarian assistance would cement India's attempts to prove itself as a worthy partner for the Middle Eastern countries' (Tekir, 2023, 5). Indeed, scholars have examined how countries across the globe—including Russia, Turkey, Qatar, United Arab Emirates, and Indonesia—have leveraged disaster response toward these and similar ends (Hasimi, 2014; Donelli, 2017; Barakat, 2019; Setiawan & Hamka, 2019; Gökalp, 2020; Robinson, 2022; Altunisik, 2023; Zakharova & Burghart, 2023).

Unsurprisingly, the factors that drive governmental funding for humanitarian assistance and disaster response include not only genuine perceptions of humanitarian need but also other elements that include geopolitical interests and internal bureaucratic politics. A growing body of research focused on this issue has generated mixed findings. Numerous studies have found a misalignment between levels of governmental donorship and humanitarian needs resulting from disasters. Possibly relevant factors driving funding patterns include governments' economic interests, geographic proximity between giving and receiving countries, preexisting geopolitical alignments, identity characteristics, ideological alignments, and levels of news coverage related to the disaster in question (Fink & Redaell, 2010; Raschky & Schwindt, 2012; Fielding, 2014;

Wei & Marinova, 2016; Nohrstedt & Baekkeskov, 2018; Eichenauer et al., 2020; Cheng & Minhas, 2021; Dellmuth et al., 2021; Bommer etal, 2022; Scott et al., 2022). Still, not all studies have found geopolitical, economic, or security interests to be the prime factors shaping governmental funding patterns (Becerra et al., 2014). Additionally, scholars have also probed more nuanced dimensions of the question, examining, for example, how a government's aid allocation dynamics can vary across different sectors (e.g., the social infrastructure and services sector, the economic infrastructure and services sector and the production sector) or across different agencies within the same government, based on factors that include the level of bureaucratic independence from executive power (Arel-Bundocket al., 2015; Guillon & Mathonnat, 2020).

The general takeaway from analyses of donor data is to highlight the role that governmental interests play in impacting how the humanitarian system engages in 'triage' (as one scholar has dubbed it), meaning the process of making choices about priorities amidst limited resources (Krause, 2014). The normative ideal of humanitarianism is that the principle of impartiality should shape triage, meaning that disaster responders should prioritize the most urgent cases based solely on needs. The existing body of research in this area suggests a plethora of other factors (as outlined above) that drive the reality of humanitarianism away from this ideal.

Moving beyond how governmental interests can 'pollute' approaches to disaster response, one can find a less cynical linkage between disaster response and other foreign policy objectives encapsulated in the concept of 'disaster diplomacy.' This notion—quite distinct from 'humanitarian diplomacy'—points toward the possibility disaster management can inadvertently fuel processes of conflict de-escalation or even peace (Kelman, 2006). Notable examples include cooperation (albeit in most cases short-lived) between various geopolitical foes, including Greece and Turkey after the earthquake in 1999, India and Pakistan after earthquakes in 2001 and 2005, the United States and Iran after the 2003 Iranian earthquake, the United States and Cuba in relation to monitoring hurricanes, South Korea and North Korea in the context of famine beginning in the 1990s, China and Japan after the 2011 earthquake and tsunami, and Ethiopia and Eritrea during droughts between 1999 and 2002 (Ker-Lindsay, 2000; Streich & Mislan, 2014; Kelman et al., 2018; Yeger & Podoler, 2022).

Southeast Asia has proved particularly potent for disaster diplomacy. The Association of Southeast Asian Nations has served as a forum for promoting international coordination and cooperation in response to largescale disasters (Cook, 2021). To illustrate with one example, a study of Cyclone Nargis, which caused widespread destruction in Myanmar in 2008, concludes 'that an extremely severe natural disaster is capable of promoting regional cooperation despite the actual cold political environment' (Papp, 2023, 25). Moreover, the history of Quadrilateral Security Dialogue (known as the 'Quad')—involving Australia, India, Japan, and the United States—illustrates how international cooperation on humanitarian assistance and disaster relief can play a role in forging security ties. The 'Quad' owes its lineage to the Tsunami Core Group formed in the wake of the 2004 Indian earthquake and tsunami (Grossman, 2005; Fraser, 2023). The focus of the 'Quad' is strategic security dialogue, but cooperation on humanitarian assistance and disaster relief remains a key strand in 'Quad' members' engagements with one another (Chen, 2022).

A final dimension relevant to this discussion is state engagement on devising and adopting international legal frameworks that govern humanitarian assistance and disaster response. A growing body of history and political science scholarship examines how state interests have shaped international negotiations on relevant legal and normative instruments, including the Geneva Conventions and their Additional Protocols, as well as refugee law (Finnemore, 1996; Moorehead, 1998; Mantilla, 2020; Grace, 2022; Van Dijk, 2022). This line of research can inform our understanding of relevant current and future efforts in this area, such as ongoing efforts to negotiate a new pandemic treaty in response to the COVID-19 pandemic (Phelan, 2023). These types of negotiations—as opposed to diplomacy that relates to operational issues—are important to consider as part of the broader 'diplomacy-disaster aid nexus.' Indeed, this dimension points toward the ways that states—guided by a cacophony of considerations that include geopolitical interests and genuine concern for addressing humanitarian needs—shape the broad institutional, legal, and normative structures within which humanitarian assistance and disaster relief unfolds.

The Domestic Politics of Disaster Response

Just as governments leverage international disaster response as a means of pursuing various foreign policy goals, governments' engagement in domestic disaster response constitutes a form of domestic public diplomacy that can be crucial to state leaders' efforts to gain and maintain domestic political support. This section touches on three key dimensions of the domestic leg of the 'diplomacy-disaster aid nexus.' One dimension is that domestic constituents judge leaders, at least in part, on how well they manage disasters. Consider one notable historical example: the impact of the Great Bhola Cyclone that struck East Pakistan (now Bangladesh, then united with West Pakistan) in 1970, killing over 250,000 people. The West Pakistan government's failed response and evident disregard for the victims of the cyclone 'marked the end of any remaining political legitimacy and the moral claim that West Pakistan had over the governance of East Pakistan,' operating 'as a catalyst in shaping the political reconfiguration of united Pakistan,' feeding into the process that led to the birth of Bangladesh as an independent country, as one study concludes (Islam et al., 2023, 240 and 247).

Similarly, in contexts around the globe, inadequate government responses have fueled domestic discontent, including after Hurricane Katrina in the United States (Atkeson & Maestas, 2012; Van Klinken, 2021). The consequences of failed domestic disaster response can be so severe that, as previous research has found, after instances of widely criticized domestic disaster responses, governments can be incentivized to escalate international hostilities to distract local attention from the government's domestic failures (Lee et al., 2022). In this sense, disaster response constitutes a mechanism through which governmental leaders can build and sustain (or when it goes poorly, tarnish) relationships with key domestic constituencies.

A second dimension, which flows directly from the first element just discussed, is the role that domestic politics plays in shaping governmental domestic disaster relief priorities. This aspect is the domestic corollary to the previous section's examination of the role that political considerations play in influencing governmental funding patterns for international disaster relief. To illustrate with one example, a study of disaster risk financing in Mexico concludes 'that voters punish politicians for the occurrence of natural disasters but reward them for the allocation of post-disaster aid,' incentivizing 'politicians to overspend on post-disaster aid, in particular in election years' (Boudreau, 2015, 1). Just as vulnerability constitutes one of many factors shaping patterns of international humanitarian response, the same is true at the domestic level, where disaster response, as a politicized public diplomacy mechanism, strays far from needs and vulnerabilities as the key metric for shaping policies and decision-making (Garret & Sobel, 2003; Depoorter, 2006; Shughart, 2006; Sylves & Búzás, 2007; Daniels, 2013; Sainz-Santamaria & Anderson, 2013; Husted & Nickerson, 2014; Gasper, 2015; Kahan, 2015; Davies, 2017).

A third dimension—the domestic corollary to the previous section's examination of disaster diplomacy in the international domain—is the role that domestic disaster response can play in promoting peace in the context of civil wars. Research findings suggest that disasters can influence the trajectory of civil wars to some degree, although these impacts depend on various contingencies, including whether or not the disaster impacted one or both sides of the conflict (Nemeth & Lai, 2022). It also appears to be the case, as with disasters and international conflict, that impacts tend to be short-lived (Gaillard et al., 2008; Haer & RezaeeDaryakenari, 2022). Conversely, disasters can also fuel political tensions and even prompt increasingly repressive practices from authoritarian regimes (Barceló et al., 2022; Chung & Rhee, 2022; Passeri, 2022). As these comments suggest, this leg of the 'diplomacy-disaster aid nexus' can point in different directions, possibly enabling or straining more peaceful domestic relations. Nevertheless, the overall takeaway is that, as in the international domain, disasters can shape diplomatic relations at the domestic level between leaders and constituents, and even between warring combatants, further highlighting the importance of the broader political landscape within which disaster response occurs.

Engaging Non-State Armed Groups during Disaster Response

All dimensions of the 'diplomacy-disaster aid nexus' examined throughout this chapter also apply to NSAGs. Researchers have directed a great deal of analytical attention toward understanding interactions between humanitarian organizations and NSAGs related to gaining and maintaining access for response operations, complying with IHL, and promoting civilian protection (e.g., Clements, 2019). This strand of work encompasses a growing evidence base that has documented NSAGs' perspectives on humanitarian organizations, as well as the depth and breadth of NSAGs' knowledge of IHL and humanitarian principles, which can vary greatly from group to group and across organizational levels within groups (Jackson & Giustozzi, 2012; Jackson & Aynte, 2013; Mosel & Jackson, 2013; Jackson & Davey, 2014; Jackson, 2016; Terry & McQuinn, 2020).

Additionally, NSAGs (just as states do) use disaster response as a diplomatic tool for furthering their objectives with international and local stakeholders. A wealth of political science scholarship has probed a range of possibly relevant factors that can shape NSAGs' interests and behavior related to issues of humanitarian concern. On the one hand, some NSAGs seek to obstruct humanitarian access, attack aid workers, and victimize civilians for a variety of reasons, including to signal resolve or strength (especially when battling a comparatively stronger government) or to terrorize civilian population as part of a broader strategy of undermining popular support for the state (Azam & Hoeffler, 2002; Narang & Stanton, 2017). On the other hand, other factors might incentivize NSAGs to enable humanitarian access. For example, NSAGs that have consolidated territorial control might view humanitarian response as aligned with broader governance objectives, and hence, might exhibit a greater propensity to grant and facilitate humanitarian access (Breslawski, 2021). Moreover, one could draw a distinction between legitimacy-seeking NSAGs (those needing to garner and sustain political support from international and local audiences) and NSAGs that are more predatory, seeking economic advantages for its members (Jo, 2015). The notion is that legitimacy-seeking NSAGs (i.e., groups that depend on external support from human rights-conscious third-party governments and/or need local support to further a political agenda) are more likely to refrain from civilian victimization than other types of NSAGs.

Researchers have begun to explore how these, and other, factors shape the dynamics of NSAG engagement in disaster response (Grace et al., 2023). Evidence points to the possibility that NSAGs, by granting and enabling access for humanitarian organizations, can help boost their reputations with local populations (Breslawski, 2023). A notable example is the COVID-19 pandemic, during which NSAGs across the globe—including the Restoration Council of Shan State in Myanmar, armed gangs in El Salvador, and the Taliban in Afghanistan—contributed to the response in numerous ways, including spreading public health information; enforcing movement restrictions;

and delivering food, medicine, and health infrastructure (Breslawski, 2021). NSAGs also can play a vital role in responses to hurricanes, earthquakes, and other natural hazards. To illustrate with one example from the United States, one author writes of the far-right militia, the Oath Keepers, 'Disasters provide the Oath Keepers with opportunities to fundraise and gain the trust of people who might not otherwise be sympathetic to their anti-government cause. By arriving to crisis zones before federal agencies do, the Oath Keepers take advantage of bureaucratic weaknesses, holding a hand out to people in desperate circumstances' (Teirstein, 2023). For NSAGs, disaster response can be an important tool of outreach, political mobilization, and diplomacy.

Conclusion

This chapter has offered an overview of the current state of thinking and research on the 'diplomacy-disaster aid nexus.' The overall conclusion is that state and NSAG interests are ever-present threads across the domain of disaster response. The key for moving forward in this field—in terms of policy, practice, and research—is to understand that one cannot so easily extract this web of interests from our understanding of how humanitarianism operates. One should not take the 'humanitarian' out of humanitarian response but one also cannot ignore the interests of key relevant stakeholders. Indeed, a core task of the field of humanitarian diplomacy and humanitarian negotiation is to cultivate modes of understanding the interests of one's counterpart, set alongside the full range of a humanitarian negotiator's social, cultural, and relational negotiation capital (Grace, 2020a). Nevertheless, there is much more work to be done to push toward a deeper synthesis of the two legs of the 'diplomacy-disaster aid nexus.' A wealth of research examines the practice of humanitarian diplomacy and humanitarian negotiation. There is also a wealth of social scientific research that has probed what states and NSAGs seek to gain from engaging in disaster response, facilitating humanitarian access, and protecting civilians. Key to pushing our understanding forward will be to redouble efforts to explore how these two strands of research and thinking can inform one another.

References

Altunisik, Meliha. "Humanitarian Diplomacy as Turkey's National Role Conception and Performance: Evidence from Somalia and Afghanistan." *Southeast European and Black Sea Studies*, vol. 23, issue 3, 2023, 657–673.

Arel-Bundock, Vincent; Atkinson, James; and Potter, Rachel Augustine. "The Limits of Foreign Aid Diplomacy: How Bureaucratic Design Shapes Aid Distribution." *International Studies Quarterly*, vol. 59, issue 3, September 2015, 544–556.

Atkeson, Lonaa Rae and Maestas, Cherie D. *Catastrophic Politics: How Extraordinary Events Redefined Perceptions of Government*. Cambridge University Press, 2012.

Azam, Jean-Paul and Hoeffler, Anke. "Violence against Civilians in Civil Wars: Looting or Terror?" *Journal of Peace Research*, vol. 39, issue 4, July 2002, 461–485.

Barakat, Sultan. "Priorities and Challenges of Qatar's Humanitarian Diplomacy." Chr. Michelsen Institute, CMI Brief, 2019, https://www.cmi.no/publications/6906-priorities-and-challenges-of-qatars-humanitarian-diplomacy.

Barceló, Joan; Kubinec, Robert; Cheng, Cindy; Rahn, Tiril Høye; and Messerschmidt, Luca. "Windows of Repression: Using COVID-19 Policies against Political Dissidents?" *Journal of Peace Research*, vol. 59, issue 1, 2022, 73–89.

Becerra, Oscar; Cavallo, Eduardo A.; and Noy, Ilan. "Foreign Aid in the Aftermath of Large Natural Disasters." *Review of Development Economics*, vol. 18, issue 3, 2014, 445–460.

Bogatyreva, Olga. "Humanitarian Diplomacy: Modern Concepts and Approaches." *Herald of the Russian Academy of Sciences*, vol. 92, issue 14, 2022, 1349–1366.

Bommer, Christian; Dreher, Axel; and Perez-Alvarez, Marcello. "Home Bias in Humanitarian Aid: The Role of Regional Favoritism in the Allocation of International Disaster Relief." *Journal of Public Economics*, vol. 208, April 2022, 1–16.
Boudreau, Laura. "Discipline and Disasters: The Political Economy of Mexico's Sovereign Disaster Risk Financing Program." Fondation Pour Les Études et Recherches sur le Développement International, Policy Brief, 2015, https://ferdi.fr/dl/df-xpNsUiGwTXFBebzSuXdiBhmW/ferdi-b128-discipline-and-disasters-the-political-economy-of-mexico-s.pdf.
Breslawski, Jori. "Territorial Control & Armed Group Cooperation with Humanitarians." 2021, https://papers.ssrn.com/sol3/papers.cfm?abstract_id=4083723.
Breslawski, Jori. "Can Insurgents Capture Aid through Credit Claiming? Evidence from Afghanistan." *Journal of Global Security Studies*, vol. 8, issue 4, December 2023, 1–10.
Centre of Competence on Humanitarian Negotiation. "What Is Humanitarian Negotiation?" Centre of Competence on Humanitarian Negotiation, n.d., https://frontline-negotiations.org/what-is-humanitarian-negotiation/.
Centre of Competence on Humanitarian Negotiation. "CCHN Field Manual on Frontline Humanitarian Negotiation." 2019, https://frontline-negotiations.org/wp-content/uploads/2019/11/CCHN-Field-Manual-EN.pdf?hsCtaTracking=45723d72-16e4-4711-949f-8453d17ad182%7Cdc4930fd-ad4a-48f4-a3da-53b714fa12b2.
Chen, Christopher. "The Quad and HADR Operations: Prospects for Cooperation with Southeast Asia." S. Rajaratnam School of International Studies, RSIS Commentary, November 2022, https://reliefweb.int/report/world/quad-and-hadr-operations-prospects-cooperation-southeast-asia.
Cheng, Cindy and Minhas, Shahryar. "Keeping Friends Close, But Enemies Closer: Foreign Aid Responses to Natural Disasters." *British Journal of Political Science*, vol. 51, 2021, 940–962.
Chung, Eunbin and Rhee, Inbok. "Disasters and Intergroup Peace in Sub-Saharan Africa." *Journal of Peace Research*, vol. 59, issue 1, 2022, 58–72.
Clements, Ashley Jonathan. *Humanitarian Negotiations with Armed Groups: The Frontlines of Diplomacy*. Routledge, 2019.
Colona, Silvia. "Humanitarian Principles: A Bridge and a Compass for Civil-Military Coordination." *A Civil-Military Response to Hybrid Threats*, edited by Eugenio Cusumano and Marian Corbe. Palgrave Macmillan, 2017, 123–143.
Cook, Alistair DB. "Humanitarian Diplomacy in ASEAN." *Asian Journal of Comparative Politics*, vol. 6, issue 3, 2021, 188–201.
Daniels, R. Steven. "The Rise of Politics and the Decline of Vulnerability as Criteria in Disaster Decisions of the United States, 1953–2009." *Disasters*, vol. 37, issue 4, 2013, 669–694.
Davies, Gareth. "Lyndon Johnson and Disaster Politics." *Presidential Studies Quarterly*, vol. 47, issue 3, September 2017, 529–551.
Dellmuth, Lisa M.; Bender, Frida A.-M.; Jönsson, Aiden R.; Rosvold, Elisabeth L.; and von Uexkull, Nina. "Humanitarian Need Drives Multilateral Disaster Aid." *PNAS*, vol. 118, issue 4, 2021, https://www.pnas.org/doi/full/10.1073/pnas.2018293118.
Depoorter, Ben. "Horizontal Political Externalities: The Supply and Demand of Disaster Management." *Duke Law Journal*, vol. 56, 2006, 101–125.
Donelli, Federico. "Features, Aims and Limits of Turkey's Humanitarian Diplomacy." *Central European Journal of International and Security Studies*, vol. 11, issue 3, 2017, 59–83.
Eichenauer, Vera Z; Fuchs, Andreas; Kunze, Sven; and Strobl, Eric. "Distortions in Aid Allocation of United Nations Flash Appeals: Evidence from the 2015 Nepal Earthquake." *World Development*, vol. 136, December 2020, 1–17.
Fielding, David. "The Dynamics of Humanitarian Aid Decisions." *Oxford Bulletin of Economics and Statistics*, vol. 76, issue 4, August 2014, 536–564.
Fink, Günther and Redaell, Silvia. "Determinants of International Emergency Aid—Humanitarian Need Only?" *World Development*, vol. 39, issue 5, 2010, 741–757.
Finnemore, Martha. *National Interests in International Society*. Cornell University Press, 1996.
Fraser, Dominique. "The Quad: A Backgrounder." Asia Society Policy Institute, May 16, 2023, https://asiasociety.org/policy-institute/quad-backgrounder.
Gaillard, Jean-Christophe; Clavé, Elsa; and Kelman, Ilan. "Wave of Peace? Tsunami Disaster Diplomacy in Aceh, Indonesia." *Geoforum*, vol. 39, issue 1, January 2008, 511–526.
Garret, Thomas A. and Sobel, Russel S. "The Political Economy of FEMA Disaster Payments." *Economic Inquiry*, vol. 41, issue 3, July 2003, 496–509.

Gasper, John T. "The Politics of Denying Aid: An Analysis of Disaster Declaration Turndowns." *Journal of Public Management & Social Policy*, vol. 22, issue 2, 2015.

Gökalp, Deniz. "The UAE's Humanitarian Diplomacy-Claiming State Sovereignty, Regional Leverage and International Recognition." Chr. Michelsen Institute, CMI Working Paper, 2020, https://www.cmi.no/publications/7169-the-uaes-humanitarian-diplomacy-claiming-state-sovereignty.

Gong, Lina. "China's Disaster Diplomacy in Southeast Asia." *China Review*, vol. 21, issue 4, November 2021, 77–107.

Grace, Rob. "The Humanitarian as Negotiator: Developing Capacity across the Aid Sector." *Negotiation Journal*, vol. 36, issue 1, Winter 2020a, 13–41.

Grace, Rob. "Humanitarian Negotiation with Parties to Armed Conflict: The Role of Laws and Principles in the Discourse." *Journal of International Humanitarian Legal Studies*, vol. 11, 2020b, 68–96.

Grace, Rob. "Incrementalism in International Lawmaking: The Development of Normative Frameworks of Protection for Forcibly Displaced Persons." *International Law and Time: Narratives and Techniques*, edited by Klara Polackova Van der Ploeg, Luca Pasquet, and León Castellanos-Jankiewicz. Springer, 2022, 135–152.

Grace, Rob; Alejandria, Maria Carinnes; Bates, Madison; Boland, Samuel T.; Nylen, Alexandria; Tayyeb, Zein; and Levine, Adam C. "Moving Humanitarian-Military Relations Forward: A New Typology." *Journal of International Humanitarian Action*, vol. 8, issue 2, 2023, 1–15.

Grace, Rob and Lempereur, Alain, "Four Dilemmas of Acceptance: Insights from the Field of Humanitarian Negotiation." Global Interagency Security Forum, 2021, https://gisf.ngo/wp-content/uploads/2021/10/Four_Dilemmas_of_Acceptance_insights_from_the_field_of_humanitarian_negotiation.pdf.

Grossman, Marc. "The Tsunami Core Group: A Step toward a Transformed Diplomacy in Asia and Beyond." *Security Challenges*, vol. 1, issue 1, 2005, 11–14.

Guillon, Marlène and Mathonnat, Jacky. "What Can We Learn on Chinese Aid Allocation Motivations from Available Data? A Sectorial Analysis of Chinese Aid to African Countries." *China Economic Review*, vol. 60, April 2020, 1–21.

Haer, Roos and RezaeeDaryakenari, Babak. "Disasters and Civilian Victimization: Exploring the Dynamic Effect in Africa, 1997–2017." *Journal of Peace Research*, vol. 59, issue 1, 2022, 43–57.

Hasimi, Cemalettin. "Turkey's Humanitarian Diplomacy and Development Cooperation." *Insight Turkey*, vol. 16, issue 1, 2014, 127–145.

Husted, Thomas and Nickerson, David. "Political Economy of Presidential Disaster Declarations and Federal Disaster Assistance." *Public Finance Review*, vol. 42, issue 1, 2014, 35–57.

Islam, Mohammad Mozahidul; Chatterjee, Niladri; and Basar, Muhammad Asiful. "The Final Straw? Bhola Cyclone, 1970 Election, Disaster Politics, and the Making of Bangladesh." *Contemporary South Asia*, vol. 31, issue 2, 2023, 236–250.

Jackson, Ashley. "In Their Words: Perceptions of Armed Non-State Actors on Humanitarian Action." Geneva Call, 2016, https://reliefweb.int/report/world/their-words-perceptions-armed-non-state-actors-humanitarian-action.

Jackson, Ashley and Aynte, Abdi. "Humanitarian Negotiations with Al-Shabaab in Somalia." Humanitarian Policy Group, December 2013, https://www.refworld.org/pdfid/5346928c4.pdf.

Jackson, Ashley and Davey, Elanor. "From the Spanish Civil War to Afghanistan: Historical and Contemporary Reflections on Humanitarian Engagement with Non-State Armed Groups." Humanitarian Policy Group, May 2014, https://cdn.odi.org/media/documents/8974.pdf.

Jackson, Ashley and Giustozzi, Antonio. "Talking to the 'Other Side': Humanitarian Engagement with the Taliban in Afghanistan." Humanitarian Policy Group, December 2012, https://cdn.odi.org/media/documents/7968.pdf.

Jo, Hyeran. *Compliant Rebels: Rebel Groups and International Law in World Politics*. Cambridge University Press, 2015.

Kahan, Jerome. "Future of FEMA – Preparedness or Politics?" *Homeland Security & Emergency Management*, vol. 12, issue 1, 2015, 1–21.

Kelman, Ilan. "Acting on Disaster Diplomacy." *Journal of International Affairs*, vol. 59, issue 2, Spring/Summer 2006, 215–240.

Kelman, Ilan; Field, Jessica; Suri, Kavita; and Bhat, Ghulam M. "Disaster Diplomacy in Jammu and Kashmir." *International Journal of Disaster Risk Reduction*, vol. 31, 2018, 1132–1140.

Ker-Lindsay, James. "Greek-Turkish Rapprochement: The Impact of 'Disaster Diplomacy'?" *Cambridge Review of International Affairs*, vo. 14, issue 1, 2000, 215–232.

Krause, Monika. *The Good Project: Humanitarian Relief NGOs and the Fragmentation of Reason*. University of Chicago Press, 2014.
Kurtzer, Jacob D. "Denial, Delay, Diversion: Tackling Access Challenges in an Evolving Humanitarian Landscape." Center for Strategic and International Studies, September 2019, https://csis-website-prod.s3.amazonaws.com/s3fs-public/publication/Kurtzer_DenialDelayDiversion_WEB_FINAL.pdf.
Labbé, Jérémie and Daudin, Pascal. "Applying the Humanitarian Principles: Reflecting on the Experience of the International Committee of the Red Cross." *International Review of the Red Cross*, vol. 97, issue 897/898, 2016, 183–210.
Lee, Bomi K.; Mitchell, Sara McLaughlin; Schmidt, Cody J.; and Yang, Yufan. "Disasters and the Dynamics of Interstate Rivalry." *Journal of Peace Research*, vol. 59, issue 1, 2022, 12–27.
Lempereur, Alain. "Humanitarian Negotiation to Protect: John Rabe and the Nanking International Safety Zone (1937–1938)." *Group Decision and Negotiation*, vol. 25, January 2016, 663–691.
Lempereur, Alain. "Negotiation Power: How Humanitarian Frontliners Get Things Done with Hard Bargainers." *Negotiation Journal*, 2023, vol. 39, issue 4, 1–24.
Magone, Claire; Neuman, Michael; and Weissman, Fabrice, editors. *Humanitarian Negotiations Revealed: The MSF Experience*. Hurst, 2012.
Mancini-Grifoli, Deborah and Picot, André. "Humanitarian Negotiation: A Handbook for Securing Access, Assistance and Protection for Civilians in Armed Conflict." HD Centre for Humanitarian Dialogue, 2004, https://hdcentre.org/insights/humanitarian-negotiation-a-handbook-for-securing-access-assistance-and-protection-for-civilians-in-armed-conflict/.
Mantilla, Giovanni. *Lawmaking under Pressure: International Humanitarian Law and Internal Armed Conflict*. Cornell University Press, 2020.
Marrus, Michael. *The Unwanted: European Refugees from the First World War through the Cold War*. Temple University Press, 2002.
Mc Hugh, Gerard, and Bessler, Manuel. "Humanitarian Negotiation with Armed Groups: A Manual for Practitioners." United Nations, January 2006, https://resourcecenter.undac.org/wp-content/uploads/2020/12/HumanitarianNegotiationswArmedGroupsManual.pdf.
Mercy Corps. "Playbook: Negotiating for Humanitarian Access." November 2018, https://www.mercycorps.org/sites/default/files/2019-11/NegotiatingForHumanitarianAccessPlaybook.pdf/.
Minear, Larry and Smith, Hazel, editors. *Humanitarian Diplomacy: Practitioners and Their Craft*. United Nations, 2007.
Moorehead, Caroline. *Dunant's Dream: War, Switzerland and the History of the Red Cross*. Carroll & Graf Publishers, 1998.
Mosel, Irina and Jackson, Ashley. "Talking to the 'Other Side': Humanitarian Negotiations in Southern Kordofan and Blue Nile, Sudan." Humanitarian Policy Group, July 2013, https://cdn.odi.org/media/documents/8591.pdf.
Narang, Neil and Stanton, Jessica A. "A Strategic Logic of Attacking Aid Workers: Evidence from Violence in Afghanistan." *International Studies Quarterly*, vol. 61, issue 1, 2017, 38–51.
Nash, George H. *The Life of Herbert Hoover: The Humanitarian, 1914-1917*. W.W. Norton & Company, 1988.
Nemeth, Stephen and Lai, Brian. "When Do Natural Disasters Lead to Negotiations in a Civil War?" *Journal of Peace Research*, vol. 59, issue 1, 2022, 28–42.
Nohrstedt, Daniel and Baekkeskov, Erik. "Political Drivers of Epidemic Response: Foreign Healthcare Workers and the 2014 Ebola Outbreak." *Disasters*, vol. 42, issue 1, 2018, 41–61.
Papp, Bendegúz. "An Analysis of ASEAN's Cyclone Nargis 2008 Disaster Diplomacy." *Asian International Studies Review*, vol. 24, 2023, 25–46.
Passeri, Andrea. "Authoritarian Rule and the Weaponisation of Natural Disasters: The Case of Myanmar from Cyclone Nargis to the COVID-19 Pandemic (2008–2021)." *Journal of Asian Security and International Affairs*, vol. 9, issue 2, 2022, 280–300.
Pejic, Jelena. "Strengthening Compliance with IHL: The ICRC-Swiss Initiative." *International Review of the Red Cross*, vol. 98, issue 1, 2016, 315–330.
Phelan, Alexandra L. "The World Health Organization's Pandemic Treaty." *BMJ*, vol. 380, 2023, 463–464.
Powell, Colin L. "Remarks to the National Foreign Policy Conference for Leaders of Nongovernmental Organizations." 2001, https://avalon.law.yale.edu/sept11/powell_brief31.asp.
Raschky, Paul A. and Schwindt, Manijeh. "On the Channel and Type of Aid: The Case of International Disaster Assistance." *European Journal of Political Economy*, vol. 28, issue 1, March 2012, 119–131.

Régnier, Philippe. "The Emerging Concept of Humanitarian Diplomacy: Identification of a Community of Practice and Prospects for International Recognition." *International Review of the Red Cross*, vol. 93, issue 884, December 2011, 1211–1237.

Robinson, Jonathan. "Russian Foreign Humanitarian Assistance: Identifying Trends Using 15 Years of Open-Source Data." *Expeditions with MCUP*, 2022, 1–28.

Sainz-Santamaria, Jaime and Anderson, Sarah E. "The Electoral Politics of Disaster Preparedness." *Risks, Hazards and Crisis in Public Policy*, vol. 4, issue 4, 2013, 234–249.

Schenkenberg van Mierop, Ed. "Coming Clean on Neutrality and Independence: The Need to Assess the Application of Humanitarian Principles." *International Review of the Red Cross*, vol. 97, issue 897/898, 2016, 295–318.

Scott, Martin; Bunce, Mel; and Wright, Kate. "The Influence of News Coverage on Humanitarian Aid: The Bureaucrats' Perspective." *Journalism Studies*, vol. 23, issue 2, 2022, 167–186.

Setiawan, Asep and Hamka Hamka. "Role of Indonesian Humanitarian Diplomacy toward Rohingya Crisis in Myanmar." 2019, https://eudl.eu/pdf/10.4108/eai.5-11-2019.2292481.

Shughart, William F. "Katrinanomics: The Politics and Economics of Disaster Relief." *Public Choice*, vol. 127, 2006, 31–53.

Streich, Philip A. and Mislan, David Bell. "What Follows the Storm? Research on the Effect of Disasters on Conflict and Cooperation." *Global Change, Peace, and Security*, vol. 26, issue 1, 2014, 55–70.

Sutton, Rebecca and Rhoads, Emily Paddon. "Empathy in Frontline Humanitarian Negotiations: A Relational Approach to Engagement." *Journal of International Humanitarian Action*, vol. 7, issue 23, 2022, 1–13.

Sylves, Richard and Búzás, Zoltán I. "Presidential Disaster Declaration Decisions, 1953–2003: What Influences Odds of Approval?" *State and Local Government Review*, vol. 39, no. 1, 2007, 3–15.

Teirstein, Zoya. "When Disaster Strikes, Far-Right Groups See An Opportunity To Gain Trust." HuffingtonPost, May 17, 2023, https://www.huffpost.com/entry/far-right-groups-fema-disaster-relief_n_6463af3de4b005be8ff66084#:~:text=For%20nearly%20a%20decade%2C%20the,meals%20and%20doing%20construction%20work.

Tekir, Gökhan. "Humanitarian Diplomacy as a Tool for India's Middle East Strategy." *Communication and Diplomacy*, vol. 10, 2023, 3–22.

Terry, Fiona and McQuinn, Brian. "The Roots of Restraint in War." International Committee of the Red Cross, 2020, https://www.icrc.org/en/publication/4352-roots-restraint-war.

Van Dijk, Boyd. *Preparing for War: The Making of the Geneva Conventions*. Oxford University Press, 2022.

Van Klinken, Gerry. "Typhoon Disaster Politics in Pre-1945 Asia: Three Case Studies." *Disaster Prevention and Management*, vol. 31, issue 1, 2021, 35–46.

Wei, Jiuchang and Marinova, Dora. "The Orientation of Disaster Donations: Differences in the Global Response to Five Major Earthquakes." *Disasters*, vol. 40, issue 3, 2016, 452–475.

Williamson, Jamie A. "Using Humanitarian Aid to 'Win Hearts and Minds': A Costly Failure?" *International Review of the Red Cross*, vol. 93, issue 884, December 2011, 1035–1061.

Yeger, Shay and Podoler, Guy. "The North Korean Famine and Inter-Korean Relations—Disaster Diplomacy Perspective." *International Journal of Disaster Risk Reduction*, vol. 79, 2022.

Zakharova, Liudmila and Burghart, Sabine. "Russia—A Re-Emerging Donor in North Korea." *European Journal of East Asian Studies*, vol. 22, 2023, 121–149.

12

FLOW AND MANAGEMENT OF DISASTER RELIEF

The Case of Bangladesh

Shitangsu Kumar Paul, Md. Shohel Khan, and Shehan Tawsif

Introduction

Bangladesh is one of the most populated countries in the world. This densely populated country faces momentous disaster risk, such as floods that inundate about 70% of the landmass of the country (Sovacool, 2017; Islam et al., 2020). Bangladesh also faces further natural and anthropogenic disasters, such as cyclones, tidal surges, storm surges, water logging, heatwaves, drought, fire incidents, saline intrusion, arsenic contamination, and various types of pollution (Paul, 2013; Guha-Sapir et al., 2017; Mallick et al., 2017; Sai et al., 2018; Seddiky et al., 2020; Tawsif et al., 2023). About 6.9 million people died, 3.6 billion were impacted, and $550 billion USD of economic losses were incurred between 1900 and 2012 due to coastal flooding in Bangladesh (Kugler & De Groeve, 2007; Jonkman & Vrijling, 2008; Kundzewicz et al., 2014; Kabir & Hossen, 2019). In the early 1960s, about 50,000 people died across six cyclones in Bangladesh, and an estimated 300,000 fatalities were accounted for in a single cyclone that struck the country in 1970 (Harun-Or-Rashid et al., 2022). Another cyclone, Sidr, caused 4,406 deaths, injured 55,009, and caused $839 million USD in economic losses in 2007 (Arif et al., 2023). Additionally, COVID-19 caused 29,477 deaths as of 6 December 2023 in Bangladesh (DGHS, 2023).

The World Risk Report's latest World Risk Index ranked Bangladesh as the ninth most disaster risk prone among 193 countries based on overall risk determined by exposure, susceptibility, and coping and adaptation capacities (WRR, 2023). Bangladesh also confronts numerous challenges in disaster risk reduction due to a lack of capacity building among various stakeholders and supporting organizations (Alam et al., 2011), insufficient policies (Choudhury et al., 2019), poor governance (Bhuiyan, 2015), inappropriate distribution of scarce funds (Mallick, 2014; Islam, 2015), and anomalies between coordination and collaboration (Sovacool et al., 2018; Paul et al., 2022). Out of Bangladesh's 64 districts, 53 are denoted as vulnerable to flood (Barua et al., 2016). A major source of floodwater comes upstream from India as part of the massive Ganga-Brahmaputra-Meghna river system that inundates the riverbanks of these districts and disrupts the livelihoods of their inhabitants. Disaster risk reduction includes preparedness, response and rescue, recovery, and rehabilitation—all of which connect with government and nongovernment entities, their institutional frameworks, along with their collaborations and partnerships, acquisition and usage of funds, and relief distribution and development activities. Several guidelines (e.g., Hyogo

DOI: 10.4324/9781003414834-17

Framework 2005–2015 and Sendai Framework 2015–2030) have been adopted to reduce losses and increase resilience to disasters in Bangladesh (Mallick, 2014).

Disasters have negatively impacted access to livelihoods in Bangladesh, which is one of the world's most vulnerable nations due to its geographic location, close proximity to the sea, dense population, low level of awareness, and impoverished economic standing (Juran & Trivedi, 2015; Paul, 2019; Rahman & Islam, 2019; Tasnuva et al., 2021). Management of disaster events has evolved from its initial emphasis on relief and rehabilitation to a multifaceted mission encompassing responsibilities ranging from catastrophe prediction and preparedness to long-term recovery and rebuilding to future disaster risk reduction. As a result, disaster management in Bangladesh has become an intricate, complex, and vital national endeavor (Sabur, 2012). Given this context, this research aggregates the results of several studies conducted in different areas of Bangladesh at different periods of time. The overarching objective is to better understand the flow of relief to and within the disaster-prone country of Bangladesh.

Methods

Study Area

Bangladesh comprises 64 administrative districts, of which only six are declared as less hazard prone by the Government of Bangladesh (GoB) in the Bangladesh Delta Plan 2100. The plan reports that Bangladesh is divided into six hotspots based on the magnitude of natural hazard vulnerabilities and hydrology. For this study, nine villages and one city corporation area were purposively selected from eight districts (Barguna, Patuakhali, Gopalganj, Sirajganj, Rajshahi, Satkhira, Khulna, and Bagerhat) (Figure 12.1). Among these study areas, Rajshahi City Corporation (RCC) is considered an urban slum while Tatulbaria, Angulkata, and Charkashem villages are located on the shoreline of the Bay of Bengal, approximately 30 km from the coast, and on an offshore island, respectively. Bannabari village of Tungipara Upazila in Gopalganj District and Suvagacha of Kajipur Upazila in Sirajganj District are situated roughly 200 and 180 km from the capital city of Dhaka, respectively. Southkhali (both north and south) and Rajoir are the shoreline villages of coastal Bagerhat District, while Gangarampur and Ganapatipur villages are the interim and inland areas of Khulna and Satkhira Districts, respectively. Other than RCC, all of the study areas are rural and the majority of household respondents are primarily dependent upon agriculture, either directly or indirectly, for livelihood. RCC is categorized into three slum zones (inner, middle, and outer).

Data Collection

Data were collected from primary sources including reconnaissance surveys, household surveys, key informant interviews (KII), and focus group discussions (FGD). In total, 1,304 households, 131 FGD respondents, and 101 KII participants were selected randomly from the study areas (Table 12.1). The questionnaires (separate for each study) were finalized through reconnaissance surveys in each respective study area. Household heads (at least 30 years old) were interviewed from each household. FGDs comprising 8–13 participants (36–70 years old) with at least three females were conducted with predetermined questions to both supplement and validate questionnaire data. KIIs were also conducted with predetermined questions among individuals directly engaged in rescue, relief distribution, coordination, damage estimation, and reconstruction. Each FGD lasted 50–60 minutes, while each KII required 25–30 minutes. All data were analyzed using SPSS version 26.0.

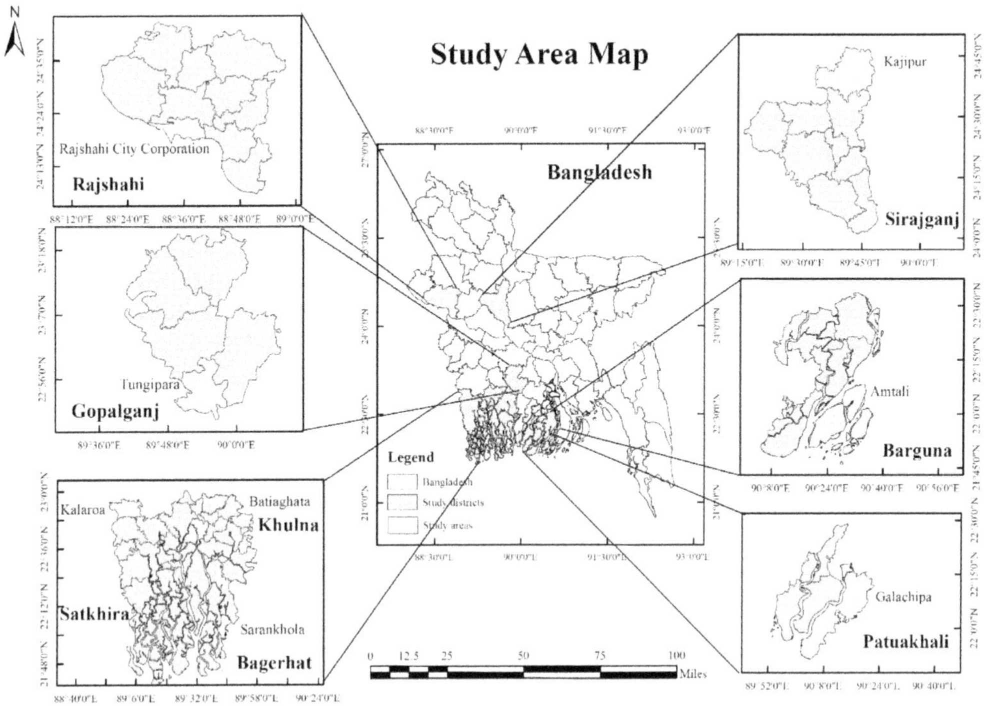

Figure 12.1 Map of the Study Areas

Source: Created by Authors.

Table 12.1 Study Areas, Sampling, and Times of Data Collection

Study area	*Upazila and district*	*Total HHs*	*Sampled HHs*	*FGD respondents*	*KIIs*	*Data collection*
Angulkata	Amtali, Barguna	575	198	9	19	December 2010
Tatulbaria	Amtali, Barguna	170	90	10	10	December 2010
Charkashem	Galachipa, Patuakhali	43	43	11	9	December 2010
Bannabari	Tungipara, Gopalganj	150	35	8	12	January 2005
Suvagacha	Kajipur, Sirajganj	350	59	13	15	January 2005
Southkhali	Sarankhola, Bagerhat	1,138	180	21 (two FGDs)	7	January–June 2011
Rajoir	Sarankhola, Bagerhat	1,518	184	12	8	July–August 2022
Gangarampur	Batiaghata, Khulna	581	71	9	4	June–July 2022
Ganapatipur	Kalaroa, Satkhira	690	83	11	4	May–June 2022
RCC	RCC, Rajshahi	4,010	361	29 (three FGDs)	13	January–March 2022
Total		**8,087**	**1,304**	**131**	**101**	

Source: Table from authors.

Results and Discussion

Relief Providing Agencies

There has been a notable shift in disaster management approaches and relief-providing agencies in Bangladesh over the past few decades. Dwivedi et al. (2018) report that GoB utilizes two types of supply chains to assist individuals during and after crises. One is to provide consistent, year-round assistance to the impoverished regardless of any particular calamity, which is termed as humanitarian aid. The second is used as emergency funds, which are distributed during disasters by GoB through local governments. Humanitarian aid is the regular, ongoing supply chain for aid distribution (Shareef et al., 2019). The goal of improving the lives and livelihoods of impoverished populations is being pursued by numerous nongovernment organizations (NGOs), such as Grameen Bank, BRAC (Bangladesh Rural Advancement Committee), Rangpur Dinajpur Rural Service, Association for Social Advancement, CARE, CARITAS, UNICEF, Oxfam, and various microfinance institutions. GoB uses these networks to manage the entire emergency supply chain for planning, acquiring, and distributing relief in the event of a disaster (DMB, 2015; Shareef et al., 2019). GOs include union parishad offices (smallest administrative units in Bangladesh), Department of Agricultural Extension offices, Bangladesh Krishi Bank, Bangladesh Army, Bangladesh Navy, and Upazila Nirbahi Office (UNO), which collectively work together to distribute relief materials in distressed areas. GOs are the prime authority for preparing relief budgets and distribution, and they collaborate with NGOs to transport relief supplies to affected areas, although this is not always easy to achieve. In addition, NGOs may receive donations directly from agencies around the world (Chen & Zhao, 2021) while international NGOs receive relief goods from international agencies which are further distributed to disaster-affected areas. However, NGOs sometimes have difficulty performing their activities in an efficient and equitable manner, which means that the delivery system may leave certain areas distressed while others may receive an abundance of relief supplies (Chen, 2021). These scenarios have been observed during cyclones and floods as well as during the COVID-19 situation in RCC.

Need Assessment and Institutional Frameworks

In Bangladesh, the most common practice is to distribute relief to affected populations when aid is received from domestic and international sources. The task of identifying populations in need is often difficult because needs are not always properly identified by data collectors and/or may be biased by local politicians, demonstrating the need for highly experienced field staff and more standardized needs-based assessments (Aase, 2020). Moreover, the effectiveness of GoB policies, particularly their capacity to safeguard the country's millions of poor and insecure residents from the regular crises of existence and subsistence, remains a critical factor in determining the state's legitimacy (Ali et al., 2021). NGOs operating in underprivileged communities have the chance to bias and facilitate their microcredit membership by identifying needy individuals regardless of level of poverty, infrastructure, and access to facilities. Together with NGOs, GoB has tried to better locate underprivileged communities to identify the mitigation, readiness, and preventive tactics that can be mainstreamed at local levels (Seddiky et al., 2020).

In 2009, cyclone Aila hit the coast of Bangladesh and severely impacted coastal livelihoods. In terms of relief assessment, the International Federation of Red Cross and Red Crescent Society (IFRCS) made an emergency appeal for $2.2 million USD for nine months to help 36,800 households in spite of GoB's lack of an official appeal for aid (Paul & Chatterjee, 2019). Bangladesh Red Crescent Society (BDRCS) sent funds to each of its units to buy dry meals within hours of the catastrophe as part of emergency efforts. Furthermore, BDRCS and IFRCS collaborated to mobilize assessment teams for additional emergency response. The two organizations also provided hygiene kits and galvanized corrugated sheeting to 10,000 households that were shipped from stock that had been prepositioned. In order to gather potable drinking water from treatment plants, Oxfam, BDRCS, and IFRCS jointly worked in the impacted zones to supply 11,000 water jugs (Paul & Chatterjee, 2019). BRAC also took part in cyclone Aila relief efforts, along with the United Nations Development Program and World Bank.

In Bangladesh, establishing a legislative framework to guide the massively expansive activities related to disaster management is a difficult task (Sabur, 2012). By changing the paradigm from a relief-oriented mechanism to a proactive, comprehensive approach, the GoB, development partners, and other humanitarian actors have significantly reduced the number of casualties caused by disasters in Bangladesh over time (Hasan et al., 2019). Still, stakeholders must work collectively to respond to disasters, and the absence of effective partnerships among stakeholders is the primary weakness in the institutional mechanism of disaster management in Bangladesh (Azad et al., 2019). Thus, GoB has now institutionalized a dynamic, holistic system to effectuate integrated and comprehensive disaster management (Figure 12.2).

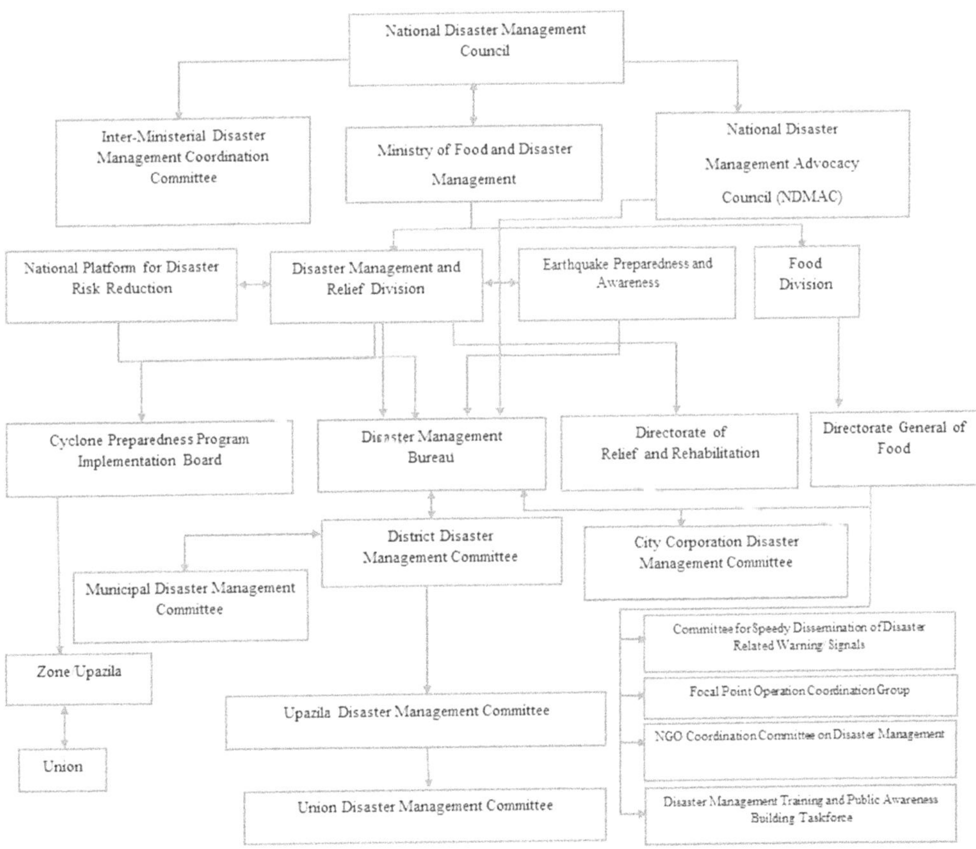

Figure 12.2 Disaster Management Institutional Framework in Bangladesh

Source: Adapted from DMB (2015).

Relief

Emergency Help and Rescue

Emergency help and rescue operations are required during and immediately after disasters in order to provide support to affected populations. The World Food Programme reports that relief items can be categorized as food, non-food, and cash (WFP, 2017). Slum dwellers in RCC received relief items during the COVID-19 pandemic when a lockdown disrupted sources of income to combat food insecurity. Findings show that, overall, 78.7% of slum dwellers received relief items during the pandemic, which came from various organizations. Geographically, 82.2% (inner), 84.2% (middle), and 72.2% (outer) of respondents received relief. A chi-square test ($\chi^2 = 6.621$) further revealed significant differences among the three zones in terms of receiving relief. The outer zone received less relief than the inner and middle zones because it is located farther away from the central business district. Apparently, relief providers emphasized the inner zones over the peripheral zone, perhaps due to ease of distribution or political influence. In terms of sources of relief, 70.1%, 19%, and 18% of respondents claimed that they received relief from GOs, NGOs, and other institutions or individuals, respectively. GOs mainly supplied rice, pulses, vegetables, potatoes, sugar, oil, salt, mangoes, vermicelli, soap, musk, and similar food items. It is also worth noting that, sometimes, NGOs preferred families who have good relationships with NGO workers. The relief receiving trend was reduced from the central business district moving toward the periphery among all of the relief suppliers. These results were significant across the study areas when considering GOs ($\chi^2 = 6.20$), NGOs ($\chi^2 = 29.59$), and other individuals and institutions ($\chi^2 = 21.51$). Geographically, 32% (inner), 21% (middle), and 8% (outer) of respondents received relief (e.g., food, money, and safety equipment) from BDRCS, different donors, and wealthy individuals during the lockdown. Again, the outer zone was neglected in relief distribution. Furthermore, significant differences (F test) exist among the slum zones in terms of rice, pulse, flour, oil, sugar, salt, mango, vegetables, vermicelli, musk, soap, sanitizer, detergent, and cash relief (Table 12.2).

Table 12.2 Relief (Average) Received by Respondents during COVID-19 Pandemic in RCC, Bangladesh

Relief items	*Inner*	*Middle*	*Outer*	*Overall*	*ANOVA (F)**
Rice (kg)	1.89	1.57	1.22	1.50	12.833
Pulse (kg)	1.11	0.82	0.72	0.85	7.318
Flour (kg)	0.87	0.34	0.36	0.48	19.894
Oil (l)	0.92	0.46	0.48	0.58	15.949
Sugar (kg)	0.66	0.12	0.13	0.25	34.712
Salt (kg)	0.44	0.09	0.08	0.18	23.903
Mango (kg)	2.44	0.17	0.99	1.08	38.189
Vegetables (kg)	0.89	0.42	0.12	0.41	34.392
Vermicelli (kg)	0.68	0.25	0.14	0.31	39.956
Musk (number)	0.43	0.13	0.09	0.19	21.793
Soap (number)	2.22	0.50	0.46	0.91	32.138
Sanitizer (ml)	0.38	0.03	0.05	0.12	38.380
Detergent (kg)	0.50	0.03	0.04	0.15	47.925
Cash money (BDT.)	583	499	186	413	8.830

Notes: *Significant at 95% confidence level; Source: Field survey, 2022.
Source: Field survey, 2022.

Flood conditions rendered the availability and accessibility of food to affected populations during and after the 2004 flood in Suvagacha and Bannabari villages. About 50% of children and women in these areas consume low-quality food due to inadequate food supply and are thus more likely to fall into malnutrition and sickness. In Bannabari, 62.9% of households had adequate food stock during and after the flood because they stockpiled food beforehand, but in Suvagacha, the figure was only 39% because most of the respondents are functionally landless and had no capability to stockpile food. This result is consistent with Salam et al.'s (2019) study on the flood affected *char* lands of the Jamuna River in Siranganj but inconsistent with Khan et al.'s (2021) report on flash floods in Tanguar *Haor* in Sunamganj. Chi-square tests show significant differences in the availability of food between two villages. Overall, about 29% respondents had their own food source followed by 27.7% getting assistance from NGOs, 23.4% from GoB, and 15.2% having to purchase from the market. At the village scale, 72.1% and 16.3% of respondents had their own source of food, and 18.6% and 14.2% had buying capacity from the local market in Bannabari and Suvagacha villages, respectively. Similarly, 36.2% and 30.5% respondents from Suvagacha received food assistance from NGOs and GoB during and after flood, but such assistance was totally absent in Bannabari. Thus, the study reveals that Bannabari villagers are dissatisfied with both governmental and nongovernmental food assistance. On the other hand, respondents in Suvagacha are neutral regarding government-provided food assistance and satisfied with assistance from NGOs. Note that due to the severity and intensity of floods in Suvagacha, GoB had to deploy the army to rescue people and provide assistance.

The study in Barguna and Patuakhali Districts found that organizational assistance for rescuing cyclone victims significantly varied among the three villages. In Angulkata, Charkashem, and Tatulbaria villages, 37.9%, 32.6%, and 10% of respondents received rescue assistance, respectively, as GoB and NGO personnel were able to access Angulkata and Charkashem more easily. Major organizations involved in rescue operation were BDRCS and union parishad members, while NGOs such as Save the Children, Oxfam, and Songram also operated rescue activities at a limited scale. Across the villages, 100% (Tatulbaria), 72.3% (Angulkata), and 46.67% (Charkashem) of respondents claimed that BDRCS volunteers actively performed their duties during and after cyclones. Such findings unveil that rescue operations are primarily the responsibility of local GOs and BDRCS volunteers. In many cases, NGOs are reluctant to provide such services apart from their main duties, which revolve around microcredit. GoB's Standing Orders on Disasters provides strict guidelines to local governments and BDRCS volunteers for providing primary medical facilities to people injured from cyclones through mobile medical teams (Alam, 2023). This study reports that 62% overall and 97%, 58.1%, and 47.5% of respondents from Tatulbaria, Charkashem, and Angulkata, respectively, have received first aid. However, significant difference exists among the villages. The total number of deaths and injured population in sampled households are 30 and 58 in Tatulbaria, 12 and 87 in Angulkata, and 1 and 31 in Charkashem, respectively. This study also reveals that 65% of respondents from Tatulbaria village received first aid from NGOs, followed by Angulkata (56%) and Charkashem (50%). First aid and post-cyclone healthcare facilities are also furnished by medical teams from different donor countries. In addition, the Bangladesh Army and Navy as well as doctors from public hospitals also carried out first aid but at a limited scale. Therefore, activities of such organizations could be expanded to provide first aid services immediately after a cyclone to help reduce mortality.

Assistance for Relief, Rehabilitation, and Rebuilding Livelihoods

GoB's Standing Orders on Disasters reports that the major responsibilities of local government entities include rescue and relief operations, rapid loss and damage assessment for receiving relief, sinking tube wells on an emergency basis, removing saline water from ponds and wells, ensuring fair and quick distribution of relief and other aid, coordination and supervision of relief and rehabilitation activities by voluntary organizations, implementation of post-disaster rehabilitation schemes, and reestablishing communication networks through quick maintenance. Relief is considered to be partial assistance to disaster-affected populations to save lives, reduce suffering, facilitate the livelihood recovery process, and enhance livelihood securities.

FGD and KII respondents from Angulkata and Tatulbaria reported that some people received relief one day after Cyclone Sidr. The study finds that overall, nearly 99% of respondents have received relief from various sources immediately after Sidr. Some respondents in Charkashem also received food from neighboring people (most cases from relatives) one day after Sidr. Union parishad, Agricultural Extension Office, Bangladesh Krishi Bank, Bangladesh Army, Bangladesh Navy, and UNO provided immediate response regarding rescue and relief distribution. Union parishad provided rice, flattened rice, and similar items. GoB Army and Navy distributed dry food, drinking water, oral saline, and water purifying tablets. Likewise, UNO played a crucial role in the reconstruction of houses in all three villages. Respondents received BDT 5,000 for each damaged house and BDT 20,000 for each death in a family. Apart from this, UNO coordinated relief and rehabilitation activities in the respective upazila. NGOs distributed goods such as rice, puffed rice, flattened rice, household utensils, potable water, potatoes, fire boxes, clothing, medicine, and water purifying tablets. Lutheran Healthcare, a USA-based NGO, operated a 'food for work' program among the cyclone affected populations in Tatulbaria. The organization provided 30 kg of rice (varied according to family size) per month per household for three months after Sidr. World Food, another NGO, operated a similar program in Angulkata and Tatulbaria, while Oxfam did the same in Charkashem. According to respondent's opinions, such food for work programs were the most effective for sustaining the livelihoods after the cyclone.

Some NGOs supported farmers by providing seeds and agricultural and fishing accessories to assist rebuilding post-cyclone livelihoods. Agricultural Extension Office provided limited seeds and fertilizers in Angulkata while Ahasania Mission (domestic NGO) provided support to fishermen to support livelihoods. The organizations formed cooperatives among the fishermen and provided a boat with the condition that fishermen must purchase an engine for the boat with their own funds. However, the effort was not successful. BRAC also distributed some cows among its members in Angulkata. However, despite having limitations, a majority of the respondents in all three villages reported that relief materials given by various government and NGOs were sufficient to save their lives after Cyclone Sidr (97%, 87.8%, and 51.5% in Angulkata, Tatulbaria, and Charkashem, respectively).

Institutional Assistance for Enhancing Coping and Livelihood Security

Relief is important for assisting affected populations in the short-term period after a disaster. However, if such assistance is provided for a longer period of time, it may develop relief dependency (Paul & Routray, 2011). Almost all cyclone-affected respondents have received relief from various organizations and donor agencies within a period of three months—most frequently varying from one to five times in the form of cash, rice, pulses, cooking oil, household utensils, and clothes. However, relief agencies have not provided any support services, money, or materials for rebuilding the local economy or reviving livelihoods (e.g., agriculture and fishing). Even three years after Sidr, respondents strongly expressed their desire to receive further support for rebuilding livelihoods. Such findings demonstrate that post-cyclone relief operations are crisis-oriented, reactive, and not usually proactive in an effort to support rebuilding livelihoods. Therefore, it is necessary and more sensible to provide assistance to produce food and productive materials that might enhance the coping ability of disaster-affected populations. The traditional form of relief operations (e.g., food, blankets, and clothes) seems less effective in the study area. Rather relief should acknowledge the real needs raised by affected populations such as agricultural inputs, fishing equipment, and employment opportunities. Assistance should be delivered in a way that helps affected populations to secure their livelihoods and regain or even exceed their pre-disaster status. It is also important to provide relief to those who experience greater losses and those at the bottom of the society who are unlikely to recover without external help. Moreover, if relief is made available immediately after a cyclone, it should be based on priority and should aim to maintain community assets and prevent them from reaching the final stages of destitution.

Humanitarian Assistance

Humanitarian assistance has garnered increased attention worldwide due to the increasing rate and magnitude of natural and anthropogenic disasters. Relief inflows and formalized distribution systems have played a dominant role in responding to disasters in Bangladesh. As disasters occur frequently, a systematic top-down relief distribution and management system has been essential to 'build back better'. GoB determines humanitarian assistance based on attaining at least four criteria from a total set of 12 criteria (see details in DMB, 2015). Humanitarian support includes providing food items as well as non-food items like corrugated iron sheets, blankets, tents, and cash (Shareef et al., 2019). These support systems are operated throughout the year to assist poor people.

After Cyclone Sidr in the southern part of Bangladesh, nearly 98% of households of Southkhali (north and south) received sanitary latrines donated by the NGO Muslim Aid. Other supports from NGOs included cash (92%), a new house (78%), radio for hearing weather information during fishing (66%), bundle of corrugated sheets (21%), life jacket (17%), fishing net (18%), and domestic animals (7%). Furthermore, almost 98% of respondents in Southkhali (north and south) received BDT 15,000 for repairing their house by Rupantor, an NGO, immediately after Cyclone Sidr. Respondents in the study area also received BDT 5,000 for fully destroyed houses and BDT 1,000 for moderately destroyed houses from the Bangladesh Army. However, it must be noted that poor construction materials were provided by donor agencies, and supplied materials were not always properly used to reconstruct the houses. New houses were not properly distributed to the original victims, and some houses are not resilient to future disasters. Respondents were very pleased for receiving their new houses, but the relatively wealthy were unhappy because the new houses could not be properly constructed with the supplied materials and would not last long.

Realities and Practicalities

Previously, GoB was often criticized for corruption and irregularities in distributing relief, particularly for relief not being distributed among actual affected populations (Smith, 1992). For example, GoB was highly criticized for irregularities and widespread corruption after the 1974 floods that resulted in a nationwide famine (Paul, 2003). Therefore, donor agencies and international NGOs sought to provide assistance and post-disaster emergency aid through alternative methods (Lewis, 1992). Moreover, many NGOs had proven their competency and impartiality in the distribution of relief assistance during this time. This led to challenges in emergency aid distribution through NGOs during the 1970s to mid-1980s in Bangladesh (Paul, 1998, 2003). Since that time, NGOs are again playing a crucial role in relief distribution, although reports of corruption and nepotism are still found in cases of relief distribution by local governments, such as by union parishad members. Despite the constraints, NGOs are efficiently and successfully distributing relief among affected populations. However, findings of this study are consistent with several earlier studies that NGO coverage is lower compared to GoB and that NGO members receive more relief than non-members—meaning that membership in an NGO plays a vital role in receiving relief (Paul, 1998; Mustafa, 2003; Paul et al., 2022). Therefore, NGOs have sometimes first fulfilled their primary mission of running their activities (e.g., microcredit) and been biased in their distribution of relief. These findings are consistent with Paul (2003), who states that the relief and rehabilitation activities of NGOs are often limited to the places where their microcredit projects are already implemented. Additionally, union parishad offices do not document the lists of relief receivers and sometimes they do not know how many NGOs are performing activities during/after a disaster, so there exists mismanagement and a lack of coordination for relief distribution.

Our study from Gangarampur and Ganapatipur depicts that union parishad members are corrupt in most cases when selecting populations for relief distribution, and that NGOs also give privilege to their members. In Rajoir, however, respondents reported that relief was properly distributed, and the poor and affected populations received equitable relief. Almost all of the respondents agreed that relief and donations (e.g., latrines, pond sand filters, plastic drums, and ferrocement filters for water conservation) were received—although the materials did not last long in most of the cases, which is consistent with Tawsif et al. (2023) and Khan and Paul (2023). Overall, respondents appealed that they want to be informed what is supposed to be allotted and to then make sure that such assistance is actually received, including materials to support rehabilitation and reconstruction.

Distribution, Coordination, Biases, and Misuse

In Bangladesh, the Department of Disaster Management (DDM) usually receives the official budget from the Ministry of Disaster Management and Relief in the July of each calendar year. With the budget, DDM purchases blankets, corrugated iron sheets, and other supporting non-food items in September. These items are stockpiled in warehouses in Dhaka, Chattagram, and Khulna. GoB normally procures rice and wheat through the Ministry of Food, which are warehoused in each upazila food store. Considering the area and population, DDM sends an allotment letter for rice to the District Relief and Rehabilitation Officer (DRRO), which then sends a requisition for rice to the upazila food warehouse to distribute. Cash distributions are straightly handled by the DRROs. Sometimes, extra food or even cash is provided to each district in order to purchase other relief items (e.g., dry food) (DMB, 2015). During an emergency response, the Project Implementation Officer assesses the relief requirements with the help of volunteers. The union parishad chairman/members, under the direction of the respective UNO, forward these requests to the DRRO. DRRO generally adjusts the requests and compiles the data before sending them to the DDM head office in Dhaka. DDM analyzes the requests received from DRRO, reduces the quantity, and then sends the allotment order to each respective district. The relief procurement (goods), storage (warehouse), and distribution (prepared from bottom-up but allotted by top-down) system in Bangladesh is declared as a push system. Administrative districts are grouped as A, B, and C by the Disaster Management Bureau based on the number of upazilas and unions, and these districts are allotted a fixed BDT 200,000 and 100 metric tons of rice, BDT 150,000 and 75 metric tons of rice, and BDT 100,000 and 50 metric tons of rice every year, respectively. Non-food items are distributed based on two criteria, population and poverty index, each weighted at 50%. Shareef et al. (2019) claim that in 2016, BDT 500,000 was distributed to every district for purchasing dry food for distressed and extremely poor people. Relief distribution from the district level is downstream, but collecting information for relief requirements from wards or unions is primarily aggregated and placed at the upazila level and then forwarded to district level. Therefore, this management system of GoB is a combination of push-pull system.

In many cases, NGOs are biased in relief distribution, and UNO does not always coordinate relief activities during emergencies because each organization, including NGOs, works independently immediately after disaster. As a result, some people receive relief multiple times while others are totally deprived. For example, an FGD in Rajoir revealed that almost 8% of respondents received relief more than once during and/or after disaster. Thus, relief is sometimes distributed inequitably, sometimes misused by local populations, and sometimes the distributing authorities (e.g., union parishad officials) personally benefit.

Conclusion

GoB has well-established disaster management institutions in all departments and administrative tiers. Community risk assessment and participatory risk reduction plans are some of the successful initiatives that have involved communities in the disaster risk reduction process. Similarly, reducing risk factors (e.g., cyclone shelters, *killas* (generally, a safe place higher than normal flood level to shelter humans and protect livestock), warning signal dissemination, coastal afforestation, and introducing disaster management issues in the education system) have proven impactful in recent years.

Despite having success in disaster management, some limitations have been identified such as ambiguities in cyclone warning signals, timely dissemination of warnings, lack of relief coordination, and nepotism during relief distribution. Officially remote *char* and islands are given special priority, but this study finds that island inhabitants still lag behind and are more vulnerable to impending disasters and to achieving post-disaster livelihood security. Similarly, disaster preparedness and the post-disaster relief activities of government and NGOs are concentrated on central business districts or headquarters and well-connected areas. Post-disaster income generating activities and livelihood rebuilding need to be prioritized, as the study found that 'food/cash for work' was the most effective to save populations from starvation. This study also found that both NGOs and GoB are more efficient in distributing relief compared to earlier times; corruption and irregularities have also decreased but still exist. Finally, NGOs are sometimes biased and seek to continue their programs (e.g., microcredit activities) as their main goal. Lastly, zero tolerance regarding nepotism and strict coordination among relief distributing authorities—both national and international—should be followed for the proper rehabilitation and rebuilding of disaster-affected people's livelihoods.

References

Aase, Maren. "Listing for change? Exploring the politics of relief lists in Bangladesh after Cyclone Sidr." *Disasters* 44.4 (2020): 666–686.

Alam, Edris. "Factors of cyclone disaster deaths in coastal Bangladesh." *Heliyon 9* (2023): e18417.

Alam, Khurshid, Md Shamsuddoha, Thomas Tanner, Moshahida Sultana, Muhammad Jahedul Huq, and Sumaiya S. Kabir. "The political economy of climate resilient development planning in Bangladesh." *IDS Bulletin* 42.3 (2011): 52–61.

Ali, Tariq Omar, Mirza Hassan, and Naomi Hossain. "The moral and political economy of the pandemic in Bangladesh: Weak states and strong societies during Covid-19." *World Development* 137 (2021): 105216.

Arif, Md Saidul Islam, Ibnul Mahdi, Md Adil Rafi, Saadmaan Jubayer Khan, and Md Mostafizur Rahman. "Cyclone exposure mapping in coastal Bangladesh: A multi-criteria decision analysis." *Heliyon* 9.10 (2023).

Azad, M. Abul Kalam, M. Salim Uddin, Sabrina Zaman, and Mirza Ali Ashraf. "Community-based disaster management and its salient features: A policy approach to people-centered risk reduction in Bangladesh." *Asia-Pacific Journal of Rural Development* 29.2 (2019): 135–160.

Barua, Uttama, M. Shammi Akhter, and Mehedi Ahmed Ansary. "District-wise multi-hazard zoning of Bangladesh." *Natural Hazards* 82 (2016): 1895–1918.

Bhuiyan, Shahjahan. "Adapting to climate change in Bangladesh: Good governance barriers." *South Asia Research* 35.3 (2015): 349–367.

Chen, Yingzhen. "Relief goods distribution problem: Considering donation strategies, fairness, and interventions." *Progress in Disaster Science* 12 (2021): 100198.

Chen, Yingzhen, and Qiuhong Zhao. "Donation-based relief goods distribution problem." *Computers & Industrial Engineering* 152 (2021): 106990.

Choudhury, Mahed-Ul-Islam, M. Salim Uddin, and C. Emdad Haque. "Nature brings us extreme events, some people cause us prolonged sufferings: The role of good governance in building community resilience to natural disasters in Bangladesh." *Journal of Environmental Planning and Management* 62.10 (2019): 1761–1781.

Director General of Health Services (DGHS). "COVID-19 Dynamic Dashboard for Bangladesh." Government of the People's Republic of Bangladesh, Dhaka (2023).

Disaster Management Bureau (DMB). "National Plan for Disaster Management, Standing Orders on Disaster." Government of the People's Republic of Bangladesh, Dhaka, Bangladesh (2015).

Dwivedi, Yogesh K., Mahmud Akhter Shareef, Bhasker Mukerji, Nripendra P. Rana, and Kawaljeet Kaur Kapoor. "Involvement in emergency supply chain for disaster management: A cognitive dissonance perspective." *International Journal of Production Research* 56.21 (2018): 6758–6773.

Guha-Sapir, Debby, Philippe Hoyois, Pascaline Wallemacq, and Regina Below. "Annual Disaster Statistical Review 2016." The Numbers and Trends (2017): 1–91.

Harun-Or-Rashid, G. M., Juan Jose Castro, Mithun Chakrabartty, and Md. Kamruzzaman. "Proposal of resilient housing features for the disaster affected community of Bangladesh." *Sustainability* 14. 21 (2022): 14061. https://doi.org/10.3390/su142114061

Hasan, Md Robiul, Mahbuba Nasreen, and Md Arif Chowdhury. "Gender-inclusive disaster management policy in Bangladesh: A content analysis of national and international regulatory frameworks." *International Journal of Disaster Risk Reduction* 41 (2019): 101324.

Islam, Mohammad Mozahidul. "The politics of the public food distribution system in Bangladesh: Regime survival or promoting food security?" *Journal of Asian and African Studies* 50.6 (2015): 702–715.

Islam, Shafiqul, Cordia Chu, Leong Liew, and James CR Smart. "Distributing flood shelters for disaster risk reduction: Exploring the practices in Bangladesh from a political economy perspective." *Disaster Prevention and Management: An International Journal* 29.3 (2020): 322–339.

Jonkman, Sebastiaan N., and Johannes K. Vrijling. "Loss of life due to floods." *Journal of Flood Risk Management* 1.1 (2008): 43–56.

Juran, Luke, and Jennifer Trivedi. "Women, gender norms, and natural disasters in Bangladesh." *Geographical Review* 105.4 (2015): 601–611.

Kabir, Md Humayain, and Md Nazmul Hossen. "Impacts of flood and its possible solution in Bangladesh." *Disaster Advances* 12.10 (2019): 48–57.

Khan, Md Shohel, Mizanur Rahman Apu, Sameena Begum, and Mohammad Maruf Billah. "Sectoral impacts of flash flood in Tanguar Haor in Sunamganj of Bangladesh." *Asian Journal of Geographical Research* 4.2 (2021): 55–64.

Khan, Md Shohel, and Shitangsu Kumar Paul. "Fresh water management in coastal Bangladesh: Preparedness and adaptation." *Discover Water* (2023) 3:27.

Kugler, Zsófia, and Tom De Groeve. "The Global Flood Detection System." JRC Scientific and Technical Reports (2007): 1–45.

Kundzewicz, Zbigniew W., Shinjiro Kanae, Sonia I. Seneviratne, John Handmer, Neville Nicholls, Pascal Peduzzi, Reinhard Mechler, Laurens M. Bouwer, Nigel Arnell, Katharine Mach, Robert Muir-Wood, G. Robert Brakenridge, Wolfgang Kron, Gerardo Benito, Yasushi Honda, Kiyoshi Takahashi, and Boris Sherstyukov. "Flood risk and climate change: Global and regional perspectives." *Hydrological Sciences Journal* 59.1 (2014): 1–28.

Lewis, David. "Catalysts for Change? NGOs, Agricultural Technology and the State in Bangladesh". ODI, London No. 2291-2019-4442 (1992).

Mallick, Bishawjit. "Cyclone shelters and their locational suitability: An empirical analysis from coastal Bangladesh." *Disasters* 38.3 (2014): 654–671.

Mallick, Bishawjit, Bayes Ahmed, and Joachim Vogt. "Living with the risks of cyclone disasters in the southwestern coastal region of Bangladesh." *Environments* 4.1 (2017): 13.

Mustafa, Daanish. "Reinforcing vulnerability? Disaster relief, recovery, and response to the 2001 flood in Rawalpindi, Pakistan." *Global Environmental Change Part B: Environmental Hazards* 5.2 (2003): 71–82.

Paul, Bimal Kanti. "Coping mechanisms practised by drought victims (1994/5) in North Bengal, Bangladesh." *Applied Geography* 18.4 (1998): 355–373.

Paul, Bimal Kanti. "Relief assistance to 1998 flood victims: A comparison of the performance of the government and NGOs." *Geographical Journal* 169.1 (2003): 75–89.

Paul, Shitangsu Kumar. "Prevalence of nutritional insecurity in the central coast of Bangladesh: Evidence from Cyclone Sidr'." *The Journal of Geo-Environment* 10 (2013): 22–35.

Paul, Shitangsu Kumar. (2019). "Post cyclone household food security in coastal Bangladesh." In Khatun, H., Baqee, A. and Kabir, H. (Eds), *People at Risk: Disaster and Despair*, 185–209. Disaster Research Training and Management Centre, University of Dhaka.

Paul, Bimal Kanti, and Subarna Chatterjee. "Climate Change-Induced Environmental Hazards and Aila Relief Measures Undertaken to Sundarbans in Bangladesh and India." The Sundarbans: A Disaster-Prone Eco-Region: Increasing Livelihood Security (2019): 469–490.

Paul, Shitangsu Kumar, Jannatul Naim, and Mousumi Akando Mou. "Assessing efficiency of disaster management institutions at local level in the coastal region of Bangladesh." *Nature-Based Solutions* 2 (2022): 100024.

Paul, Shitangsu Kumar, and Jayant K. Routray. "Household response to cyclone and induced surge in coastal Bangladesh: Coping strategies and explanatory variables." *Natural Hazards* 57 (2011): 477–499.

Rahman, Md Siddiqur, and Abu Reza Md Towfiqul Islam. "Are precipitation concentration and intensity changing in Bangladesh overtimes? Analysis of the possible causes of changes in precipitation systems." *Science of the Total Environment* 690 (2019): 370–387.

Sabur, AKM Abdus. "Disaster management system in Bangladesh: An overview." *India Quarterly* 68.1 (2012): 29–47.

Sai, Fabio, Lydia Cumiskey, Albrecht Weerts, Biswa Bhattacharya, and Raihanul Haque Khan. "Towards impact-based flood forecasting and warning in Bangladesh: A case study at the local level in Sirajganj district." *Natural Hazards and Earth System Sciences Discussions* (2018): 1–20.

Salam, Mohammed Abdus, Md Shohel Khan, and Md Abu Hena Sarker. "Impacts and coping techniques to flood: A case study on two adjacent char land of Jamuna River in Sirajganj District, Bangladesh." *International Research Journal of Environmental Sciences* 8.3 (2019): 53–61.

Seddiky, Md Assraf, Helen Giggins, and Thayaparan Gajendran. "International principles of disaster risk reduction informing NGOs strategies for community based DRR mainstreaming: The Bangladesh context." *International Journal of Disaster Risk Reduction* 48 (2020): 101580.

Shareef, Mahmud Akhter, Yogesh K. Dwivedi, Rafeed Mahmud, Angela Wright, Mohammad Mahboob Rahman, Hatice Kizgin, and Nripendra P. Rana. "Disaster management in Bangladesh: Developing an effective emergency supply chain network." *Annals of Operations Research* 283 (2019): 1463–1487.

Smith, Keith. (1992). *Environmental Hazards: Assessing Risk and Reducing Disaster*. Routledge.

Sovacool, Benjamin K. "Don't let disaster recovery perpetuate injustice." *Nature* 549.7673 (2017): 433–433.

Sovacool, Benjamin K., May Tan-Mullins, and Wokje Abrahamse. "Bloated bodies and broken bricks: Power, ecology, and inequality in the political economy of natural disaster recovery." *World Development* 110 (2018): 243–255.

Tasnuva, Anjum, Md Riad Hossain, Roquia Salam, Abu Reza Md Towfiqul Islam, Muhammad Mainuddin Patwary, and Sobhy M. Ibrahim. "Employing social vulnerability index to assess household social vulnerability of natural hazards: An evidence from southwest coastal Bangladesh." *Environment, Development and Sustainability* 23 (2021): 10223–10245.

Tawsif, Shehan., Shitangsu Kumar Paul, and Md Shohel Khan. "Changing pattern of livelihood capitals of urban slum dwellers during COVID-19 Pandemic". *International Journal of Human Capital and Urban Management* 9.1 (2023): 101–118.

World Food Programme (WFP). "Forecast-Based Emergency Preparedness Project Bangladesh" (2017). www.ifrc.org. Accessed on 12 June 2023.

WRR (World Risk Report). 2023. ISBN 978-3-946785-16-3. Available at: https://weltrisikobericht.de/wp-content/uploads/2023/10/WRR_2023_english_online161023.pdf?fbclid=IwAR3UJhyAxBWho6cSmKTnyX4Z8CMBn3zsQBc6mwqhO-IzuoA5hPs8Gr4hd84. Accessed on 22 January 2024.

13

HUMANITARIANISM AND VOLUNTEERISM IN DISASTER RESPONSE

The Case of Sikh Volunteers during the 2022 Australia Floods

Sweta Baniya and Badal Pokharel

Introduction

In recent years, when disaster strikes, we have witnessed many humanitarian and voluntary actions performed by communities across the globe. Citizens may resort to humanitarian aid from voluntary sectors due to a lack of confidence in government support (Lai et al., 2019). Subsequently, the world's recent big disasters like the Nepal Earthquake 2015 also witnessed similar kinds of transnational humanitarian actions performed by various local and global communities via social media (Baniya, 2022). Humanitarian actions and volunteerism are part of any disaster relief and are very important for saving the lives of the most marginalized communities in the world. Anthropology scholar Miriam Ticktin (2014) argues that humanitarianism is hard to define. She states, "In its dominant characterization, humanitarianism, is one way to 'do good' or to improve aspects of the human condition by focusing on suffering and saving lives in times of crisis or emergency" (Ticktin, 2014, 274). With the rise in the use of social media, there is an upsurge in humanitarian actions led by various volunteers via multiple platforms such as Facebook, X (formerly Twitter), and WhatsApp. In Australia, where the frequency of disasters has been rising rapidly, the collaboration of humanitarian actors and volunteers to address the impact of disasters has been significant. Showcasing humanitarianism and volunteerism in the Australian floods of 2022 conducted by Sikh Volunteers, we argue that rapid and effective disaster response is successful with collaboration and coordination among humanitarian actors, volunteers, and the community. This is accomplished by presenting an in-depth case study of Sikh Volunteers, a volunteer-based humanitarian organization in Australia that has contributed significantly to various disasters in Australia.

The February 2022 floods in Australia severely impacted the northern river areas and Hawkesbury-Nepean catchment in New South Wales (NSW), Australia. The intense weather in Southern Queensland and northern NSW received more than a year's worth of rainfall in one week. This led to a declaration of national emergency in Australia in March 2022. Hundreds of thousands of people were affected by the floods. Over the past decades, floods have become a frequently occurring geohazard all over the globe. With its extreme weather patterns, Australia is one of the most

DOI: 10.4324/9781003414834-18

flood-affected countries. The variable climate, coupled with the influence of La Niña, led to increased rainfall across Australia, impacting various communities (Ummenhofer et al., 2015). The adverse flooding events in 2011, 2021, and 2022 are linked to La Niña, associated with warmer Pacific Ocean waters intensifying storminess along the east coast. Over the last two years, La Niña-induced extreme rainfall caused widespread flooding, particularly affecting marginalized communities.

Australia has faced various disasters throughout its history, and the resilience of its communities has played a crucial role in responding to and recovering from these challenges. From bushfires and floods to cyclones and droughts, Australians have come together to support one another and contribute to rebuilding efforts. Many researchers have highlighted the role of the public in the response to floods in Australia. Jacqueline K Wachira and A. John Sinclair (2005) noted that members of the public are often the first responders during a flood, and they engage in activities such as moving furniture and physical assets and participating in neighborhood committees to reduce the damage. Andrew Gissing, Chas Keys, and Steve Opper (2010) studied how the Victoria State Emergency managed the response to the 2010 Victorian Spring Floods event in close partnership with other emergency survival agencies. Likewise, Fiona Roberts, Frank Archer, and Caroline Spencer (2021) studied the role of Australian non-profit organizations (NPOs) in strengthening the community's resilience toward disasters in Victoria, Australia. They highlighted NPOs had significant contributions in disaster-affected communities pre-, during, and post-disasters. In the face of ongoing environmental challenges, such as climate change, communities in Australia continue to adapt and build resilience.

Researchers, including the Intergovernmental Panel on Climate Change (IPCC), project future increases in climate change-induced rainfall and flooding (Pörtner et al., 2022). While the Australian government has been working to support the survivors of the floods, many humanitarian actions were being led by various local-level organizations including the volunteers. United Nations Office for the Coordination of Humanitarian Affairs (OCHA) describes humanitarian actions as "Humanitarian action is premised on the fact that, along with having their basic needs met, affected people often also need protection from violence, abuse, coercion, and deprivation" (IASC, 2019, 16). Organizations that worked during the Australian floods focused on meeting community needs during the floods and working closely with many volunteers who contributed their time during the disaster. In this article, we provide a case study of the Sikh Volunteers and an analysis of their activities from their website and Facebook, followed by an analysis of an in-depth interview with one of the leaders from Sikh Volunteers Australia about their work that focused primarily on two different floods in Australia. Our study reflects that volunteers play a pivotal role in effective crisis communication and inclusivity in flood hazard risk management and fostering resilience within communities. The findings of this study aspire to contribute vital insights for effective disaster response strategies, inclusive policies, and the recognition of the invaluable contribution of volunteers amid escalating climate challenges.

Literature Review

Initial responders in a humanitarian crisis such as a disaster are typically regular citizens, as has been researched by various scholar (Wachira and Sinclair, 2005; Roberts et al., 2021; Radice, 2022). Individuals play a crucial role in helping impacted people during the response and recovery phase, which is often a valuable support to the concerned authorities. During any humanitarian crisis, there is need of humanitarianism which plays a pivotal role in disaster response, embodying the compassionate spirit of collective support and aid in adversity. In times of crisis, be it natural/human-induced disasters or conflict-driven emergencies, the essence of humanitarianism shines through as individuals and organizations mobilize to alleviate suffering and restore a sense of normalcy. In the world of natural disasters, being humanitarian is not a simple idea—it's quite complicated. Erica Bornstein and Peter Redfield (2011) argue that humanitarianism, development, and human rights all broadly seek to ameliorate and improve aspects of the human condition and hence, it is crucial to not only help with physical problems but also pay attention to how people feel inside. Likewise, Henry Radice (2022) argues that, when we try to help, we should recognize that the people affected by the disaster have their voices and opinions about what should happen. Oftentimes times with blanket approaches to disaster response, people's voices might be subdued. Andreev Anatoliy and Svetlana Artemyeva (2020) also suggest we should focus more on individual people when doing humanitarian work. Focusing on humanitarianism allows us to understand the understanding the bigger social and cultural picture in post-disaster situations.

Furthermore, technology has become a powerful force in shaping and enhancing humanitarian efforts across the globe. In recent years, innovations in various technological fields have played a crucial role in improving the efficiency, reach, and impact of humanitarian initiatives. One notable advancement is the use of data analytics and artificial intelligence, enabling organizations to analyze vast amounts of information quickly. This facilitates more informed decision-making in areas such as disaster response, resource allocation, and risk assessment. For example, the use of digital technologies, like drones, to assist in natural disaster situations is gaining popularity, as highlighted by Ning Wang (2020). Furthermore, some experts, such as Irni Eliana Khairuddin et al. (2022), suggest adopting decentralized blockchain technology as a way to improve communication and coordination among different groups involved in humanitarian efforts. Despite these technological strides, there remains a challenge in making sure everyone understands the seriousness of natural disasters and the importance of taking steps to prevent and lessen their impact, as pointed out by Robert Hamilton (2000). The introduction of eco-drones as a tool for humanitarian purposes, especially in the face of severe flooding, raises important questions about how well the government is managing the situation and the responsibilities of the drone operators, according to J.D. Schnepf (2020). So, even though we have these advanced technologies, there is still work to be done to ensure widespread awareness and collaboration when it comes to dealing with the challenges posed by natural disasters.

The exploration of social media in responding to disasters has been widespread, mainly focusing on its capacity to improve the spread of information and identify needs (Muniz-Rodriguez et al., 2020; Baniya, 2022). In the current world, humanitarian action on the ground is supported by volunteering and humanitarian organizing that happens on various social media. Christian Ehnis and Deborah Bunker (2012), who studied Queensland floods of 2011, argue that the full potential of social media for engaging in two-way communication and collaboration with the public is often not fully utilized. However, coming to the context of the 2016 Louisiana floods, social media proved to be crucial for communication when regular cell services were down, but its effectiveness in reaching out to government agencies was limited, according to Daniel Bosch (2017). The actual and potential applications of social media in emergencies are varied, encompassing extending emergency response efforts, fostering social unity, and advancing research, as highlighted by David E Alexander (2014).

Moreover, the significance of community involvement and active participation in disaster resilience and recovery has been highlighted by many studies (Cutter et al., 2008; Joseph et al., 2020; Roberts et al., 2021). Australia's National Disaster Risk Reduction Framework and National Strategy for Disaster Resilience (NSDR) emphasize the value of incorporating the voices and engagement of the community in the implementation of the framework. Similarly, community involvement has been stressed by international disaster risk reduction and management policies like the United Nations Hyogo Framework and Sendai Framework. These policies and frameworks also reflect on the importance of public involvement irrespective of how resourceful governments are. All of these actions during a disaster are important as volunteerism in natural disaster response embodies the spirit of solidarity and community resilience. Volunteers play a vital role in providing immediate aid, support, and comfort to those affected by disasters. Terry Fulmer et al. (2007) focus on how university volunteers step up during disasters, showing that they are eager to help and have useful skills. Likewise, John Twigg and Irina Mosel (2017) highlight that when disasters happen in cities, people often come together on their own to help out. Similarly, Lilla Yumagulova et al. (2021) also talk about how volunteers in Indigenous communities play a role during emergencies and point out the cultural factors that make volunteering easier or harder. Furthermore, Nayomi Kankanamge et al. (2019) researched how regular people can use their smartphones and computers to help with disaster planning, especially by sharing important information during and after disasters.

The scope of digital volunteerism in natural disasters is expanding, marked by the rise of movements like Digital Humanitarian Volunteers (DHV) that play a significant role as argued by Jaziar Radianti and Terje Gjøsæter (2019). However, they argue that challenges exist in ensuring the accessibility of digital tools and platforms for these volunteers, potentially affecting their effectiveness (Radianti and Gjøsæter, 2019). The collaboration between digital volunteer networks and formal response organizations is influenced by technical and organizational factors, with information technology and a shared understanding of disaster situations being crucial factors, according to Chul Hyun Park and Erik Johnston (2019). The motivations and contributions of digital volunteers are also significant, emphasizing a preference for non-monetary measures of success and a need for feedback and impact measurement (Barrera and Yang, 2019). All these studies highlight how important it is to include these volunteer groups in official plans for dealing with disasters.

Overall, the integration of technology into humanitarian efforts marks a significant step forward in addressing global challenges and responding more effectively to crises. Voluntary efforts grounded in the humanitarian principles are important for emergency preparation and disaster management as they establish a link between the official response and the public, encourage community involvement, and advance the common welfare (Alexander, 2014; Baniya, 2022).

Sikh Volunteers in Action during the 2022 Australian Floods

In this section, we provide a detailed case study of Sikh Volunteers on how their work incorporates humanitarianism and volunteerism. We situate our case study within a volunteer-based humanitarian organization that has worked during Australian floods in 2021. For our research, we wanted to find out how humanitarian organizations that are primarily volunteer-based in Australia launched disaster response during the 2021 floods. Within our research, we sought to understand the effectiveness of the humanitarian response and how the community members themselves have responded to the crisis. Along with this, we wanted to understand the humanitarian and volunteerism as a response to this crisis. We conducted an ethically cognizant study (IRB approval in both the US and Australia) to understand how humanitarian response was launched as a part of our larger study where we interviewed n=15 individuals. Due to space and limitations, we only will focus on our interview with a leader of Sikh Volunteers and we also present an in-depth analysis of the organization.

We identified the organization from community members in Australia who have directly benefited from Sikh Volunteers during the fire as well as during the floods. We also accessed information on the Sikh Volunteers from their open-access website and Facebook. First, we emailed and then we reached out via phone to set up an appointment to talk with the member. We conducted an in-depth semi-structured interview with a member of Sikh Volunteers (identity kept anonymous due to the IRB regulations). Our interview process was semi-structured as we wanted to hear the narratives of how volunteerism and humanitarianism were conducted during the Australian floods. One set of questions focused on volunteering and humanitarian work. The interview lasted around 80 minutes and we had an in-depth and detailed conversation about the participant's experience of being a first responder in the floods and distributing the food to the needy. By interviewing the leader of Sikh Volunteers, we found that humanitarianism and volunteerism are implemented via various systems within Australia. In our analysis, we looked for specific details about humanitarianism and volunteerism as a focus of this specific article.

According to their website, Sikh Volunteers is a "registered charity and a non-profit volunteer organization focusing on reducing distress in the community by providing free food to disadvantaged and needy individuals/families" (Sikh Volunteers). Their major work focuses on providing free food and drinking water to the communities that are affected by the emergencies. To conduct this task, the organization also recruits volunteers who support their endeavors of distributing food and working in collaboration with other various non-profits. The organization currently situated in Victoria, Australia is an organization that embraces diversity and supports any community that needs their immediate needs. In the past, the organization has worked during bushfires and COVID-19, and they were the very first responders to reach the communities. During the 2022 floods, the organization worked for 22 days providing 15,000 cooked meals and distributing them to needy communities. Along with the major emergency work, the organization also distributes free food for four days a week and their volunteers also deliver food if needed for certain community members. More importantly, they also engage with various schools and colleges to provide insight into their work. For their work, Sikh A has been awarded by various city councils and governmental organizations in Australia. Recently, they were awarded Australian Human Rights Heroes in 2020.

With our interview with one of the members of the Sikh Volunteers, we were able to get an in-depth insight on how on a day-to-day basis, they contributed to the floods and how humanitarian actions were launched by the volunteers. The analysis included the content of the website as well as their Facebook profile and interviews. This varied type of data allowed us to understand the humanitarianism and volunteerism work of Sikh Volunteers in more detail. For the analysis, we took the guidance provided by the United Nations High Commissioner for Refugees' emergency handbook. The guidebook says, "The principal motivation of humanitarian action is to save lives and alleviate suffering in a manner that respects and restores personal dignity" (Handbook for the United Nations, 2019). They list four different principles: humanity, impartiality, neutrality, and independence (Table 13.1). Various organizations that work in disaster, war, or conflict put these principles in action. Likewise, in a recent white paper commissioned by the Secretariat of the Plan of Action to Integrate Volunteering into the 2030 Agenda for the Global Technical Meeting on Volunteering 2020, author Chris Millora (2020) revisits the basic typology of volunteering, which provides a diverse and broader framework of understanding the work of volunteering. Millora (2020) argues that "the rapid and widespread changes in the social, political, economic, technological landscapes throughout the world have also created causes for volunteers, new tools they can use and new channels for volunteering" (5). He revisits the typology of volunteering from 1999 typology and expands it into five categories: mutual aid, service, campaigning, participation, and leisure. In the following Tables 13.1 and 13.2, we showcase how Sikh Volunteers have been using humanitarian principles and the volunteering categories.

Table 13.1 Humanitarian Principles

Humanitarian principles	*Definition by UNHCR*	*Example from Sikh Volunteers*
Humanity	Humanity is the principal driver for any response to a crisis, whether caused by conflict, violence, or natural or human-made disaster	The work of Sikh Volunteers is dedicated to humanity with a focus on food distribution during disaster, which is one of the major needs
Impartiality	Impartiality means that humanitarian action is based solely on need, with priority given to the most urgent cases irrespective of race, nationality, gender, religious belief, political opinion, or class	Sikh Volunteers implements a non-discriminatory policy as they openly state that they serve their community regardless of race, color, religion, or origin of country
Neutrality	The neutrality of humanitarian action is further upheld when humanitarian actors refrain from taking sides in hostilities or engaging in political, racial, religious, or ideological controversies	In providing free food to various disaster-affected communities, Sikh Volunteers refrain from taking sides—rather they engage in collaboration with various other entities
Independence	Independence requires humanitarian actors to be autonomous; they are not to be subject to control or subordination by political, economic, military, or other non-humanitarian objectives	Sikh Volunteers is an autonomous organization; during the floods they traveled 34 hours to reach the flooded area and distributed food—they are driven by their own desire for service

Source: Table from author.

As our interviewee shared since the inception of the organization in 2014, their main aim has been to promote multiculturalism and reach out to any member of the community who is in need. They have also been active in other events than floods such as bushfire and COVID-19. As our participant recalls,

> In 2019 there was a bushfire very close to where we live in Bunyip. Bushfire was not of, you can say a massive scale, it was detained very quickly. It was controlled very quickly, but still, it kept on going for three days. So that was the first time that we supported anyone in a disaster situation because Bunyip is like a 3-hour drive. When we got to know about this bushfire, we thought that we could take food to the people who were in the evacuation center and support them.

In situations of disaster, Sikh Volunteers is always ready to mobilize their volunteers and do their humanitarian work. The humanitarian organization is willing to go to various parts of the country and collaborate with people on the ground to provide services the food to the community in need. Likewise, even in 2020 in the second bushfire, Sikh Volunteers was equally active. These two events according to our interviewee provided them with a lot of lessons regarding the cultural needs of people and training the volunteers and managing resources. During the time of COVID-19, they started a home food delivery service. Hence, in disasters that Australian communities have faced, Sikh Volunteers has been part of humanitarian response specifically providing food to the community in need. During the Australian floods, Sikh Volunteers went to serve the community in NSW in the Taree region in 2021, and in 2022 they served in the Lismore area and

Table 13.2 Volunteering Principles

Typology of volunteering (expanded)	*Definition by Chris Millora (2020)*	*Example from Sikh Volunteers*
Mutual Aid/ Self-Help	Mutual aid is the wealth of informal, person-to-person helping activities embedded in community and cultural practices; people gather and volunteer together as a response to a shared need or issue	Sikh Volunteers is a volunteer-driven organization and anyone who is willing to contribute is welcome; they do regular volunteer training and prepare volunteers to work in high-stakes disaster situations
Service	Service volunteering is where volunteers respond to the perceived needs of another person or community	Sikh Volunteers deploy their volunteers based on the needs of the community; they focus on service through food delivery, which is a need during disasters
Campaigning	Campaigning usually involves the collective action of a group or an individual to amplify "marginalized" voices and to change the status quo	Sikh Volunteers campaign using their website and Facebook, where they encourage volunteers as well as non-volunteers to share information specifically about the needs of the community
Participation	Participation is where volunteers give time and effort to engage with governance and decision-making mechanisms at different levels	Individuals engaged in volunteering via Sikh Volunteers dedicate their time to driving and reaching out to various needy communities with warm home-cooked meals
Leisure	Volunteer activities that express personal interests or passions such as in the arts, culture, and sports; these activities contribute to wider wellbeing and cohesion	During times of non-crisis, Sikh Volunteers is involved in community engagement work and reaches out to schools, colleges, and universities to share their work and encourage volunteerism

Source: Table Created by author.

in 2023 in Victoria. Sikh Volunteers has served several thousands of communities in Australia with a grounding in humanitarian principles.

Furthermore, during the NSW floods, which is 12 hours from where Sikh Volunteers is located, our participant mentioned that people started texting them on social media if they will come and serve. Hence, they decided to drive 12 hours and sometimes they have driven more than 34 hours to serve communities. According recalled by our participant, they also involved various volunteers on the way while they were serving:

> In 2021 we were six volunteers who originally started from Victoria. When we reached Coffs Harbor, fortunately, there was a Sikh family living in the area. They came up and offered us their garage and their home as well. So, there were nine family members who volunteered… We frequently contact them. In 2022 there were four volunteers who started originally from Victoria where I was one of them. Again, because the Coffs Harbor family was in constant contact. So, we made our first stop at Coffs Harbor. We reached their house at around 8:30 in the evening and we prepared our first meal there. We loaded it into a food van and we started towards Woolgoolga.

In doing their service to the community, Sikh Volunteers welcomes other potential volunteers who are willing to serve their communities. They do a lot of partnerships with the local community and local volunteers who support their endeavor. The most important thing they do is collaborate with a community who need their support. This collaboration allows them to understand the needs of the community, the situation after the disaster, and how many meals to prepare for many community members. Their volunteering actions are driven by humanitarian principles, and shared by our interlocutor:

> In 2022 during the floods, we started four from here, from Victoria. When we reach Lismore, you can say the support came from complete strangers. Complete stranger people. When we reach there, the level of support is such that when we reach in Lismore there is a ration cap for the people of the Lismore. I have never seen such a thing in my life. The Woolworths of the Lismore were taking out the bread, and loaf and one person can purchase only six slices. Six slices of bread, two potatoes, one onion, that is one person, that's it. No more. That is the level of the condition in Lismore when we reach them. But despite this condition, the farmers, the people who were in the area whose crops are not gone, their stock is still with them. They were serving the community and they were bringing the ration to us as well for cooking. We have two vans with us. One van is filled with gas, utensils, and everything else. One van was filled with groceries. Each, but when we started cooking the next day in the Lismore from that garage, we realized that our ration was going to go away only in two days. In the morning we started cooking up around 1,100 meals.

The above narrative shared by the participant showcases the intersection of humanitarianism and volunteerism and how it interplays not just within their organization but also people who are completely strangers to the organization. While the condition at Lismore due to floods was dreadful and people were limited with ration, the Sikh Volunteers still were able to provide their services, work with the community, and partner with various other people to provide warm meals to people and they served 1,100 meals during that time.

Overall, our analysis reveals that there is an intrinsic link between humanitarianism and volunteerism within the work of Sikh Volunteers. Humanitarian actions in the Australian floods have been launched via Sikh Volunteers, which incorporates the actions launched by various volunteers who practice and implement the principles of humanitarianism. The organization invites various volunteers to participate.

Conclusion

In times of disasters, synergy between humanitarianism and volunteerism plays a crucial role in easing suffering and helping affected communities recover. This dynamic relationship is exemplified by volunteer-based organizations during crises. In our study, we presented the case study of Sikh Volunteers during the NSW floods in 2021–22 that provided a vivid example of how the principle of humanitarianism and volunteerism work in action. When the floods occurred and expanded, Sikh Volunteers prioritized the basic needs of communities, aligning with the core principles of humanity—saving lives and alleviating suffering while respecting personal dignity. The volunteering categories, as explained by Chris Millora (2020) find practical expression in the Sikh Volunteers' diverse approach, covering mutual aid, service, campaigning, participation, and leisure.

The case study highlights the adaptability of volunteer-driven efforts in responding to different disasters, from bushfires to the COVID-19 pandemic and the 2021–22 floods. Sikh Volunteers demonstrate a flexible and responsive form of volunteerism, reaching distant places and collaborating with communities in highly impacted places like Lismore and Taree in northern NSW, showcasing both geographical reach and a commitment to meeting various community needs. Importantly, the personal story shared by a member of Sikh Volunteers emphasizes the humanitarian values guiding their actions. The organization's collaborative efforts, even in situations with limited resources like ration-capped Lismore, showcase their resilience and ability to form meaningful partnerships with diverse communities. This intersection of humanitarianism and volunteerism, witnessed in Sikh Volunteers' work, not only aids in disaster relief but also contributes to building resilient and compassionate communities. In essence, the case study of Sikh Volunteers sheds light on the vital connection between humanitarian principles and volunteer-driven actions. It illustrates their collective effectiveness in responding to disasters and fostering a culture of service within different communities.

Acknowledgments

We would like to acknowledge the College of Liberal Arts and Human Sciences at Virginia Tech and Dr. Farida Jalalzai for providing us with a Global Partnership Project Grant and the School of Civil and Environmental Engineering at the University of New South Wales for their support to conduct this research.

References

Alexander, D. 'The Voluntary Sector in Emergency Response and Civil Protection: Review and Recommendations'. International Journal of Emergency Management, vol. 7, no. 2, Inderscience Publishers, 2010, p. 151, https://doi.org10.1504/ijem.2010.033654.

Alexander, D. E. 'Social Media in Disaster Risk Reduction and Crisis Management'. Science and Engineering Ethics, vol. 20, no. 3, Springer Science and Business Media LLC, Sept. 2014, pp. 717–733, https://doi.org10.1007/s11948-013-9502-z.

Andreev, A., and Svetlana, A. 'Humanitarian Approach as a Scientific Problem'. Applied Psychology and Pedagogy, vol. 5, no. 3, Infra-M Academic Publishing House, July 2020, pp. 28–38, https://doi.org10.12737/2500-0543-2020-28-38.

Baniya, S. "Transnational Assemblages in Disaster Response: Networked Communities, Technologies, and Coalitional Actions During Global Disasters." Technical Communication Quarterly, Jan. 2022, DOI.org (Crossref), https://doi.org/10.1080/10572252.2022.2034973.

Barrera, G., and Yang, M. 'Evaluation of Digital Volunteers Using a Design Approach: Preferences and Motivations in Disaster Response'. Information Systems for Crisis Response and Management, ISCRAM, vol. 2019, 2019, pp. 906–915.

Bornstein, E., and Redfield, P. Forces of Compassion: Humanitarianism between Ethics and Politics, Sar Press, 2011.

Bosch, M. P. A. D. 'Use of Social Media and E-Government in Disasters: 2016 Louisiana Floods Case Study'. Journal of Emergency Management (Weston, Mass.), vol. 15, no. 6, Weston Medical Publishing, Nov. 2017, pp. 391–405, https://doi.org10.5055/jem.2017.0347.

Cutter, S. L., Barnes, L., Berry, M., Burton, C., Evans, E., Tate, E., Evans, E., and Webb, J. 'A Place-Based Model for Understanding Community Resilience to Natural Disasters'. Global Environmental Change: Human and Policy Dimensions, vol. 18, no. 4, Elsevier BV, Oct. 2008, pp. 598–606, https://doi.org10.1016/j.gloenvcha.2008.07.013.

Ehnis, C., and Bunker, D. 'Social Media in Disaster Response: Queensland Police Service – Public Engagement during the 2011 Floods'. ACIS 2012: Proceedings of the 23rd Australasian Conference on Information Systems, 2012.

Fulmer, T., Portelli, I., Foltin, G. L., Zimmerman, R., Chachkes, E., and Goldfrank, L. R. 'Organization-Based Incident Management: Developing a Disaster Volunteer Role on a University Campus'. Disaster Management & Response: DMR: An Official Publication of the Emergency Nurses Association, vol. 5, no. 3, Elsevier BV, July 2007, pp. 74–81, https://doi.org10.1016/j.dmr.2007.06.001.
Gissing, A., Keys, C., and Opper, S 'Towards Resilience Against Flood Risks'. The Australian Journal of Emergency Management, vol. 25, no. 2, April 2010, pp. 39–45.
Hamilton, R. M. 'Science and Technology for Natural Disaster Reduction'. Natural Hazards Review, vol. 1, no. 1, American Society of Civil Engineers (ASCE), Feb. 2000, pp. 56–60, https://doi.org10.1061/(asce)1527-6988(2000)1:1(56).
Handbook for the United Nations Resident and Humanitarian Coordinator (2019). Retrieved February 6, 2024, from https://rchc-handbook.unocha.org/chapter-a.html
"Humanitarian Principles." UNHCR, 22 July 2019, https://emergency.unhcr.org/protection/protection-principles/humanitarian-principles.
Inter-Agency Standing Committee (IASC). Leadership in Humanitarian Action: Handbook for the UN Resident and Humanitarian Coordinator, United Nations Office for the Coordination of Humanitarian Affairs, 2019.
Joseph, J. K., et al. 'Community Resilience Mechanism in an Unexpected Extreme Weather Event: An Analysis of the Kerala Floods of 2018, India'. International Journal of Disaster Risk Reduction: IJDRR, vol. 49, no. 101741, Elsevier BV, Oct. 2020, p. 101741, https://doi.org10.1016/j.ijdrr.2020.101741.
Kankanamge, N., Yigitcanlar, T., Goonetilleke, A., and Kamruzzaman, M. 'Can Volunteer Crowdsourcing Reduce Disaster Risk? A Systematic Review of the Literature'. International Journal of Disaster Risk Reduction: IJDRR, vol. 35, no. 101097, Elsevier BV, Apr. 2019, p. 101097, https://doi.org10.1016/j.ijdrr.2019.101097.
Khairuddin, I. E., Uzir, N. A., Zaini, M. K., and Ghazali, A. K. 'Decentralized Distribution of Humanitarian Aid for Natural Disaster Relief'. Environment-Behaviour Proceedings Journal, vol. 7, no. SI10, e-IPH Ltd., Nov. 2022, pp. 233–239, https://doi.org10.21834/ebpj.v7isi10.4127.
Lai, C., She, B., and Ye, X., "Unpacking the Network Processes and Outcomes of Online and Offline Humanitarian Collaboration." Communication Research, vol. 46, no. 1, SAGE Journals, Feb. 2019, pp. 88–116, https://doi.org/10.1177/0093650215616862.
Millora, C. 21st Century Volunteering Practices. 6 July 2020, https://www.unv.org/publications/21st-century-volunteering-practices.
Muniz-Rodriguez, K., Ofori, S. K., Bayliss, L. C., Schwind, J. S., Diallo, K., Liu, M., Yin, J., Chowell, G., and Fung, I. C. 'Social Media Use in Emergency Response to Natural Disasters: A Systematic Review with a Public Health Perspective'. Disaster Medicine and Public Health Preparedness, vol. 14, no. 1, Cambridge University Press (CUP), Feb. 2020, pp. 139–149, https://doi.org10.1017/dmp.2020.3.
Park, C. H., and Johnston, E. 'Determinants of Collaboration between Digital Volunteer Networks and Formal Response Organizations in Catastrophic Disasters'. International Journal of Organization Theory and Behavior, vol. 22, no. 2, Emerald, May 2019, pp. 155–173, https://doi.org10.1108/ijotb-07-2018-0088.
Pörtner, H., Roberts, D. C., Tignor, M. M. B., Poloczanska, E. S., Mintenbeck, K., Algeria, A., Craig, M., Langsdorf, S., Löschke, S., Möller, V., Okem, A., Rama, B. Climate Change 2022: Impacts, Adaptation and Vulnerability. Contribution of Working Group II to the Sixth Assessment Report of the Intergovernmental Panel on Climate Change, 2022.
Radianti, J., and Gjøsæter, T. 'Digital Volunteers in Disaster Response: Accessibility Challenges'. Universal Access in Human-Computer Interaction. Multimodality and Assistive Environments, Springer International Publishing, 2019, pp. 523–537, https://doi.org10.1007/978-3-030-23563-5_42. Lecture Notes in Computer Science.
Radice, H. 'Humanitarianism as Civic Practice? Humanity, Politics and Humanitarian Activism'. Journal of Civil Society, vol. 18, no. 2, Informa UK Limited, Apr. 2022, pp. 142–160, https://doi.org10.1080/17448689.2022.2121296.
Roberts, F., Archer, F., and Spencer, C. 'The Potential Role of Nonprofit Organisations in Building Community Resilience to Disasters in the Context of Victoria, Australia'. International Journal of Disaster Risk Reduction: IJDRR, vol. 65, no. 102530, Elsevier BV, Nov. 2021, p. 102530, https://doi.org10.1016/j.ijdrr.2021.102530.
Schnepf, J. D. 'Flood from Above: Disaster Mediation and Drone Humanitarianism'. Media+Environment, vol. 2, no. 1, University of California Press, July 2020, https://doi.org10.1525/001c.13466.
Sikh Volunteers Australia Inc. (n.d.). Retrieved February 6, 2024, from https://sikhvolunteersaustralia.org/

Tester, K. 'Humanitarianism'. International Journal of Cultural Studies, vol. 13, no. 4, SAGE Publications, July 2010, pp. 375–389, https://doi.org10.1177/1367877910374787.
Ticktin, M. "Transnational Humanitarianism." Annual Review of Anthropology, vol. 43, no. 1, 2014, pp. 273–89. Annual Reviews, https://doi.org/10.1146/annurev-anthro-102313-030403.
Twigg, J., and Mosel, I. 'Emergent Groups and Spontaneous Volunteers in Urban Disaster Response'. Environment and Urbanization, vol. 29, no. 2, SAGE Publications, Oct. 2017, pp. 443–458, https://doi.org10.1177/0956247817721413.
Ummenhofer, C. C., Sen Gupta, A., England, M. H., Taschetto, A. S., Briggs, P. R., and Raupach, M. R. 'How Did Ocean Warming Affect Australian Rainfall Extremes during the 2010/2011 La Niña Event?' Geophysical Research Letters, vol. 42, no. 22, 2015, pp. 9942–9951.
Wachira, J. K., and Sinclair, A. J. 'Public Participation in the Emergency Response Phase of Flooding: A Case Study of the Red River Basin'. Canadian Water Resources Journal/Revue Canadienne Des Ressources Hydriques, vol. 30, no. 2, Informa UK Limited, Jan. 2005, pp. 145–158, https://doi.org10.4296/cwrj3002145.
Wang, N. '"We Live on Hope…": Ethical Considerations of Humanitarian Use of Drones in Post-Disaster Nepal'. IEEE Technology and Society Magazine, vol. 39, no. 3, Institute of Electrical and Electronics Engineers (IEEE), Sept. 2020, pp. 76–85, https://doi.org10.1109/mts.2020.3012332.
Yumagulova, L., Phibbs, S., Kenney, C. M., Woman-Murno, D. Y. O., Christianson, A. C., McGee, T. K., and Whitehair, R. 'The Role of Disaster Volunteering in Indigenous Communities'. Environmental Hazards, vol. 20, no. 1, Informa UK Limited, Jan. 2021, pp. 45–62, https://doi.org10.1080/17477891.2019.1657791.

Section IV

Disaster Response Issues and Visions

14

BIASES AND VULNERABLE POPULATIONS IN RELIEF DISTRIBUTION

Emily Kaplan, Chandra Pauline Daniel, and Gifty Immanuel

Introduction to Relief Distribution Biases

Allocation of relief in emergent situations can be impacted by biases. Humanitarian aid disbursement may be affected by two types of biases, namely, extrinsic and intrinsic. Extrinsic biases are attitudes, preferences, and beliefs of which people are aware (Vela et al., 2022). Implicit biases are unconscious mental processes that affect decision-making beyond deliberate thinking and reflection. Such unconscious biases are difficult to recognize and address, as noted by the late U.S. Supreme Court Justice Ginsburg (Weisberg, 2020). Unconscious biases include set, preconceived ideas about what something is like (e.g., stereotypes). As biases disproportionately affect actions and outcomes, 'bias literacy' is essential when formulating a just relief structure. Insight into equitable decision-making processes is therefore practical and pertinent to disaster relief providers.

Bias management is recognized and integrated in standards, policy, and legislation, including disaster response. However, complete eradication of biases in decision-making remains elusive. During disaster response, it is functionally necessary to use a utilitarian approach, to do the most good for the most people, while ensuring vulnerable populations are protected and prioritized. The Americans with Disabilities Act (ADA) and the Rehabilitation Act provide examples of protective legislation for vulnerable populations that have been affected by intentional or unintentional biases. In recent years, biases have been demonstrated to influence decision-making among administrators (Zaksek and Arvai, 2004; Arvai et al., 2006), emergency management administrators (Roberts and Wernstedt), and health care providers serving vulnerable populations (Vela et al., 2022). There are many intersections among the varied roles and capacities of distribution relief providers. For scope and purposes of this chapter, each is simply referred to as a relief provider. In addressing biases, relief providers can reduce inequities in distribution and help achieve important global sustainable development goals.

Vulnerable Populations

A vulnerable population refers to the reduced capacity of a group of individuals sharing a feature to resist the effects of, or adequately respond to, a disaster. These populations require additional considerations in relief distribution as vulnerable groups are the most exposed to harm and/or are

DOI: 10.4324/9781003414834-20

at higher risk of experiencing harm. Harm during relief may occur via the direct and/or indirect actions of providers as well as institutionally or systemically through the design of supply chains, service delivery, and products. A particular group may be more or less vulnerable under differing circumstances. Children, for example, are quite vulnerable during armed conflict due to loss of family or homes but are comparatively less vulnerable during periods of peace. Supply chains disrupted by disaster, such as for pharmaceuticals, may disproportionally result in harm to vulnerable populations. Appropriate products may assist vulnerable populations, such as the provision of ramps for the less mobile at housing locations, and withholding of the same may make relief resources inaccessible.

As a critical part of sustainable relief distribution, it is essential to identify potentially vulnerable groups and classify their challenges, thereby empowering relief providers on potential provider-biases in relief distribution. Relief providers may identify vulnerable groups in multiple ways, such as those described by the U.S. Centers for Disease Control and Prevention categorized by (1) socioeconomic status (SES), (2) age, (3) gender, (4) medical issues, functional access needs, and disabilities, (5) race and ethnicity, and (6) limited language proficiency (Centers for Disease Control (CDC) and Prevention, 2015). Another example of vulnerable populations by UNAIDS (2014a,b) identifies 12 categories, including HIV positive, displaced, transgender, prison, drug user, and sex worker populations.

Vulnerable Populations in Disaster Relief

Vulnerable population considerations should be integrated in each phase of disaster relief distribution, including preparedness, response, mitigation, and recovery. Those considerations should carry through monitoring, validating information, and actions. The needs of vulnerable populations must be considered in each aspect of relief distribution, including but not limited to feeding, sheltering, distribution of emergency supplies, and reunification services. Vulnerable populations include the non-exhaustive list of above categories and the consideration of specific needs such as service animals, unique dietary needs, medical considerations, and functional access needs such as for sheltering (e.g., wheelchair accommodations and limited technological access). The detrimental effects of lower SES and low access to social determinants of health (SDH), such as education and being part of a vulnerable population, add double jeopardy through biases, norms, and social relations, all of which increase the vulnerability of a population.

Social vulnerability may be assessed through a variety of measures, including the Social Vulnerability Index (SoVI). Indices such as the SoVI can capture location-based socioeconomic and demographic measures of a population. Measures may include, for example, the percentage of people in poverty, households with single parents, those who are uninsured, and households with individuals with a disability. Social vulnerability measures can be adapted for country-specific use. These measures are used in conjunction with hazard vulnerability analysis to identify geographic areas requiring dedicated attention by emergency responders. Inclusion of social vulnerability indices in modeling extends the capacity of hazard vulnerability analyses beyond place and situational-based analyses (e.g., models based solely on likelihood of floods, fires, or tsunamis).

Inclusion of SoVI measures adapted for country-specific needs (e.g., built environment, access to medical services, and language barriers) have been applied to identify the social vulnerability of populations at various scales. Modeling social vulnerability may yield insights not apparent when modeling includes only limited data such as meteorological or topographic considerations.

Populations in an area less prone to earthquakes compared to other areas in a country may be counterintuitively more socially vulnerable in the face of an earthquake due to lack of earthquake-resistant infrastructure (Aksha et al., 2019).

Various other location-specific considerations should be considered when providing relief to vulnerable populations. In many countries, a gender-disparity exists in access to food, health services, employment, and education, which is often driven by political or social ideology. In many instances, females are therefore more vulnerable to the effects of disasters than males and therefore should be treated with additional consideration in disaster response. In Bangladesh, the interplay among the physical environment, frequent natural disasters, and gender norms is evident in the comparatively increased vulnerability of women (Juran and Trivedi, 2015) as well as among Rohingya refugee populations (Nasar et al., 2022). A variety of measures of gender disparity may be used, such as the gender inequality index (GII), gender disparity index (GDI), gender equality measure (GEM), and the Gender Status Index (GSI). Expertise should be sought in the adoption of indices, as they are often either tailored for specific geographic regions or may be 'off-the-shelf' in nature and thus require geographic and cultural contextualization.

Discrimination

An overt outcome of any type of bias may be discrimination toward vulnerable populations. Discrimination is typically considered as treatment of a group that is unjust. Discrimination toward groups should be considered along four dimensions: practical discrimination, formal discrimination, societal or implicit discrimination, and hidden groups. Figure 14.1 portrays the broad spectrum of discrimination resulting from extrinsic and intrinsic biases toward vulnerable populations.

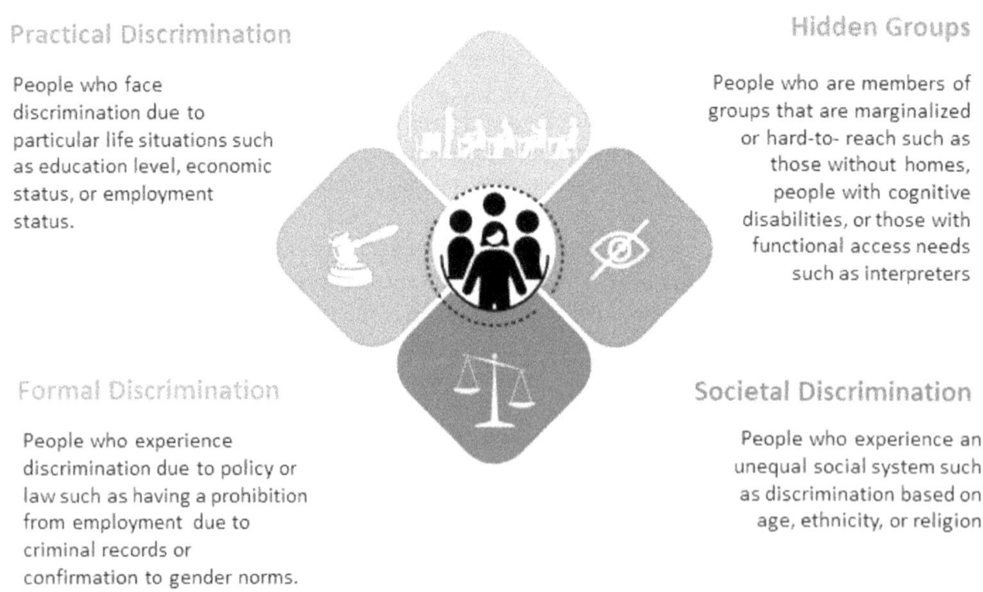

Figure 14.1 Discrimination Categories of Vulnerable Groups

Source: Image from authors. Adapted from Scott, Kelly, Pires, and Bennici (2023). Adapted with permission.

The 'Double Jeopardy Effect'

Whether acting unknowingly or knowingly, any type of bias may cause inefficient or absent relief distribution. When biases are introduced into systems, vulnerable populations are in a position of increased jeopardy. The population is negatively affected by the intersection of the disaster, then subsequently by biases toward the group. As a result, the effects of inefficient or absent relief are amplified and concentrated in the population already struggling.

Risk

Relief providers work in environments filled with risk. With every consideration and subsequent decision, unconscious or conscious, there is corresponding risk. How do we know what we know about the world around us? Is what we think we know actually the case? How do we know what we do not know? Risk perception is egocentric; it is our own minds making sense of our world. Not only do relief providers respond to perceived risk, so do the populations they serve. We must examine ourselves and our decisions to ensure we are making as sound, bias-free judgments as possible. Being aware of our biases and mitigating them potentially eliminate the reinforcement of vulnerable populations and discrepancies in relief distribution.

Types of Biases

Decision-making is often thought to be the result of a rational process based on available information and preferential arrangements of choices (Kahneman, 2011). However, it is well demonstrated that people engage in decision-making by relying on cognitive biases. Kahneman (1997) and Kahneman and Tversky (1979) describe how humans are biologically wired to use different pathways when thinking 'Fast' (System 1) or 'Slow' (System 2). When 'thinking fast,' people use heuristics, learned patterns, rules, or mental maps to quickly arrive at a solution. Khanmen and Tversky (1979) developed 'prospect theory,' a model demonstrating that decisions are made in relationship to an individual's unique perspective, including the role of biases, at a particular point in time. Unintentional and automatic biases are termed cognitive biases. Cognitive biases occur as a result of unconsciously learned patterns over a lifetime. Cognitive biases are systems of thinking humans follow but do not align with their logical reasoning or assessments of plausible scenarios. Again, humans are biologically wired to use heuristics to quickly arrive at a solution; heuristics is favored by the brain when an individual encounters high-risk, high-stress, or information-incomplete environments. A ready example of mental map use is the Stroop effect. There is a delay in the time it takes the mind to process a scenario that agrees with itself than when a scenario does not align with learned patterns. For example, try reading the word 'green' when printed in the color green versus red. It takes longer for most people to read the word green when it is printed in the color red than when it is printed in the color green (Scarpina and Tagini, 2017).

Presented in Table 14.1 is a non-exhaustive list of biases, examples for providers, and remedies to combat each respective bias. Remediation measures that are effective in mitigating one type of bias are often effective in the mitigation of the effects of several other types of biases. In describing remediation toward types of bias, a 'whole community' approach is encouraged throughout the disaster management process. Encouraging the local people and businesses affected by disasters makes everyone a partner in relief distribution, self-governance, and self-protection. Whole community approaches are more likely to effectively engage a wider range of resources, which results in affected populations reporting a more positive experience throughout disaster response and recovery (see Table 14.1).

Table 14.1 Examples of Common Biases and Potential Methods of Remediation

Confirmation bias
Definition: Occurs when people accept information that is consistent with and strengthens their own attitudes toward a specific subject (i.e., conforming to existing norms). Emergency managers and responders may select to evaluate only data confirming their own beliefs, resulting in a skewed picture of a situation. **Example:** The plan for typical seasonal flu management if extrapolated for a novel virus will not be effective. However, confirmation bias leads one to expect that if it worked once, it would work again. **Remediation:** Educate and reinforce education for donor(s), relief provider(s), or recipient population(s) on pertinent novel information. Continually reassess the situation.
Fallacy of planning bias
Definition: Underestimating the risk, time, and costs and disregarding the lived experience of risk, time, and costs in completing a prior similar task. **Example:** Relief providers believe because they have planned, practiced a plan, and thought through every possible iteration (e.g., flu pandemic planning) that the plan will be effective when executed with similar time and costs (e.g., responding to the novel COVID-19 virus). **Remediation:** Routinely revisit plans and recognize that all plans must be flexible and change given the circumstances.
Expectation bias
Definition: An individual accepts data or information that is congruent with what they believe or anticipate. **Example:** The Federal Emergency Management Agency uses a placeholder estimate that less than 10% of the population would seek housing and shelter absent additional information; however, vulnerable populations may disproportionally seek shelter compared to the general population depending on the disaster situation. If the initial assessment reveals that 4% of a population would seek shelter and the decision maker is advised 4% of the vulnerable population would therefore seek shelter, the information would be accepted as congruent with traditional estimates. However, the acceptance of congruent information may result in overlooking the type of disaster and thus result in underestimating the number of individuals in a vulnerable population likely to seek shelter. During a power failure, for example, persons dependent on power for breathing equipment may preferentially seek shelter, which may be overlooked due to expectation bias. **Remediation:** Apply an objective and unbiased approach in collecting and analyzing data. Include multiple sources of reliable data.
Normalcy bias
Definition: Predisposes individuals to discard information of minor abnormalities as normal. **Example:** Notification of an earthquake in an earthquake prone area may result in failure to take precautions or not selecting to evacuate, presuming the likelihood of a severe earthquake to be lower. **Remediation:** Acknowledge disasters will occur and have contingency plans. Include public stakeholders in planning emergency measures. Issue unambiguous and clear warnings from public officials and individuals respected in communities (Frandsen and Johansen, 2020).
Risk perception bias
Definition: One's tendency to believe in their own perspective of events during emergency and disaster. Risk perception is egocentric in nature. **Example:** While communicating during disaster relief, responders and planners tend to believe that everyone understands what they know. The severity of a biological incident may be perceived as severe by providers who presume members of the community perceive the severity to be the same. **Remediation:** Use simple language to communicate facts. Consider alternatives and seek community feedback. Use models or algorithms to assist in decision-making (Kahneman et al., 2019).

(*Continued*)

Table 14.1 (Continued)

Errors of omission and commission
Definition: *Errors of omission* occur when inaction, and its attendant results, is perceived as less serious than action. Conversely, in an *error of commission*, an emergency manager may choose to act to minimize disturbances in relief distribution and response and to incur any associated costs, be it economic, social capital, or loss of political will. • **Type I:** False-positive occur when an event is incorrectly concluded to be occurring. • **Type II:** False-negative conclusions are when the situation is predicted but does not occur. **Example:** During Hurricane Floyd, evacuation errors were issued across Florida due to the predicted path of the storm. The storm spared Florida while affecting omitted areas such as much of South Carolina, which subsequently issued evacuation orders. A total of 3.5 million evacuees caused interstate traffic gridlock, business, and school closures. Households with a physically disabled family member were less likely to evacuate due to perceived lack of access to services and functional access needs during the evacuation. These households also experienced greater costs due to storm damage (Frieden, 2006). **Remediation:** Obtain an accurate estimation of the likelihood of an imminent event. Engage in post-interpretation review to check for errors and misinterpretation. Use residual analysis to test the difference between a model and the data from an actual event.
Anchoring bias
Definition: Tendency to prioritize the first piece of information and evaluate subsequent information against the first information received. **Example:** The initial arbitrary overview of an outbreak (stomach bug), if initially perceived to be mild, will continue to be viewed as mild (by the health care provider) even when evidence is on the contrary (severe dehydration). **Remediation:** Engage in constant re-evaluation and updating information. Utilize real-time data and modify interventional strategies accordingly.
Recency bias
Definition: Tendency to overemphasize the importance of recent experiences or the latest information possessed when estimating future events. Recency bias misleads providers to expect that the most plausible scenarios in predicting future events are directly related to the most recent disaster experienced. **Example:** In earthquake- and tsunami-prone areas, such as Japan, tsunamis may be perceived to be the most likely disaster for a period of time after a tsunami disaster. In 2011, the Tōhoku Tsunami resulted in not only a tsunami and flooding disaster but also a nuclear disaster due to an electrical failure at the Fukushima Daiichi nuclear power plant, increasing bias that the next most likely disaster would be a tsunami. **Remediation:** Engage in consistent evaluation of the population dynamics (Population Situation Analysis), update information prior to relief distribution, and evaluate changing threats over time.
Availability of heuristics
Definition: In the absence of complete information and when under stress, the brain favors the use of known patterns to create actionable information (Kahneman, 2011; Kahneman et al., 2019). The mind may skip over, or fill in, information due to a lack of complete information or consideration, which may result in a decision-maker inadvertently rendering gaps in services delivered. **Example:** In quickly estimating population size for relief, providers may use census information and fail to consider 'hidden populations' such as those without homes and not counted in a standard census. **Remediation:** Review the relevant information, avoid underestimation, use multiple data sources, avoid decision-making based on superficially available information, and use decision-making algorithm aids (Kahneman and Tversky, 1979; Kahneman, 2011; Kahneman et al., 2019).

(*Continued*)

Table 14.1 (Continued)

Overconfidence effect
Definition: The overreliance and credible belief in one's own information/responses without any validation based on the outcomes; it is an epistemic vice. **Example:** Emergency managers are routinely asked for information and must often act on limited information. They believe and have confidence in the accuracy and efficacy of the information they provide. The Emergency Manager may be overly confident 10,000 homes are affected by power loss. Subsequently, it is determined with the given information that only 8,000 homes are affected, though the manager strongly believed their 'call' was correct. **Remediation:** Test data gathering systems to ensure, as best possible, valid, reliable data is being provided. Ask others for their evaluation of the data at hand. Consider one's level of confidence with a sense of humility.
Political bias
Definition: The altering of information or actions to be guided by a political position or those guided by political accommodations. **Example:** Elected officials may pressure providers to distribute relief preferentially by constituency or subgroups of constituencies served (Roberts and Wernstedt, 2019). Providers may then prioritize relief distribution for an elected official's home constituencies. **Remediation:** Diversify aid distribution augmented by seeking professional advice. Avoid concentration of risk where any single political leader or entity maintains fiscal or operational control. Be prepared to explain to political leadership one's decisions pertaining to relief distribution.
Home bias
Definition: 'Home country' bias or regional favoritism. Preferential investment in domestic aid rather than diversifying foreign aid and reducing impartiality in aid distribution. **Example:** Following a disaster, people and those leading aid relief may preferentially send money to their countries or leaders of their birth country or ethnicity. In an opposing viewpoint, leaders may choose to refuse aid from those perceived as 'foreign' due to potential fears as appearing, or being, subordinate to the country providing aid. **Remediation:** Increase memorandums of understanding between regions and countries and avoid clustering of aid flow within a geographic area.
Social media bias
Definition: Social media platforms can abuse and distort information on an ongoing disaster, which can have profound effects on disaster relief outcomes. **Example:** Inequitable access to online platforms (digital divide), social influences, and social networks can shape human behavior, modify information in disaster response, and negatively affect outcomes. **Remediation:** Creating awareness on the prevalence of abuse and distortion of messages on social media platforms and technological devices. Teaching social media literacy and information evaluation. Assessment of technological access by populations.

Source: Table Created by authors.

Both providers and recipients of relief act according to biases. People perceiving risk to be high are more likely to act in a risk-averse manner. When providing relief, understanding risk promotes positive actions by individuals engaged in relief endeavors and the populations they serve. For example, those with a higher perception of risk, such as persons aged over 65 in response to COVID-19, were more likely to seek vaccination (Dawood et al., 2022). When public planners issue hurricane evacuation warnings, the public is more likely to evacuate the impact area if perceived risk is higher. Emergency assistance and disaster relief personnel also triage relief services according to perceived risk. To appropriately assess and communicate risk to vulnerable populations, it is necessary to assess the population from more than one viewpoint.

Bias Mitigation and System Design

Vela et al. (2022) offer a conceptual model within the health care system that is applicable to relief distribution. The model demonstrates that provider-level implicit bias interventions should be accompanied by interventions that systematically change structures inside and outside of relief distribution. A variety of activities and structural changes may be implemented to mitigate the impact of biases, and these changes can be engineered into system design in order to mitigate the impact of biases.

Instructional Interventions and System Design

Current interventions include instruction on the existence and harmful role of bias in perpetuating health disparities, as well as skills training for the management of bias. These interventions can raise awareness of biases and engage relief providers, including health care providers, in establishing egalitarian goals for care delivery and relief distribution. However, instruction alone is not demonstrated to create substantial change in behavior in the clinical, emergency management, or learning environments (Vela et al., 2022). Unfortunately, the effectiveness of these interventions may be hampered by providers' work and social environments, which are rife with discriminatory practices that sustain the very biases providers are seeking to diminish. Similarly, due to the intrinsic nature of heuristic and risk-perception biases, skills and training may only minimally reduce acting on biases. A holistic approach to bias mitigation therefore should include instructional interventions *as well as* a strong emphasis on system engineering and design.

Policy, Best Practices, and Other Considerations in System Design

Policy is an imperative in relief distribution as it provides (binding) expectations, roles, and protocols to reduce the inclusion of biases in relief distribution and ensure inclusion on task forces and working groups (i.e., mainstreaming). During disaster response and relief provision, mass care and emergency assistance task forces charged with providing relief should include a specific working group whose members ensure that the needs of vulnerable populations are met. In the United States, Emergency Support Function (ESF) #6—Mass Care, Emergency Assistance, Housing and Human Services, is coordinated at the federal level by the Federal Emergency Management Agency (FEMA). FEMA and the American Red Cross are the primary agencies responsible for execution and are supported by multiple federal agencies and non-governmental organizations (NGOs). ESF #6 states that functional needs of special populations must be addressed and not overlooked (Federal Emergency Management Agency [FEMA], 2016). Systems design can also employ an itemized approach, such as checklists, to ensure cognitive biases do not incidentally result by overlooking vulnerable populations, care, or associated tasks in relief provision (Gawande, 2010). Standards in relief distribution and humanitarian aid are promulgated by both governments and NGOs. A commonly used model for humanitarian aid standards is provided by The Humanitarian Standards Partnership, which publishes the Sphere Handbook. Sphere includes a commonly adopted humanitarian charter, principles of protection, and considerations for categories of aid provision. The charter recognizes rights to life with dignity, to humanitarian aid, and to protection and security inclusive of vulnerable populations (Sphere Association, 2018). Using effective standardized approaches, such as Sphere, leads to better outcomes with less variation.

Task Forces and Working Groups can help reduce the incidental inclusion of biases in relief distribution, especially among vulnerable populations. During disaster response and relief provision, mass care and emergency assistance task forces and working groups should be created and charged with providing relief according to ethical and equitable standards. Again, in the United States, ESF #6 ensures, via policy, that functional needs of special populations are addressed. This model is replicated at state and local levels during disaster response, and similar models are utilized in the international setting.

Liaisons dedicated to populations with special needs are provided to incident management assistance teams, such as those responsible for sheltering and feeding. Liaisons may also be embedded in mobile distribution teams to ensure vulnerable population needs are represented and met. Liaisons reduce the introduction of bias into relief response and can work across working groups and at different scales of relief distribution. Similarly, representatives from a broad range of public agencies and service providers that interface with vulnerable populations should be included in relief activities, such as planning, tabletop, and functional exercises.

Social Norms and societal beliefs, including biases toward vulnerable populations, may obstruct relief distribution. This is due to the institutionalization of social biases within political, educational, and economic systems (Webster et al., 2022), for example, through policies that favor or disfavor particular groups. This can be assessed through the comparison of measures, such as the Gender Social Norms Index (GSNI), against efforts to address gender-based violence and the provision of health care services (Daniel, 2023). Deliberative efforts to reduce social bias and discrimination can be deployed via modeling and queues, social pressure, and policy in order to affect systemic changes.

Mental Health deserves a special note as there often exists bias toward and stigmatization of those with reduced mental health or mental health diseases. Psychological first aid is integral when responding to and fostering the resilience of affected individuals, particularly among vulnerable populations. Children who experience disasters, for example, are at increased risk for mental health disorders over a lifetime (McFarlane and Van Hoof, 2009). Policy and practice must include provision of mental health and trauma-competent caregiving in planning, response, and recovery. In the United States, a variety of resources are available to relief providers through state (e.g., offices of mental health), federal (e.g., crisis counseling training programs), and local and community-based agencies and businesses for crisis counseling. International mental health relief providers (e.g., StrongMinds Mental Health Africa and Doctors Without Borders) may liaise with task forces to provide services focused on, and with special expertise in, vulnerable populations.

Engineering Informed Systems

Preparedness and Community Needs Assessments

Preparedness activities include community needs assessments of vulnerable populations. Community needs assessments collect data and engage a wide range of stakeholders in the community to understand gaps. In the context of vulnerable populations, stakeholders representing vulnerable populations, or the organizations that serve them, should be included. Widely sharing the information, needs, and best practices uncovered from needs assessments presents an opportunity to reduce bias and improve perspectives in disaster planning.

Community Partner Inclusion

Federal, state, local, as well as national and international voluntary organizations with expertise in vulnerable populations are encouraged to participate during all phases of disaster relief. In the United States, examples include having public agencies, such as Offices of the Aging, Public Health Departments, Health and Human Services, Departments of Homeless Services, Departments of Community Affairs, and Offices of Disability Services, along with The ARC (the largest national community organization serving individuals with disabilities), share information and network during preparedness planning.

Including community partners provides opportunities to ensure the wide variety of vulnerable population needs are met. Listening sessions and in-person events allow community partners to ask questions and share concerns. Public sessions must be held in multiple formats with appropriate accommodations for the vulnerable population, such as ADA compliant facilities, the provision of translation services, and hosting sessions in non-threatening environments. It is always important to note that community partners may provide information not otherwise considered, stressing the importance of listening and learning during the process. In the United States, the American Red Cross is tasked with the authority to provide disaster relief and has expertise in serving a variety of vulnerable populations, including children and the elderly. Inclusion of agencies such as the U.S. Public Health Service may expand services, and service considerations, for vulnerable populations such as sexual and reproductive health services. Agencies ensuring security of the public are also included at the federal, state, and local levels with expertise in ensuring safety of vulnerable populations such as the Federal Bureau of Investigation's Crimes Against Children and Human Trafficking Unit. Internationally, agencies with expertise in vulnerable populations should be sought out such as the United Nations Population Fund (UNFPA) with expertise in care for women and mental health care support, Joint United Nations Program on HIV/AIDS, and United Nations Children Fund (UNICEF), along with their local chapters.

No one is immune to bias. Even subject matter experts have biases, and, in fact, they favor different decision-making patterns and use of heuristics more than novices (Chi et al., 1981). Experts may be more comfortable relying on limited data as their experience informs their next steps as part of a mental model in which they do not deliberate over each individual step and consideration. They may even be more overconfident in their abilities and have a higher risk tolerance. While this underscores the significance that experts too must be cognizant of their thoughts and actions, it should not negate the use of expert knowledge to support vulnerable populations.

Leveraging Technology for Bias Mitigation

Valid and Reliable Data

Valid and reliable data is necessary to make informed decisions. In disaster relief activities, data availability is frequently restricted, patchwork, and arrives in uneven flows. It is necessary to ensure available data is inclusive of vulnerable populations and their needs to reduce the introduction of biases. This holds during obtaining, organizing, analyzing, and mentally processing the data. For example, housing provides a ready example of valid data that may result in the introduction of bias. The U.S. census reports include the number of residences and individuals residing in a residence in a particular area. This is valid, accurate data. However, it may not be reliable in disaster relief when attempting to measure the actual number of individuals residing in a particular area. Several vulnerable populations are hard-to-count (HTC) and therefore may be hidden. Undocumented immigrants, people not living in traditional housing, elderly individuals living with relatives or alone, and young children are often undercounted (Stempowski, 2023). Appropriate data collection techniques unique to each vulnerable population must be used. For example, while admittedly imperfect, methods such as counting those in nontraditional housing via 'midnight counts' of individuals living in shelters and other areas can help to account for homeless, transient, and other 'invisible' populations. Vulnerable population subject matter experts should be consulted for appropriate data capture techniques. Statistical modeling and analysis may also be considered in light of both validity and reliability (Daoud and Dubhashi, 2023).

Computer Modeling

Many modeling and simulation systems are used in all phases of disaster response, of which two are presented below. Geographic information system (GIS) science may be leveraged to map and model the needs, locations, and clusters of vulnerable populations (Aksha et al., 2019). This may be predictive in nature, or real-time. Tools such as ArcGIS and QGIS are capable of providing real-time visuals of disaster infrastructure, which can also be considered in access and distribution of relief. These tools also can assess proximity to health care or other services as well as evacuation routes and thus should be taken into consideration in scenario modeling (Bazargani et al., 2022). Such models are normative and can assist in making trade-off and location-allocation decisions. Other forms of modeling include agent-based models (ABMs), which simulate interactions among people, things, places, and time—including considerations of vulnerable populations. Agents are people who engage in behaviors producing mass quantities of micro-data. ABM can therefore be used to assess how vulnerable populations are self-organizing and behaviors that *a priori* were hidden among small, hard-to-reach groups (Macal and North, 2005). Spikes in pharmaceutical purchases (i.e., a behavior) and increased interactions with pharmacy staff (i.e., relationships) ahead of the notice of an event may portend to a trend of fear among medically vulnerable individuals or 'agents.' Emergency responders may use this information to guide further analysis, identify affected populations, and tailor interventions. These types of interactions are increasingly able to be tracked and predicted. While imperfect and always subject to human errors, enhanced and accurate modeling capacity reduces information in incomplete environments and can be employed to avert and reduce bias.

Artificial Intelligence and Machine Learning

Increasingly, artificial intelligence (AI), artificial neural networks, and machine learning tools combined with big data and spatial data sciences are increasingly used to improve response in disaster relief. Such tools connect data across systems; however, it is necessary to ensure and test that the system is learning with data inclusive of vulnerable populations. Due to cognitive biases in the programming of machine learning, vulnerable populations may be underrepresented (Kliegr et al., 2021). Comparing known data about vulnerable populations against AI-driven data periodically can help to prevent relief distribution gaps. For example, if social media messages are mapped against disaster-damaged areas, an uptick may be seen and used to inform relief. It is necessary to overlay these maps and data sources against inclusive social vulnerability assessments to ensure vulnerable populations are not masked during relief distribution.

Empowering Individuals and Communities to Act

When biases or attendant discrimination are present, they reduce the ability of a community to care for itself; thus, empowerment is a powerful tool. 'Whole community' approaches include all community members and allow community members the power to act, thereby improving preparedness and resiliency by enabling their voice and already inherent agency to take root. Much of relief distribution is accomplished by neighbors helping neighbors. Providing communities, including vulnerable populations, the means to self-assist during disaster improves relief distribution efforts (Kennedy et al., 2008). This may include recognizing and acting to mitigate biases within a community or toward a different community. If a bias is present, for example, toward a population that is homeless, the community may be less likely to assist the homeless population. Empowering a community (e.g., by reducing biases) may enable that community to both increase its own resilience and ability to engage in assisting vulnerable populations. Community members without bias toward vulnerable groups may be more likely, for example, to provide food, shelter, or assistance to such populations. Reduction of biases toward vulnerable populations is supported through forming community working groups, platforms for shared discussion, and inclusive activities, which provide the opportunity to reduce biases among community members when engaging in relief distribution.

Conclusion

Biases toward vulnerable populations are evident in relief distribution. Inadvertent or intentional inclusion of biases reduces the efficiency, inclusivity, and equity of relief distribution. These biases include:

- Explicit biases: Conscious and intentional expressed attitudes toward a group;
- Implicit biases: Unintentional attitudes toward a group inclusive of cognitive patterns such as risk perception.

To maximize efficient and equitable relief distribution, integrating bias mitigation is imperative in the development of policies, hard and soft distribution systems, creation and actions of working groups, production and use of information systems, and service delivery mechanisms writ large. Acting to reduce the unintended effects of biases is a force multiplier in relief distribution. A wide variety of bias mitigation measures may be adopted, including ensuring data accuracy, educational programs, whole community approaches, and policy changes. Due to the biological imperative to use heuristics along with the social constraints in which we live, biases will not be eradicated, though their attendant influence may be mitigated. Conscious acknowledgement of biases and their roles in decision-making empowers disaster relief responders to ensure no vulnerable population is left behind.

References

Aksha, Sanam K., Luke Juran, and Yang Zhang. "An Analysis of Social Vulnerability to Natural Hazards in Nepal Using a Modified Social Vulnerability Index." *International Journal of Disaster Risk Science*, vol. 10, no. 1, 2019, pp. 103–16, https://doi.org/10.1007/s13753-018-0192-7

Arvai, Joseph, Robin Gergory, Dan Ohlson, Bruce Blackwell, and Robert Gray. "Letdowns, Wake-up Calls, and Constructed Preferences: People's Responses to Fuel and Wildfire Risks." *Journal of Forestry*, vol. 104, no. 4, 2006, pp. 173–81, Supplemental Index, https://doi.org/10.1093/jof/104.4.173

Bazargani, Jalal Safarii, Mostafa Zafar, Abolghasesm Sadeghi-Niaraki, and Soo-Mi Choi. "A Survey of GIS and AR Integration: Applications." *Sustainability*, vol. 14, no. 16, 2022, pp. 10134.

Centers for Disease Control (CDC) and Prevention. *Planning for an Emergency: Strategies for Identifying and Engaging at-Risk Groups; a Guidance Document for Emergency Managers*, first edition. CDC, 2015.

Chi, Michelene T. H., Paul J. Feltovich, and Robert Glaser. "Categorization and Representation of Physics Problems by Experts and Novices." *Cognitive Science*, vol. 5, no. 2, 1981, pp. 121–52.

Daniel, Chandra P. Socio-economic Determinates of Gender Based Violence: SDG Analytics of the Global GBV Scenario with Special Reference to Technology-Facilitated Gender-Based Violence and Adolescent Birth Rates. 2023. New York Medical College, DrPH dissertation.

Daoud, Adel, and Devdatt Dubhashi. "Statistical Modeling: The Three Cultures." *Harvard Data Science Review*, vol. 5, no. 1, 2023. https://doi.org/10.1162/99608f92.89f6fe66

Dawood, Shaimaa Samir, Rasha Salah Eweida, Zohour Ibrahim Rashwan, Maha Mohammed Abd Elmoniem, and Nancy Mahmoud Elsaky. "Older Adults' Perceived Risk Regarding COVID-19 and Intention to Vaccinate: Role of Health-Related Literacy." *Research in Gerontological Nursing*, vol. 15, no. 2, 2022, pp. 57–67. https://doi.org/10.3928/19404921-20220218-02

Federal Emergency Management Agency (FEMA). Emergency Support Function #6 – Mass Care, Emergency Assistance, Temporary Housing and Human Services Annex. 2016.

Frandsen, Finn, and Winnie Johansen. Crisis Communication. vol. 00023, De Gruyter Mouton, 2020. Handbooks of Communication Science.

Frieden, Lex. "The Impact of Hurricanes Katrina and Rita on People with Disabilities: A Look Back and Remaining Challenges." National Council on Disability, 2006. https://files.eric.ed.gov/fulltext/ED496270.pdf. Accessed November 11, 2023.

Gawande, Atul. "The Checklist Manifesto: How to Get Things Right/Atul Gawande." 1st ed., Metropolitan Books, 2010. New York Medical College Library Catalog, https://research.ebsco.com/linkprocessor/plink?id=78bd9eac-e119-320a-ac02-08e11c792a74

Juran, Luke, and Jennifer Trivedi. "Women, Gender Norms, and Natural Disasters in Bangladesh." *Geographical Review*, vol. 105, no. 4, 2015, pp. 601–11. https://doi.org/10.1111/j.1931-0846.2015.12089.x

Kahneman, Daniel. "New Challenges to the Rationality Assumption." *Legal Theory*, vol. 3, 1997, pp. 105–24. HeinOnline, https://research.ebsco.com/linkprocessor/plink?id=c891d28b-bb65-3574-9c5f-98cac59ebd09

Kahneman, Daniel. *Thinking, Fast and Slow*. New York: Macmillan, 2011.

Kahneman, Daniel, and Amos Tversky. "Prospect Theory: An Analysis of Decision under Risk." *JSTOR*, vol. 47, no. 2, 1979, pp. 263–91.https://doi.org/10.2307/1914185..

Kahneman, Daniel, Dan Lovallo, and Oliver Sibony. "A Structured Approach to Strategic Decisions." *MIT Sloan Management Review*, vol. 60, no. 3, 2019. Gale Business: Insights, https://research.ebsco.com/linkprocessor/plink?id=46f995d2-88c4-33c0-b4aa-07e1951fe5ac

Kennedy, Jim, Joseph Ashmore, Elizabeth Babster, and Ilan Kelman. "The Meaning of 'Build Back Better': Evidence from Post-Tsunami Aceh and Sri Lanka." *Journal of Contingencies and Crisis Management*, vol. 16, 2008, pp. 24–36. https://doi.org/10.1111/j.1468-5973.2008.00529.x

Kliegr, Tomáš, Štěpán Bahník, and Johannes Fürnkranz. "A Review of Possible Effects of Cognitive Biases on Interpretation of Rule-Based Machine Learning Models." *Artificial Intelligence*, vol. 295, 2021, pp. 103458.

Macal, Charles M., and Michael J. North. "Tutorial on agent-based modeling and simulation." *Proceedings of the Winter Simulation Conference, 2005*. IEEE, 2005.

McFarlane, Alexander C., and Miranda Van Hoof. (2009). Impact of Childhood Exposure to a Natural Disaster on Adult Mental Health: 20-Year Longitudinal Follow-up Study. *The British Journal of Psychiatry, 195*(2), 142–48.

Nasar, S., Raz, S., Parray, A. A., Hossain, M. R., Sultana, R., Nadim, A. S. M., ... & Rahman, M. S. (2022). An assessment of gender vulnerability in the humanitarian crisis in Cox's Bazar, Bangladesh: Developing a gender-based vulnerability index in the Rohingya and Host community contexts. *International Journal of Disaster Risk Reduction, 81*, 103246.

Office of the Federal Register, National Archives and Records Administration. 122 Stat. 3553 – ADA Amendments Act of 2008. U.S. Government Publishing Office, https://www.govinfo.gov/app/details/STATUTE-122/STATUTE-122-Pg3553

"*Planning for an Emergency: Strategies for Identifying and Engaging at-Risk Groups; a Guidance Document for Emergency Managers*." edited by Centers for Disease Control and Prevention (CDC), first edition. CDC, 2015. http://www.cdc.gov/nceh/hsb/disaster/default.htm

Roberts, Patrick S., and Kris Wernstedt. "Decision Biases and Heuristics among Emergency Managers: Just Like the Public They Manage For?" *The American Review of Public Administration*, vol. 49, no. 3, 2019, pp. 292–308.
Scarpina, Federica, and Sofia Tagini. "The Stroop Color and Word Test." *Frontiers in Psychology*, vol. 8, 557, 12 Apr. 2017, https://doi.org/10.3389/fpsyg.2017.00557
Scott, Kelly, Céline da Graça Pires, and Maria Bennici. *Respecting the Rights of Vulnerable Groups during the COVID-19 Pandemic*. Business for Social Responsibility. 2023.
Sphere Association. *The Sphere Handbook: Humanitarian Charter and Minimum Standards in Humanitarian Response*, fourth edition. 2018. www.spherestandards.org/handbook
Stempowski, Deborah. "Counting Every Voice: Understanding Hard-to-Count and Historically Undercounted Populations." United States Census Bureau, 2023. Accessed November 11, 2023.
Tversky, Amos, and Daniel Kahneman. "The Framing of Decisions and the Psychology of Choice." *Science*, vol. 211, no. 4481, 1981, pp. 453–58.
UNAIDS. "12 Populations Left behind." 14 July 2014a. https://www.unaids.org/en/resources/infographics/20140716keypopulations. Accessed February 2, 2023.
UNAIDS. *The Gap Report*. Joint United Nations Program on HIV/AIDS (UNAIDS), 2014b, https://www.unaids.org/sites/default/files/en/media/unaids/contentassets/documents/unaidspublication/2014/UNAIDS_Gap_report_en.pdf
Vela, Monica, Amarachi I. Erondu, Nichole Smith, Monica E. Peek, James N. Woodruff, and Marshall H. Chin. "Eliminating Explicit and Implicit Biases in Health Care: Evidence and Research Needs." *Annual Review of Public Health*, vol. 43, 2022, p. 477, Gale Academic OneFile, https://doi.org/10.1146/annurev-publhealth-052620-103528
Webster, Craig S., Saana Taylor, Courntney Thomas, and Jennifer M. Weller. "Social Bias, Discrimination and Inequity in Healthcare: Mechanisms, Implications and Recommendations." *British Journal of Anaesthesia Education*, vol. 22, no. 4, 2022, pp. 131–37, ScienceDirect, https://doi.org/10.1016/j.bjae.2021.11.011
Weisberg, Jessica. "Remembering Ruth Bader Ginsburg in Her Own Words." Elle. Hearst Magazine Media. 2020. https://www.elle.com/culture/career-politics/interviews/a14788/supreme-court-justice-ruth-bader-ginsburg/#. Accessed October 1, 2023.
Zaksek, Melissa, and Joseph L. Arvai. "Toward Improved Communication About Wildland Fire: Mental Models Research to Identify Information Needs for Natural Resource Management." *Risk Analysis: An International Journal*, vol. 24, no. 6, 2004, pp. 1503–14, Health Business Elite, https://doi.org/10.1111/j.0272-4332.2004.00545.x

15

MYTHS RELATED TO DISASTER RESPONSE

Thomas Evan O'Shea

Introduction

Mass panic, looting, martial law, flock evacuations, 'disaster shock,' and general antisocial behaviour following disasters have all been attributed to a 'disaster mythology' (Constable, 2008). All have variously been debunked (Drabek, 1996), yet much like the related idea of superstition (Sjöberg & Wåhlberg, 2002), there is evidence to suggest that those engaged in contemporary disaster management and emergency planning assume the prevalence of such myths to *actually be true* (Birkland, 2016; Nogami, 2018; Wood & Miller, 2021). The full scale and extent to which this negatively affects management and planning for disasters are difficult to establish. However, such myths have been shown to generate unhelpful dynamics related to historically significant natural disasters (Tierney, Bevc & Kuligowski, 2006; Constable, 2008; Jacob et al., 2008; Nöggerath, Geller & Gusiakov, 2011; Nunn, 2014; Aguirre, 2020). Largely reinforced by mass media amplification (Heide, 2004; Nogami, 2018), these myths have cultured negative relationships between people, the environment, and those responsible for managing the dynamics between both entities (Easthope, 2018), presenting unnecessary obstacles to sustainability and resilience-building strategies. In this chapter, disaster myths related to the concept of disaster response will be explored through their origin, scale, extent, and impact to disentangle reality from myth. The result is that we may renew our viewpoint of the real issues and further orientate ourselves towards the most effective pathways of response for future disasters.

Origins of Myths

Most people hold three fundamental assumptions:

1 The world is benevolent,
2 The world is meaningful,
3 The self is worthy.

Contrary to some current evidence, and while caution should be applied in extrapolating these assumptions towards all instances, they are commonly held in good faith (Janoff-Bulman, 2010), offering a first-order reference for the functional relationships between people and their environment.

 DOI: 10.4324/9781003414834-21

However, to these assumptions, we can begin tracing the origins of dysfunction around cause and effect, particularly related to disasters (Etkin, 2014). Disaster planning is 'only as effective as the assumptions on which it is based' (Heide, 2004, pg. 337). While models and interpretations of disasters (e.g., how we plan for and respond to them) are undoubtedly motivated by a desire to improve the effectiveness of planning and are numerous to this effect, the multi-faceted nature of disasters has historically confounded attempts to unify them into a general theory or universal explanatory power (Alexander, 2002). As a result, we are unable to confidently apply simple first-order assumptions, like those given at the outset of this chapter, with confidence towards a positive outcome in every relevant instance (Jia et al., 2023). However, disasters do, regardless of origin, encompass all human dimensions. Thus, they have become enmeshed within our social and cultural fabric, enduring in our many cultural histories and societal structures (Figure 15.1), which has presented us with

Figure 15.1 A 19th-Century Woodblock Depiction of Namazu, a Giant Wels Catfish and Deity in Japanese Mythology Responsible for Earthquakes as well as the Subsequent Redistribution of Wealth. The cultural influence of the catfish as a symbol in this context is why the Japanese Meteorological Agency has adopted it as the symbol for their domestic Earthquake Early Warning System

Source: Unknown Artist. 'Namazu.' World History Encyclopaedia. Last modified June 09, 2017. https://www.worldhistory.org/image/6748/namazu/

multiple means of better understanding the universality of disasters, as well as the many dimensions of response to such events.

Myths, or 'traditional oral tales' (Nunn, 2014), are as much a fundamental component in the history of human society as natural dynamics are in the planetary environment (Svenson, 2009; Liritzis, Westra & Miao, 2019). Therefore, it is not surprising that the two are so innately entwined, such that floods, storms, fires, and volcanic eruptions have all historically been fertile source material for mythological parables, narratives, and, indeed, entire mythologies (West & Smith, 1997; Cashman & Cronin, 2008; Nunn, 2014). Natural events pervade the timeline of human development in a rhizomatic manner (Tyler & Swartz, 2012); in terms of the myths around these, they have been taken on as a means of rationalising loss and change (see Figure 15.1), which has historically been on a biblical scale (Kempe, 2003; Gluchman, 2016). However, how myths extend in the other direction, outwardly from the human system into the planetary environment, to determine and shape our responses to natural events is less obvious (Constable, 2008; Jacob et al., 2008). Given the extent to which natural events have been so influential during the evolution of society (Macklin & Lewin, 2015), it would be reasonable to assume that our interpretations and rationalisation of such events determine how we have, and will continue to, receive them (Ng, 2023). Thus, this dynamic determines the scale of the event's impact on the various intricacies of our entwined, and ever more complicated, global environment (Hewitt, 1995; Sun & Faas, 2018).

Evidence has suggested that disasters related to these natural events can, like traditional myths, challenge our beliefs about the world and our place within it, perpetuating potential shifts in entire worldview. Part of this shift lies in the consolidation of cognitive dissonance around such events, leading to well-intentioned disaster management messaging producing an undesirable public reaction (Wood & Miller, 2021). Mapping such shifts can be difficult, with avoidance and denial also playing an important role in maintaining psychological paradigms (Beck, 2009), leading to maladaptation, such as the resettling of historically disaster-prone areas, including flood plains (Lazzarin, Defina & Viero, 2023; Riaux et al., 2023), and the adherence to failed risk-reduction strategies (Kondo & Lizarralde, 2021). This dissonance further enables misconceptions to perpetuate through system scales, from individuals up to the levels of national and international governance, intensifying through the extension of antecedent societal misconceptions. Disasters therefore essentially strip back mitigating routines and perceptions of abundance to exacerbate existing points of socio-economic difficulty (Heide, 2004; Singh, Eghdami & Singh, 2014). Acting in parallel to the proliferation of such misconceptions can also be a sense of opportunity, or even gain, for those whose business it is to involve themselves with such events, including emergency planners, journalists, and politicians (Etkin, 2014; Rowthorn & Maciejowski, 2020).

Boundaries and Spaces

At the forefront of disasters are people, and before the realisation of a disaster, the management and effective communication of this management are fundamental to protecting people from the myriad impacts of a disaster (Marlowe et al., 2018; Fakhruddin et al., 2020). Variously, this dynamic has historically sat within the very broad reach of the term 'risk'; latterly reorientated under the auspices of Disaster Risk Reduction (DRR) (Wachinger et al., 2013; Marshall, 2020), which is facilitated and managed by a collective of disaster professionals (Nogami, 2018). In categorisation of this risk and its ambivalences, Ulrich Beck (2009) described how classical distinctions, particularly in the sociological categories of risk, merge into relatively 'greater or lesser degrees of risk' that enable it to be managed to some degree over time. Critically, however, in the instance of a disaster, Beck describes how risk functions like an 'acid bath' in which venerable classical distinctions are dissolved, essentially making risk very difficult to manage during and after the disaster. Here, the 'binary coding' of society within the horizon of risk (e.g., permitted or forbidden, legal or illegal, right or wrong, us and them) does not exist, with *all* becoming a relative scale of risk. Within the context of disaster response, an instance of extreme risk is essentially realised, whereby a complex space of moral and political responsibility is opened, and, as Beck further describes, the meanings of proximity, reciprocity, dignity, justice, and trust are transformed beyond their recognisable form. Exactly what this transformation looks like is made more complex by the specificity of each disaster as well as the socio-cultural modes implicated. However, despite this, and even in the smallest conceivable microcosm, the relation between the decision-maker and recipient always endures, be it positive or negative, to define the outcome of the disaster itself for all concerned.

By unpacking disaster risk in this way, Beck (2009) emphasised the crucial role that communication within this enduring relationship plays in the proliferation and mitigation of risk. Also, this approach demonstrates how, when applied generally, risk can generate a problematic space within which there is room for misinterpretations and assumptions around the fundamental roles and relationships in society, particularly at the complex nexus of nature, a community at risk, and the professionals managing that risk. Indeed, this problem was recognised by the United Nations Office for Disaster Risk Reduction (UNDRR) in the Sendai Framework (Aitsi-Selmi et al., 2015) and was a primary motivation for why Beck sought to define society through risk. The aim of doing this is to dispel the space of ambiguity between *what risk is* and what it can be *portrayed as*, so that some sense of equilibrium between risk and catastrophe may be maintained until change is unavoidable. This concept is particularly relevant from a social standpoint as it should theoretically enable the layperson to make better-informed decisions about their risk in parallel with professionals and, theoretically, advance discourse around the social dimensions of risk so that its future management is more effective. This critical lens, cast over the space between the assumed and reality, is also crucial if we are to maintain a critical sense of what is and is not a myth, or indeed develop a sense of separation between the traditional, historical myth and the more insidious, destabilising form of disaster myth (Srinivas & Nakagawa, 2008; Gotham, 2014; Berariu et al., 2015).

Giving definition to the complex space described by Beck, from which such negative myths find their origin, is also a fundamentally difficult endeavour. In *Nine Fallacies of Natural Disaster* (1977), Michael Glantz outlined nine generalisations that proved to be particularly problematic for responding to a period of prolonged drought in the West African Sahel, also with some general applicability to environmental problems. Notably in this address, Glantz substantiates the difficulties of responding to drought in the region through a lack of progress over the 60 years leading up to the droughts and famines across West Africa in the 1970s, establishing the first fallacy which is that 'people learn from their mistakes.' The second fallacy is the assumption that 'things will have to change,' with change having notably been described as an ordeal for people who fear its unknown qualities (Hoffer, 1967). Whilst the fallacies are identified from a case study of relatively unique groups and populations across the Sahel region, these two fallacies are an indication of generalisations that still endure today, some time beyond that specific case (Simpson & Hancock, 2009; Sivapalan, Savenije & Blöschl, 2012; Sivapalan, 2015; Pande & Sivapalan, 2017; Madani & Shafiee-Jood, 2020; Leal Filho et al., 2023). These fallacies suggest that the complex space of disaster myth is aligned with assumptions made in good faith, but which do not endure in any meaningful way at a fundamental level, generating continued difficulties in response to disasters that extend far beyond the original event.

Examples of Myths

An important example, regularly referenced in the disaster myth literature, is Hurricane Katrina (Aguirre, 2020). Despite having happened in 2005, the scars from the event remain today (Vollen & Ying, 2023), with some communities across the southern half of the United States still to fully recover and many having been completely devastated (Raker et al., 2019; Raker & Woods, 2023). The psychosocial impact of the event remains complex, resulting from the compound factors of immense natural force and failure of the storm and flood protection structures, most notably in New Orleans (Yarnal, 2007; Colten et al., 2008; Link, 2010) with significant negative impact on proximal states also being well documented (Chuan et al., 2019; Wood et al., 2021). The negative impacts of the event also reverberated through the system of US governance, putting it under visibly significant pressure and destabilising the authority of incumbent President George W. Bush (Sylves, 2006). Such far-reaching impact was broadly brought about through internal disorganisation of the bodies responsible for managing and responding to the event (US Congress, 2006), and external pressures resulting from an overall lack of preparation as well as a lack of clarity around accountability across the spectrum or management and response further problematised the context (Moynihan, 2009).

Beyond the confusion and ire that immediately followed Katrina was Hurricane Rita, a secondary large-scale event that further compounded the poor response to the initial event. Around these two events, assumptions made by key actors in management and governance (Schneider, 2005) further perpetuated a distorted local, national, and international response to Katrina (Eikenberry, Arroyave & Cooper, 2007). This increasingly complex situation was then further exacerbated by conflicting concerns and responsibilities in the response efforts, amplified through intense, often misrepresentative, mass media coverage (Sommers et al., 2006; Voorhees, Vick & Perkins, 2007), leading to very prominent responses by celebrities, rather than professionals, in the expanse of time between disaster onset and official response (Torres, 2006).

Jacob et al. (2008) dissected the Katrina event directly through the discourse of disaster myth, which led to a wholesale re-writing of the doctrine for public health impact, from disasters, descended from the World Health and Pan American Health Organisations. Key to understanding the role that disaster myths played in perpetuating the negative impacts of Katrina, Jacob et al.'s review outlines ten myths and their general impacts on public health. Subsequently, they contextualise the myths and impacts based on their respective roles in the development of the Hurricane Katrina disaster. There are certainly similarities between the ten myths of Katrina and the nine fallacies (Glantz, 1977) described some 30 years prior for the droughts in the Sahel region. Glantz particularly highlighted the need for a systematic approach (fallacy 6), rather than a piecemeal and sequential approach, for addressing preparation and response to disasters. This issue is facilitated by a lack of clear political leadership (fallacy 7) and the myth of interdependence (fallacy 8)—the latter having been described as a 'lack of distinction between *what is* and *what ought to be*' in terms of the relational ties from governance to resources, the effects of which are passed on to the individuals in the disaster area. Jacob et al.'s (2008) ten myths certainly echoed elements of these latter fallacies described by Glantz, with specific impacts on public health in the instance of the Katrina disaster. Assumptions made around the infallibility of interconnectedness can, and did, lead to what Glantz referred to as a 'crisis of expectations,' particularly when socio-economic connectedness fails and expectations are not met during an event, perpetuating both immediate and long-term detriment to social and economic systems. This was certainly a forewarning of 'the complex space of moral and political responsibility,' described in the Risk Society presented some three decades later by Ulrich Beck (2009), now since further expanded through the ten myths for Katrina shortly after that (Jacob et al., 2008).

This suggests that contemporary disaster myths have origins in the complex space of unaccountability which replicates in a cyclical pattern, from which we are not learning and therefore continue to make assumptions in 'good faith' that things will be better next time, or worse still, that we are safe(r). Drawing parallels and contrasting the assessments and reflections of a hurricane with a drought, some 30 years apart, does present some obvious discrepancies. Yet, similarities, in terms of impacts, do seem to exist on the same spectrum, and the overall messages for improved response to the specific type of natural hazard, and disasters more generally, have endured in like fashion. Indeed, the current sentiment is that these messages have echoed an increasing urgency because of the growing pressures of climate-driven change and increasing socio-environmental complexity (Radunsky & Cadman, 2021). Therefore, it's reasonable to assume that the gap, between *what is* and *what ought to be*, has only grown, increasing the space for disaster myths to grow and perpetuate further in an increasingly insidious fashion over time.

There are notable recent examples to suggest this might be the case for historically disaster-prone regions, stemming from assumptions based on myths of cohesion, mutual aid, and community consolidation (Aguirre, 2020), which are seemingly positive concepts by comparison to those outlined at the outset of this chapter. These behaviours, when linked to disaster events, have been directly misrepresented as mechanisms of resilience and recovery in recent times, in some instances as a substitute for effective state-led recovery (Lam, 2023). One such example of this is the spirit of *bayanihan*, the myth-cum-social value, affiliated with community resilience in the Philippines, which became particularly affiliated with those who were so severely impacted by Typhoon Haiyan in 2013 (Su et al., 2023). *Bayanihan* received international attention following the 2013 disaster, primarily because Barack Obama directly referred to it as being a fundamental quality of resilience and recovery for the Philippine people in his address to them following the event. By contrast, when its supposed values are used as criteria for assessing community resilience across the whole timeframe of Haiyan (i.e., pre-event preparedness, during, and post-event response), the reality is that *bayanihan* has better served as a mechanism for social analysis, enabling insight beyond the zeitgeist of disaster myth stemming from assumptions made about community cohesion and an innate resilience to such events.

By referencing the principles of *bayanihan* to the communities affected by Haiyan, researchers could, alongside the affected communities, benchmark social practices and perceptions of the responses to Haiyan as well as their ideas of each other (Bonifacio & Epe, 2023). This is an analogue for some ethnographic techniques, which represent a useful tool for gauging how influential community cohesion and mutual aid can be in the outcome of such events (Handwerker, 2001). By this measure, the communities affected by Haiyan indicated that the principles of *bayanihan* were indeed prevalent before the onset of Haiyan: albeit non-specifically, everyone got along, people shared, and people generally helped each other when necessary. Further accounts have suggested that these qualities became somewhat elevated during the event but then receded following the event (Bonifacio & Epe, 2023). This post-event recovery period was, from the perspective of affected communities with *bayanihan* as their benchmark, punctuated with large degrees of suspicion and an overall reduction in the sense of community cohesion. The initial trend is what one might expect to learn about following a disaster on the scale of Haiyan; however, the reporting and political amplification of *bayanihan* as an intangible but intrinsic fallback for the disaster-afflicted community's resilience and rebuilding, independent of municipal and independent support, was effectively untrue. Indeed, reports since Haiyan have indicated that external perceptions of the worst affected group's need for support were misled to such an extent that it extended the waiting time for provisions that were very much needed. This was in stark contrast to reports at the time which indicated that such communities were recovering well without the explicit need for these provisions (Bonifacio & Epe, 2023).

Communication around *bayanihan* suggests that prominent figures are just as liable to believe in the myth as anyone else, and in such a complex space, this can further consolidate misconceptions. Historical analyses of such events are also important mile markers for how disaster myths can—through their bases in assumptions—abstract perceptions of such events in seemingly subtle ways but with tangibly negative effects that reverberate well beyond the initial disaster event. Earmarked as one of the overall greatest challenges to DRR in the UN sustainability agenda, the need for improved communication for disaster management at the socio-environmental interface (Izumi et al., 2019) has generated increased focus and elevated research output on disaster communication over the last decade (Agyepong & Liang, 2023). However, the result of this renewed focus and how it has translated to tangible benefit at the forefront of DRR has been mixed (Mercer, 2010; Mercer et al., 2016; Alexander, 2022). For example, the 2023 Turkey-Syria earthquake, where many thousands of fatalities were directly attributed to building collapse, despite supposedly built to advanced anti-earthquake protocol based on extensive DRR research (D'Ayala, 2023).

The Turkey-Syria earthquake is also a fresh example that serves to highlight the particularly damaging effects of the 'interconnectedness' or 'Nirvana' fallacy (Coyne, 2006). The assumed provision of international aid in the aftermath of a disaster event speaks to a deeper and more complex interplay between historical socio-political forces that can act to expand the dimensions or the space of unaccountability (Beck, 2009). This further highlights the damaging impact of an assumption that broad response and provisions can be distributed despite significant damage to regional and local infrastructure and communication networks (Kolivand et al., 2023). It is now often repeated that effective pre-event planning and preparedness are the best measures for ensuring a better disaster response for those who inhabit hazardous regions of the globe (Twigg, 2004, 2015; Cook & Zurita, 2016). For pre-event measures to be successful, following Glantz (1977), largely requires that some lessons are learned from past events and, following Jacob et al. (2008) for Katrina, constructively developed towards a more holistic, objective-based response methodology. A critical lens on past events relative to current measures and practices supplements such a methodology, for example, revisiting building codes for earthquake-affected regions in respect of the recent Turkey-Syria case would be a positive aspect of overall response (D'Ayala, 2023). Opening avenues of critical communication around these examples to co-create with the affected communities (like that undertaken in the Philippines following Haiyan) while using a clear and defined terminology would further reconcile *what is* with *what ought to be* for laypeople and professionals alike (McClure, 2020; McClure et al., 2023).

Refocusing and Realignment

Enquiries into the mode of disaster myth perpetuation have shown that disaster professionals and laypeople alike are subject to patterns of belief that shape their perceptions of disasters and thus determine patterns of response to them. With certain myths, there are patterns to suggest that mass media is much more influential on lay people's perceptions of disasters compared to professionals, yet professionals are not necessarily more likely to be insightful on myths, including crime and panic, than lay people (Nogami, 2018). This analysis is important as it indicates that myths do not necessarily have a discernible origin other than via mass media and that professionals should focus on what they do best, which is enhancing measures of preparedness to support the laypeople within communities (Nogami, 2018). The complexities of this can be brought back to the nature of communication and language, as well as the assumptions made around preparedness, which when taken in isolation against specific events, has patterns that require considered strategy.

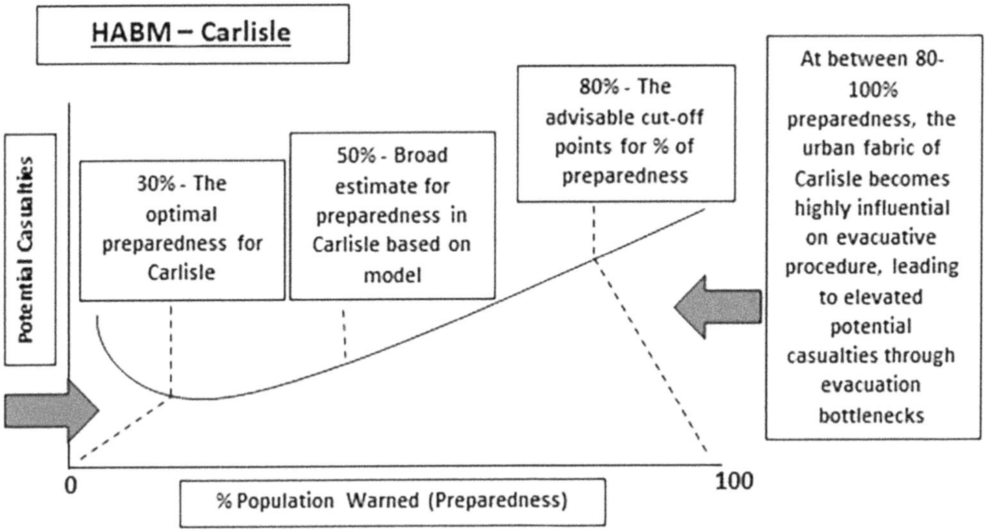

Figure 15.2 An Output of the Hydro Agent-Based Model (HABM) Simulated for the 2005 Carlisle Flood in the United Kingdom. In both instances (Figures 15.2 and 15.3), a preparedness protocol was simulated across many hundreds of thousands of simulations for disaster events. Illustrated are patterns of optimality for such events that are relative to advised optimality for evacuation measures. This further suggests that being 100% prepared for these events can lead to an elevated likelihood of fatality and overall disaster, an outcome far worse than in the imagined realm of disaster myths

Source: O'Shea, Bates and Neal (2020).

Figures 15.2 and 15.3 were generated by simulations of evacuative responses to a flood event and a tsunami, respectively. They illustrate a simplification of modelled results for fatalities generated by an agent-based simulation of responsive actions prior to, during, and after the events, with the affected population's behaviours having been calibrated with survey data and guidelines for action based on retrospective analyses (O'Shea, Bates & Neal, 2020). The modelling of these dynamics sought to help identify finer patterns of preparedness and optimal strategies for informing decision-making processes by simulating the events with different styles of warning alongside incrementally diversified public behaviours. Revealing some intricacies of the socio-environmental interface, this methodology bridges the discourse of disaster professionals, encompassing strategic preparedness and the real-life actions enacted by an affected public in response to such events.

The use of these, and similar methods (Comfort et al., 2004; Merz et al., 2020, Shults, 2023), can move us beyond the space of assumption and miscommunication inherent in disaster myths. Indeed, discourse analysis of public thought on disasters and climate has demonstrated that affected laypeople can form semantically relevant perceptions of the complexities of disaster events (Nogami, 2018). Therefore, these approaches offer a positive step towards closing the space of assumption and ambiguity between what is and ought to be, and where disaster myths take form, by offering a new medium of layperson and specialist interpretations alike. This is particularly pertinent if we are to, as the Risk Society concept similarly does, assume that the action(s) encompassing being prepared for disaster exist on a spectrum of actions that can also lead to us becoming

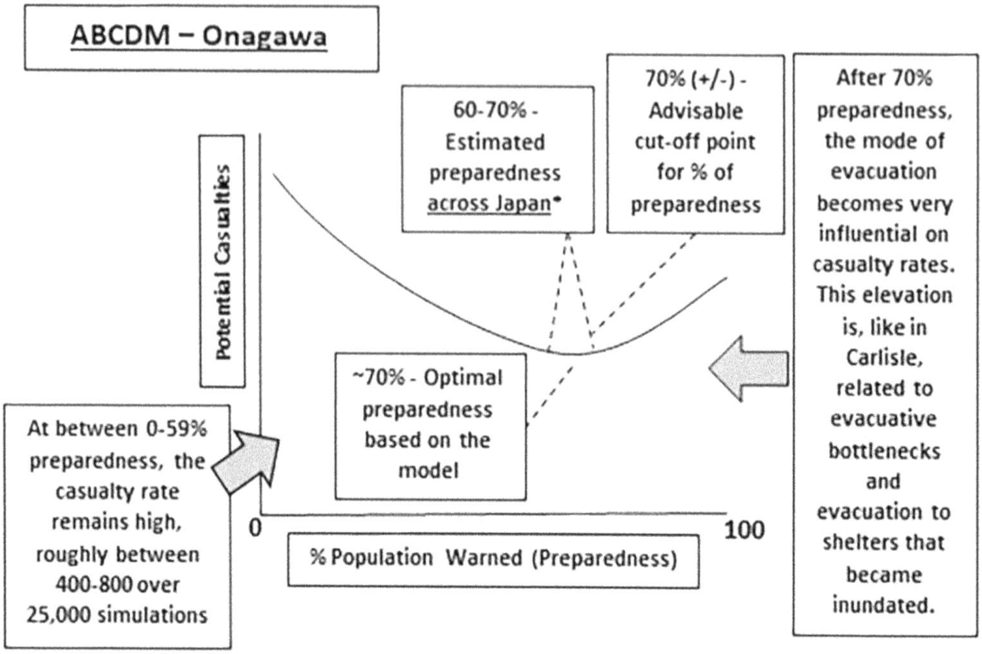

Figure 15.3 An Output of the Agent-Based Coastal Disaster Model (ABCDM) Simulated for the 2011 East Asian Tsunami in Onagawa, Japan. By comparison to a similar illustration shown in Figure 15.2 for similar disaster dynamics, the optimal preparedness for a tsunami is quite different compared to that for a flood event

Source: Author's own image.

more vulnerable in the event of a disaster. If we are to understand such hidden possibilities better, then we must equip ourselves with the appropriate modes of inquisition to be able to dispel the possibilities of myths arising from an adversarial space of unaccountability, extending the reach of the disaster, and reducing capacity of meaningful action to reduce continued negative impact. Equipped with such modes of enquiry and the accountability they can generate, we might better understand the roles and accountabilities of each other in such events, consolidating our resilience to them in the process (Easthope, 2018).

Discussion and Conclusions

Ultimately, myths are, much like the disasters they can be associated with, a socially produced phenomenon (Sun & Faas, 2018). The finding that mass media influences the layperson's perceptions of disaster myths up to five times more by comparison to a disaster professional indicates that the pathway from disaster to media to layperson is one with significant potential to open the space within which negative disaster myths find form (Nogami, 2018). Yet, laypeople in disaster-prone regions of the globe have been responsible for communicating knowledge of the environment and nature in forms that construct culturally significant narratives and stories, some passed down over hundreds of years and across many different communities through cultural practices (Nunn, 2014; Prawoto & Octavia, 2021). This is a fundamental practice for extending the memories of communities in historically disaster-prone areas towards contemporary resilience-building frameworks to support a continued understanding and communication, around the workings of the environment, between those who experience them and those who now manage them. This is a historically well-defined space within which the role of the disaster professional is not so apparent, and this could very well be the reason why the space of uncertainty is given dimension through modern disaster myths, particularly in ever-changing and complex urban environments like cities.

Similar discussion on the social production of disasters and disaster social constructs (Tierney, 2003; Sun & Faas, 2018) has indicated that such intricacies need to be disassembled and reframed to promote effective resilience within the ever-changing space of socio-environmental dynamics. This discussion has alluded to some of the benefits inherent in community-based storytelling as an effective, resilience-building technique that has maintained communities in disaster-prone areas for centuries. The difficulty now is the translation of such effective techniques across to less disaster-prone areas or into more mobile, contemporary, communities where attachments to cultural practices of storytelling and listening are less tangible (Easthope, 2018; Akbar, Abubakar & Bouregh, 2020). Global urban populations are projected to continue growing and expanding into disaster-prone areas (Elmqvist et al., 2021) exposing ever-growing concentrations of people and assets to new and elevated risks. An important element of this adaption to a new future will be to deconstruct the myth of safety, which, in historical instances, has been found to obfuscate honest discussions of risks (Nöggerath, Geller & Gusiakov, 2011) that cannot simply be addressed by one single measure, such as the technocentric ambitions of all people living in one futuristic megalopolis (Angelo & Wachsmuth, 2020).

Indeed, advancements in the understanding of disaster memories have led us to better understand the non-linear effects of disasters, including complex post-traumatic stress disorder (PTSD), depression, and loss-related grief (Janoff-Bulman, 2010; Sandhu & Kaur, 2013; Sveen et al., 2018). These are examples of effects that are not mythical; they are very human, and they have been shown to impede resilience building at various scales—particularly at the systematic scale of the group and individual—well beyond the initial event impact (Shaw, Scully & Hart, 2014). As such, the complex health and psychological effects of being impacted by a disaster take effect on scales that vary from local to global. This emphasises the pressing need for a holistic extension of community memories of such events to inform a better long-term preparedness, that is generated at the intersection of cognitive health, scientific explanation, and popular culture (Prawoto & Octavia, 2021).

Often, the literature concerned with this subject matter closes with an open message of perseverance, of continued vigour, and the extension of good practice out into the fog of a very uncertain future. This chapter will instead close with a message that emphasises renewal and re-orientation because sometimes, even when there is a will, there isn't always a way (Glantz, 1977, fallacy 9)—and in these instances, we must make a new space to fully consider what is needed to establish effective pathways leading to a resilient, equitable, and sustainable future reality for which we are all accountable for the outcomes, imagined or otherwise.

References

Aguirre, B.E., 2020. The myth of disaster myths. *Oxford research encyclopedia of politics*. Retrieved 22 Nov. 2024, from https://oxfordre.com/politics/view/10.1093/acrefore/9780190228637.001.0001/acrefore-9780190228637-e-1616

Agyepong, L.A. and Liang, X., 2023. Mapping the knowledge frontiers of public risk communication in disaster risk management. *Journal of Risk Research*, *26*(3), pp.302–323.

Aitsi-Selmi, A., Egawa, S., Sasaki, H., Wannous, C. and Murray, V., 2015. The Sendai Framework for disaster risk reduction: Renewing the global commitment to people's resilience, health, and well-being. *International Journal of Disaster Risk Science*, *6*, pp.164–176.

Akbar, N., Abubakar, I.R. and Bouregh, A.S., 2020. Fostering urban sustainability through the ecological wisdom of traditional settlements. *Sustainability*, *12*(23), p.10033.

Alexander, D.E., 2002. *Principles of emergency planning and management*. Oxford University Press.

Alexander, D.E., 2022. On the meaning of impact in disaster risk reduction. *International Journal of Disaster Risk Science*, *13*(5), pp.822–827.

Angelo, H. and Wachsmuth, D., 2020. Why does everyone think cities can save the planet?. *Urban Studies*, *57*(11), pp.2201–2221.

Beck, U., 2009. *World at risk*. Polity.

Berariu, R., Fikar, C., Gronalt, M. and Hirsch, P., 2015. Understanding the impact of cascade effects of natural disasters on disaster relief operations. *International Journal of Disaster Risk Reduction*, *12*, pp.350–356.

Birkland, T.A., 2016. Policy process theory and natural hazards. *Oxford research encyclopedia of natural hazard science*. Retrieved 22 Nov. 2024, from https://oxfordre.com/naturalhazardscience/view/10.1093/acrefore/9780199389407.001.0001/acrefore-9780199389407-e-75

Bonifacio, G. T., & Epe, R. B. (Eds.). (2023). *Disasters in the Philippines: Before and After Haiyan* (1st ed.). Bristol University Press. https://doi.org/10.2307/jj.2487425

Cashman, K.V. and Cronin, S.J., 2008. Welcoming a monster to the world: Myths, oral tradition, and modern societal response to volcanic disasters. *Journal of Volcanology and Geothermal Research*, *176*(3), pp.407–418.

Chuang, W.C., Eason, T., Garmestani, A. and Roberts, C., 2019. Impact of Hurricane Katrina on the coastal systems of southern Louisiana. *Frontiers in Environmental Science*, *7*, p.68.

Colten, C.E., Kates, R.W. and Laska, S.B., 2008. Community resilience: Lessons from New Orleans and Hurricane Katrina. *CARRI Report*, *3*, pp.2–4.

Comfort, L.K., Dunn, M., Johnson, D., Skertich, R. and Zagorecki, A., 2004. Coordination in complex systems: Increasing efficiency in disaster mitigation and response. *International Journal of Emergency Management*, *2*(1–2), pp.62–80.

Constable, M., 2008. Disaster mythology: Looting in New Orleans. *Disaster Prevention and Management: An International Journal*, *17*(4), pp.519–525.

Cook, B.R. and Zurita, M.D.L.M., 2016. Planning to learn: An insurgency for disaster risk reduction (DRR). *International Journal of Disaster Risk Reduction*, *19*, pp.265–272.

Coyne, C.J., 2006. Reconstructing weak and failed states: Foreign intervention and the nirvana fallacy. *Foreign Policy Analysis*, *2*(4), pp.343–360.

D'Ayala, D., 2023. Commentary: Reflections on the Turkey–Syria earthquakes of 6 February 2023. *Proceedings of the Institution of Civil Engineers-Structures and Buildings*, *176*(6), pp.478–481.

Drabek, T.E., 1996. *The social dimensions of disaster: Instructor guide*. FEMA, Emergency Management Institute.

Easthope, L., 2018. *The recovery myth: The plans and situated realities of post-disaster response*. Springer.

Eikenberry, A.M., Arroyave, V. and Cooper, T., 2007. Administrative failure and the international NGO response to Hurricane Katrina. *Public Administration Review*, *67*, pp.160–170.

Elmqvist, T., Andersson, E., McPhearson, T., Bai, X., Bettencourt, L., Brondizio, E., Colding, J., Daily, G., Folke, C., Grimm, N. and Haase, D., 2021. Urbanization in and for the Anthropocene. *npj Urban Sustainability*, *1*(1), p.6.

Etkin, D., 2014. *Disaster theory: An interdisciplinary approach to concepts and causes*. Butterworth-Heinemann.

Fakhruddin, B., Clark, H., Robinson, L. and Hieber-Girardet, L., 2020. Should I stay or should I go now? Why risk communication is the critical component in disaster risk reduction. *Progress in Disaster Science*, *8*, p.100139.

Glantz, M.H., 1977. Nine fallacies of natural disaster: The case of the Sahel. *Climatic Change*, *1*(1), pp.69–84.

Gluchman, V., 2016. Moral theory and natural, or social, disasters. *Human Affairs*, *26*(1), pp.3–7.

Gotham, K.F., 2014. Reinforcing inequalities: The impact of the CDBG program on post-Katrina rebuilding. *Housing Policy Debate*, *24*(1), pp.192–212.

Handwerker, P.W., 2001. *Quick ethnography: A guide to rapid multi-method research*. Rowman Altamira.

Heide, E.A., 2004. Common misconceptions about disasters: Panic, the disaster syndrome, and looting. *The first 72 hours: A community approach to disaster preparedness*, Margaret O'Leary (ed.). iUniverse, Inc. p. 340–380.

Hewitt, K., 1995. Excluded perspectives in the social construction of disaster. *International Journal of Mass Emergencies & Disasters*, *13*(3), pp.317–339.
Hoffer, E., 1967. *The ordeal of change*. Perennial Library.
Izumi, T., Shaw, R., Djalante, R., Ishiwatari, M. and Komino, T., 2019. Disaster risk reduction and innovations. *Progress in Disaster Science*, *2*, p.100033.
Jacob, B., Mawson, A.R., Marinelle, P. and Guignard, J.C., 2008. Disaster mythology and fact: Hurricane Katrina and social attachment. *Public Health Reports*, *123*(5), pp.555–566.
Janoff-Bulman, R., 2010. *Shattered assumptions*. Simon and Schuster.
Jia, C., Zhang, C., Li, Y.F. and Li, Q.L., 2023. Joint pre-and post-disaster planning to enhance the resilience of critical infrastructures. *Reliability Engineering & System Safety*, *231*, p.109023.
Kempe, M., 2003. Noah's Flood: The Genesis story and natural disasters in early modern times. *Environment and History*, *9*(2), pp.151–171.
Kolivand, P., Kivi, H.K., Hasheminezhad, S.F., Saberian, P. and Shamspour, N., 2023. The presence of international relief teams in the 2023 Turkish earthquake: Challenges, strengths, and lessons learned. *Prehospital and Disaster Medicine*, *38*(3), pp.1–2.
Kondo, T. and Lizarralde, G., 2021. Maladaptation, fragmentation, and other secondary effects of centralized post-disaster urban planning: The case of the 2011 "cascading" disaster in Japan. *International Journal of Disaster Risk Reduction*, *58*, p.102219.
Lam, L.M., 2023. Asymmetrical power relationships in disaster reduction remain the key barrier to sustainable recovery: Neoliberalism and Nepal's post-earthquake recovery experiences. *Sustainable Development*, *32*(2), pp. 1485–1495.
Lazzarin, T., Defina, A. and Viero, D.P., 2023. Assessing 40 years of flood risk evolution at the micro-scale using an innovative modeling approach: The effects of urbanization and land planning. *Geosciences*, *13*(4), p.112.
Leal Filho, W., Trevisan, L.V., Rampasso, I.S., Anholon, R., Dinis, M.A.P., Brandli, L.L., Sierra, J., Salvia, A.L., Pretorius, R., Nicolau, M. and Eustachio, J.H.P.P., 2023. When the alarm bells ring: Why the UN sustainable development goals may not be achieved by 2030. *Journal of Cleaner Production*, *407*, p.137108.
Link, L.E., 2010. The anatomy of a disaster, an overview of Hurricane Katrina and New Orleans. *Ocean Engineering*, *37*(1), pp.4–12.
Liritzis, I., Westra, A. and Miao, C., 2019. Disaster geoarchaeology and natural cataclysms in world cultural evolution: An overview. *Journal of Coastal Research*, *35*(6), pp.1307–1330.
Macklin, M.G. and Lewin, J., 2015. The rivers of civilization. *Quaternary Science Reviews*, *114*, pp.228–244.
Madani, K. and Shafiee-Jood, M., 2020. Socio-hydrology: A new understanding to unite or a new science to divide? *Water*, *12*(7), p.1941.
Marlowe, J., Neef, A., Tevaga, C.R. and Tevaga, C., 2018. A new guiding framework for engaging diverse populations in disaster risk reduction: Reach, relevance, receptiveness, and relationships. *International Journal of Disaster Risk Science*, *9*, pp.507–518.
Marshall, T.M., 2020. Risk perception and safety culture: Tools for improving the implementation of disaster risk reduction strategies. *International Journal of Disaster Risk Reduction*, *47*, p.101557.
McClure, A., 2020. *Inclusive, participatory and reflexive learning processes for climate resilience: Key lessons from FRACTAL*. FRACTAL Working Paper 9. Retrieved from http://www.fractal.org.za/wp-content/uploads/2020/04/FRACTAL-learning-workin-paper_layout.pdf.
McClure, A., Patel, Z., Ziervogel, G. and Hardman, J., 2023. Exploring the role of transdisciplinary learning for navigating climate risks in African cities: The case of Lusaka, Zambia. *Environmental Science & Policy*, *149*, p.103571.
Mercer, J., 2010. Disaster risk reduction or climate change adaptation: Are we reinventing the wheel? *Journal of International Development: The Journal of the Development Studies Association*, *22*(2), pp.247–264.
Mercer, J., Gaillard, J.C., Crowley, K., Shannon, R., Alexander, B., Day, S. and Becker, J., 2016. Culture and disaster risk reduction: Lessons and opportunities. In *Natural hazards and disaster risk reduction* (pp. 4–25). Routledge.
Merz, B., Kuhlicke, C., Kunz, M., Pittore, M., Babeyko, A., Bresch, D.N., Domeisen, D.I., Feser, F., Koszalka, I., Kreibich, H. and Pantillon, F., 2020. Impact forecasting to support emergency management of natural hazards. *Reviews of Geophysics*, *58*(4), p.e2020RG000704.
Moynihan, D.P., 2009. *The response to Hurricane Katrina* (pp.27–45). International Risk Governance Council.

Ng, S.L., 2023. The role of risk perception, prior experience, and sociodemographics in disaster preparedness and emergency response toward typhoons in Hong Kong. *Natural Hazards*, *116*(1), pp.905–936.
Nogami, T., 2018. Disaster myths among disaster response professionals and the source of such misconceptions. *Journal of Contingencies and Crisis Management*, *26*(4), pp.491–498.
Nöggerath, J., Geller, R.J. and Gusiakov, V.K., 2011. Fukushima: The myth of safety, the reality of geoscience. *Bulletin of the Atomic Scientists*, *67*(5), pp.37–46.
Nunn, P.D., 2014. Lashed by sharks, pelted by demons, drowned for apostasy: The value of myths that explain geohazards in the Asia-Pacific region. *Asian Geographer*, *31*(1), pp.59–82.
O'Shea, T., Bates, P. and Neal, J., 2020. Testing the impact of direct and indirect flood warnings on population behaviour using an agent-based model. *Natural Hazards and Earth System Sciences*, *20*(8), pp.2281–2305.
Pande, S. and Sivapalan, M., 2017. Progress in socio-hydrology: A meta-analysis of challenges and opportunities. *Wiley Interdisciplinary Reviews: Water*, *4*(4), p.e1193.
Prawoto, E. and Octavia, L., 2021. Disaster Storytelling: Extending the Memory of the Community Toward Disaster Preparedness from Myth, Scientific Explanation, and Popular Culture. *Journal of Disaster Research*, *16*(2), 228–233.
Radunsky, K. and Cadman, T., 2021. Addressing climate change risks: Importance and urgency. In *Handbook of climate change management: Research, leadership, transformation* (pp. 1405–1431). Springer International Publishing.
Raker, E.J., Lowe, S.R., Arcaya, M.C., Johnson, S.T., Rhodes, J. and Waters, M.C., 2019. Twelve years later: The long-term mental health consequences of Hurricane Katrina. *Social Science & Medicine*, *242*, p.112610.
Raker, E.J. and Woods, T., 2023. Disastrous Burdens: Hurricane Katrina, federal housing assistance, and well-being. *RSF: The Russell Sage Foundation Journal of the Social Sciences*, *9*(5), pp.122–143.
Riaux, J., Kuper, M., Massuel, S. and Mekki, I., 2023. Riding the waves of discomforts: Reflecting on the dialogue of hydrologists with society. *Journal of Hydrology*, *626*, p.130189.
Rowthorn, R. and Maciejowski, J., 2020. A cost–benefit analysis of the COVID-19 disease. *Oxford Review of Economic Policy*, *36*(Supplement_1), pp.S38–S55.
Sandhu, D. and Kaur, S., 2013. Psychological impacts of natural disasters. *Indian Journal of Health and Wellbeing*, *4*(6), p.1317.
Schneider, S.K., 2005. Administrative breakdowns in the governmental response to Hurricane Katrina. *Public Administration Review*, *65*(5), pp.515–516.
Shaw, D., Scully, J. and Hart, T., 2014. The paradox of social resilience: How cognitive strategies and coping mechanisms attenuate and accentuate resilience. *Global Environmental Change*, *25*, pp.194–203.
Shults, F.L., 2023. Simulation, science, and stakeholders: Challenges and opportunities for modelling solutions to societal problems. *Complexity*, *2023*(1), p.1375004..
Simpson, N.C., and Hancock, P.G., 2009. Fifty years of operational research and emergency response. *Journal of the Operational Research Society*, *60*(sup1), pp.S126–S139.
Singh, S.R., Eghdami, M.R. and Singh, S., 2014. The concept of social vulnerability: A review from disasters perspectives. *International Journal of Interdisciplinary and Multidisciplinary Studies*, *1*(6), pp.71–82.
Sivapalan, M., 2015. Debates—Perspectives on socio-hydrology: Changing water systems and the "tyranny of small problems"—Socio-hydrology. *Water Resources Research*, *51*(6), pp.4795–4805.
Sivapalan, M., Savenije, H.H. and Blöschl, G., 2012. Socio-hydrology: A new science of people and water. *Hydrological Processes*, *26*(8), pp.1270–1276.
Sjöberg, L. and Wåhlberg, A.A., 2002. Risk perception and new age beliefs. *Risk Analysis: An International Journal*, *22*(4), pp.751–764.
Sommers, S.R., Apfelbaum, E.P., Dukes, K.N., Toosi, N. and Wang, E.J., 2006. Race and media coverage of Hurricane Katrina: Analysis, implications, and future research questions. *Analyses of Social Issues and Public Policy*, *6*(1), pp.39–55.
Srinivas, H. and Nakagawa, Y., 2008. Environmental implications for disaster preparedness: Lessons learnt from the Indian Ocean Tsunami. *Journal of Environmental Management*, *89*(1), pp.4–13.
Su, Y., Mangada, L. and Pacoma, A.J., 2023. Beyond bayanihan: Overcoming myths of community resilience in typhoon Haiyan post-disaster recovery. In *Disasters in the Philippines* (pp. 267–286). Bristol University Press.
Sun, L. and Faas, A.J., 2018. Social production of disasters and disaster social constructs: An exercise in disambiguation and reframing. *Disaster Prevention and Management: An International Journal*, *27*(5), pp.623–635.

Sveen, J., Bergh Johannesson, K., Cernvall, M. and Arnberg, F.K., 2018. Trajectories of prolonged grief one to six years after a natural disaster. *PLoS One*, *13*(12), p.e0209757.
Svensen, H., 2009. *The end is nigh: A history of natural disasters*. Reaktion Books.
Sylves, R.T., 2006. President Bush and Hurricane Katrina: A presidential leadership study. *The ANNALS of the American Academy of Political and Social Science*, *604*(1), pp.26–56.
Tierney, K., 2003. Disaster beliefs and institutional interests: Recycling disaster myths in the aftermath of 9–11. In *Terrorism and disaster: New threats, new ideas* (Vol. 11, pp. 33–51). Emerald Group Publishing Limited.
Tierney, K., Bevc, C. and Kuligowski, E., 2006. Metaphors matter: Disaster myths, media frames, and their consequences in Hurricane Katrina. *The Annals of the American Academy of Political and Social Science*, *604*(1), pp.57–81.
Torres, J.A., 2006. *Hurricane Katrina and the devastation of New Orleans, 2005*. Mitchell Lane Publishers, Inc.
Twigg, J., 2004. *Disaster risk reduction: Mitigation and preparedness in development and emergency programming*. Overseas Development Institute (ODI).
Twigg, J., 2015. *Disaster risk reduction*. Overseas Development Institute (ODI).
Tyler, J.A., and Swartz, A.L., 2012. Storytelling and transformative learning. In *The handbook of transformative learning: Theory, research, and practice* (pp.455–470). Jossey-Bass.
U.S. House of Representatives Select Bipartisan Committee to Investigate the Preparation for and Response to Hurricane Katrina, 2006. *A failure of initiative: Final report of the select bipartisan committee to investigate the preparation for and response to Hurricane Katrina* (Vol. 109, No. 377). US Government Printing Office.
Unknown Artist. (2017, June 09). Namazu. World History Encyclopedia. Retrieved from https://www.worldhistory.org/image/6748/namazu/'
Vollen, L. and Ying, C. eds., 2023. *Voices from the storm: The people of New Orleans on Hurricane Katrina and its aftermath*. Haymarket Books.
Voorhees, C.C., Vick, J. and Perkins, D.D., 2007. 'Came hell and high water': The intersection of Hurricane Katrina, the news media, race and poverty. *Journal of Community & Applied Social Psychology*, *17*(6), pp.415–429.
Wachinger, G., Renn, O., Begg, C. and Kuhlicke, C., 2013. The risk perception paradox—Implications for governance and communication of natural hazards. *Risk Analysis*, *33*(6), pp.1049–1065.
West, B. and Smith, P., 1997. Natural disasters and national identity: Time, space, and mythology. *The Australian and New Zealand Journal of Sociology*, *33*(2), pp.205–215.
Wood, E. and Miller, S.K., 2021. Cognitive dissonance and disaster risk communication. *Journal of Emergency Management and Disaster Communications*, *2*(01), pp.39–56.
Wood, Z.B., Wang, D.C., Lee, C., Barnes, H., Abouezzeddine, T. and Aten, J., 2021. Social support, religious coping, and traumatic stress among Hurricane Katrina survivors of southern Mississippi: A sequential mediational model. *Traumatology*, *27*(4), p.336.
Yarnal, B., 2007. Vulnerability and all that jazz: Addressing vulnerability in New Orleans after Hurricane Katrina. *Technology in Society*, *29*(2), pp.249–255.

16

OUTBREAK OF EPIDEMICS DURING DISASTER RESPONSE

Laila Shahzad

Introduction

Major disasters cause significant loss of life and injuries on a large scale. Globally, disasters have increased over the past decades (Alexander, 2015). For example, the number of disasters increased from 44 in year 1960 to 611 in year 2019, representing a more than a 13-fold increase (Chen et al., 2021). The surge in disasters may be linked to global climate change, but human involvement such as urbanization, population growth, and infrastructure development have increased the severity of these events (Shahzad et al., 2019). According to Ahmad et al. (2022), every country is affected by some form of disaster every year. Some communities are more vulnerable than others due to issues such as lack of resources, inequality, poverty, low adaptive capacity, and lack of political stability (Shahzad et al., 2021).

When an area is affected by a disaster, post-disaster challenges such as access to clean water, shelter, and food along with the outbreak of communicable and non-communicable diseases increase the cases of mortality and morbidity. In impact phases (0–4 days), rescue of victims and treatment of injuries takes place. In the post-impact phase (four days to four weeks), water-, food-, and vector-borne diseases spread. After an incubation period (over one month), diseases such as malaria and yellow fever spread in the recovery phase, as depicted in Figure 16.1. A disaster can increase the risk of communicable diseases by increasing the transmission of local pathogens and by introducing new pathogens in the area (Lee et al., 2020). Post-disaster epidemics often include respiratory infections, infections due to wounds and injuries, infections due to contaminated food and water, and vector- and insect-borne diseases. The nature of the disaster typically determines the risk and impact of communicable diseases. Floods are the disasters that cause the most death and displacement of people worldwide every year (Erickson et al., 2019), and contaminated floodwater poses a serious risk to human health (Hammer et al., 2019). Meanwhile, geo-disasters such as volcanic eruptions and earthquakes might not directly lead to the spread of diseases, but the post-disaster disruption of the health system and daily life facilities can increase the risk of diseases spreading.

In the 21st century, due to higher frequency and magnitude, disasters are generating severe health, social, environmental, and economic consequences (Erickson et al., 2019). Health impacts upon affected communities come in the form of mortality or morbidity and can cause

DOI: 10.4324/9781003414834-22

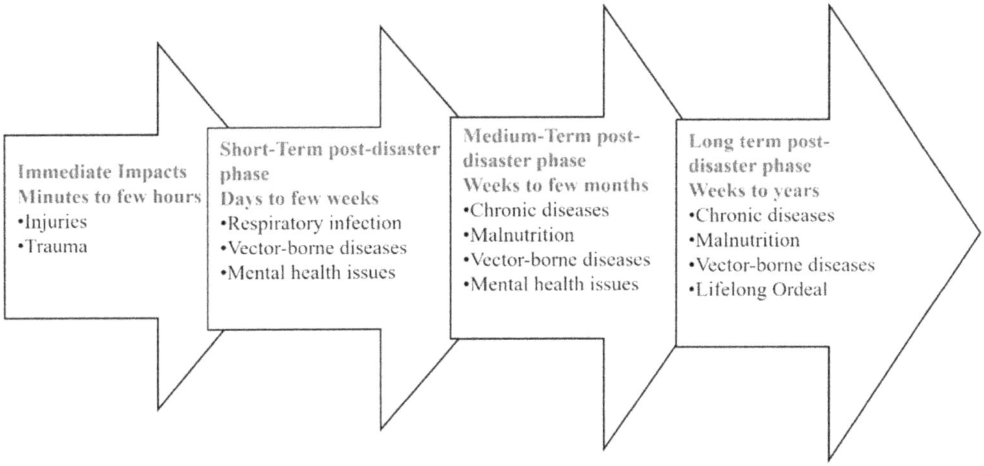

Figure 16.1 Timeframe of Diseases and Morbidities During Disaster Recovery
Source: Image from author.

socioeconomic damage to countries (Ahmad et al., 2022). The health effects of disasters can be seen as both short- and long-term costs for individuals, households, and economies—and low-income countries are often more impacted due to a lack of health resources and infrastructure available. This chapter provides an account of historical epidemics in the post-disaster phase, a relationship which is well documented but not fully understood due to the chaotic situation and simultaneous favorable grounds for both the biological existence and socio-physical transmission of communicable diseases.

Examples of Disasters and Epidemics

Communicable diseases spread due to a lack of health resources and conditions in the post-disaster response stage. The damages are felt by both high-income and lesser developed countries, but the latter may be more affected due to lack of preparedness plans, lack of infrastructure and resources, and inefficient institutional arrangements (Shahzad et al., 2021). Table 16.1 provides a non-exhaustive list of disasters that resulted in outbreaks of disease.

Disaster Types and Disease Outbreaks

The factors contributing to outbreaks of communicable diseases following disasters are predominantly linked to displacement of the population. Numerous elements such as the access to clean water, sanitary facilities, availability of soap and hygiene resources, population density in relief and transitional settings, and the overall health condition and behaviors of the affected population intertwine within disease environment of the area to impact the susceptibility to and transmission of communicable diseases. These conditions increase not only the spread of diseases but also mortality rates due to the lack of health resources and stress/surge in the health system. The remainder of this section discusses epidemics as they relate to floods, earthquakes, and drought.

Table 16.1 Historically Prominent Disasters and Associated Disease Outbreaks

Disaster	*Year*	*Location*	*Disease Outbreak(s)*	*References*
Volcanic eruption	1999	Tungurahua, Ecuador	• Tuberculosis in shelters	Tobin and Whiteford (2002)
Indian Ocean tsunami	2004	India, Indonesia, Malaysia, Maldives, Myanmar, Sri Lanka, Seychelles, Thailand, Somalia	• Measles • Hepatitis A and E • Acute gastroenteritis • Post-traumatic stress disorder	Clasen et al. (2006) Juran (2012)
Earthquake	2005	Pakistan	• Hepatitis E • Upper respiratory tract infections • Diarrhea	Watson et al. (2007)
Hurricane Katrina	2005	US Gulf Coast	• Diarrhea • Gastrointestinal infections • Skin infections • COVID-19	Murthy and Christian (2010)
Samoa earthquake and tsunami	2009	Samoa	• Upper respiratory tract infections • Pneumonia • Tsunami lung	Mavrouli et al. (2021)
Haiti earthquake	2010	Haiti	• Cholera • Typhoid • Malaria • Respiratory tract infections	Hadeed et al. (2023) Magloire et al. (2010)
Japan earthquake and tsunami	2011	East Japan	• Respiratory tract infections • Tetanus • Measles	Takahashi et al. (2012)
Typhoon Haiyan	2013	Philippines, China, Vietnam, Hong Kong	• Hypertension • Acute respiratory infections • Diarrhea	Salazar et al. (2017)
Gorkha earthquake	2015	Nepal	• Scrub typhus	Asokan and Vanitha (2017)
Hurricane Harvey	2017	Texas, US	• Respiratory tract infections • Gastrointestinal illnesses	Shukla et al. (2018)
Hurricane Maria	2017	Puerto Rico, US	• Leptospirosis • Dengue • Malaria • Diarrhea	Shin et al. (2017)
Hurricane Irma	2017	Barbuda and parts of the US and British Virgin Islands	• Respiratory illnesses • Dengue • Zika virus • West Nile virus • Cholera and dysentery	Deresinski (2017)
Earthquake	2017	Pohang, South Korea	• Anxiety • Stress-related disorders	Han (2022)

(*Continued*)

Table 16.1 (Continued)

Disaster	*Year*	*Location*	*Disease Outbreak(s)*	*References*
Cyclones Idai and Kenneth	2019	Mozambique	• Cholera • Malaria • Diarrhea	Kandawasvika et al. (2021)
Beirut explosion	2020	Beirut, Lebanon	• Depression • Anxiety • Post-traumatic stress disorder	El Khoury Malhame et al. (2021)
Flood	2022	Pakistan	• Malaria • Dengue • Diphtheria • Measles	Lama and Tatu (2022)
Earthquake	2023	Morocco	• Cholera • Typhoid • Diarrhea	Albuquerque (2023)
Flood	2023	Libya	• Diarrhea • Cholera • Malaria	Taylor (2023)

Source: Table Created by author.

Floods

Floods represent the most prevalent form of disaster on a global scale and the most common weather-related disaster. The immediate injuries and fatalities resulting from flood disasters typically arise from drowning and blunt force trauma (Howard et al., 1996). The public health implications of flooding encompass disease outbreaks triggered by the displacement of individuals into overcrowded temporary shelters and the contamination of water sources with fecal matter and hazardous chemicals (Watson et al., 2007). Furthermore, flooding often leads to the reproduction of mosquitoes, increasing mosquito-borne diseases such as malaria, chikungunya, dengue fever, yellow fever, West Nile, Zika, Japanese encephalitis, and others. In comparison to flood disasters, the public health consequences of tropical cyclones (i.e., hurricanes or typhoons) and tornadoes, particularly in terms of infectious disease outbreaks, have received less attention but can result in similar consequences (Qadri et al., 2005).

In low- and middle-income countries, people are more likely to get sick from both infectious and noninfectious diseases when floods occur (Shahzad et al., 2021). This is because such places may already have the existence or prevalence of certain diseases, and the health systems are often under-resourced. Other issues such as pollution of and damage to the environment, poverty, and political issues all make it worse when floods come. For example, in Bangladesh, infectious diseases often occur every year during and/or after heavy rains and floods. This is due to standing floodwater, interacting with the floodwater, people living close together in crowded places, low quality housing, and of course lack of potable water and functioning sewage systems—especially in the slum areas of the capital city of Dhaka (Donner & Rodríguez, 2011).

In the heavily populated South Asian countries of Bangladesh, Pakistan, Nepal, and India, many people live in flood-prone areas. Floods occur nearly every year, especially during the monsoon season, leading to frequent outbreaks of gastrointestinal and diarrheal diseases as well as leptospirosis. For example, in July 1998, a severe flood in the Malda district of West Bengal, India, promulgated a cholera outbreak. It quickly affected more than 16,000 people, mostly children, who were residing in unsanitary conditions where human waste and garbage were not properly removed, thereby contaminating water sources and the local environment (Espadale et al., 2018). In Bangladesh during the years 1988, 1998, and 2004, outbreaks of diarrheal diseases occurred after monsoon rains and related floods. The primary pathogens responsible were *Vibrio cholerae*, enterotoxigenic *Escherichia coli* (ETEC), and rotavirus (Qadri et al., 2005). All three of these disease-causing agents not only became more common but also constituted a larger proportion of the pathogens found in diarrheal stools, indicating that the viruses and bacteria have a high potential for causing epidemics (Gertler et al., 2015).

In Ethiopia, devastating floods in October 2006 caused more than 40,000 people to suffer from severe diarrhea. Similarly, in January 2007, Eastern Angola had an outbreak of cholera after heavy flooding, which had severe impacts. There was also an outbreak of typhoid fever during floods in Jakarta, Indonesia, in 2001, and in 1988 in Khartoum, Sudan, there was an outbreak of hepatitis E after flooding (Brown & Murray, 2013).

Studies have found that when there is excess rainfall or flooding, there can be more diseases spread by mosquitoes (Ruiz et al., 2018; Coalson et al., 2021). This happens because floods create more places where mosquitoes can lay their eggs, like in cases of blocked drains, saturated ditches, and pools of water. For example, after a large flood in Mozambique in 2000, there was a four-to-five-fold increase in malaria cases. After the 2004 Indian Ocean tsunami, the South Andaman and Nicobar Islands of India experienced a high incidence of malaria as there were more mosquitoes due to the salty water. Furthermore, outbreaks of Rift Valley fever have occurred over the past 48 years in Kenya after heavy rainfall periods (Walker, 2018), and a study in Thailand found that due to heavy rainfall cases of dengue hemorrhagic fever (DHF) rose (Zasada et al., 2014).

Floods increase the risk of both infectious and noninfectious diseases (Huang et al., 2016), and the impacts of flooding on human health can be divided into short- and long-term impacts. Short-term health impacts include death, injuries, trauma, and communicable diseases. Non-communicable diseases, psychosocial issues, and malnutrition are the long-term health impacts of floods (Dayrit et al., 2018). Some of common morbidities are discussed below.

Dermatological diseases frequently occur in the aftermath of flooding. Floods do not directly cause infections, but wounds and injuries, if not treated properly, can cause skin infections (Bandino et al., 2015). Moreover, floodwater contains pollutants, including sewage and chemicals; contact with these pollutants can cause irritation, infections, and allergic reactions. For example, after the 2010 floods in Pakistan, dermatological diseases spiked with greater than 70,000 cases of skin inflammation and infections (Parker et al., 2022).

Respiratory infections are common due to overcrowding in temporary shelters. Acute upper respiratory infections have been reported after floods in the US and in the Pakistan floods of 2010, 2012, and 2013 (Fendrick et al., 2023; Sharif et al., 2024). Consumption of contaminated water can also introduce pathogens that cause sore throats, coughs, and rhinitis. Post-flood tuberculosis (TB) is another concern, particularly in low-income countries with limited resources (Shokri et al., 2020). Similarly, cases of COVID-19—with respiratory symptoms, including productive and non-productive cough, sore throat, nasal obstruction, rhinorrhea (runny nose), and dyspnea (short breath)—can increase during disasters due to displacement and crowding in temporary facilities (Alzahrani et al., 2023).

Gastrointestinal diseases are highly associated with floods. Floodwater is contaminated with sewerage and microbes that causes diarrhea. The risk of infection can persist even weeks or months after the floods due to the incubation period of pathogens (Nichols et al., 2018).

Leptospirosis is a bacterial infection that is transmitted to humans when they come into contact with the urine of infected animals, either directly or through soil or water. After a flood, interactions between humans and animals (e.g., rats) increase, leading to outbreaks (Radi et al., 2018).

Mosquito-borne illnesses increase post-flood as the standing floodwater creates new and expanded breeding sites for mosquitoes. As already mentioned, the incidence of malaria, West Nile, and DHF often increases after floods. It has been highlighted that in order to control mosquito-borne illnesses after floods, effective control measures should be taken not only near temporary shelters but also near empty houses and flood-affected areas where disaster-affected populations return (Adekunle et al., 2019). Methods such as chemicals or 'fogging,' mosquito nets, personal body sprays (e.g., those that contain DEET), and public education and behavior change programs can help mitigate the occurrence of mosquito-borne illnesses after floods and other hydrometeorological disasters.

Floodwater causes contamination of drinking water sources and increases the probability of consuming a range of bacteria, viruses, protozoa, and helminths. *Typhoid* and *hepatitis A* and *E viral infections* are common. For example, Pal et al. (2016) documented an outbreak of hepatitis A among children in a flood rescue camp in Uttarakhand, India. In Denmark in August 2010, when seawater mixed with sewage due to heavy rain that filled up the sewer system, there was an outbreak of diarrhea and vomiting among people who partook in a triathlon. This was the result of three different pathogens, *E. coli*, *Campylobacter jejuni*, and *Giardia lamblia*, with a total of 42% of triathlon participants getting ill (Ota & Kawai, 2012).

In terms of non-communicable diseases, increases in *post-traumatic stress disorder* (PTSD) are associated with disasters. According to Fontalba-Navas et al. (2017), severe flooding in Almeria, Spain, in 2012 led to an increase in cases of PTSD, and it should be noted that mental health issues associated with flooding may be more severe among the elderly (Bei et al., 2013). Floods also cause chronic health issues. An analysis of the health burden after a flood event in China revealed that there was an increase in mortality rates and the number of years of potential life lost due to exposure to floods (Alderman et al., 2012).

Earthquakes

When earthquakes occur, the primary causes of mortality and morbidity are related to building and structural collapses (Forzieri et al., 2017). However, disease outbreaks are also common. As a case in point, after an earthquake in Italy in April 2009 there was a salmonella outbreak among children. This was likely a result of the earthquake altering the subsurface geology and groundwater, which caused water from a local spring to get contaminated. People used the water to grow crops and 155 children got sick, with the highest number of cases among two-year-olds. As a result, 44 children had to seek medical attention at a hospital (Behbod et al., 2017).

Another study on Italy's earthquake investigated the increase in cases of infectious diseases. Before the earthquake, 7.41% of people were admitted to the hospital for infectious diseases. However, in the two months after the earthquake, the rate rose to 27.18%. There were also more people diagnosed with infectious diseases who did not need to be admitted to the hospital, with the rate increasing from 12.04% to 27.29% (Hattori et al., 2022).

In Spain, there were 163 cases of chickenpox in Lorca in the eight weeks before an earthquake in May 2011. After the earthquake, 1424 people had to go to a temporary medical camp to control the spread of chickenpox (Pacheco Barzallo et al., 2018). While there was no direct relation between the earthquake and epidemic, the emergency situation arose and was exacerbated due to the loss of health facilities, which served to concentrate and increase the number of cases.

In February 2018, an earthquake with multiple aftershocks caused severe damage to nearly 65% of health facilities in the Hela and Southern Highlands provinces of Papua New Guinea. The United Nations Children's Fund (UNICEF) and World Health Organization (WHO) raised concerns and explicitly cautioned communities that there was increased risk of waterborne and vaccine-preventable disease outbreaks in the region, which they noted would be problematized as a result of the closed and inaccessible healthcare facilities. In addition, predominant hygiene behaviors, poor sanitation infrastructure and sewage treatment, and relatively low immunization rates among children in the marginalized Highland region contributed to the health threat (Floret et al., 2006).

Drought

Droughts occur when seasonal precipitation drops below the normal or long-term average. Human activities problematize this process through land use changes and increasing temperatures due to anthropogenic climate change. Drought-related health impacts vary widely depending on the severity of the drought, existing health condition of the baseline population, and availability of resources to mitigate these impacts. Health impacts include diseases, food insecurity (potentially leading to mal- and under-nutrition), population displacement, and loss of livelihood. The health impacts of drought may appear over a longer period of time; therefore, managing the loss is a hard-hitting task (Mera, 2018). According to recent data, an estimated 55 million people are affected by drought every year and 700 million people are at risk of displacement by 2030 as a result (Lieber et al., 2022).

The risk of cholera outbreak increases with droughts due to reduced availability of water for hygiene purposes. Rieckmann et al. (2018) found that cholera outbreaks in sub-Saharan Africa occur four times more frequently during droughts compared to floods. Short rains after a long period of drought cause contamination of water resources with microbial pathogens and increase the risk of waterborne diseases. Groundwater tables drop during periods of drought and the flow of surface water also diminishes. In the UK, an outbreak of *Cryptosporidium* was recorded due to contamination of boreholes when heavy rainfall occurred after an extended dry period (Levy et al., 2016).

Droughts disrupt food supply and accessibility, as there is less water available to produce and process it. A shortage of food can lead to malnutrition and weaken the immune system, rendering individuals more susceptible to infectious diseases (Lieber et al., 2022). Drought can also affect the transmission of vector-borne diseases as changes in temperature and the availability of water affect the habitat and breeding processes of vectors such as mosquitoes, ticks, and flies, which can transmit diseases malaria, DHF, West Nile, and Lyme disease, among others (Brown et al., 2014). Droughts can cause the migration of people from affected areas in search of water, livelihood, and other resources. Displacement often forces individuals to live in overcrowded shelters or urban slums, and infectious diseases may spread due to poor living conditions and close human contact (Asmall et al., 2021).

Prolonged droughts can disturb mental health through stress induced by food insecurity, trauma from increased conflict, and other fluctuations that disrupt everyday life. Edwards et al. (2015) studied the impacts of drought on farmers and concluded that it adversely affected the mental health of farmers in Australia and caused long-term traumatic symptoms. Drought also increases the risk of respiratory diseases by increasing air dust particulate pollution. Wang et al. (2021) reported that seasonal drought in China increases the clinical visits of children and adolescents due to increased respiratory illness. Wildfires are more common during drought, which degrade air quality and intensify particulate matter in the air, thus perturbing cases of asthma and TB and increasing the risk of respiratory illness (Machado-Silva et al., 2022).

Common Routes of Disease Transmission

After a disaster, the outbreak of infectious diseases poses a serious threat to human health. Extreme weather events can create an environment in which infectious diseases of humans, animals, and zoonoses (diseases that are able to spread directly from humans to animals and vice versa) grow and spread. On the basis of mode of transmission, communicable diseases and their routes are discussed below.

Transmission through Water

Waterborne diseases primarily include gastrointestinal, diarrheal, and respiratory diseases as well as leptospirosis—which are caused by water contaminated with pathogenic microorganisms (Smith et al., 2013). Disasters damage water, sanitation, and drainage infrastructures, pollute clean water supplies, and often result in standing water. Thus, access to clean drinking water is a key challenge after a disaster. According to Shokri et al. (2020), extreme events disrupt water supply systems (e.g., broken pipes and infiltration of contaminants into pipes) and sanitation infrastructure (e.g., backflows and damaged septic tanks) resulting in contamination of water with human and animal excreta (e.g., *E. coli*) as well as other pollutants, thereby increasing the risk of waterborne diseases.

Vector-Borne Transmission

The relationship between disasters and vector-borne diseases (VBD) is influenced by multiple factors which may include the lifecycle of the vector, growth of an infectious agent within a vector, environmental determinants of human and vector interface, susceptibility of the host, and individual and community health behaviors. Disasters typically cause changes in environmental conditions that may alter the transmission rate by vectors in human beings. Standing water and hydrological perturbations can act as suitable breeding grounds for mosquitoes (and other vectors) and can pose long-term health risks well after the initial impacts of the environmental change (Arshad et al., 2022). VBDs such as dengue fever, chikungunya, malaria, yellow fever, Zika, Japanese encephalitis, and West Nile have been found to increase in flood-affected areas. Regionally, sub-Saharan Africa is already at risk to numerous VBDs (malaria being the major case in point) (Zerbo et al., 2020), Southeast Asia is reported to have become hyperendemic in terms of dengue (Bhatia et al., 2014), and millions of Americans live in areas demonstrating historical risk to yellow fever, dengue, and Zika (Lehman et al., 2007)—and locally acquired cases of malaria were just confirmed in the US in 2023 (Murthy & Christian, 2010). These existing contexts will have to grapple with further changes in water, vegetation, environmental conditions, and human behaviors (e.g., increased migration and human-wildlife interfaces) as disasters and anthropogenically induced climate change continue to unfold.

Transmission due to Overcrowding

Amplified occurrences of communicable diseases such as acute (lower) respiratory infections, increased transmission of measles, and incidence of TB have been linked to the crowded conditions that prevail in evacuation centers, cyclone shelters, and temporary housing. Overcrowded shelter environments also allow sandflies to thrive and increase the chances of disease transmission among vulnerable populations (Mavrouli et al., 2021).

Vulnerable Groups in Disaster Response

Children, elderly populations, women, disabled populations, migrants and displaced persons, and unvaccinated individuals are more at risk of infection than others (The Sphere Handbook, 2018). It has been observed globally that more women and children die in disasters compared to men (Juran & Trivedi, 2015). In a survey of evacuees after Hurricane Katrina in Louisiana, 44% of parents reported new behavioral and emotional problems among at least one of their children (Brodie et al., 2006). Children are more susceptible to injuries and psychological trauma both during and after disasters. Roughly 1 billion children are at high risk due to climate change and climate-related disasters worldwide, so the issue is widespread (Yasin et al., 2023). The perturbation of family structure by disasters renders them more vulnerable due to their dependence on adults. Disasters also place pressure on children in terms of absence from school.

Elderly individuals often require special assistance in evacuation and during disasters due to mobility issues. Elderly are more likely to have chronic and degenerative health issues, making them more susceptible to infectious diseases (Banerjee, 2020). Weakened immune systems and preexisting health conditions also render elderly populations more vulnerable to post-disaster health issues and infectious diseases.

Women have special health considerations during disasters, such as menstrual hygiene management. Access to feminine hygiene products during and after disasters (e.g., in transitional settlements) is disrupted and can lead to infections and skin irritation. Further, pregnant women have special health needs as disasters worsen their health and access to prenatal and hospital care (Juran, 2012). Additionally, issues related to dehydration and nutrition can affect lactation among new mothers, and mothers who lose infants in disasters may suffer from engorged breasts or need to express milk.

Disasters negatively affect disabled and differently abled populations. Their dependence upon others makes more and prone to infectious diseases and mental trauma (Levy & Patz, 2015). Finally, socioeconomic inequalities also increase the vulnerability of low-income communities during disasters and situate them more susceptible to injuries, morbidities, and mortality (Aksha et al., 2019).

Underlying Causes of Disease Outbreaks in Low- and Middle-Income Countries

After a disaster, both developed and lesser developed countries face the challenge of epidemics, but low- and middle-income countries often face a greater burden of diseases due to prevailing resource mismanagement as well as weak institutional arrangements. Low-income countries typically have fragile construction and weak infrastructure, and the disaster-affected populations often reside in areas more vulnerable to disasters due to limited resources. Such contexts lack the financial resources to invest in better infrastructure, which is further compounded by disasters and climate change (Yasin et al., 2023). According to the World Bank, 91% of deaths due to climate-related hazards occurred in low- and middle-income countries. Healthcare systems in lesser developed countries often lack adequate equipment, resources, training, and infrastructure to deal with emergency situations. In the event of disaster, an already weak health system fails to provide medical assistance to those in need (Shahzad et al., 2019). Access to clean water and sanitation is the main challenge confronted by people after a disaster; yet in order to prevent diseases, clean drinking water and proper sanitation services are essential (Vaishali et al., 2020). As relatively poorer countries are already struggling to provide these services, disasters serve to worsen the situation. Evacuation of people from disaster-affected areas, rapid deployment of search and rescue teams, and transportation of injured people to hospitals all require stable transportation networks. However, poor transportation infrastructure, lack of digital mapping of networks and road infrastructure, and environmental degradation and mismanaged land use add to the complexity of emergency response (Shahzad et al., 2021).

Strategies for the Prevention of Disease Spread

In the response phase of a disaster, the mitigation of disease spread is aided by practical preplanned preventive measures. Communicable diseases spread mostly because of unhygienic conditions combined with high population crowding. The emergency response plan in the post-disaster phase requires a clear evacuation plan that encompasses the provision of shelter, water, and sanitation needs, a safe food supply, and a healthcare plan and medical supplies (Basaria et al., 2023). Following are a few steps that act as control measures in cases of post-disaster displacement.

Provision of Shelter, Water, and Sanitation Facilities

As the most critical point of entry, providing appropriate shelter or identifying a site for evacuation can reduce the chance of disease spreading. The size of the shelter site should be according to the number of affected persons as well as the need to have access to other necessary facilities. The size of shelter place is important as it can reduce crowding and the chance of disease spread as communicable diseases have a higher prevalence in congested and less ventilated places. Many diseases, such as TB, diarrhea, measles, and respiratory infections, are reported due to high population pressure in overfilled settlements in the disaster response phase (Jafari et al., 2011). Furthermore, the provision of safe water and sanitation facilities is essential for the control of communicable diseases. Water is the most common medium of disease transmission in shelters as many basic needs are linked to water availability such as drinking, cooking, cleaning, and self-hygiene. Each of these processes introduces opportunities for the introduction of pathogens, and all interact with the availability of hygiene infrastructure (e.g., toilets and drainage) and resources (e.g., water and soap).

Preparation and Safe Supply of Food

The preparation and supply of food to disaster-affected populations is an effective control measure to reduce foodborne contamination. The World Health Organization (2005) has recommended a guideline for ensuring the safe supply of food to disaster-affected populations. This involves using clean water for cooking, preserving the food to prevent the growth of microorganisms, cooking food at high temperatures to kill any microorganisms, and separating raw and cooked food. Hygienic storage vessels, serving and eating utensils, and plates are also required, not to mention soap for washing utensils and hands.

Healthcare and Medical Supplies

Another key control measure for preventing disease outbreaks is the timely provision of medical facilities and healthcare to those who are sick, injured, or vulnerable to developing an ailment. This should involve setting up a medical clinic at the site of a shelter (Jafari et al., 2011). Temporary health clinics are useful for providing initial and regular checkups and basic medicines to control

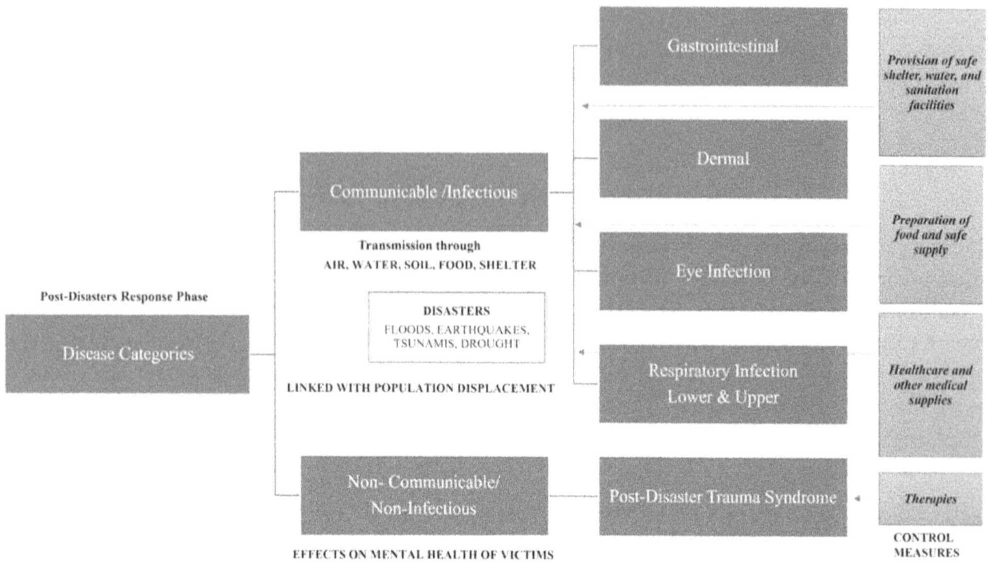

Figure 16.2 Schematic Diagram of Different Disease Types and Control Measures

Source: Image from author.

secondary mortality or morbidity at disaster sites. Today, with advancements in technology, a mobile health service can also provide effective control. Along with these, running a vaccination campaign for more vulnerable groups like children and the elderly can be an effective measure.

On the basis of a literature review, the following flowchart (Figure 16.2) was developed to conceptualize the types of diseases and control measures after a disaster. It is imperative to implement a monitoring and reporting system to assess a problem and provide a prompt solution. Identifying a solution within a short period of time is also key to making a control measure effective. Creating awareness about good hygiene practices, proper disposal of human and solid waste, and vector-borne diseases helps pave the way to controlling disease transmission. Finally, it is important to involve local communities in disease control measures and the prevention of disease outbreaks. Local communities can be engaged in the distribution of hygiene supplies, maintenance of sanitation facilities, and monitoring of water quality. Such roles can help usher in long-term behavioral change and a feeling of stewardship.

Conclusion

The escalating frequency and severity of disasters pose substantial threats to human life and well-being on a global scale. This chapter is an effort to demonstrate the nexus between disasters and the risk of communicable diseases, which underscores the imperative for robust policy implementation and resource planning in disaster response. A proactive and comprehensive approach to disease control is essential, not only to mitigate immediate morbidity and mortality but also to address the long-term health consequences that often accompany such catastrophic events. By prioritizing and implementing effective measures, societies can reduce their vulnerabilities, enhance their resilience, and ultimately minimize the adverse health impacts associated with disasters.

References

Adekunle, A. I., Adegboye, O. A., & Rahman, K. M. (2019). Flooding in Townsville, north Queensland, Australia, in February 2019 and its effects on mosquito-borne diseases. *International Journal of Environmental Research and Public Health, 16*(8), 1393.

Ahmad, J., Ahmad, M. M., Su, Z., Rana, I. A., Rehman, A., & Sadia, H. (2022). A systematic analysis of world-wide disasters, epidemics and pandemics associated mortality of 210 countries for 15 years (2001–2015). *International Journal of Disaster Risk Reduction, 76*, 103001.

Aksha, S. K., Juran, L., Resler, L. M., & Zhang, Y. (2019). An analysis of social vulnerability to natural hazards in Nepal using a modified social vulnerability index. *International Journal of Disaster Risk Science, 10*(1), 103–116.

Albuquerque, D. C. (2023). Here's how to help earthquake-hit Morocco reduce the risk of disease. Aljazeera. https://www.aljazeera.com/opinions/2023/9/11/heres-how-to-help-earthquake-hit-morocco-reduce-the-risk-of-disease#:~:text=So%20what%20can%20we%20do,drums%2C%20filters%20and%20purification%20tablets.

Alderman, K., Turner, L. R., & Tong, S. (2012). Floods and human health: A systematic review. *Environment International, 47*, 37–47.

Alexander, D. E. (2015). *Disaster and Emergency Planning for Preparedness, Response, and Recovery*. Oxford University Press.

Alzahrani, M. M., Alaraifi, A. K., Aldosari, L. H., Hijazi, L. O., & Alsaab, F. A. (2023). Clinical manifestations of COVID-19 versus other upper respiratory tract infections in pediatric patients. *Saudi Medical Journal, 44*(1), 74–79. https://doi.org/10.15537/smj.2023.44.1.20220439

Arshad, T., Wajahat, A., Jabeen, A., & Ali, S. H. (2022). Malaria and dengue outbreaks during a national disaster in Pakistan: A rising concern for public health. *Journal of Global Health, 12*.

Asmall, T., Abrams, A., Röösli, M., Cisse, G., Carden, K., & Dalvie, M. A. (2021). The adverse health effects associated with drought in Africa. *Science of the Total Environment, 793*, 148500.

Asokan, G. V., & Vanitha, A. (2017). Disaster response under One Health in the aftermath of Nepal earthquake, 2015. *Journal of Epidemiology and Global Health, 7*(1), 91–96.

Bandino, J. P., Hang, A., & Norton, S. A. (2015). The infectious and noninfectious dermatological consequences of flooding: A field manual for the responding provider. *American Journal of Clinical Dermatology, 16*, 399–424.

Banerjee, D. (2020). The impact of Covid-19 pandemic on elderly mental health. *International journal of geriatric psychiatry, 35*(12), 1466.

Basaria, A. A., Ahsan, A., Nadeem, A., Tariq, R., & Raufi, N. (2023). Infectious diseases following hydrometeorological disasters: Current scenario, prevention, and control measures. *Annals of Medicine and Surgery, 85*(8), 3778–3782.

Behbod, B., Leonardi, G., Motreff, Y., Beck, C. R., Yzermans, J., Lebret, E., & Pirard, P. (2017). An international comparison of the instigation and design of health registers in the epidemiological response to major environmental health incidents. *Journal of Public Health Management and Practice: JPHMP, 23*(1), 20.

Bei, B., Bryant, C., Gilson, K. M., Koh, J., Gibson, P., Komiti, A., & Judd, F. (2013). A prospective study of the impact of floods on the mental and physical health of older adults. *Aging & Mental Health, 17*(8), 992–1002.

Bhatia, R., Ortega, L., Dash, A. P., & Mohamed, A. J. (2014). Vector-borne diseases in South-East Asia: Burdens and key challenges to be addressed. *WHO South-East Asia J Public Health, 3*(1), 2–4. DOI:10.4103/2224-3151.206878

Brodie, M., Weltzien, E., Altman, D., Blendon, R. J., & Benson, J. M. (2006). Experiences of Hurricane Katrina evacuees in Houston shelters: Implications for future planning. *American Journal of Public Health, 96*(8), 1402–1408.

Brown, L., Medlock, J., & Murray, V. (2014). Impact of drought on vector-borne diseases–how does one manage the risk? *Public Health, 128*(1), 29–37.

Brown, L., & Murray, V. (2013). Examining the relationship between infectious diseases and flooding in Europe: A systematic literature review and summary of possible public health interventions. *Disaster Health, 1*(2), 117–127.

Chen, Y. E., Li, C., Chang, C. P., & Zheng, M. (2021). Identifying the influence of natural disasters on technological innovation. *Economic Analysis and Policy, 70*, 22–36.

Clasen, T., Smith, L., Albert, J., Bastable, A., & Fesselet, J. F. (2006). The drinking water response to the Indian Ocean tsunami, including the role of household water treatment. *Disaster Prevention and Management: An International Journal, 15*(1), 190–201.
Coalson, J. E., Anderson, E. J., Santos, E. M., Madera, G. V., Romine, J. K., Dominguez, B., Richard, D. M., Little, A. C., Hayden, M. H., & Ernst, K. C. (2021). The complex epidemiological relationship between flooding events and human outbreaks of mosquito-Borne diseases: A scoping review. *Environment Health Perspective, 129*(9), 96002. DOI:10.1289/EHP8887.
Dayrit, J. F., Bintanjoyo, L., Andersen, L. K., & Davis, M. D. P. (2018). Impact of climate change on dermatological conditions related to flooding: Update from the international society of dermatology climate change committee. *International Journal of Dermatology, 57*(8), 901–910.
Deresinski, S. (2017). Extreme weather and infectious diseases: What will follow Hurricanes Harvey and Irma? *Infectious Disease Alert, 37*(1).
Donner, W., & Rodríguez, H. (2011). Disaster Risk and Vulnerability: The Role and Impact of Population and SocietyEm-DAT (2016). Em-DAT: The international disaster database. Centre for Research on the Epidemiology of Disasters. Universite Catholique de Louvain. Brussels. Retrieved July, 2016.
Edwards, B., Gray, M., & Hunter, B. (2015). The impact of drought on mental health in rural and regional Australia. *Social Indicators Research, 121*, 177–194.
El Khoury Malhame, M., Harajli, D. A., Reykowska, D., & Ohme, R. (2021). Traumatic effect of Beirut port explosion on Lebanese people's experiences. *Available at SSRN 3929408.*
Erickson, T. B., Brooks, J., Nilles, E. J., Pham, P. N., & Vinck, P. (2019). Environmental health effects attributed to toxic and infectious agents following hurricanes, cyclones, flash floods and major hydrometeorological events. *Journal of Toxicology and Environmental Health, Part B, 22*(5-6), 157–171.
Espadale, E., Pinchbeck, G., Williams, N. J., Timofte, D., McIntyre, K. M., & Schmidt, V. M. (2018). Are the hands of veterinary staff a reservoir for antimicrobial-resistant bacteria? A randomized study to evaluate two hand hygiene rubs in a veterinary hospital. *Microbial Drug Resistance, 24*(10), 1607–1616.
Fendrick, A. M., Monto, A. S., Nightengale, B., & Sarnes, M. (2023). The economic burden of non-influenza-related viral respiratory tract infection in the United States. *Archives of Internal Medicine*, 24;*163*(4), 487–494.
Floret, N., Viel, J. F., Mauny, F., Hoen, B., & Piarroux, R. (2006). Negligible risk for epidemics after geophysical disasters. *Emerging Infectious Diseases, 12*(4), 543–548. DOI:10.3201/eid1204.051569. PMID: 16704799; PMCID: PMC3294713.
Fontalba-Navas, A., Lucas-Borja, M. E., Gil-Aguilar, V., Arrebola, J. P., Pena-Andreu, J. M., & Perez, J. (2017). Incidence and risk factors for post-traumatic stress disorder in a population affected by a severe flood. *Public Health, 144*, 96–102.
Forzieri, G., Cescatti, A., e Silva, F. B., & Feyen, L. (2017). Increasing risk over time of weather-related hazards to the European population: A data-driven prognostic study. *The Lancet Planetary Health, 1*(5), e200–e208.
Gertler, M., Dürr, M., Renner, P., Poppert, S., Askar, M., Breidenbach, J., & Wilking, H. (2015). Outbreak of *Cryptosporidium hominis* following river flooding in the city of Halle (Saale), Germany, August 2013. *BMC Infectious Diseases, 15*, 1–10.
Hadeed, S. J., Broadway, K. M., Schwartz-Watjen, K. T., Tigabu, B., Woodards, A. J., Swiatecka, A. L., … & Wu, A.. (2023). Notional spread of cholera in Haiti following a natural disaster: Considerations for military and disaster relief personnel. *Military Medicine, 188*(7–8), e2074–e2081.
Hammer, C. C., Brainard, J., & Hunter, P. R. (2019). Risk factors for communicable diseases in humanitarian emergencies and disasters: Results from a three-stage expert elicitation. *Global Biosecurity, 1*(1), 1–14.
Han, C. (2022). Induced seismicity and acute development of anxiety and stress-related mental disorders: Findings from the 2017 Pohang earthquake. *Environmental Health Perspectives, 130*(6), 067701.
Hattori, T., Chagan-Yasutan, H., Koga, S., Yanagihara, Y., & Tanaka, I. (2022). Seminar lessons: Infectious diseases associated with and causing disaster. *Reports, 5*(1), 7.
Howard, M. J., Brillman, J. C., & Burkle, F. M. (1996). Infectious disease emergencies in disasters. *Emergency Medicine Clinics, 14*(2), 413–428.
Huang, L. Y., Wang, Y. C., Wu, C. C., Chen, Y. C., & Huang, Y. L. (2016). Risk of flood-related diseases of eyes, skin and gastrointestinal tract in Taiwan: A retrospective cohort study. *PloS One, 11*(5), e0155166.
Jafari, N., Shahsanai, A., Memarzadeh, M., & Loghmani, A. (2011). Prevention of communicable diseases after disaster: A review. *Journal of Research in Medical Sciences: the Official Journal of Isfahan University of Medical Sciences, 16*(7), 956.

Juran, L. (2012). The gendered nature of disasters: Women survivors in post-tsunami Tamil Nadu. *Indian Journal of Gender Studies, 19*(1), 1–29.
Juran, L., & Trivedi, J. (2015). Women, gender norms, and natural disasters in Bangladesh. *Geo. Review, 105*(4), 601–611.
Kandawasvika, G. Q., Thabethe, N., Dzangare, J., & Manangazira, P. (2021). Impacts of tropical cyclones Idai and Kenneth on public health in Southern Africa. *Cyclones in Southern Africa, 3: Implications for the Sustainable Development Goals*, 69–79.
Lama, A. P., & Tatu, U. (2022). "Climate change and infections: Lessons learnt from recent floods in Pakistan." *New Microbes and New Infections, 49*.
Lee, J., Perera, D., Glickman, T., & Taing, L. (2020). Water-related disasters and their health impacts: A global review. *Progress in Disaster Science, 8*, 100123.
Lehman, J. A., Hinckley, A. F., & Kniss, K. L. (2007). Effect of hurricane Katrina on arboviral disease transmission. *Emerging Infectious Diseases, 13*(8), 1273–1274. DOI:10.3201/eid1308.061570
Levy, B. S., & Patz, J. A. (2015). Climate change, human rights, and social justice. *Annals of Global Health, 81*(3), 310–322.
Levy, K., Woster, A. P., Goldstein, R. S., & Carlton, E. J. (2016). Untangling the impacts of climate change on waterborne diseases: A systematic review of relationships between diarrheal diseases and temperature, rainfall, flooding, and drought. *Environmental Science & Technology, 50*(10), 4905–4922.
Lieber, M., Chin-Hong, P., Kelly, K., Dandu, M., & Weiser, S. D. (2022). A systematic review and meta-analysis assessing the impact of droughts, flooding, and climate variability on malnutrition. *Global Public Health, 17*(1), 68–82.
Machado-Silva, F., Libonati, R., de Lima, T. F. M., Peixoto, R. B., de Almeida Franca, J. R., Magalhães, M. D. A. F. M., … & DaCamara, C. C.. (2022). Drought and fires influence the respiratory diseases hospitalizations in the Amazon. *Ecological Indicators, 109*, 105817.
Magloire, R., et al. (2010). "Launching a National Surveillance System after an earthquake-Haiti, 2010." *Morbidity and Mortality Weekly Report, 59*(30), 933–938.
Mavrouli, M., Mavroulis, S., Lekkas, E., & Tsakris, A. (2021). Respiratory infections following earthquake-induced tsunamis: Transmission risk factors and lessons learned for disaster risk management. *International Journal of Environmental Research and Public Health, 18*(9), 4952.
Mera, G. A. (2018). Drought and its impacts in Ethiopia. *Weather and Climate Extremes, 22*, 24–35.
Murthy, S., & Christian, M. D. (2010). Infectious diseases following disasters. *Disaster Medicine and Public Health Preparedness, 4*(3), 232–238.
Nichols, G., Lake, I., & Heaviside, C. (2018). Climate change and water-related infectious diseases. *Atmosphere, 9*(10), 385.
Ota, H., & Kawai, H. (2012). An unusual case of pleural empyema in a tsunami survivor. *Asian Cardiovascular and Thoracic Annals, 20*(3), 344–346.
Pacheco Barzallo, D., Pacheco Barzallo, A., & Narvaez, E. (2018). The 2016 earthquake in Ecuador: Zika outbreak after a natural disaster. *Health Security, 16*(2), 127–134.
Pal, S., Juyal, D., Sharma, M., Kotian, S., Negi, V., & Sharma, N. (2016). An outbreak of hepatitis a virus among children in a flood rescue camp: a post-disaster catastrophe. *Indian Journal of Medical Microbiology, 34*(2), 233–236.
Parker, R., et al. (2022). "The dermatological manifestations of extreme weather events: A comprehensive review of skin disease and vulnerability." *The Journal of Climate Change and Health, 8*, 100162.
Qadri, F., Khan, A. I., Faruque, A. S. G., Begum, Y. A., Chowdhury, F., Nair, G. B., & Svennerholm, A. M. (2005). Enterotoxigenic *Escherichia coli* and Vibrio cholerae diarrhea, Bangladesh, 2004. *Emerging Infectious Diseases, 11*(7), 1104.
Radi, M. F. M., Hashim, J. H., Jaafar, M. H., Hod, R., Ahmad, N., Nawi, A. M., & Ayub, N. I. F. (2018). Leptospirosis outbreak after the 2014 major flooding event in Kelantan, Malaysia: A spatial-temporal analysis. *The American Journal of Tropical Medicine and Hygiene, 98*(5), 1281.
Rieckmann, A., Tamason, C. C., Gurley, E. S., Rod, N. H., & Jensen, P. K. M. (2018). Exploring droughts and floods and their association with cholera outbreaks in sub-Saharan Africa: A register-based ecological study from 1990 to 2010. *The American Journal of Tropical Medicine and Hygiene, 98*(5), 1269.
Ruiz, E. F., Vasquez-Galindo, C. M., Aquije-Pariona, X. M., & Torres-Roman, J. S. (2018). Outbreaks caused by *Aedes aegyptis* due to El Niño in a coastal area of Peru. *Travel Medicine and Infectious Disease, 21*, 78–79.

Salazar, M. A., Law, R., Pesigan, A., & Winkler, V. (2017). Health consequences of Typhoon Haiyan in the Eastern Visayas Region using a syndromic surveillance database. *PLoS Currents*, *9*.
Shahzad, L., Shah, M., Saleem, M., Mansoor, A., Sharif, F., Tahir, A., & Ghafoor, G. (2021). Livelihood vulnerability index: A pragmatic assessment of climatic changes in flood affected community of Jhok Reserve Forest, Punjab, Pakistan. *Environmental Earth Sciences*, *80*, 1–16.
Shahzad, L., Tahir, A., Sharif, F., Khan, W. U. D., Farooq, M. A., Abbas, A., & Saqib, Z. A. (2019). Vulnerability, well-being, and livelihood adaptation under changing environmental conditions: A case from mountainous region of Pakistan. *Environmental Science and Pollution Research*, *26*, 26748–26764.
Sharif, F., Shahzad, L., & Batool, M. (2024). The association between climatic factors and waterborne infectious outbreaks with a focus on vulnerability in Pakistan: Integrative review. *International Journal of Environmental Health Research* *9*:1–18. https://doi.org/10.1080/09603123.2024.2302040.
Shin, P., Sharac, J., Gunsalus, R., Leifer, B., & Rosenbaum, S. (2017). Puerto Rico's community health centers: Struggling to r ecover in the wake of Hurricane Maria. *RCHN Community Health Foundation*. https://geigergibson.publichealth.gwu.edu/sites/g/files/zaxdzs4421/files/2021-12/final-ggrchn-puerto-rico-health-center-report_10.31.pdf
Shokri, A., Sabzevari, S., & Hashemi, S. A. (2020). Impacts of flood on health of Iranian population: Infectious diseases with an emphasis on parasitic infections. *Parasite Epidemiology and Control*, *9*, e00144.
Shukla, M. A., Woc-Colburn, L., & Weatherhead, J. E. (2018). Infectious diseases in the aftermath of hurricanes in the United States. *Current Tropical Medicine Reports*, *5*, 217–223.
Simonovic, S. P., Kundzewicz, Z. W., & Wright, N. (2021). Floods and the COVID-19 pandemic—A new double hazard problem. *WIREs Water*, *8*(2), e1509. DOI:10.1002/wat2.1509.
Smith, J. K., Young, M. M., Wilson, K. L., & Craig, S. B. (2013). Leptospirosis following a major flood in Central Queensland, Australia. *Epidemiology & Infection*, *141*(3), 585–590.
Takahashi, T., Goto, M., Yoshida, H., Sumino, H., & Matsui, H. (2012). Infectious diseases after the 2011 Great East Japan earthquake. *Journal of Experimental & Clinical Medicine*, *4*(1), 20–23.
Taylor, L. (2023). Stretched medical aid agencies race to disaster zones in Libya and Morocco. *BMJ: British Medical Journal (Online)*, *382*, 2131.
The Sphere Handbook: Humanitarian Charter and Minimum Standards in humanitarian response. (2018). 4th Edition. ISBN: 978-1-908176-707. Sphere association 2022 Geneva Switzerland. Pages-458
Tobin, G. A., & Whiteford, L. M. (2002). Community resilience and volcano hazard: The eruption of Tungurahua and evacuation of the faldas in Ecuador. *Disasters*, *26*(1), 28–48.
Vaishali, P. J., Malik, A., Talukdar, S. J., & Singh, P. (2020). Emergence and containment of zoonoses after disasters.
Walker, M. D. (2018). Leptospirosis: The possible risk to those participating in water-based sports and activities. *British Journal of General Practice*, *68*(673), 394–339.
Wang, B., Wang, S., Li, L., Xu, S., Li, C., Li, S., ... & Luo, B. (2021). The association between drought and outpatient visits for respiratory diseases in four northwest cities of China. *Climatic Change*, *167*(1), 2.
Watson, J. T., Gayer, M., & Connolly, M. A. (2007). Epidemics after natural disasters. *Emerging Infectious Diseases*, *13*(1), 1.
World Health Organization. (2005). Ensuring food safety in the aftermath of natural disasters. http://www.who.int/foodsafety/foodborne_disease/emergency/en/index6.html.
Yasin, A., Shahzad, L., Hayyat, M. U., Ghafoor, G. Z., Saeed, S., Sharif, F., ... & Saleem, M. B. (2023). Assessing and mapping the impact of tourism and changing climatic conditions in Himalayan region of Pakistan. *Journal of Environmental Management & Tourism*, *14*(3), 605–622.
Zasada, A. A., Forminska, K., Ogrodnik, A., Gierczynski, R., & Jagielski, M. (2014). Screening for anthrax occurrence in soil of flooded rural areas in Poland after rainfalls in spring 2010. *Annals of Agricultural and Environmental Medicine*, *21*(3).
Zerbo, A., Delgado, R. F., & González, P. A. (2020). Aedes-borne viral infections and risk of emergence/resurgence in sub-Saharan African urban areas. *Journal of Biosafety and Biosecurity*, *2*(2), 58–63. https://doi.org/10.1016/j.jobb.2020.10.002.

17

TECHNOLOGY AND TECHNOLOGICAL METHODS IN DISASTER RESPONSE

Wendy Stout, Nathan Heazlewood, and Ed Cook

Introduction

Technology is integral to disaster response. Accurate and timely intelligence information is critical for decision makers to inform and understand situations, locations, and risks in both the natural and built environments. Technological methods for disasters have been foundational for decision support for many years, including telecommunications, the internet, imagery, and basic geospatial analysis. Traditionally, technology use in disaster response centered on the activities of professionals, including emergency call center operators, medical personnel, firefighters, and military personnel. Government officials have been guided on how to respond based on a general disaster management plan, which defines a chain of command in the time of a disaster, as well as an Information and Communications Technology-Business Continuity Plan (ICT-BCP), which guides government officials on how to sustain ICT facilities after a disaster. The critical difference between emergency management and other scenarios is the emphasis of knowing *where* and *when*. Using systems that do not allow this information to be visualized using location can create challenges for responders. This includes not having shared awareness of where events, tasks, and resources relate to each other. Geospatial technology addresses how location information is collected, analyzed, and disseminated. Investing in emerging geospatial software/technology, developing people and processes, understanding the ongoing needs of users, and enabling interoperable data is key to developing technology capacity that will be successfully implemented before and during a disaster.

Disasters Can Be the Catalyst for Innovative Technology Applications

Effective disaster management is intelligence-led. Geospatial technology is transforming how information is recorded, exchanged, and analyzed, creating a new paradigm for how disaster response is conducted. Conventional geospatial response generally has been led by small, centralized geographic information systems (GIS) teams that include governmental agencies, relief response organizations, and GIS vendors and may operate out of centralized emergency operations centers in or near the disaster location. These GIS units are in a support capacity and collect relevant data, damage information, create maps to support decision makers, and disseminate information, but often lack the workforce to handle large disasters. Technology can support and help overcome that deficiency. Specialized workflows over the last decade have become commoditized through

DOI: 10.4324/9781003414834-23

web-based platforms, allowing a non-specialist audience the ability to view and maintain data (Gordon & Cook, 2021). Many of the emerging geo-enabled technologies highlighted in this chapter have found applications in disasters because traditional methods were not sufficient to meet the real-time demands. Technology challenges and opportunities can be categorized under five broad headings: data, software, technology infrastructure, process, and people/organization. This chapter discusses each of these areas, which can be useful ways to understand how technology can be successfully used in emergency management.

Box 17.1: Case Study – Innovation

A severe flood in Pakistan in August 2022 caused significant agricultural losses. This flood affected ~33 million people and increased food insecurity. The damage assessment process led to a new technology to evaluate crop-specific losses. The new index, Disaster Vegetation Damage Index (DVDI), using Synthetic Aperture Radar (SAR) satellite data and optical satellites, was developed to map flood extent and detect differences in vegetation before and after the flood event to assess flood damage (2022 Pakistan Floods, 2023).

Data Challenges in Disaster Response

Important data considerations include understanding data collection, leveraging data sources, data management, data standards, awareness, data security, and data quality.

Data Collection

Understanding what data is required includes evaluating and categorizing existing data and determining required data. The categories of threats, assets, personnel, tasks, and assignments are useful to structure and communicate information with stakeholders.

What Data to Collect about Assets

Assets are defined as 'things that have value.' Some assets assist with an emergency response (e.g., vehicles, bases, staging locations, fuel reserves, and equipment). The definition also encompasses vulnerable assets that an emergency response is seeking to protect (e.g., houses, infrastructure, industrial sites, and areas of environmental or cultural value).

Elements of data to collect should include the following:

- Quantity of each asset.
- Quality or status of each asset.
- Geo-temporal attributes, such as the location of moving assets associated with a timestamp
- Vulnerabilities.
- Responsibilities or ownership.
- For certain types of assets, cost is also an important factor: for example, knowing how much it costs for each hour of flight time for an additional helicopter.
- In addition to actual values, it is also useful to enable forecasting.

What Data to Collect about Tasks and Assignments

Emergency management systems revolve around assessing threats and then creating tasks to deal with those threats. Often during emergencies, it is not possible to promptly deal with all tasks; therefore, there must be methods of prioritization. To achieve this, tasks must be matched to assets and personnel that can respond to those tasks. An emergency management system should efficiently manage the process of tracking when tasks are created, when resources are assigned, and when tasks are marked as complete. This information should also be easily accessed and understood through reporting (i.e., map-based dashboards).

What Data to Collect about Threats

Attributes, such as status, size, current location, direction of travel, and speed, are common for any type of threat (whether it is a hurricane or a wildfire or civil unrest, etc.). It is also critical to understand the rate of change and measurements at past, present, and future times. For example, knowing the location and heading of a wildfire yesterday and a different location and heading today can be important in planning for tomorrow. Threats are not just fast-moving factors during an initial event. For example, after a flood, a secondary threat might be a slow-moving outbreak of disease well after rain has stopped. This secondary threat may be deadlier than the initial threat. There are approaches such as predictive analysis and scenario planning that can assist with this.

Collecting data about threats is a priority, but it is also critical to communicate these threats effectively to the public. Alert systems can be effective, but only if they are utilized. Conversely, sometimes there is sensitivity of data about threats (especially security-related); therefore, methods to collect the area of impact should be nuanced (i.e., using movement control data such as roadblocks and cordons for defining an area impacted by a terrorism threat).

Box 17.2: Case Study – Data Challenge #1

During the 2018 Camp Fire in California, chaos broke loose when the emergency notification system the county used failed to activate for all evacuees. The program, CodeRED, had two emergency notification modes. One was for zone-by-zone evacuation, which is helpful in evacuating people and maintaining order and traffic. The second mode was an all-resident evacuation alert. Under the pressure of the moment and not sure what to do, the sheriff's office chose the zone by-zone alerts without knowing that to receive these alerts, residents had to opt in to CodeRED alerts. Fewer than half of the residents had registered for these alerts. The sheriff's office did not learn until the next day that those alerts had not been sent to everyone. The Camp Fire would become California's deadliest wildfire, killing 85 people (Comfort et al., 2019).

Leveraging Data Sources

In a disaster situation, there is a need to mobilize, collect data, and turn it around as quickly as possible. Ideally, data considerations are addressed prior to a disaster response situation, and a data plan is already in place. This could include in-house creation of data, purchasing data from a reputable vendor, partnering with other organizations for data acquisition, or obtaining open data. There must be a trust strategy in place, establishing the trust factor for each dataset. In selecting the right data, several data questions should be answered, such as: has the correct baseline data been established, is predictive data required, what kind of field data is needed, and how data can be acquired from surrounding communities. Collaborative data sharing is a particularly important consideration during a disaster. Planning and pre-approving data sharing among agencies prior to an emergency can save a lot of time (and lives).

Authoritative and Open Data

Many governmental agencies have a repository of authoritative data on roads, addresses, impervious surfaces, elevation data, and many other valuable data layers that form a baseline picture of the location where a disaster is taking place. Government, scientific, academic, and increasingly commercial sector organizations often publish data to the internet that can be downloaded for free as open data. Some particularly useful examples of this can be found within the ArcGIS Living Atlas of the World.

Remote Sensing Data

Remote sensing is a technical term defined as the process of gathering information about the Earth's surface, atmosphere, or waters from a distance, typically using various sensors and instruments on satellites, aircraft, drones, boats, submersibles, or other platforms. There are a wide variety of sensors that can be attached to these platforms to gather different measurements of electromagnetic radiation (such as visible light, infrared, microwave, nuclear radiation, sound, and other wavelengths) or measuring distances, shapes, and reflectivity through return signals from radar or LiDAR. Data from these sensors may be available as open data from websites such as the NASA Open Data Portal https://data.nasa.gov/. In other cases, commercial companies make this type of data available to purchase. If defined areas or specifications are required, modern platforms such as satellites can be 'tasked' to focus repeatedly on a specific area at a specific time, such as an area recovering from an emergency.

LiDAR Scanners and Global Positioning Systems

LiDAR scanners are devices that use laser light to measure distances and create detailed maps of surfaces. LiDAR scanners are used in topographic mapping, self-driving vehicles, and environmental monitoring. Global Positioning System (GPS) devices use triangulation from satellites to determine the coordinates of the device on the face of the Earth. These devices may be embedded alongside the telemetry devices in vehicles, as stand-alone devices, or within modern mobile phones. The GPS coordinates can be recorded and/or transmitted alongside other data. For example, location coordinates can be included with a photograph, with field-collected data, or with a social media post (if that function is enabled).

Big Data, Temporal Data, and Streaming Data

Responding to disasters requires many data sources, but each successive disaster also provides a staggering amount of data. Using data mining techniques and big data analytics to extract information from previous disasters, decision makers can collect insights that facilitate better forecasting of future incidents. Any data that contains time elements is called temporal data. Often temporal data is related to the 'Internet of Things' (IoT), which refers to a network of interconnected physical devices, objects, and sensors that are embedded with technology devices to collect, exchange, and transmit data over the internet or other communication networks. This data can be combined with advancements such as smart cities and digital twins and streaming data to provide real-time visualizations.

Data Management

The input-process-output (IPO) model is a basic and self-explanatory tool for data management commonly used to understand how data passes through any system. Simply put, this model assesses:

- Where will data in a system come from?
- What processing needs to be done with that data?
- What will the outputs from the system be?

Expanding on the data source concept can be a useful way to examine the data an organization is currently using and to determine whether there may be additional sources they should be utilizing. Another important way the IPO model can be used to manage data is to contribute to understanding the data lifecycle. It is important to plan how data will be created or acquired, how it will be updated or refreshed, and how it will be archived or deleted. Two common problems are stale data (a dataset that has not been refreshed frequently) and zombie data (data that was collected a long time ago, but the circumstances of its collection have been long forgotten). Failing to deal with these issues can cause confusion over which datasets to use, misunderstanding of the suitability of different datasets, and budget wasted on storing unnecessary data.

One crucial element of a good data life-cycle process is to create and maintain good metadata. Metadata is defined as data about the data (e.g., information about the source, age, definitions, quality, and completeness of the data). One important element of metadata to plan for is change logging, i.e., recording what changes have been made to the data, when, by whom, and in some cases even why a change was made.

Data Standards

Determining which data standards to follow is critical for planning and maintaining any system. The creation of a system, often a common-operational picture, is achieved through connecting different systems among different agencies through common standards rather than through a single software solution (Phyn, 2018). In addition to considering the appropriate type of standard, the standards can vary depending on the scale, ranging from international down to the organization level. It is important to note that in some cases standards will vary among different jurisdictions. It is critical to understand those differences when one jurisdiction is assisting another. For example, the differences in units of measurement can cause dangerous risks when using another county's data in a foreign aid situation.

Data Awareness

Ideally, the response team identifies and collects all the data needed for decision support well before a disaster strikes. These common information requirements are either persistent/standing information requirements or situational/priority information requirements (New Zealand Institute of Intelligence Professionals, 2023). However, even the most robust data collection is not valuable if users are unaware the data is available or do not know how to access or use it. Training in available data is essential. Users should have a good understanding of the potential data limitations and risks and the channels to use if they have questions, concerns, or if they want to request additional data.

Data Security, Authorization, Access, and Sharing Protocols

Data security encompasses a range of practices and measures and a proactive approach to counter unauthorized access. Risk assessments should be regularly conducted to identify vulnerabilities and potential threats to sensitive data. There should be well-defined protocols for authorization, limiting who can access and modify data. Robust encryption techniques should be in place to secure data both at rest and in transit. Data classification schemes facilitate determining who can share data with whom and under what conditions. Role-based access control (RBAC) ensures individuals have access only to the data necessary for their roles. Implementing multi-factor authentication (MFA) adds an additional layer of security to user access.

Sensitive data includes Personally Identifiable Information (PII), such as names, addresses, Social Security numbers, or biometric information. Security-sensitive data includes facility access information and the location or movements of security personnel or assets. Commercially sensitive data pertains to information that, if disclosed, could negatively impact an organization's competitiveness or financial stability, such as trade secrets, intellectual property, or confidential financial data. Best practice includes maintaining an inventory of all sensitive data, specifying its purpose and custodianship, ensuring secure handling throughout its lifecycle, and securing compliance with relevant privacy laws and regulations.

Data Quality

There are many aspects of quality of data that should be assessed for disaster response well before an event occurs. Metadata should be in place to document the lineage to data's origin, transformation, and use. It is important to know the extent to which data consistently covers a target population or geographic region. Depending on purpose, accuracy may need to be high, such as floorplans for first responders to a school shooting. Alternatively, lower accuracy could be acceptable, such as aggregated rainfall totals to make a visualization for public safety awareness to caution against driving through high water.

Box 17.3: Case Study – Data Challenge #2

The UK Covid dashboard team learned an important lesson in 2021. On March 7th, Covid daily death count had not been processed by the usual deadline for the dashboard. The team decided to input the data they did have for other categories but left a zero in place of a daily death count. They knew this number would seem absurd, so the team uploaded an explanation for the data discrepancy on their website. The National media noticed the zero deaths figure without the clarifying note. Misinformation and distrust spread quickly while the team corrected the figure online. The team revised their data practices, such as delaying posting data in gaps, creating automatic feeds, and direct subscriptions to their website rather than relying on media reporting. The team saw a 35% increase in trust in their service after implementation (Peeples, 2022).

Software Application Challenges for Disaster Response

Functional Requirements and Non-Functional Requirements

In the context of IT and software development, both functional requirements (FRs) and non-FRs (NFRs) are essential components of a project's specification and play a crucial role in defining what a system should do and how it should perform. They should provide a clear roadmap for developers, testers, and other stakeholders to understand what the system should achieve and how it should perform. Properly defining and managing these requirements is essential to delivering a product that meets user expectations and business objectives.

FRs describe what the system is supposed to do and outline its expected behavior in response to various inputs or interactions. FRs might include features like data input forms, report generation, map design, and any other specific interactions or processes the software must support.

NFRs define the characteristics that describe how a software system should perform rather than what it should do. NFRs focus on the overall system performance/speed, scalability, security, and other qualities that are essential for the system's success.

System Performance

One critically important NFR is system performance. This relates to the speed of data storage and retrieval. Achieving speed of processing in IT systems requires a holistic approach that encompasses retrieval optimization, concurrency control for multiuser scenarios, streamlined data ingestion and export capabilities, and continuous monitoring and improvement efforts. These factors collectively contribute to a responsive and high-performing IT system.

Data Storage and Retrieval Efficiency

The efficiency of data storage and retrieval heavily depends on the database design. Properly structured databases with optimized schemas can significantly improve data access times. Techniques such as indexing and partitioning can enhance query performance. The choice of storage technology, whether it is traditional relational databases, in-memory databases, or distributed storage systems, can have a significant impact on data retrieval speed. For example, the Data Store component used with the ArcGIS system is built on a PostgreSQL database that has been optimized for performance. Caching frequently accessed data can dramatically improve retrieval speed. Caching mechanisms store copies of frequently used data in memory or faster storage, reducing the need to fetch data from slower, persistent storage devices.

Performance Optimization

This involves fine-tuning queries and database operations to execute as efficiently as possible. This includes using appropriate indexes, minimizing the number of queries, and avoiding resource-intensive operations. Distributing workloads across multiple computers through load balancing can prevent bottlenecks and ensure the system can handle a high volume of requests without degradation in performance. The ability to dynamically increase computing resources (vertical or horizontal scaling) based on demand is essential for maintaining performance during peak usage periods. Continuous monitoring and profiling of the system's performance can identify bottlenecks and areas for improvement. Tools like performance analytics, application performance monitoring (APM) systems, ArcGIS Monitor, and profiling tools can help with this.

Multiuser Editing

Multiuser editing scenarios require robust concurrency control mechanisms to prevent conflicts and ensure data consistency. Techniques like locking, optimistic concurrency control, or versioning can be employed to manage concurrent access to shared data. For real-time collaboration, technology is available to provide instant updates and notifications, ensuring changes made by one user are immediately visible to others.

Technology Infrastructure Challenges for Disaster Response

Addressing technology infrastructure challenges for emergency management ensures systems remain robust, reliable, and accessible during crises, ultimately aiding in effective emergency response and management. High availability is crucial for uninterrupted access to essential systems during emergencies. Employing redundant computer servers, load balancers, automatic failover mechanisms, and data centers can minimize downtime. In the event of a regional disaster or infrastructure failure, services can be set up to continue from a distant location. Cross-regional hosting involves distributing infrastructure across different geographic regions, enhancing resilience by mitigating the impact of localized emergencies. Continuous support is vital for emergency management systems, as crises can occur at any time. Utilizing remote monitoring and alerting systems to detect and respond to problems in real-time can ensure immediate assistance during critical situations. Integrating technology within business continuity processes and programs ensures dependencies are identified and their impact on activities is mitigated during unplanned disruptions.

Leveraging alternative power sources, such as generators, solar panels, or battery backups, can maintain system operation. Off-grid fallback ensures technology services remain operational during power outages, a common occurrence during emergencies.

Alternative internet connections should also be considered, such as satellite-based access. Offline modes in GIS applications allow users to access essential data and functionality when connectivity is lost. Users can pre-download maps or data for offline use during fieldwork or in areas with limited connectivity. A regular process of producing analog products (i.e., paper maps), if required in austere conditions response (i.e., loss of printer access) should be considered a failsafe.

Box 17.4: Case Study – Infrastructure Challenges

The 2019 Venezuelan Blackouts stand as an example of failed off-grid fallback after a fire overheated the lines of the primary hydroelectric power plant at the Guri Dam. The overheated lines triggered load rejection mechanisms that protected the lines to the dam. That momentary loss of power caused turbines to increase speed, creating an electrical system overload. While safety control systems were activated, the system became uncontrollable and forced operators to disconnect the safety generators, causing overloads at two other power plants. Over a dozen backup plants powered by natural gas or diesel failed to function during the outage due to lack of fuel (Molina Guzman, 2019).

Infrastructure Hosting

The decision of where to host technology for emergency management depends on factors like performance requirements, budget, scalability needs, data security, and compliance considerations.

Organizations must weigh the advantages and disadvantages of on-site, cloud, hybrid, or containerized hosting to best meet their specific technology requirements. Hosting technology 'on the premises' means deploying servers and infrastructure within an organization's physical location. This approach provides maximum control over hardware and network resources, which can lead to high performance if the system is effectively tuned to the operations of the organization. It can be suitable for organizations with strict data security requirements or specific compliance needs, although on-site hosting may require higher initial capital investments and ongoing maintenance costs.

Conversely, crowd hosting relies on remote data centers managed by cloud service providers. These providers offer a wide range of resources that can be scaled to meet performance needs because they tend to have a large capacity of computing power distributed across multiple localities and sites. This allows the system to handle one location experiencing an unexpected spike more easily in computing power requirements, catastrophic hardware failure, crippling cyber-attack, or power outage. Cloud hosting offers flexibility and a pay-as-you-go model, which can be cost-effective. It is well-suited for remote access and can quickly adapt to changing circumstances during emergencies.

A hybrid approach combines on-site and cloud hosting. Organizations can keep sensitive data on the premises while using the cloud for scalability and remote access. This approach provides a balance between control and flexibility. It can be advantageous for organizations with diverse requirements.

Containerization, such as using Kubernetes containers, allows applications to run consistently across different environments, whether on-site or in the cloud. It simplifies deployment, scaling, and management. Containerization enables flexibility in choosing where to host specific components of a system.

Physical Technology Infrastructure Considerations

The physical safety and resilience of technology infrastructure are crucial to maintain operational readiness during crises. Protecting technology infrastructure involves safeguarding server rooms against floods, fires, earthquakes, and other risks. These criteria should apply regardless of whether the server room is on the premises or in the cloud. If hosted in the cloud, then assurances should be sought from the hosting provider that their facilities provide protection for unforeseen circumstances. Server rooms may require flood barriers, raised floors, and sump pumps to prevent water damage. Fire suppression systems, such as sprinklers and fire-resistant walls, should be implemented to protect servers from fire hazards. In earthquake-prone regions, server rooms should be equipped with seismic bracing and reinforced structures to withstand ground motion. Securing desktops and screens as well as using ruggedized mobile devices with waterproofing and durability features are critical to ensure reliability and safety of technology assets in emergency situations.

People and Process Challenges for Disaster Response

Technological methods for disaster response primarily focus on data, software, and technology; however, people and processes are an essential element of successful disaster response.

People and Organization Challenges

Ensuring the right skills and training through workforce planning and development can help address the challenges of preparing responders for disaster response. Roles and responsibilities should be well-defined and understood, including any changes to delegated authority due to an emergency. Governance must provide the right level of oversight and assurance. It is important to have data available to determine which personnel are available, where they are, what skills they have, and their status. High-pressure industries like emergency management must deal with staff attrition as an ongoing readiness challenge. A capability development program to build competencies and skills is recommended, and this should include frequent system use to retain knowledge.

Process Challenges

Planning for change management and change configuration management should be done before disaster strikes. During disaster response, the efficiency of service requests can make a difference in how well issues that arise are addressed in real time. Adequate training drills of the technology designed for process support is important so that it runs smoothly during an actual disaster event. An example of this is a communications primary, alterative, contingency, and emergency (PACE) plan; these identify a PACE process for maintaining a network of voice communication (PACE Plan, 2019). Applying this to information management systems, identifying how an activity/process can be achieved if one component is unavailable, allows a controlled level of reliance on a chosen technology and alternatives when that technology is unavailable.

Geospatial Information Systems (GIS) for Disaster Response

GIS is a particularly beneficial technology for disaster response. GIS allows communities to build resilience and be better prepared for a disaster, respond more efficiently and effectively with real-time information to map disaster response, and recover faster from disaster events using mobile tools in the field. The GIS vendor company Esri has created a useful tool to aid with understanding high-level FRs of a successful GIS system. Many of these patterns are also relevant to other branches of technology. The tool is named 'Common Patterns of GIS Use,' as seen in Figure 17.1.

Mapping and Visualization

Mapping and visualization utilize visual representations to convey location and relationships effectively to enhance understanding. By using 2D and 3D maps and charts, data gains context and becomes easily communicable and understandable. Providing maps enhances comprehension far beyond lists, text blocks, or spreadsheets by providing understanding of what things look like in reality. One common example is that emergency responders visualize emergency sites remotely as a briefing before they go into a site with anticipated risks (for example, a factory that has been damaged in an earthquake).

Additionally, visualizations of the past guide optimal decisions for the future. Geospatial data and changes to that data can have the dates and times that changes were automatically recorded. By recording 'timestamps' on changes to geospatial data, it is possible to go back to the state the data was in and view it at a specific time, e.g., at the time of an event. Alternatively, a sequence can be viewed in a video-like replay (at different speeds and with the ability to reverse, pause or replay, etc.). This can be used for determining what decisions were made during an emergency and what impact those decisions had. Future predictions can also be simulated to inform planning and decisions.

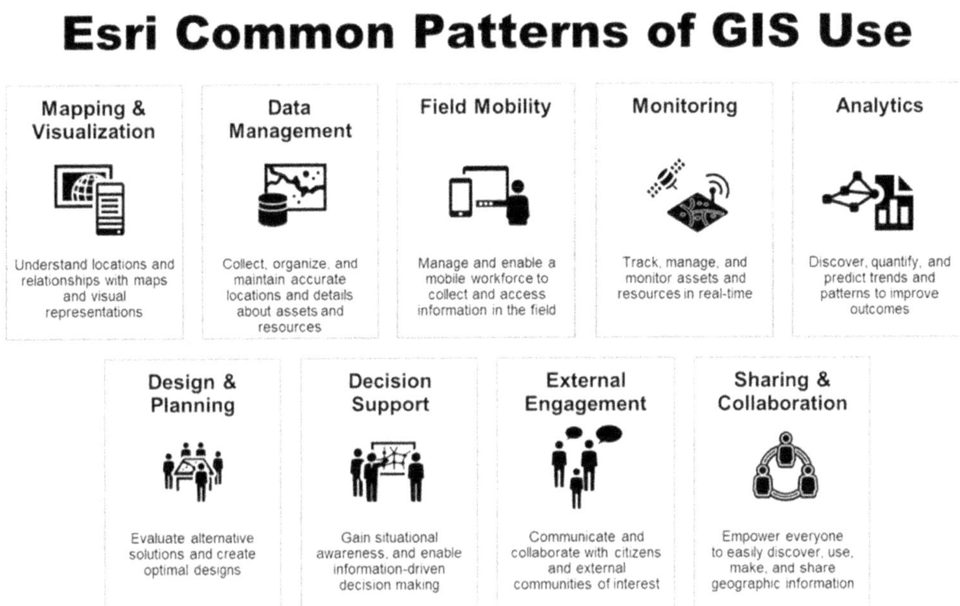

Figure 17.1 The Common Patterns of GIS Use

Source: From Law (2018).

Data Management

Data management involves collecting, organizing, and maintaining accurate asset and resource information. Geographic referencing provides a reliable and permanent indexing method for business data. By using the underlying geography as a common referencing system, this ensures data retrieval and processing, even as landscapes and street addresses change. GIS tools, including standard mobile devices, automate manual data handling and capture processes, seamlessly incorporating location data. This reduces manual labor, enhancing data utility. An essential GIS strength lies in its capacity to source and integrate external data, enriching an organization's datasets.

Many organizations are still using processes that rely on paper forms that then need to be entered into a computer system or passing unstructured emails from one team to another and collecting location data (such as street addresses or about storage sites) without realizing those processes could be more efficient by using *location aware* mobile devices or processes.

'Data is the new gold' – it is extremely valuable to organizations. With geospatial technology, master data from within the organization or authoritative/useful data from other organizations is widely available. Often organizations are unaware of the abundance of data that is available they could and should be making better use of from geospatial open data sites (that are often free of charge).

Data can be validated using specialized GIS tools. Geospatial tools and best practices support data quality. There are many processes or methods to help prevent errors from being introduced into the data, including data models, industry-specific editing templates, attribute rules, and data-specific editing tools. There are also many tools that can be used to identify errors that exist within a dataset.

Field Mobility

Field mobility in GIS and emergency management involves efficiently managing and empowering a mobile workforce to collect and access vital information while on the go. This serves to enhance operational visibility, optimize workforce scheduling, mitigate issues stemming from outdated data, and equip personnel with the necessary information for seamless task execution in the field.

By employing diverse methods to connect mobile devices with enterprise databases, field personnel can seamlessly view and edit location data. This facilitates field crews in using mobile devices to gain insights into work requirements, coordinate and dispatch resources effectively, and even access optimized routes, all in real-time. These personnel can instantly update records on-site and promptly receive their next task and destination, ensuring a streamlined workflow.

Monitoring

Monitoring enables real-time tracking and management of assets and resources. Organizations can analyze and display data from various sensors, devices, and social media feeds simultaneously. Real-time location tracking enhances safety and operational efficiency by ensuring awareness of nearby tasks or assets. GPS, transponders, mobile phones, and QR codes facilitate location transmission and visualization, helping monitor real-time movements and historical tracking.

Analytics

Analytics employs data analysis techniques to convert information into actionable insights. Geospatial tools enable users to uncover, quantify, and predict trends and patterns, enhancing decision-making and business outcomes. Key strengths include the ability to analyze relationships between datasets based on geography, eliminating the need for shared identifier codes. This facilitates effortless data integration for more comprehensive analysis. GIS excels in complex tasks such as route optimization, traffic analysis, leak detection in water pipes via flowmeter data, and 3D modeling for flood level estimation or understanding terrain during a wildfire.

Design and Planning

Design and planning patterns involve evaluating solutions for optimal designs. GIS facilitates collaborative design decision-making across organizations and communities. Geospatial tools empower users to analyze data not only in space but also over time, offering insights into historical changes and future projections. Users can create and compare multiple versions of data, including predictive data for demonstrating future events, such as flood spread, disease outbreaks, or suburban housing developments. These tools also enable the combination of various datasets to visualize and predict cumulative impacts, such as the convergence of flooded rivers, storm surges, and king tides resulting in coastal inundation.

Decision Support

Decision support streamlines data visualization on maps and dashboards to enhance comprehension of activities, projects, and operations. It amalgamates data, metrics, and operational layers to provide decision makers with valuable insights. For instance, field crews and managers often lack a comprehensive overview of nearby activities, leading to missed efficiency opportunities. Utilizing location-based management, activities become more efficient through resource sharing and eliminating redundancy. Location awareness can similarly aid operational decisions concerning tasks, issues, personnel, vehicles, and assets in many ways.

External Engagement

Engagement facilitates two-way information sharing with the public, partner agencies, and external stakeholders. This enhances collaboration, fosters informed decision-making, and improves community awareness of situations and decision processes. Visualizing data on maps simplifies communication, making information comprehensible even to non-GIS experts. Targeted communication can be localized to specific areas of interest, down to hyper-local neighborhoods. Consistent engagement can occur through various channels, including web, mobile, or print, ensuring accessibility.

Sharing and Collaboration

Sharing and collaboration empower users to self-serve, discover, utilize, create, and distribute maps, extending location information's value across organizations and communities. Even those with minimal GIS knowledge can configure web and mobile apps effortlessly using 'drag and drop' functionality, avoiding the need for coding.

Recent Technological Advances in Disaster Response

Disaster response has witnessed numerous revolutionary technological advancements in recent years. Combining previously disparate sources into a ready-to-use real-time picture of a response using web platforms has lessened the burden on classic intelligence collection and processing, allowing more time for analysis (Maunder, 2018). Advanced technologies such as machine learning (ML) and artificial intelligence (AI) are being employed to improve tracking and forecasting. Virtual and augmented reality (VR and AR) scenario simulations support field exercises that prepare first responders prior to a real disaster. The use of predictive tools such as sensors and digital twins can optimize disaster preparedness by predicting flooding and other dangers, thus leading to more efficient disaster practices. Additionally, improvements in photogrammetry, satellite technology, unmanned aerial vehicles (UAVs), and advanced GPS systems have enhanced the accuracy of location analysis and tracking, contributing to better decision-making. Digital tools such as dashboards have been used to enhance traceability and improve communication with citizens. Social media and citizen mapping efforts have led to increased engagement and empowerment of communities faced with disaster. Recent advances in various technologies are crucial in the context of emergency management for enhancing preparedness, response, and recovery efforts:

Machine Learning/Artificial Intelligence

AI and ML using GIS, called GeoAI (geospatial artificial intelligence), are being used to rapidly create and classify data. One example is analyzing satellite imagery or aerial photography to automatically identify, measure, and create 3D records of buildings, trees, or rivers. Another example is using terrestrially captured video along with GPS and video direction telemetry to automatically read street signs and assign a location to each of them. ML/AI Edge detection is being used in analyzing the gradient or variations in pixel values across the image to identify changes. The result is a visual representation of the edges within the image. These can be used for various purposes, such as object recognition, image segmentation, and geospatial feature extraction. Edge detection is a fundamental step in many image analysis and computer vision tasks, as it helps to simplify complex images and extract meaningful information for further processing or interpretation. One use of this technique is AI-driven change monitoring of satellite imagery to detect changes in disaster-affected regions, aiding damage assessment. A different AI technique is to assess social media to analyze posted text to identify crisis events, assess public sentiment, and provide real-time information.

Smart Cities and Internet of Things

Smart city technologies offer a real-time geospatial perspective on urban infrastructure, including traffic, utilities, and environmental conditions, improving situational awareness during emergencies. IoT sensors provide data on temperature, humidity, air quality, and more, helping monitor disaster conditions and assess risks. Traffic, radiation, temperature, and gas and chemical sensors are just a few of the IoT datasets that provide decision makers with the ability to monitor how a city is functioning during disaster response. Additionally, closed-circuit television cameras and/or movement sensors, fire and smoke sensors, seismographs, and building access systems provide additional information for situational awareness. In coastal or inland flooding related disasters, water quality sensors, wave and tide gauges (for coastal inundation and tsunami situations) and river level/flood gauges can be used to predict and monitor with the goal of protecting citizens.

Box 17.5: Case Study – Technology Solutions

The 2019 European heat wave struck France the hardest, with Paris ranking number 1 for heat-related mortality out of 854 European cities. Paris responded by creating 800 'cool islands' spread throughout the city that average 2°C–4°C cooler than surrounding areas. These islands include things like parks, water features, and museums, which can be found on the mapping app EXTREMA. The city created the app to guide residents to cool islands, and it gathers feedback from residents to monitor the use of the spaces and to help make the islands more inclusive and monitor the use of the spaces. Paris created a real-time geospatial application for urban infrastructure and invited the public to participate in its maintenance for future heat waves (C40 Cities Climate Leadership Group, 2019).

Virtual Reality/Augmented Reality

VR and AR mirror the physical world but represent the intangible world of data and information through computer-generated environments. These simulations enable immersive training for emergency responders and can help plan evacuation routes or simulate disaster scenarios. 3D mapping and data visualization of everything from buildings, vehicles, and the natural landscape are the basis for VR or AR capabilities. 3D can be augmented with time-enabled sequences to provide video-like '4D' visualization, providing responders with a geospatial perspective on the duality of reality. The digital mirroring of the real world is commonly described as a digital twin. Location-based data can be combined with spatial analytics tools to provide real-time geospatial analytics that help in simulating and predicting different scenarios using digital twins. For example, a digital twin of a wind farm could simulate wind patterns that could predict power generation scenarios for turbines and how they could be affected during a hurricane or typhoon.

Box 17.6: Case Study – Data Integration

The 2020 Vizag LG Gas Leak in India, which killed 12 people and hospitalized hundreds, demonstrates the need for data integration. Investigations have revealed lessons have not been learned since the 1984 Bhopal gas tragedy, which killed at least 2,500 people at the time and thousands more in the years post-disaster. After this disaster, several laws and guidelines were implemented but were not supported with effective data, so the response to chemical disasters such as Vizag falls short of protecting people. Post-disaster analysis has revealed data integration could have protected citizens. This could include air quality sensors, dispersion models, and graphs of indoor and outdoor pollutant concentrations to develop threat zones for evacuation plans, and maps that show roads, facilities, schools, homes, and response assets overlaid with predicted hazard zones (Gettleman et al., 2020).

Building Information Modeling

Building information modeling (BIM) provides detailed 3D models of structures, aiding in pre-planning for response and recovery operations. Designing buildings with resilience planning can minimize the impact of disasters. The advantages of BIM include knowledge of the building's design, such as systems, materials, and locations of exits.

Linked Data (Graph Databases)

Graph databases connect disparate datasets, enabling better understanding of complex relationships and information sharing during emergencies. It enables a deeper understanding of objects and systems by connecting spatial attributes and exploring these interactions. The Geocoded Disasters (GDIS) dataset provides spatial geometry for almost 10,000 worldwide disasters between 1960 and 2018. Connecting the EM-DAT database to other data sources through linked data allows for rigorous empirical analysis.

Drones/UAVs

Drone 3D mapping is a powerful technology for disaster response by creating digital models of the real world from aerial imagery, or LiDAR, and other data. Drones can be used both to gain ground truth information during a disaster and can be the cause of the disaster as well. For instance, Russian drone attacks in Ukraine in 2023 resulted in the destruction of nearly 40,000 tons of grain. Conversely, UAV and imagery footage can show time stamp information and proof to the world of post-attack damage. While many people voluntarily are willing to forgo privacy for convenience and safety, the power dynamics of government or other organizations using drones for data collection during a disaster event should be done with extreme caution and deference to information compliance regarding individual privacy rights.

Conclusion

Disaster relief relies heavily on technology to provide critical information to decision makers. Critical areas where technology plays an important role include gathering intelligence, enhancing communication, analyses to support decision-making, and the ability to help reduce risks. Technology is rapidly advancing to provide new tools and information about where and when events occur, what the impacts are, and to prioritize actions that should be taken. It is important that these technological advances are monitored to assess what benefits have become available for different situations. Careful planning is essential to reduce risks and maximize benefits with the use of technology. It is strongly recommended that this planning take place before these tools are needed and not left until emergency situations are occurring.

References

2022 Pakistan Floods. *Center for Disaster Philanthropy*, 6 Sept. 2023, https://disasterphilanthropy.org/disasters/2022-pakistan-floods/.

C40 Cities Climate Leadership Group. *2019 Cities100: 100 City Projects Making the Case for Climate Action*. C40 Cities Climate Leadership Group, 2019. https://c40.my.salesforce.com/sfc/p/#36000001Enhz/a/1Q000000MfJq/jNN04dftc8c7DuStWPE2ouYg1EiOkDP9Fdjo5PxnJm4

Comfort, Louise, Soga, Kenichi, Stacey, Mark, McElwee, Millard, Ecosse, Chiara, Dressler, Jillian, & Zhao, Bingyu. *Collective Action in Communities Exposed to Recurring Hazards*., 2019. https://hazards.colorado.edu/quick-response-report/collective-action-in-communities-exposed-to-recurring-hazards. Accessed 1 Dec. 2023.

Gettleman, Jeffrey, Raj, Suhasini, Schultz, Kai, & Yasir, Sameer. "Gas Leak in India at LG Factory Kills 11 and Sickens Hundreds." *The New York Times*, 7 May 2020. *NYTimes.com*, https://www.nytimes.com/2020/05/07/world/asia/india-lg-gas.html.

Gordon, Teresa, & Cook, Ed. CDEM Geospatial Strategy: 2021–2026 (p. 30, Rep.). Taranaki CDEM Group, 2021. https://www.civildefence.govt.nz/assets/Uploads/documents/resilience-fund/2019-20/03/2019-03-GIS-Investigation-Final-Report.pdf

New Zealand Institute of Intelligence Professionals. *Intelligence Practitioners Handbook*, New Zealand Institute of Intelligence Professionals, 2023. https://nziip.org.nz/wp-content/uploads/2024/01/nziip-practitioners-handbook.pdf

Law, Derek. "ArcGIS: Common Patterns of Use Poster." *Esri Community*, 6 Apr. 2018, https://community.esri.com/t5/arcgis-online-documents/arcgis-common-patterns-of-use-poster/ta-p/915296.

Maunder, Jeff. *The Geospatial Intelligence Continuum during Sudden Onset Disaster Response*. In Kristin Stock, & Deborah Bunker (Eds.), *Proceedings of ISCRAM Asia Pacific 2018: Innovating for Resilience – 1st International Conference on Information Systems for Crisis Response and Management Asia Pacific* (pp. 246–253). 2018. Auckland: Massey University.

Molina Guzman, Julio. *Failure Analysis of the National Electrical System in Venezuela – Total Loss of the System*. Universidad Central De Venezuela, 2019, pp. 1–8, https://es.scribd.com/document/401835067/Origen-de-La-Falla-v2.

PACE Plan Reference. *FM 6-02 Signal Support to Operations*. Headquarters, Department of the Army, 2019.

Peeples, Lynne. (2022). "Lessons from the COVID data wizards," vol. 603, no. 7902, Mar. 2022, pp. 564–67. www.nature.com, https://doi.org/10.1038/d41586-022-00792-2.

Phyn, Derek. *New Zealand GIS for Emergency Management (NZGIS4EM): Making GIS and Its Practitioners Integral to Emergency Management in New Zealand*. Massey University, 2018, pp. 223–32.

PART II

Disaster Recovery

Section V

Disaster Recovery: Key Elements and Activities

18

KEY ELEMENTS AND ACTIVITIES OF DISASTER RECOVERY

Rico Kongsager

Introduction

Disasters can occur unexpectedly and cause extensive disruption to communities and infrastructure, but recovery efforts can restore societal functions and provide a chance to improve equity and resilience (Phillips et al., 2021). The process of recovery, which is designed to put the community back together (Schwab et al., 2016), is a major aspect of disaster and its management greatly impacts citizens (Johnson & Olshansky, 2016). Recovery refers to the coordinated efforts and actions undertaken to restore essential services, infrastructure, and community functions following a disaster. It involves the systematic process of recovery planning, implementation, and evaluation to minimize the disruption caused by the disastrous event and to increase resilience to future events. The process of disaster recovery encompasses a range of activities and elements aimed at reviving affected areas that occur through a process of repairing, rebuilding, and reconstruction (March & Kornakova, 2017). The aim is not only to return a system to a functional state (Blaikie et al., 2003; Coppola, 2020) but also to return the community to a less vulnerable state (Alexander, 1993) and at the same time create new opportunities for future development. Recovery follows the response phase and thereby completes the disaster management cycle (although the cycle never actually ends or begins), and if a reoccurrence of hazards is expected, it is recommendable to have an overlap with the prevention/mitigation phase (March & Kornakova, 2017).

Understanding disaster recovery is essential for effective disaster management, as understanding the key elements and activities involved can better prepare organizations for disasters and contribute to a swift and successful recovery. The process aims to bring back a community, political territory, or geographic area to its pre-disaster state following a catastrophic event. However, disaster recovery is not just about rebuilding physical structures—it also involves rehabilitating affected individuals (along with their livelihoods, social networks, and trauma) and revitalizing the economy. Recovery is a multidimensional process that requires a comprehensive approach to address the various aspects, and thus, it is often considered to be the most challenging of the four stages of the disaster cycle. To be effective, there is a need for close collaboration among professionals, agencies, and interest groups from a diverse range of disciplines and perspectives (March & Kornakova, 2017).

DOI: 10.4324/9781003414834-26

Goals and Objectives

The primary goal of disaster recovery is to facilitate the immediate and effective restoration of affected areas. This entails ensuring the safety and well-being of individuals, rebuilding damaged infrastructure, and reestablishing vital community services. Additionally, disaster recovery aims to enhance community resilience and improve preparedness for future disasters. To achieve these goals, disaster recovery activities are guided by several objectives, including:

- Assessing the extent of damage and identifying priority areas for recovery efforts;
- Coordinating the mobilization of resources, including financial, human, and material assets;
- Developing and implementing recovery plans that align with community needs and priorities;
- Establishing partnerships and collaborations between stakeholders to support recovery efforts;
- Ensuring the equitable distribution of resources and services to affected individuals and communities; and
- Integrating risk reduction and resilience-building measures into recovery strategies to mitigate the impact of future disasters.

Temporal Aspects of Recovery

The process of recovery usually begins only a few hours after a disaster, following the immediate response stage, though there are some pre-disaster recovery activities. The duration of the recovery phase can extend over several years depending on the extent of the damage and the resources and capabilities that the community has at its disposal (Alexander, 2002; Blaikie et al., 2003). Survivors may endure economic stagnation, weakened social ties, and declining health services in the post-disaster period, and while physical recovery often only spans for years, the psychological effects can persist for decades (Johnson & Olshansky, 2016). This extended nature of the recovery phase often leads to certain planning complexities, and a common challenge that arises during this phase is the balance between instant action to address immediate concerns and the achievement of long-term disaster risk reduction objectives (March & Kornakova, 2017).

An issue often creating a challenge is that recovery managers feel pressed to return a community to normal as quickly as possible (Phillips et al., 2021). Consequently, communities tend to be rebuilt very similarly compared to the pre-disaster state (March & Kornakova, 2017). However, this might not be favorable in the long run, because it is important to note that disasters can provide opportunities for infrastructure and economic upgrades, among others, that potentially can lead to a better quality of life and improved long-term prospects for residents (Coppola, 2020). This aspect usually encompasses what is commonly referred to as the window of opportunity, which is a limited period after a disaster with significant potential for enhancing and improving measures to prevent disasters (Alexander, 1993; Mileti, 1999). During this period there is a willingness of many community members to radically change approaches, however; the window closes slowly as a strong desire to return to pre-disaster conditions gradually returns (March & Kornakova, 2017).

Thus, critical timing of disaster recovery activities must balance speed and caution as it is a challenge to meet immediate needs while also focusing on long-term goals of reducing vulnerability and improving development. This can be accomplished by dividing the recovery activities into sub-phases determined by time: pre-disaster, short-term, and long-term actions. Pre-disaster recovery is a proactive approach that takes place before a disaster happens, which involves anticipating future recovery issues to build capacity to improve recovery outcomes. Short-term recovery involves damage assessment, infrastructure restoration, and temporary housing, allowing communities to resume daily routines. Long-term recovery involves planning for improved rebuilding. The process starts with damage assessments that inform short-term priorities and lay the groundwork for long-term planning (Phillips et al., 2021). The transition between these sub-phases is dependent on the specific conditions of the disaster aftermath and the capabilities of the affected government and participating agencies (Coppola, 2020). The key elements of these three sub-phases and their related activities are described in the following sections.

Pre-Disaster Recovery: A Proactive Approach

Pre-disaster recovery actions are an integral part of disaster management, providing a roadmap for reducing disaster risk and enhancing community resilience. By planning for recovery before a disaster occurs, communities can respond more effectively and efficiently when a disaster strikes and their recovery efforts can better contribute to long-term disaster risk reduction and sustainable development. It involves a series of measures taken before a disaster strikes to foster a robust response and smooth transition to the recovery phase. This not only helps to minimize the initial impact of disasters but also facilitates the ability of communities to 'build back better,' creating a safer and more resilient future (International Recovery Platform, 2012).

Pre-disaster recovery planning (PDRP) is a proactive approach that communities can adopt to prepare for potential disasters. It involves considering how to manage important recovery issues, such as locating temporary housing, planning rebuilding efforts, and reestablishing essential economic activity. One of the key aspects of PDRP is ensuring the continuity of government and essential services as, in the event of a disaster, the government must continue to function and provide essential services to its citizens. This includes everything from emergency services to utilities and infrastructure (FEMA, 2017; March & Kornakova, 2017; Coppola, 2020).

However, it must be noted that in many places the political appetite and resources often are not sufficient and strong enough to support pre-disaster planning processes (Alexander, 2002), and having a well-functioning pre-disaster plan in place is rare in many locations (Phillips et al., 2021). Consequently, unplanned activities will almost always occur as the nature of disasters often is unforeseen and/or unpredictable (March & Kornakova, 2017)—but pre-disaster planning can reduce the risk of haphazard rebuilding during the post-event recovery phase. For instance, many needs are common to all disaster types, and many processes they generate are similar and can therefore be identified and studied in advance, such as debris removal, housing reconstruction, environmental remediation, and infrastructure repair.

Although it is difficult to foretell exactly how a disaster will affect a community, recovery is a process that is performed within a time-constrained setting and upon which victims' lives and livelihoods directly depend (Coppola, 2020). Accordingly, it is essential to have a comprehensive pre-disaster recovery plan in place that can guide the recovery process and make sure that the community is better prepared and more resilient in the face of future disasters. The disaster recovery plan is a comprehensive program that addresses a wide range of scenarios. This plan should be flexible and adaptable, taking into account the unique characteristics and needs of the community as well as the specific nature and impacts of different types of disasters. It should also be regularly reviewed and updated to reflect changes in the community and advances in disaster management practices and technologies (Twigg, 2004). The plan outlines the steps the organization will take to protect and recover from a disaster and includes a clear definition of organizational roles and responsibilities, including identification of who is responsible for which actions during the recovery phase. This is vital for effective disaster recovery, as a well-defined structure ensures that everyone knows their role and can act quickly and efficiently when disaster strikes.

The United Nations Office for Disaster Risk Reduction and the International Recovery Platform has developed a guidance note on PDRP. The guidance note presents a sequence of steps to initiate and sustain PDRP and to assist communities in building back better following a disaster. The process involves developing a planning framework, considering key steps at a broad level, presenting relevant examples, and providing recommendations based on the experiences of others. The goal of PDRP is to strengthen future recovery efforts and outcomes, and it is intended to help government officials, whether they are working in small communities or across an entire nation, to define and follow a process to pre-plan for disaster recovery (International Recovery Platform, 2012).

One important consideration in PDRP, and generally in recovery, is housing. Disasters often result in significant damage to homes, leaving many people without a place to live. PDRP involves identifying potential locations for temporary and transitional housing as well as developing plans for rebuilding permanent homes. Economic recovery is also a critical component of PDRP. Disasters can significantly impact the local economy, disrupting businesses and leading to job losses and even depopulation of the affected area. Consequently, PDRP involves developing strategies to reestablish essential economic activity and support businesses in the aftermath of a disaster (Eadie et al., 2001). These are only some of the important issues to consider in PDRP. The following subsections describe recovery decisions and activities (mainly based on Coppola, 2020; FEMA, 2017; International Recovery Platform, 2012; Twigg, 2004) that can be planned before a disaster. It is important to note that the specific elements of a plan can vary depending on the nature of the community and the potential risks it faces.

Establishing a Disaster Recovery Team

Deciding roles and participation of the team that will handle disaster recovery efforts is imperative. This team is responsible for developing and implementing disaster recovery strategies, which also include mechanisms for coordination and information sharing. The team's tasks include identifying critical services, assessing potential risks, and formulating contingency plans. The team comprises individuals with diverse skills and expertise, including local government officials, emergency services personnel, and representatives from key community organizations (including volunteers). Having a disaster recovery team established enables swift restoration of essential services post-disaster, thereby minimizing disruption and facilitating recovery.

Identifying Critical Services

It is imperative to determine which services are essential for the community's functioning and survival. Critical services like healthcare, electricity, water, wastewater, and transportation are vital for community functioning. The aim is to establish their swift restoration post-disaster to minimize disruption and facilitate recovery. This process requires thorough risk assessment, infrastructure evaluation, and contingency and redundancy planning. By identifying these services and planning for disasters beforehand, communities can prioritize resources, develop robust recovery strategies, and enhance their resilience to future disasters.

Developing a Communication Plan

Deciding on how information will be disseminated before, during, and after a disaster is an important element of recovery planning. This plan assists in keeping all stakeholders informed about the recovery process. This could involve setting up hotlines, websites, or using local media, and it also includes internal communication within the organization and external communication with the public and other stakeholders.

Planning for Temporary Housing

Identifying potential locations for temporary housing and developing plans for their quick deployment is part of successful disaster recovery. Planning for temporary housing involves strategizing for adequate and safe shelter for individuals displaced by disasters and includes identifying potential locations, resources, and logistics for setting up temporary or transitional housing. It seeks to ensure that those affected have access to necessities and a safe space for all populations (e.g., elderly and differently-abled groups) during the recovery phase. By anticipating housing needs, communities can minimize the disruption caused by displacement and provide immediate relief to those affected.

Determining Financial Resources

Successful recovery efforts are fostered when sources of funding are identified and systems are equipped to receive and manage large influxes of funds in advance of disasters occurring. Financial planning makes certain that sufficient funds are available for immediate relief measures and long-term recovery initiatives. This includes resources for rebuilding critical services, the provision of temporary housing, and supporting affected individuals. It also involves exploring various sources of supplemental funding, including government aid, insurance, community donations, and international government and non-government organizations. By designating financial resources in advance, communities can respond more promptly and effectively to disasters, thereby minimizing disruption and facilitating a smoother recovery process.

Planning for Infrastructure Restoration

Developing plans for the quick restoration of critical infrastructure is paramount in disaster recovery. This involves strategizing for the swift restoration of critical infrastructure such as healthcare facilities, electricity, water, wastewater, transportation, and other systems post-disaster. Infrastructure restoration planning includes assessing the vulnerability of existing infrastructure, identifying necessary resources, formulating contingency plans, and arranging for redundancy of services (e.g., with a nearby administration or through 'circuited' infrastructural configurations). Addition ally, the nodes or junction points of important services (e.g., pump stations, treatment centers, transducer stations, and transit nodes) and their networks should be mapped geospatially and have government or equivalent community entities tasked with their management in times of disaster. By anticipating infrastructure needs, communities can minimize disruption and expedite the recovery process.

Business Continuity Planning

Helping local businesses develop plans to continue operations during and after a disaster is a crucial element of disaster recovery. Business continuity planning involves developing strategies to ensure that essential business functions can continue and includes identifying critical business processes, assessing potential risks, planning for supply chain and delivery interruptions, and formulating contingency plans. By planning for potential disruptions, businesses can minimize downtime, maintain consumer trust, and enable a quicker return to normal operations. This proactive approach not only safeguards businesses from financial losses but also contributes to community resilience as businesses and economic activities play a vital role in supporting household and community recovery as well as rebuilding efforts.

Establishing Mutual Aid Agreements

It is important to form agreements with neighboring communities or regions (including firefighters, law enforcement, search and rescue and medical staff, utility experts, and contractors) for assistance during a disaster. These agreements involve collaborations to provide reciprocal assistance during a disaster and safeguard efficient resource allocation, sharing of expertise, and coordinated response efforts. By pooling resources and capabilities, communities can enhance their resilience and recovery capacity. Furthermore, mutual aid agreements not only facilitate immediate relief but also support long-term recovery initiatives. Therefore, they play a crucial role in disaster management by fostering collective resilience and promoting a cooperative approach to disaster recovery.

Planning for Debris Management

It is necessary to develop plans for the quick and environmentally friendly removal and disposal of debris caused by a disaster. Effective debris management is essential in clearing paths for emergency and recovery services, preventing further damage, reducing risks to public health, and allowing for the restoration and reconstruction of utilities, infrastructure, and buildings. Debris management includes identifying potential disposal sites, heavy machinery, equipment, resources, and logistics for debris removal. By anticipating challenges through pre-disaster debris management plans, communities can expedite the recovery process and restore functionality.

Developing a Post-Disaster Redevelopment Plan

Creating a long-term plan for how the community will be rebuilt in a way that reduces future disaster risk is necessary. This involves strategizing for the reconstruction and redevelopment of communities in the aftermath of a disaster. A redevelopment plan includes actionable plans for infrastructure restoration, business continuity, community rebuilding, and utilizing government resources and land to aid in redevelopment and reconstruction. Many stakeholders representing multiple sectors and communities should be involved in the process due to both the scope of redevelopment planning and its focus on affecting long-term change. The aim is to not just rebuild what was lost, but to build back better and enhance resilience and sustainability while addressing issues of pre-disaster inequities. By anticipating and planning for redevelopment challenges, communities can help generate more successful recovery processes and thus a more resilient future.

Short-Term Recovery: Relief and Temporary Features

Short-term recovery, also termed relief, is the immediate response to a crisis or setback. It may potentially be initiated while emergency response operations are still ongoing. Short-term recovery involves a series of steps aimed at restoring some kind of normalcy and supporting the well-being of those affected. This process is crucial in the aftermath of any disaster: it forms a bridge from emergency response and helps set the stage for long-term recovery and rebuilding. The primary objectives of a short-term recovery plan are multifold. It aims at reinstating essential services, such as electricity and water, healthcare services, and transportation facilities. The disruption of these services can have a significant impact on the daily lives of people and repairing them is of utmost importance. Furthermore, a short-term recovery plan focuses on providing immediate relief to those affected by the crisis. This could involve offering food, water, medical and humanitarian aid, temporary housing, and financial relief schemes. In many cases, people may have lost their homes or access to basic necessities. Providing immediate relief can help alleviate their suffering and provide them with the support they need during this challenging time. Additionally, a short-term recovery plan includes assessing the damage caused by the crisis, as thorough assessment is crucial to understand the extent of the crisis and plan the recovery process accordingly. This could involve estimating the cost of damage, evaluating the structural integrity of buildings, identifying possible land or other opportunities for redevelopment, and identifying the resources needed for recovery.

Short-term recovery actions tend to be temporary and often do not directly contribute to the community's actual long-term development (Coppola, 2020; March & Kornakova, 2017). Nonetheless, a short-term recovery plan addresses immediate needs and serves as a first step toward long-term recovery and rebuilding; thus, it lays the initial foundation for generating a resilient community that can withstand future crises. Based on input from various sources (Alexander, 2002; Coppola, 2020; Eadie et al., 2001; FEMA, 2017; March & Kornakova, 2017; Phillips et al., 2021), the following sections elaborate on some of the key elements of short-term recovery. However, it is always important to remember that every crisis is unique, and the specific steps involved in a short-term recovery plan may vary depending on both the nature and scale of the disaster as well as the population it affects.

Emergency Relief and Shelters

Providing emergency relief and temporary housing to affected individuals is of utmost importance following a disaster. The primary objective is to guarantee the safety, health, and well-being of those impacted. Emergency relief efforts include providing access to clean water, food, essential supplies, medical assistance, and often specialized health services such as machines for dialysis, access to prescription drugs and sanitary napkins, and assistance for populations with disabilities and chronic illnesses. Furthermore, temporary shelter units equipped with basic amenities to meet the immediate needs of the affected population must be arranged to accommodate individuals and families who have lost their homes or are unable to return due to safety concerns. Temporary housing is established to provide immediate refuge to displaced populations until they can return to their homes, and it may occur in hotels, schools, large public or private buildings, and through mobile or modular structures.

Damage and Needs Assessment

A critical element of disaster recovery is conducting comprehensive damage and post-disaster needs assessments (PDNAs) to evaluate the extent of the destruction caused by the disaster. This involves evaluating the structural integrity of buildings, infrastructure, housing, public facilities, and natural resources to determine the level of repair, reconstruction, or replacement required, as well as to estimate the overall cost of the disaster. This information is crucial in guiding the recovery process and helping to prioritize resources and develop logistical solutions. The damage assessment process is multifaceted and involves multiple stakeholders, including government agencies, private contractors, insurance agencies, non-governmental organizations (NGOs), and community organizations. The assessment process should be systematic, data-driven, and inclusive, ensuring that the needs and priorities of affected communities are adequately addressed. The assessment is necessary for the development of comprehensive recovery plans and policies, which should outline strategies for rebuilding infrastructure and addressing housing needs, among others. Additionally, documentation and estimates of damages are typically used to 'declare' disaster status and are required in order to be eligible for national and international financial assistance. Furthermore, the assessment will be needed when working with insurance companies and capital groups, and to anticipate the amount of debt that may be incurred or require offsetting with bonds or funding from the tax base or other budgets.

Clearing of Debris and Restoration of Essential Services

Clearing debris and making the area safe for inhabitants is an important part of the short-term recovery process. This involves clearing access roads, removing fallen trees and other debris (e.g., solid waste, soil, and vegetative waste, construction and demolition materials, household goods, and hazardous waste), and sealing off damaged buildings to allow people to safely move around in damaged areas. Following this is to restore essential services. This could involve repairing damaged power lines and water pipes, reopening hospitals and schools, and ensuring that public transportation is operational. Reinstating these services reestablishes people's basic needs, returns their functional lives, and enables their livelihood activities.

Resource Mobilization

Disaster recovery requires the mobilization of various resources, including financial, human, and material assets. Governments, NGOs, and international bodies play a pivotal role in coordinating the allocation of resources to support recovery efforts, and these efforts are increasingly assisted by supply chain logistics and technologies such as artificial intelligence. Financial resources are essential for funding reconstruction projects, while human resources are needed for labor-intensive tasks and expertise in specialized areas. Effective resource mobilization requires careful planning, coordination, and communication among stakeholders and often with neighboring jurisdictions. Clear guidelines and protocols should be established for the allocation of resources, ensuring that they are distributed equitably and efficiently.

Stakeholder Engagement and Collaboration

Engaging with the community and keeping them informed about the recovery process is an essential part of any recovery plan, and this should begin at the earliest possible stage. This involves providing regular updates about the progress of recovery efforts, addressing any concerns or questions that people may have, and ensuring that the needs of the community are being met. Moreover, successful disaster recovery relies on the active participation of affected communities (i.e., representative participation that shapes the recovery process and allows for the utilization of contextualized approaches whenever possible). Collaborations with multiple stakeholders, such as government agencies, community organizations, businesses, and individuals, are required to develop comprehensive recovery plans, share up-to-date information, and coordinate multi-objective efforts on numerous fronts. Stakeholder engagement helps to align recovery initiatives with the needs and priorities of affected communities, and it is a process that should be inclusive, transparent, and responsive to community feedback. A successful approach establishes partnerships and collaborations among stakeholders that can facilitate resource mobilization, knowledge sharing, and the development of innovative recovery strategies that are sensitive to the local physical and social geography.

Long-Term Recovery: A Path to Resilience

Long-term recovery is the process of operationalizing rebuilding and rehabilitation after a disaster and planting the seeds for future development. It goes well beyond immediate response and short-term recovery to focus on generating permanent infrastructural, economic, and social systems that are less vulnerable to future disasters. This includes the (re)introduction of long-term infrastructure, improved public services, and targeted programs that help individuals and communities return to normal or hopefully bounce forward.

The primary objectives of a long-term recovery plan are to restore, improve, and strengthen the social, economic, and physical fabric of a community. This involves not only rebuilding what was lost but also improving upon pre-disaster conditions with a focus on resilience and sustainability (Coppola, 2020; March & Kornakova, 2017; Phillips et al., 2021). Long-term post-disaster reconstruction provides opportunities for improvement and resolution of long-standing issues as it allows for the renewal of infrastructure, reconfiguration of land use, enhancement of construction and design standards, economic reinvention, the inclusion of and integration of underrepresented communities, and governance improvement. Furthermore, it presents an opportunity to mitigate the impact of future disasters through improved construction quality, avoidance of hazardous locations, and enhanced community awareness and preparedness. Consequently, long-term recovery presents significant opportunities for reducing vulnerability and fostering development (Johnson & Olshansky, 2016).

Long-term recovery is a complex and ongoing process that requires careful planning, coordination, and sustained effort over time—and it is often faster and more convenient to restore and rebuild what existed before the disaster than it is to reconstruct differently in the wake of a disaster (Johnson & Olshansky, 2016). For instance, economic renewal can be a lengthy process, often requiring a community to adapt and reinvent itself while attempting to preserve its original culture and social systems (Coppola, 2020). For major disasters, the long-term recovery process persists for years (Coppola, 2020), and it can take 10 years or longer to complete major reconstruction (Johnson & Olshansky, 2016). A study on long-term recovery narratives following major disasters in Southeast Asia found that the recovery process can continue to unravel and interact for a decade or more (Thomalla et al., 2018). Thus, the temporal scale of long-term recovery after a major disaster can vary greatly depending on the nature and severity of the disaster.

Regarding funding, long-term recovery often attracts more than any other phase of emergency management. This phase includes development assistance, insurance payouts, and disaster fund payments. Financial mechanisms such as insurance, financial aid, or buyback schemes can, if implemented wisely, compensate for property and other losses. However, the success of long-term recovery operations hinges on effective coordination and planning to utilize the funding effectively and efficiently (Coppola, 2020). Additionally, ongoing communication with the public and various social rehabilitation programs are an important part of long-term recovery efforts (March & Kornakova, 2017). The following sections elaborate further on the key long-term recovery elements emphasized in the literature (Alexander, 2002; Blaikie et al., 2003; Coppola, 2020; Eadie et al., 2001; Johnson & Olshansky, 2016; March & Kornakova, 2017; Phillips et al., 2021). The elements illustrate the breadth and depth of long-term recovery efforts, which span multiple sectors and involve a variety of stakeholders.

Planning and Coordination

Long-term recovery is a complex and challenging process that requires careful planning and coordination among stakeholders. Planning and coordination demands the strategic organization and management of numerous recovery activities across government agencies, the private sector, non-profit organizations, and the affected community. The planning phase identifies the needs of the community, sets priorities, and develops a recovery plan that outlines the steps to be taken. This plan should be comprehensive, addressing all aspects of recovery including infrastructure rebuilding, economic recovery, social services and housing restoration, and community resilience building. Coordination, on the other hand, is about ensuring that all recovery activities are carried out in a synchronized, standardized, equitable, and efficient manner. This involves coordinating the efforts of various stakeholders, managing resources, and overseeing the implementation of the recovery plan. Effective coordination helps to avoid duplication of efforts and manage resources similarly and efficiently, which will assist the recovery process to be carried out smoothly, fairly, and effectively. In essence, planning and coordination serve as the backbone of long-term recovery efforts, guiding the process and making sure that it is conducted in an organized and successful manner.

Infrastructure Restoration and Reconstruction

Restoring damaged infrastructure is a critical activity in long-term disaster recovery. However, the process also moves beyond basic repairs to include the construction and installation of new infrastructure as necessary. Infrastructure includes vital facilities such as roads, bridges, hospitals, schools, utilities (e.g., electricity, water, and sanitation systems), and communication networks. This process is not only to repair or replace damaged infrastructure but to strengthen them to better withstand future disasters by adopting stricter building codes, introducing more resilient infrastructure, and incorporating green building practices. The restoration and reconstruction process seeks to resume essential services and reestablish community functions and thus involves a comprehensive assessment of the damage and a prioritization of repair and reconstruction efforts. This requires detailed and well-planned coordination among government agencies, engineering teams, construction firms, contractors, and utility providers.

Economic Revitalization

Disasters often have a significant impact on the local economy, so measures to revive the local economy and support affected businesses are a fundamental aspect of disaster recovery. Economic revitalization efforts aim to rebuild local businesses and critical economic sectors, stimulate job creation, attract new and continued investments, prevent depopulation of disaster-affected areas, and generally restore economic stability. This may encompass financial assistance programs, grants, loans, tax incentives, and zoning initiatives to support businesses and entrepreneurs. Job placement and training programs can be implemented to provide employment opportunities to the local population, and economic development plans can be devised to diversify the economy, promote sustainable growth, and reduce vulnerability to future disasters.

Social Services and Permanent Housing

Long-term recovery also involves providing and improving social services, such as healthcare, education, and social welfare programs. These services are crucial for the well-being of the community and their restoration entails reopening hospitals and schools, continuing and perhaps adapting social welfare programs, and ensuring that these services are accessible to all members of the community. Providing affordable, disaster-resistant permanent housing is a key aspect of long-term recovery as many people may have lost their homes. This consists of retrofitting existing homes, rebuilding homes that were destroyed, constructing new multi-family housing units, and providing temporary housing solutions until these permanent dwellings are ready for occupancy. The goal is to make sure that everyone has a safe and secure place to live. In addition, these efforts focus on implementing higher standards for healthcare and education, implementing more effective social welfare programs, building homes and neighborhoods that are better able to withstand future disasters, and system learning from having managed the prior disaster. Thus, the aim is not just to rebuild what was lost but also to build a stronger and more resilient community writ large.

Community and Social Recovery

Recovering from a disaster involves addressing the social and community impacts on affected individuals. Community and social recovery activities focus on restoring social structures, providing psychosocial support, social capital formation, and promoting community cohesion. Psychosocial support is essential for addressing the emotional and psychological well-being of those affected by the disaster. Counseling services, mental health support, and trauma-informed care should be provided to individuals and communities to help them cope with the aftermath of the disaster. Community engagement programs, support groups, and recreational activities can be organized to facilitate social interaction and the rebuilding of community networks.

Environmental Recovery

Environmental recovery refers to the process of restoring and improving the health of ecosystems that have been damaged or degraded. This can be due to disasters, human activities, or a combination of both. Environment recovery involves identifying the impacts and risks (accounting for feedbacks and compounding and cascading effects) associated with the event/activity and includes management of environmental health issues (e.g., debris, waste and pollution, hazardous materials, and zoonoses) to protect air quality, water quality, soil and groundwater conditions, landscapes, ecosystems, and wildlife. This process can last years and focuses on the long-term restoration of natural and human ecologies. It includes a wide range of activities such as reforestation, soil remediation, water purification, (re)creating natural buffers, biodiversity conservation, and monitoring various parameters (e.g., pollutants and wildlife populations). A key aspect of environmental recovery is the concept of a 'green recovery,' which aims to rebuild economies in a way that enhances environmental sustainability. This includes accelerating action on climate change, biodiversity loss, and pollution crises and it acknowledges intricate links between human and environmental health.

Historic and Cultural Resources

Historic and cultural resources include landmarks, temples, relics, local practices, and traditions, all of which contribute to the cultural identity and sense of place of a community. Activities could involve the restoration of damaged cultural heritage sites and historic buildings, the preservation of ancient manuscripts, and supporting vernacular architecture, local artists, and handicrafts. Incorporating these elements into post-disaster recovery policies makes communities more safe, resilient, and sustainable. Furthermore, inclusivity and the recognition of heritage promote harmony and may foster reconciliation between groups who may have a history of conflict. Thus, long-term recovery provides an opportunity to renew these cultural resources while also addressing long-standing societal problems. In essence, the recovery of these resources is not just about rebuilding what was lost, but also about leveraging them to generate a stronger, more resilient, and culturally rich community.

Government Recovery

Recovery of the government encompasses the strategic efforts by government agencies to restore and improve public services, infrastructure, and other aspects of public life impacted by the disaster. Government plays a central role in long-term recovery, as it guides the overarching process and is responsible for ensuring that the needs of the community are met. The recovery of the government is crucial for several reasons: (1) *Leadership*: the government provides leadership and direction during the recovery process, as it sets priorities, makes key decisions, and coordinates the efforts of various stakeholders; (2) *Resource mobilization*: the government plays a key role in mobilizing resources for recovery, which includes financial resources, human resources, and material resources; (3) *Policy and regulation*: the government is responsible for establishing policies and regulations that guide the recovery process, including land use policies, building codes, zoning, and environmental regulations; (4) *Public services*: the government is responsible for reestablishing public services that have been disrupted by the disaster, including utilities, healthcare, education, and social services; (5) *Infrastructure*: the government plays a major role in rebuilding infrastructure, including roads, bridges, and public buildings; (6) *Economic recovery*: the government can implement policies and programs to support economic activities, such as providing financial aid to businesses and promoting entrepreneurship and investment; (7) *Community support*: the government can provide support to affected communities, including housing assistance, mental health services, and community development programs. These are just some of the roles the government plays, and with each disaster iteration, they should learn from and improve upon their governance strategies.

Long-Term Planning and Risk Reduction

Building resilience is a key goal of long-term recovery, so disaster recovery extends beyond the immediate aftermath and includes long-term planning and risk reduction measures. This encompasses strengthening the capacity of the community to withstand, adapt to, and recover from future disasters and incorporates lessons learned from the disaster into future planning and preparedness efforts. Risk reduction measures, such as the development of community disaster plans, early warning systems, promotion of community cohesion and social capital, improved infrastructure design, and the roll-out of community education and awareness programs, can all, if well-implemented, enhance preparedness and reduce vulnerabilities. While recovery in many cases is rebuilding what was lost in the disaster, it also presents an opportunity to 'build back better' and create a more resilient and sustainable community. The build back better or 'build back safer' principle is a systematic process to investigate the underlying reasons for failure and to incorporate lessons learned to augment the reconstruction and redevelopment process—and it does not just address the physical structures but also the people who live and work within them. Thus, structural and non-structural mitigation measures are critical components, and with careful planning and coordination, long-term recovery can pave the way for a safer and more resilient future with fewer injuries, deaths, and property loss.

Conclusion

Disaster recovery is a critical phase of disaster management that seeks to restore essential services, infrastructure, and community functions following a disaster. Moreover, it aims to improve upon the pre-disaster status quo and fashion a more resilient, sustainable, and inclusive society. Recovery involves a systematic process of planning, implementation, and evaluation to minimize disruption and future disruptions.

Disaster recovery consists of three parts. The first part is pre-disaster recovery, which is a proactive approach involving measures taken before a disaster to help fashion an efficient response and smooth transition to the recovery phase. This helps to minimize the impact of disasters and enables communities to build back better, creating a safer and more resilient future. The second part is short-term recovery, also known as relief, which is the immediate response to a crisis. It aims to return to normal and safeguard the well-being of those affected. The primary objectives of a short-term recovery plan include reinstating essential services, providing immediate relief to those affected, and assessing the damage caused by the crisis. The third and final part is long-term recovery, which focuses on rebuilding and restoring communities in the aftermath of a disaster. It goes beyond immediate response and short-term recovery efforts aiming to restore, improve, and strengthen the social, economic, and physical fabric of a community. This involves not only rebuilding what was lost but also improving upon pre-disaster conditions with a focus on resilience and sustainability. Disaster recovery, and long-term recovery in particular, provides a window of opportunity to improve the pre-disaster scenario, resolve long-standing issues, and mitigate the impacts of future disasters.

References

Alexander, D. (1993). *Natural Disasters*. UCL Press.

Alexander, D. (2002). *Principles of Emergency Planning and Management*. Oxford University Press.

Blaikie, P. M., Cannon, T., Davis, I., & Wisner, B. (2003). *At Risk: Natural Hazards, People's Vulnerability and Disasters* (2nd ed.). Routledge.

Coppola, D. P. (2020). *Introduction to International Disaster Management* (4th ed.). Butterworth-Heinemann.

Eadie, C., Emmer, R. E., Esnard, A. M., Michaels, S., Monday, J., & Philipsborn, C., & et al. (2001). *Holistic Disaster Recovery: Ideas for Building Local Sustainability after a Natural Disaster*. Natural Hazards Research and Applications Information Center, University of Colorado.

FEMA. (2017). *Pre-Disaster Recovery Planning Guide for Local Governments*. Federal Emergency Management Agency.

International Recovery Platform. (2012). *Guidance Note on Recovery: Pre-Disaster Recovery Planning*. UNISDR, IRP, and European Commission. https://www.preventionweb.net/files/31963_predisasterrecoveryweb.pdf

Johnson, L. A., & Olshansky, R. B. (2016). *After Great Disasters: How Six Countries Managed Community Recovery*. Lincoln Institute of Land Policy.

March, A., & Kornakova, M. (2017). *Urban Planning for Disaster Recovery*. Butterworth-Heinemann.
Mileti, D. (1999). *Disasters by Design: A Reassessment of Natural Hazards in the United States*. Joseph Henry Press.
Phillips, B. D., Neal, D. M., & Webb, G. R. (2021). *Introduction to Emergency Management and Disaster Science* (3rd ed.). Routledge.
Schwab, A. K., Sandler, D., & Brower, D. J. (2016). *Hazard Mitigation and Preparedness: An Introductory Text for Emergency Management and Planning Professionals* (2nd ed.). CRC Press.
Thomalla, F., Lebel, L., Boyland, M., Marks, D., Kimkong, H., Tan, S. B., & Nugroho, A. (2018). Long-term recovery narratives following major disasters in Southeast Asia. *Regional Environmental Change*, *18*(4), 1211–1222. https://doi.org/10.1007/s10113-017-1260-z
Twigg, J. (2004). Preparing for Disasters and Emergencies. In *Disaster Risk Reduction: Mitigation and Preparedness in Development and Emergency Programming*. Humanitarian Practice Network.

19
DISASTER RECOVERY CYCLE

Abdulhadi A. Al Ruwaithi

Introduction

The discourse on the disaster recovery (DR) cycle represents a critical examination of the multifaceted and dynamic processes inherent in post-disaster contexts. This exploration, grounded in the seminal works of scholars, many conducted in the US context, offers a comprehensive understanding of the complexities involved in the DR process. Disaster management is a perpetual process that encompasses readiness, prompt response, and restoration with an emphasis on acquiring wisdom and making ongoing progress. In terms of the disaster management cycle, the work of W. Nick Carter in *Disaster Management: A Disaster Manager's Handbook* provides a nuanced perspective (Carter, 2008). The text encompasses phases of prevention, mitigation, preparedness, emergency response, recovery, and national development. This conceptualization reveals that DR plans and processes must be dynamic, adapting to changing circumstances and emerging challenges.

According to recent literature, there is a growing need to reassess the four conventional phases of the disaster management cycle, of which DR is a primary phase. This re-evaluation aims to expand our comprehension of pre- and post-disaster consequences, as well as the implementation of necessary corrective measures. A number of scholars are formulating innovative theoretical models for disaster management and recovery (see Bahmani & Zhang, 2022; Bosher, Chmutina, & van Niekerk, 2021; Khan, Vasilescu, & Khan, 2008), while others contend for advancing beyond a 'conceptual stalemate' (Tierney and Oliver-Smith, 2012, 140). In practical application, various national frameworks, primarily influenced by the United States' model, are utilized to structure the process of DR. The objective of this chapter is to examine the meanings and semantics of 'disaster' and 'disaster recovery,' followed by an analysis of leading conceptual models for the disaster management cycle and DR process in both theoretical and national set-for-practice frameworks. The chapter concludes with a thorough analysis of the factors and pre-conditions that impact the progress of DR, rendering the identification of a universal recovery cycle model challenging.

This chapter begins with defining the core concepts of DR, including various interpretations and frames highlighted in both practice and the academic literature. The chapter further explores the lifespan of DR projects, emphasizing continuous planning and long-term sustainability, and

DOI: 10.4324/9781003414834-27

investigates effective management aspects necessary for successful DR. Finally, this chapter discusses the social process of recovery, acknowledging the integral role of community engagement and socioeconomic factors.

Disaster Recovery Concepts

DR is an intricate and multifaceted process that extends well beyond immediate response throughout recovery from the initial catastrophe. It encompasses a wide range of activities with the goal of restoring and, more importantly, seeking to enhance the socioeconomic and environmental conditions impacted by disasters. This section explores the concepts and definitions of what constitutes disaster and the DR process, incorporating insights from relevant literature.

It is necessary to gain a thorough comprehension of the term 'disaster' before diving into a discussion about recovery. In their review, Mayner and Arbon (2015) analyzed 128 definitions of disaster terms from various agencies and research, all written in English. Observing those definitions, disaster is considered as a state, a condition, an alteration, or an event that affects a community and causes great damage or widespread disruption. Drawing from their analysis of definitions, Mayner and Arbon (2015) propose a computer-generated definition of the disaster as "the community experiences extensive disruption and damage that surpasses its capacity to handle and overwhelms its available resources" (24).

The issue with this 'conclusive' definition lies in the combination of the words disruption and damage, which connote a physical reflection of damage. However, it is evident that pandemics can lead to disruptions without causing damage to physical infrastructure. This is independent of whether a pandemic is classified as a crisis or a disaster. Thus, to broaden the scope of inclusivity, it is imperative to incorporate the disruption of community functions or alternatively substitute 'and' with 'or' in Mayner and Arbon's (2015) definition of the disaster concept.

Considering that disaster entails significant functional disruption and/or damage to community well-being, DR can be seen as the endeavor and/or process to restore functions and community well-being to their normal or enhanced states. Typically, the DR process involves long-term, expensive, and time- and resource-intensive activities, particularly when rebuilding physical infrastructure is necessary to reestablish functionality. On certain occasions, recovery processes may experience a halt, as demonstrated by the situation in Haiti following the 2010 earthquake (Tierney & Oliver-Smith, 2012). There are also instances when the recovery of a community is not feasible in the same location, thereby necessitating relocation, as evidenced by the Fukushima Daiichi nuclear disaster. Thus, the definition of 'disaster recovery' can be described as "the differential process of restoring, rebuilding, and reshaping the physical, social, economic, and natural environment through pre-event planning and post-event actions" (Smith & Wenger, 2007, 237).

The concept of DR encompasses different aspects or scopes of activities or processes, all of which are interconnected and essential for effective recovery and the disaster management cycle writ large. The DR aspects or processes include the following:

1 **Economic Recovery**: The focus of economic recovery is to restore stability to the economy and rebuild local businesses. Small businesses and entrepreneurs, especially, encounter specific challenges in the aftermath of a disaster, which have been largely overlooked in traditional disaster research (Marshall & Schrank, 2014).
2 **Community Recovery**: The process of community recovery entails the reestablishment of social structures and networks (e.g., capacity building, social capital, and fostering community

cohesion). Both pre-disaster disparities (Tierney & Oliver-Smith, 2012) and post-disaster policies exert a significant influence on the character, speed, and inclusivity of community recovery, as demonstrated by multiple case studies (Finucane et al., 2020).

3 **Infrastructure Recovery**: The objective of infrastructure recovery is to restore critical services and facilities (e.g., utilities, bridges, and ports). Effective infrastructure recovery can impede overall DR efforts as infrastructure functionally encompasses the artery for physical DR inputs and local productive outputs to move throughout the DR process (Rouhanizadeh, Kermanshachi, & Nipa, 2020).
4 **Environmental Recovery**: This aspect refers to the restoration of natural ecologies that have been impacted or thrown off balance by disasters. The interconnection of this aspect with community and economic recovery is critical, as environmental health is part and parcel of 'one health' and all human activities rely, to some extent, on ecological features.
5 **Psychological Recovery**: Psychological recovery is centered on the mental health and well-being of those impacted. The interconnectedness of psychological recovery with other types of recovery is paramount, as mental health impacts the ability to participate in and contribute to other recovery efforts, such as economic activities and the rejuvenation of social networks and social capital.

Each seemingly chronological process of DR is intertwined, creating an intricate circuit wherein advancements in one area or attribute can impact, bolster, or suppress progress in other areas. For example, the process of economic recovery can generate the necessary resources for the rebuilding of infrastructure, thereby enabling the recovery of both the community and the environment. Similarly, the active involvement of individuals in reconstruction efforts across all domains relies heavily on their psychological well-being and ability to take ownership of and facilitate such processes.

The concept of DR involves various activities, such as reconstruction, rehabilitation, restitution, and recovery itself (Quarantelli, 1998). These terms, often used interchangeably, actually imply different aspects of the DR process. For instance, reconstruction may involve physical rebuilding, whereas rehabilitation might focus more on restoring social systems and community functions (Quarantelli, 1998). Thus, it is crucial to understand the breadth of these activities, as each plays a unique role in the collective process of recovery. Still, whether the recovery activity involves reconstructing damaged facilities, restoring agency functions, or engaging in rehabilitation, the end goal remains the same: to bring the situation back to an accepted level of functionality. This understanding of recovery provides a comprehensive perspective on the true essence of DR, rather than an idealized version. Ultimately, since definitions of disaster are rooted in community functioning, recovery becomes a social process, as contended by Nigg (1995), Quarantelli (1998), Tierney and Oliver-Smith (2012) and others.

Disaster Management Framework and Models for Disaster Recovery

In the traditional disaster management model, there are four stages that outline the management phases: disaster mitigation (and/or prevention), preparedness, response, and recovery (Figure 19.1). The aims of these phases, such as response to save lives and to prevent further damage, shape their approach. The presence of cyclical arrows implies that the process is designed to be both sequential and repeated and refined over time, rather than following a linear path. Each phase is connected, and the actions taken in one affect the outcomes of others.

Lessons learned from one phase should inform the actions taken in the others, especially when it comes to improving responses to future disasters through feedback into mitigation and preparedness (Bosher, Chmutina, & van Niekerk, 2021). This section provides an explanation of disaster

Figure 19.1 Traditional Representation of the Disaster Management Cycle

Source: Figure created by author.

management and DR processes, as discussed in Khan, Vasilescu, & Khan's (2008) paper *Disaster Management Cycle – A Theoretical Approach*. Subsequently, an alternative conceptualization will be explored, followed by an analysis of how two national DR frameworks interpret the process.

Khan, Vasilescu, and Khan (2008) emphasize DR as a critical pivot within the disaster management cycle. This cycle—typically viewed as circular but not compounding—encompasses risk reduction, preparedness, response, and recovery, with each phase linking seamlessly to the next. During risk reduction, evaluations should be conducted to mitigate potential hazards, and preparedness should involve meticulous planning and the establishment of efficient systems to ensure an effective disaster response, and so on (Khan, Vasilescu, & Khan, 2008). Once a disaster strikes, the shift from proactive to reactive measures begins in the response phase. As per the traditional management cycle, it is of utmost importance for individuals to immediately take part in activities that are focused on ensuring their personal safety and security. However, it is in the recovery phase where disaster management becomes truly transformative. DR's pivotal role is further highlighted by its emphasis on learning from previous disasters to improve future preparedness and response. According to Khan, Vasilescu, & Khan (2008), the recovery phase can be divided into two stages: rehabilitation, which seeks to restore normalcy, and reconstruction, which focuses on rebuilding and implementing preventive measures. Khan, Vasilescu, & Khan (2008) emphasize the importance of viewing the disaster management cycle as a continuous evolution, wherein each phase builds upon the experiences of the previous one. This highlights the significance of ongoing preparedness and adaptive recovery after a disaster. The phase-specific activities, as outlined in their cyclic management model, take place in three stages in relation to the timing of the disaster as follows:

1 **Pre-disaster** includes two phases: risk reduction or mitigation, and preparedness. Risk reduction/mitigation includes measures taken to reduce impacts of disasters before they occur. It encompasses risk assessments, hazard mapping, and implementing both structural and non-structural measures to mitigate potential damage. Preparedness involves planning and preparing for potential disasters. It includes contingency planning, issuing warnings, and consolidating preparations for future disasters. Preparedness seeks to ensure that systems are in place to deal with a disaster effectively and efficiently.

2 **During the disaster** is the point at which a disaster has occurred, indicating a shift from proactive to reactive measures. These involve actions to guarantee that requirements and supplies are provided to disaster-affected populations, and that their distress is alleviated. Efforts carried out during this phase are referred to as emergency response activities.
3 **Post-disaster** primarily includes two phases: response and recovery. Individual disaster response usually starts during disaster and continues afterward, which includes the immediate securing of ones' and others' safety and providing first aid assistance. Response and relief, unless the disaster scenario is expected, usually begins immediately after the disaster strikes and includes search and rescue (SAR), the provision of food, water, shelter, and emergency medical care, and other urgent services and logistics. Sometimes relief efforts are initiated before or during the disaster when the disaster is expected and sufficient preparedness programs are in place. Next, the implementation of recovery strategies that encompass rehabilitation and reconstruction begins, as in Khan, Vasilescu, & Khan's (2008) and the Asian Development Bank's (ADB's) (Carter, 2008) disaster management cycles.

The rehabilitation phase ensues after the initial response efforts, wherein the emphasis shifts from preventing further damage as a direct result of the disaster to reinstating essential services and aiding affected communities in commencing the process of returning to their usual routines. During the recovery phase of reconstruction, efforts are focused on rebuilding and restoring full services. Rebuilding processes include implementing preventive measures to reduce the impact of future disasters as well as retrofitting existing structures. This phase can extend from months to years, depending on the severity of the disaster (Khan, Vasilescu, & Khan, 2008), local community pre-disaster conditions (Himes-Cornell et al., 2018), and available capacities (Finch, Emrich, & Cutter, 2010; Tierney & Oliver-Smith, 2012).

However, alternative classifications of disaster management phases—particularly for DR processes—may exist, such as categorizations based on time or distinguishing between the short- and long-term recovery. Aligning the phases sequentially has been criticized for potentially limiting our understanding of the numerous, dynamic, and often overlapping processes (Aguirre & Lane, 2019; Bosher, Chmutina, & van Niekerk, 2021; Contreras, 2016; Tierney & Oliver-Smith, 2012). The overlapping of post-disaster phases (Bosher, Chmutina, & van Niekerk, 2021; Contreras, 2016; Tierney & Oliver-Smith, 2012) and the iterative nature of social recovery (Tierney and Oliver-Smith) challenge the traditional disaster models, including the one illustrated by Khan, Vasilescu, and Khan (2008).

In 1998, the South African Government introduced a new model for disaster management through the Green Paper process, laying the foundation for a comprehensive framework aimed at risk reduction and more effective management. The "South African Government Green Paper on Disaster Management" (South Africa, 1998) signifies a crucial transition in disaster management strategies by providing a comprehensive framework that not only addresses the South African context but also aligns with global standards. At the core of the framework lie the concepts of sustainable development and safeguarding vulnerable communities while simultaneously presenting fundamental principles to steer policies in disaster management.

One noteworthy characteristic of the South African framework is its deviation from the conventional disaster management cycle. Instead, it presents a proposed model known as the 'expand-contract' approach in which multiple strands of activities are conducted concurrently and are adaptable as required (South Africa, 1998). This model signifies a noteworthy advancement from the linear, sequential methodology conventionally utilized in disaster management cycles. It illustrates the overlapping nature of disaster management phases, attributing a certain magnitude

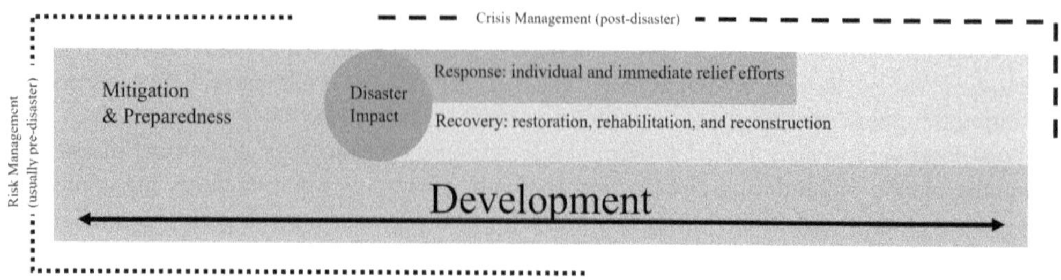

Figure 19.2 Newly Conceived Disaster Management Model Incorporating Aspects of Both (Khan, Vasilescu, & Khan, 2008) and the South African Framework (South Africa, 1998)

Source: Adapted from Khan, Vasilescu, and Khan, (2008); Created by author.

to different phases at different timings relative to the event. Figure 19.2 adapts the theoretical constructs posited by Khan, Vasilescu, and Khan (2008) through the prism of the 'expand-contract' approach of South African framework, enhancing representation of the intersecting aspects of disaster management activities.

When examining the evolution of disaster management models, it is of utmost importance to analyze the convergence and advancement from the South African model to more contemporary paradigms. Noteworthy, over two decades after the South African framework, Bahmani and Zhang (2022) introduced a more refined and explicit model of disaster management. Bahmani and Zhang's research establishes a series of specific activities within a more comprehensive framework, signifying a step forward in the conceptualization of disaster management. In contrast to Khan, Vasilescu, and Khan's (2008) approach, Bahmani and Zhang (2022) allocate space for the inclusion of all overlapping activities that are shared between the preparedness and mitigation (pre-disaster) and the emergency response and recovery phases (post-disaster). Bahmani and Zhang highlight a life-cycle approach to DR projects framed around three temporal phases: pre-disaster (including preparedness and mitigation), post-disaster (encompassing emergency response and short- and long-term recovery), and an integration phase. This last phase detailed by Bahmani and Zhang (2022) is particularly innovative, incorporating activities that span the entire spectrum of disaster management stages, from participation and communication to disaster vulnerability reduction.

The contribution of Bahmani and Zhang (2022) to the field centers around their focus on the ongoing lifespan of DR projects. Planning, design, construction, and procurement are thoroughly explored as the four critical phases of DR project management. Their comprehensive approach not only tackles the physical aspects of DR but also emphasizes the significance of meeting the social and psychological needs of affected populations and includes them within the planned activities in post-DR stages. The approach integrates continuous planning and long-term sustainability considerations, going beyond reactive measures to adopt a proactive stance toward developing resilient communities. This aligns with the broader challenge in disaster management of ensuring effective coordination among various efforts and stakeholders, as indicated by the US National Disaster Recovery Framework (NDRF) (Homeland Security, 2016).

Helical Representation

Departing from cyclical representation of disaster management and recovery process, Bosher, Chmutina, and van Niekerk (2021) introduce the Disaster Risk Management (DRM) Helix model as an approach to disaster management. The model was designed to outstrip the traditional four-phase cycle of mitigation, preparedness, response, and recovery. The helix model aims to capture the ever-changing and intricate essence of disasters, in contrast to the oversimplified and linear cyclical model. Critics argue that the cyclical model overlooks the underlying causes of disasters and perpetuates a monotonous cycle devoid of significant advancements (Bosher & Chmutina, 2017; Bosher, Chmutina, & van Niekerk, 2021;; Christoph et al., 2013).

Unlike the traditional disaster management cycle, the DRM Helix model better accounts for temporal dynamics and the adaptable allocation of resources (Bosher, Chmutina, & van Niekerk, 2021). Over time, the commitment of resources and efforts in the model changes in response to disaster risk reduction measures taken (Bosher, Chmutina, & van Niekerk, 2021). The model also allows for greater flexibility and includes multiple dimensions, accommodating various factors such as socioeconomic and political influences and the changing nature of disasters themselves. The three-dimensional helix structure visually captures the interconnected phases of disaster management, providing a more dynamic representation compared to the closed-loop cycle, thereby implying a continuous, evolving, and compounding process (Bosher, Chmutina, & van Niekerk, 2021).

The DRM Helix model integrates aspects of the life-cycle and resource/time management, as noted by Bahmani and Zhang (2022). However, a key limitation of the Helix model is the lack of specificity regarding the activities that bolster disaster risk reduction strategies within the disaster management pathway, a detail that the traditional disaster management cycle model articulates more effectively, even though both models are conceptual in nature. Additionally, it fails to accurately portray the varying resource requirements and utilization across the different types and magnitudes of disasters. Moreover, the model overlooks the cumulative impact of consecutive disasters and focuses on recurring disasters which usually require the similar activities and resources to manage (for more see the scenarios of the Helix model in Bosher, Chmutina, & van Niekerk, 2021). Thus, while the Helix model offers a macro-thematic view that provides general direction for pre-disaster (risk reduction) and post-disaster (relief, response, and recovery) activities/resources utilization and depletion, it is not clear how to provide the intricate guidance required for implementing tactical disaster risk reduction or a recovery strategy.

Ultimately, the progression of disaster management frameworks, such as the expand-contract model in the South African Green Paper (South Africa, 1998) and the more comprehensive strategies developed by Bahmani and Zhang (2022), signifies a constant effort to enhance disaster management practices. These advancements accentuate the importance of a thorough, unified approach that acknowledges the complete array of disaster management activities spanning from pre-disaster planning to the establishment of long-term recovery and the enhancement of resilience.

Disaster Recovery in Practice

The US National Disaster Recovery Framework

The practical application of disaster management in the United States, as outlined in the National Preparedness Goal and the US (NDRF 2nd edition), offers a comprehensive and coordinated approach to handling disasters. The National Preparedness Goal operationalizes the traditional phases of disaster management into five mission areas: prevention, protection, mitigation, response, and recovery (Homeland Security, 2015). By formulating mission areas, it becomes easier to take a structured approach to national preparedness, emphasizing the need for collaboration and interdependence.

In its second edition, the U.S. NDRF takes a more in-depth approach to the recovery component, categorizing DR into short-term, intermediate, and long-term recovery phases (Homeland Security, 2016). Each phase plays a distinct role, from meeting immediate basic needs to stabilizing the community and initiating long-term redevelopment. The framework highlights the crucial role of decisions made in the preparedness phase that shape the DR process, as demonstrated by scenarios such as the possible relocation of key employers due to perceived shortcomings in recovery efforts (Homeland Security, 2016). This highlights the vital role of engaging stakeholders in both pre- and post-disaster planning to guarantee timely and adequate DR.

Additionally, the U.S. NDRF acknowledges the importance of a coordinated approach, involving government entities, non-governmental organizations (NGOs), and the private sector. Such collaborations are considered crucial for successful DR, and thus, the framework outlines numerous coordinating structures and support functions (Homeland Security, 2016). It also emphasizes the significance of community engagement and adaptability, aligning with contemporary concepts in the conventional disaster management process. The DR support functions articulated in the NDRF outline the specific scope of the recovery stage, setting it apart from other stages by defining the essential capabilities needed to accomplish community objectives.

The NDRF acknowledges that preparedness plays a critical role in DR and defines the DR process as a continuum of interdependent activities that advance community recovery outcomes (Homeland Security, 2016). By illustrating the 'recovery continuum,' it becomes manifest that preparedness, response, and recovery are *interconnected and cannot be disentangled.* Response activities play a crucial role in determining successful recovery outcomes, but the groundwork for recovery planning is laid in the preparedness phase—hence the interconnectedness of all disaster management and DR processes both practically and theoretically.

The Australian Disaster Recovery Framework

On the other hand, the Australian Disaster Recovery Framework (ADRF 3rd edition) breaks down the DR process into sequential stages, with some overlap: relief (response), early recovery, long-term recovery, and ongoing community development work. With an understanding of the fluidity of the DR process, this framework acknowledges the challenges in defining specific time windows for each stage. The emphasis on the multifaceted nature of DR in this approach reflects the acknowledgment of its diverse challenges (Australian Disaster Recovery Framework (ADRF), 2022). The origins of these challenges lie in the different consequences of disasters, the subsequent occurrences that arise, the dynamics among communities, and the crucial requirement to incorporate disaster risk reduction into recovery planning (ADRF, 2022). The ADRF (2022) prioritizes recovery efforts led by the community. This approach recognizes the distinct characteristics of each community, whether it is their demographics, geography, risk factors, or hazard risk.

The ADRF categorizes recovery into four interlinked environments: social, economic, natural, and built (ADRF, 2022). This holistic perspective acknowledges the interplay between these environments, emphasizing the need for a comprehensive plan for recovery. Furthermore, to facilitate effective DR, a monitoring and evaluation framework is in place that assesses the achievement of objectives and outcomes across all four recovery environments (ADRF, 2022). Another crucial element of the Australian framework is the smooth shift from response to recovery, seeking to affect a seamless transition. Alongside the response phase, recovery initiates its course, adapting its timeline according to the event's type, severity, and resource requirements (ADRF, 2022). This transition is guided by the National Principles for Disaster Recovery, setting standards for best practices and forming the basis of emergency management plans. Last, the framework pushes toward good governance, with institutions, mechanisms, policies, and legal frameworks guiding and coordinating disaster risk reduction and related policies. This systematic approach aims to establish well-regulated DR processes, which is crucial for the efficient management of disaster impacts.

Both the United States (NDRF) and Australian (ADRF) frameworks underscore the value of a harmonized, multi-phase strategy for disaster management. The approaches highlight a shift in conceptualizing and operationalizing DR, emphasizing the importance of teamwork, readiness, and the strength of the community. While the US framework focuses on preparedness and provides a clear division of the recovery phase, the Australian framework accounts for the fluid and overlapping nature of DR processes, offering a more comprehensive guide for managing disasters.

Rethinking the Disaster Recovery Process

Since the concept of disaster is deeply connected to human communities, the DR process likewise reflects the collective efforts of the community. Modeling the recovery cycle is challenging due to the complex and unlimited factors of the community, ecology, and event-related elements, as well as post-event factors influenced by human decisions and policies (Tierney & Oliver-Smith, 2012). As others did, Tierney and Oliver-Smith (2012) critiqued earlier unidimensional and stage-oriented models of DR. They advocated for a more nuanced understanding that encompasses social variability, inequality, diversity, and disparities. In their new conceptualization of DR, emphasis was placed on the importance of pre-disaster factors, disaster impacts, immediate responses, and post-disaster variables, including governance quality, institutional capacity, and civil society-state relationships.

Perplexing Factors in Disaster Recovery

The evolving conceptualization of DR, as expounded by Tierney and Oliver-Smith (2012), signifies a paradigm shift from merely reverting to pre-disaster conditions to embracing a 'new normal' that implements and adapts to transformative changes in the aftermath of a disaster. This shift in perspective both acknowledges and aligns with ongoing social transformations, representing an interplay between DR processes and broader societal changes. In this context, it becomes challenging to differentiate the effects of exogenous variables from recovery interventions, necessitating a robust methodology.

Finch, Emrich, and Cutter's (2010) study sheds further light on the intricacy of this matter in their analysis of the effects of Hurricane Katrina on the recovery of communities in New Orleans. Their examination illuminates the connections among social vulnerability, depth of flooding, and rates at which households returned (Finch, Emrich, & Cutter, 2010). The findings show that neighborhoods with minimal flooding and lower social vulnerability witnessed higher return rates, while those with extensive flooding and heightened social vulnerability experienced significantly lower return rates.

Likewise, an examination of the consequences of tsunami recovery initiatives on women, as indicated by diverse international agencies, uncovered a variety of challenges (Jayasuriya & McCawley, 2010). Efforts to recover from the 2004 Indian Ocean tsunami brought attention to notable obstacles confronted by women, such as rigid humanitarian systems that neglected their specific requirements, hazardous conditions in internally displaced persons (IDP) camps, legal barriers, less access to information and public spaces, and limited availability of economic recovery programs (Juran, 2012; Scheper et al., 2006). Furthermore, women's grassroots organizations encountered challenges when it came to influencing the types of relief activities due to limited opportunities and leadership capacity (Scheper et al., 2006). Hence, it is essential to understand how various dimensions of diversity and inequality shape the DR process and to address it (e.g., via gender mainstreaming) in order to promote social inclusion and better recovery results for all (Juran & Trivedi, 2015; Tierney & Oliver-Smith, 2012).

Blaikie (2009) illustrates the impact of post-disaster policies on vulnerabilities, demonstrating how policies can either mitigate or exacerbate social vulnerabilities in DR. Following the 2004 tsunami, the Sri Lankan government implemented a policy disallowing rebuilding within a 200-meter buffer zone along the east coast. The discriminatory impacts of this policy were particularly evident on Tamils and Muslims, who had fled the civil war and sought shelter on vulnerable coastal lands that were subsequently decimated by the tsunami (Blaikie, 2009). The 'right to return' policy in Sri Lanka is an example of a policy meant to support IDPs in returning to their original places, either integrating locally or resettling (Blaikie, 2009). Nonetheless, implementation of the policy frequently hindered IDPs from voicing their preferences as the government prioritized its own power and interests, thereby affecting the IDPs' ability to exercise their 'right to return' (Blaikie, 2009).

Resilience as a Component of Recovery

Another layer of complexity in the DR process is the extent to which resilience is embraced to facilitate enhanced recovery outcomes for subsequent events. The concept of resilience has expanded the scope of recovery, as demonstrated by Charles, Chang-Richards, and Yiu (2022). Emphasizing the importance of resilience in post-disaster reconstruction, Charles, Chang-Richards, and Yiu (2022) highlight how communities, systems, or structures can absorb, adapt to, and recover from the impacts of a disaster. The study uncovers 24 factors that contribute to the resilience of reconstruction projects, such as political and legislative functioning, development objectives, institutional strengthening, urban governance, and other critical factors. In summary, Charles, Chang-Richards, and Yiu (2022) emphasize the role of robust governance policies, innovative technologies, comprehensive reconstruction strategies, efficient resource management, and active stakeholder engagement for post-disaster reconstruction projects.

Consecutive Disasters and Recovery

Consecutive disasters add another layer of complexity to how DR processes unfold and compound. The term 'consecutive disasters' is used to describe a situation in which two or more disasters occur in a given time period, with their direct effects overlapping in the same geographic region before recovery from the initial event is fully achieved (de Ruiter et al., 2020). The term encompasses a wide array of multi-hazard events, which include compounding events in which multiple hazards converge to produce a more pronounced effect (Zscheischler et al., 2018) as well as cascading events in which one disaster triggers another (de Ruiter et al., 2020; Pescaroli & Alexander, 2015). An example of this would be a natural hazard that causes a disaster, which then impacts industrial facilities or infrastructure, leading to further disasters. In 2011, Japan experienced a case of consecutive disasters when the Great East Japan Earthquake triggered a tsunami that flooded coastal regions. This resulted in substantial damage to the Fukushima nuclear power plant and escalated into a nuclear crisis that exacerbated the initial calamity.

The occurrence of consecutive disasters greatly complicates and prolongs the process of recovery as it adds layers of impacts and strains the already scarce availability of resources (Tarvainen, Jarva, & Greiving, 2006). The damage that overlaps—which may exceed impacts from the initial disaster—can affect various aspects such as infrastructure, economies, ecologies, and social systems, resulting in challenging decisions on resource allocation and worsening the fatigue experienced by communities engaged in DR and other disaster processes. Consecutive disasters can undermine morale and resilience, highlighting deficiencies in traditional DR planning. Therefore, a shift toward adaptive planning is necessary to more effectively manage these unique challenges. If not, even longer-term impacts such as hindered economic revival, prolonged financial struggles among public institutions, and enduring environmental damage can pose additional obstacles to DR and sustainable development (de Ruiter et al., 2020). The occurrence of consecutive disasters demands an adaptive, comprehensive, and resilient approach to DR, acknowledging the flexibility and capacity required to meet the challenges these events present.

When reflecting on DR models, it becomes clear that a multitude of factors can affect the recovery process, steering theorists and practitioners away from a linear process toward a more dynamic, inclusive, and holistic approach. Ultimately, the interplay among pre-conditions, socio-spatial disparities, pre- and post-disaster policies, community resilience, and the nature of the event emphasize the iterative nature of DR. Therefore, the comprehension of DR processes requires accounting for various pre-disaster factors that influence vulnerabilities and exposures across different communities and socio-ecological environments at different scales. DR must also consider indicators of social and economic well-being and the capacity of the government (Tierney & Oliver-Smith, 2012), and it must recognize the phenomenon of consecutive disasters. Cyclic models of disaster management are illustrative for understanding the level of institutional preparedness and interconnectedness of the various phases, especially with the new themes of recovery suggested by Bahmani and Zhang. Helical models provide a macro level picture of community progress over the course of disasters, but a gap still exists in developing a theoretical model that fully incorporates the community recovery process (Aguirre & Lane, 2019).

Conclusion

The chapter sought to highlight and emphasize the complex nature of DR, urging caution against oversimplifying recovery processes through the use of conceptual models. A disaster management strategy that encompasses a holistic, adaptable, and resilient approach is essential, and it should be firmly based on both theory and empirical evidence.

DR is a multifaceted process of restoring community infrastructure, well-being, and functions following widespread disruptions and/or substantial damage to physical infrastructure or the natural environment. This chapter explored the complexities inherent in DR, highlighting its diverse theoretical and pragmatic landscapes. Through examining various perspectives, it becomes evident that recovery processes are not uniform and can create disparities (or enlarge existing disparities), accentuating the need for a more inclusive approach. This chapter places emphasis on the importance of resilience and community participation in DR, emphasizing the necessity to move beyond the mere restoration of physical assets to encompass social, psychological, and emotional dimensions of recovery.

DR is a diverse and contested theoretical and pragmatic space; it is not a uniform process that impacts all groups within a society or community equally (Smith & Wenger, 2007). Furthermore, the DR process itself can create disparities, resulting in 'winners' and 'losers' (Tierney & Oliver-Smith, 2012). This means that assessing recovery based on aggregate or average gains can be misleading, as it (necessarily) obscures the lived experiences of different groups (Tierney & Oliver-Smith, 2012). Recovery indicators that appear positive at a community level may not hold the same positive value at the level of a displaced household or business. Thus, DR is a complex, multi-layered socio-spatial process that varies significantly across the different sectors of society, and even within communities.

The concept of recovery encompasses all activities and processes that aim to restore the human ecological milieus to an accepted, satisfied, or even improved level of functionality and capacity following the manifestation of a disaster(s). The process of DR is not uniform, highlighting the need for a more inclusive approach that encompasses the restoration of social, cultural, and economic systems in addition to traditional reconstruction of physical assets such as utilities, infrastructure, and housing (Finch, Emrich, & Cutter, 2010; Tierney & Oliver-Smith, 2012). A comprehensive, operational understanding of DR calls for a holistic approach that considers hazard risk and social vulnerabilities at different scales, the role of governance systems, and the various factors, trends, and events that coalesce to influence recovery processes and outcomes. In this vein, Bosher, Chmutina, & van Niekerk (2021) propose a helical thematic framework of DRM, arguing for a more dynamic representation of the process. Although this model provides a wider viewpoint, it fails to capture the intricate details of specific activities that enhance disaster risk reduction strategies, as seen in conventional cyclic diagrams. Nevertheless, the helix model presents a big-picture perspective that is vital for gauging the progress of disaster management.

The insights gleaned from this exploration aim to inform and enhance the effectiveness of global DR efforts, ensuring that recovery processes not only rebuild structures but also rebuild the social fabric of affected communities. In light of the complexities inherent in DR, such as the necessity for effective resource allocation, stakeholder coordination, and need for genuine community participation, this chapter highlights the criticality of implementing collaborative approaches throughout the different stages of the DR cycle. Ultimately, more extensive research is warranted in order to (1) better define and thus refine the DR process, (2) examine the factors and conditions that influence DR within communities, and (3) assess the role governance and institutional arrangements play in facilitating and hindering the alignment of DR processes with local needs and priorities.

References

Aguirre, Benigno E., and David Lane. "Fraud in Disaster: Rethinking the Phases." *International Journal of Disaster Risk Reduction*, vol. 39 (2019), pp. 101232, https://doi.org/10.1016/j.ijdrr.2019.101232

Australian Disaster Recovery Framework (ADRF). *Australian Disaster Recovery Framework*. The Australia-New Zealand Emergency Management Committee. (2022), 3rd edition.

Bahmani, Homa, and Wei Zhang. "A Conceptual Framework for Integrated Management of Disasters Recovery Projects." *Natural Hazards (Dordrecht)*, vol. 113, no. 2 (2022), pp. 859–85. https://doi.org/10.1007/s11069-022-05328-5

Blaikie, Piers M. "The Tsunami of 2004 in Sri Lanka: An Introduction to Impacts and Policy in the Shadow of Civil War." *Norsk Geografisk Tidsskrift*, vol. 63, no. 1 (2009), pp. 2–9. https://doi.org/10.1080/00291950802712061

Bosher, Lee, and Ksenia Chmutina. *Disaster Risk Reduction for the Built Environment*. John Wiley & Sons (2017).

Bosher, Lee, Ksenia Chmutina, and Dewald van Niekerk. "Stop Going Around in Circles: Towards a Reconceptualisation of Disaster Risk Management Phases." *Disaster Prevention and Management*, vol. 30, no. 4/5 (2021), pp. 525–37. https://doi.org/10.1108/DPM-03-2021-0071

Carter, W. Nick. *Disaster Management: A Disaster Manager's Handbook*. Asian Development Bank (ADB) (2008).

Charles, Shawn Hezron, Alice Yan Chang-Richards, and Tak Wing Yiu. "A Systematic Review of Factors Affecting Post-Disaster Reconstruction Projects Resilience." *International Journal of Disaster Resilience in the Built Environment*, vol. 13, no. 1 (2022), pp. 113–32. https://doi.org/10.1108/IJDRBE-10-2020-0109

Christoph Aubrecht, Dilek Ö. Aubrecht, Kierx Joachim, and Freire Sérgio. "Future-Oriented Activities as a Concept for Improved Disaster Risk Management." *Disaster Advances*, vol. 6, no. 12 (2013), pp. 1–10.

Contreras, Diana. "Fuzzy Boundaries between Post-Disaster Phases: The Case of L'Aquila, Italy." *International Journal of Disaster Risk Science*, vol. 7, no. 3 (2016), pp. 277–92. https://doi.org/10.1007/s13753-016-0095-4

de Ruiter, Marleen C., Anais Couasnon, Marc J.C. Homberg, James E. Daniell, Joel C. Gill, and Phillip J. Ward. "Why We Can No Longer Ignore Consecutive Disasters." *Earth's Future*, vol. 8, no. 3 (2020). https://doi.org/10.1029/2019EF001425

Finch, Christina, Christopher T. Emrich, and Susan L. Cutter. "Disaster Disparities and Differential Recovery in New Orleans." *Population and Environment*, vol. 31 (2010), pp. 179–202.

Finucane, Melissa L., Joie Acosta, Amanda Wicker, and Katie Whipkey. "Short-Term Solutions to a Long-Term Challenge: Rethinking Disaster Recovery Planning to Reduce Vulnerabilities and Inequities." *International Journal of Environmental Research and Public Health*, vol. 17, no. 2 (2020), p. 482. https://doi.org/10.3390/ijerph17020482

Himes-Cornell, Amber, Carlos Ormond, Kristin R. Hoelting, Natalie C. Ban, J. Jachary Koehn, Edward H. Allison, Eric R. Larson, Daniel Monson, Henry P. Huntington, and Tom A, Okey. "Factors Affecting Disaster Preparedness, Response, and Recovery Using the Community Capitals Framework." *Coastal Management*, vol. 46, no. 5 (2018), pp. 335–58.

Homeland Security (U.S. Department of). *National Preparedness Goal (United States. Federal Emergency Management Agency)*. U.S. Department of Homeland Security, FEMA (2015), 2nd edition.

Homeland Security (U.S. Department of). *National Disaster Recovery Framework (NDRF) (United States. Federal Emergency Management Agency)*. U.S. Department of Homeland Security, FEMA (2016), 2nd edition.

Jayasuriya Sisira, and Peter McCawley. *The Asian Tsunami: Aid and Reconstruction after a Disaster*. Edward Elgar (2010).

Juran, Luke. "The Gendered Nature of Disasters: Women Survivors in Post-Tsunami Tamil Nadu." *Indian Journal of Gender Studies*, vol. 19, no. 1 (2012), pp. 1–29.

Juran, Luke, and Jennifer Trivedi. "Women, Gender Norms, and Natural Disasters in Bangladesh." *Geographical Review*, vol. 105, no. 4 (2015), pp. 601–11.

Khan, Himayatullah, Laura Giurca Vasilescu, and Asmatullah Khan. "Disaster Management Cycle-a Theoretical Approach." *Journal of Management and Marketing*, vol. 6, no. 1 (2008), pp. 43–50.

Marshall, Maria I, and Holly L. Schrank. "Small Business Disaster Recovery: A Research Framework." *Natural Hazards (Dordrecht)*, vol. 72, no. 2 (2014), pp. 597–616. https://doi.org/10.1007/s11069-013-1025-z

Mayner, Lidia, and Paul Arbon. "Defining Disaster: The Need for Harmonisation of Terminology." *Australasian Journal of Disaster and Trauma Studies*, vol. 19, no. SI (2015), pp. 21–25.

Nigg, Joanne M. *Disaster Recovery as a Social Process*. Disaster Research Center, University of Delaware, Newark, (1995).

Pescaroli, Gianluca, and David Alexander. "A Definition of Cascading Disasters and Cascading Effects: Going beyond the "toppling dominos" Metaphor." *Planet@ Risk*, vol. 3, no. 1 (2015), pp. 58–67.

Quarantelli, Enrico Louis. *Disaster Recovery: Research Based Observations on What It Means, Success and Failure, Those Assisted and Those Assisting*. Disaster Research Center, University of Delaware, Newark (1998).

Rouhanizadeh, Behzad, Sharareh Kermanshachi, and Thahomina Jahan Nipa. "Exploratory Analysis of Barriers to Effective Post-Disaster Recovery." *International Journal of Disaster Risk Reduction*, vol. 50 (2020), p. 101735.

Scheper, Elisabeth, Arjuna Parakrama, Smruti Patel, and Tony Vaux. *Impact of the Tsunami Response on Local and National Capacities*. Tsunami Evaluation Coalition (2006).

Smith, Gavin P, and Dennis Wenger. "Sustainable Disaster Recovery: Operationalizing an Existing Agenda." *Handbook of Disaster Research*, edited by Havidán Rodríguez, Enrico Louis Quarantelli, and Russell Rowe Dynes. Springer (2007), pp. 234–57. https://doi.org/10.1007/978-0-387-32353-4_14
South Africa (Republic of). *Green Paper on Disaster Management*, Government Printers (1998).
Tarvainen, Timo, Jaana Jarva, and Stefan Greiving. "Spatial Pattern of Hazards and Hazard Interactions in Europe." *Special Paper-Geological Survey of Finland*, vol. 42 (2006), p. 83.
Tierney, Kathleen, and Anthony Oliver-Smith. "Social Dimensions of Disaster Recovery." *International Journal of Mass Emergencies & Disasters*, vol. 30, no. 2 (2012), pp. 123–46.
Zscheischlerr Jakob, Seh Westra, Bart J. J. M. van den Hurk, Sonia I. Seneviratne, Phillip J. Ward, Andy Pitman, Amir AghaKouchak, David N. Bresch, Michael Leonard, Thomas Wahl, and Xuebin Zhang. "Future Climate Risk from Compound Events." *Nature Climate Change*, vol. 8, no. 6 (2018), pp. 469–77. https://doi.org/10.1038/s41558-018-0156-3

20

DISASTER RECOVERY OUTCOMES

Yetta Gurtner and David King

Introduction: The Recovery Process

Disaster recovery is the process of restoring critical infrastructure and services following a disaster. It is a complex and challenging task, requiring a coordinated effort from government, businesses, and the community. The process is predicated on the ideas of resilience and a return to 'normalcy.' Emergency management strategies have adopted the concept of resilience to safeguard communities or build safe communities (COAG, 2011). This is in part a contrast to an emphasis on the reduction of vulnerability. Many aspects of vulnerability, such as low socio-economic status, are structural and outside the scope of emergency managers to exercise any control or mitigation. The primary hazard vulnerability of people and communities exists in the proximity of the natural hazard itself—living in an area prone to cyclones/hurricanes, flood-prone areas, earthquake threat, etc. Extensive mapping of hazard zones and levels of risk, inventories of community characteristics and lifelines, and integrative models using sophisticated GIS and sensor data contributed to hazard assessments of cities and regions but only underscored elements of risk exposure, susceptibility, and vulnerability. Assessments and modeling define hazard risk, but it may not be able to be mitigated any more than the social and economic vulnerability of people. Thus, a primary response to disaster risk reduction targets the positives in people and communities through education, awareness raising, and focused preparation to reduce risk and mitigate disaster impact. Vulnerability and resilience are not opposite ends of a scale—they are fluid, dynamic, and highly context dependent. It is completely plausible to be vulnerable to the threat of a disaster and resilient at the same time. Whereas vulnerability reduction is faced with intractable and structural impediments reflecting inherent weaknesses in communities and places, encouraging and developing resilience strategies targets strengths in societies and individuals. Capacity building and enhancing positive characteristics enable people to reduce their risk.

Resilience is the ability to bounce back, which suggests a return to the pre-disaster status quo. However, a place will never be the same as it was before the disaster, but the idea in many people's and agencies' minds is to get back to normal, to the pre-disaster way of life. Zhou et al. (2010) identified at least 28 definitions alone of just the concept of social resilience. Paton and Johnston (2006) argued that 'bouncing back' to a previous state is not satisfactory, because there is a changed reality to which people must adapt. Kirmayer et al. (2009) explained resilience as a

DOI: 10.4324/9781003414834-28

transformative process. The common elements of transformational definitions of resilience are the abilities of an individual, organization, or system to confront the crisis and challenges of a disaster, thence to survive and thrive despite impact and loss. In a sense, resilience may be for some a bounce forward—a transformative experience—although for others a bounce back may not be to the pre-disaster state but to a lessened or impoverished state. This variability in recovery is partially driven by the reconstructive process of recovery but is also contingent upon the capacities and experiences of the impacted community. There are consequently two primary elements operating in the recovery process: restoration of infrastructure, and community or societal recovery.

Infrastructure recovery or restoration dominates guides and manuals that are devoted to disaster recovery. For example, detailed guides are provided by Phillips (2009) and McEntire (2009) that explain the step-by-step processes of recovery. The Australia-New Zealand Emergency Management Committee (2022) and Government of South Australia (2022) provide recovery guides, although the South Australian document is a more outcomes-oriented approach. The stress of such guides is on the restoration of infrastructure with clearly defined stages in a largely linear process. Communities are acknowledged, but the tendency of an emphasis on infrastructure restoration is to view the community as a recipient or beneficiary of recovery rather than as a key stakeholder or as a driver of the process. Many models of recovery outline the processes and stages. Miles and Chang (2003) developed an urban recovery model that integrates both mathematical analyses and simulations. This model incorporates the complexity and integration of the city. The emphasis is useful on infrastructure and economy, measuring and evaluating at each stage where the outcome is a return to normal function. An extensive summary of models along with a range of case studies was subsequently presented by Davis and Alexander (2016) as an integrative text that brings together diverse locations and disasters. They present 21 models of disaster recovery and place strong emphasis on the role of the community in the recovery process. Davis and Alexander (2016) also indicate how recovery may fall short or fail to achieve successful outcomes by outlining the confusion and conflicting priorities and agendas that operate in the recovery process.

The disaster-impacted community is dependent on the restoration of infrastructure but is critically proactive throughout the recovery process (Miles et al., 2019; Patterson et al., 2010). The speed and efficiency of infrastructure recovery either drives or constrains community recovery. Thus, the factors that influence infrastructure restoration are crucial to the success of community recovery and its resilience outcomes. These factors that influence infrastructure restoration include material supplies, the supply chain and issues of distance and remoteness, the severity and breadth of the disaster impact, secondary and compounding disasters, the availability of an appropriately skilled recovery workforce, resources and institutional capacity, and the linkage of these elements to economic recovery.

When infrastructure has been restored, recovery has not necessarily been achieved; for the community it is not over. Recovery is ongoing for those who have been impacted and is further constrained by their socio-economic characteristics, social capital, and complex layers of human capacities. Community recovery both leads the process and is constrained and differentiated both by infrastructure restoration and by the nature and capitals of the community itself (Miles et al., 2019). The idea of the community is of a larger social aggregation encompassing many individuals, families, and subcommunities within a region or nation. While society suggests a higher level of complexity than community, we have adopted the term community as a surrogate for society or the public. Defining community is problematic in the context of disasters (Buckle et al., 2000; Cottrell, 2005; Stehlik, 2006).

Community and social resilience are often used interchangeably, partly because community and society are often seen as interchangeable. Mileti (1999) defined community resilience as the ability to recover based on its own resources. However, both community vulnerability and resilience are affected by the scale and definition of community, conceptualized here as a group of people with common attributes who have a sense of connection and belonging. Social resilience is a concept that operates at different scales.

Emergency managers, on the other hand, tend to consider a community in terms of a geographical locality and they assume connections through neighborliness and proximity. Such a view of community is unhelpful in urban areas where people form complex, spatially extensive communities based on networks and associations that transcend the geographical neighborhood. Communities are not spatially defined; people simultaneously are members of multiple communities (e.g., workplace, school, religious, recreational, and online, as well as their place of residence). These networks of communities contribute to social capital, which is a component of resilience (Aguirre, 2009). Some social networks within communities are well-suited to respond to a disaster, while others may reinforce their structural social disadvantages. Boon et al. (2012, 2016) adapted Bronfenbrenner's bioecological model, originally developed to understand the educational milieu of the student, to place the individual or household within a series of supporting systems. Many of these systems constitute different communities of neighborhoods, networks, memberships, and organizations, as well as institutions and infrastructural supports.

There is a further problem with the tendency to view the community as homogenous: it implies a focus and unity which is frequently not present. The reality of any community is that it is an amalgam of families, neighborhoods, organizations, networks, interest groups, minorities, and demographic and socio-economic diversity. Subgroups and networks of community are not mutually exclusive. Any one individual or family may be a member of several groups or networks and may represent a range of community characteristics. These disparate subgroups are affected very differently by the impact of a disaster and consequently recover at different rates, both temporally and spatially.

The recovery process may eventually reach a stage where the disaster impact is in the past and no more direct restoration is possible or needed. Recovery slows down and life goes on—people and communities have moved beyond the disaster, although that does not mean they are no longer affected by aspects of what happened. There is a point where the recovery process is complete or has gone as far as possible. This may be spread over a period of years—very commonly around five years, but that timescale is quite different for individual people and communities and is related to the scale of the disaster. Some people and places do not recover in the sense that we assume recovery to be a return to functions and relationships that existed before the disaster.

Resilience is a recovery concept that is conventionally defined as the capacity to bounce back, or to return to a level of normalcy. The reality is that some people bounce forward into a new reality and its opportunities, while others return to a situation that replicates the pre-disaster state (Davis and Alexander, 2016). Other people and communities remain damaged, reduced, or incapacitated. It is this indeterminate end of recovery that prompts the focus of this chapter; a consideration of societal outcomes in the post-recovery state is warranted. The purpose, intent, and direction of recovery are often not met. The outcome is a different place that may have evolved, or it may have been compromised such that it may not be a reflection of what was intended in recovery planning.

Case studies are examined to illustrate outcomes that were divergent from an expected recovery process. Emphasis is especially on population change in case study locations of New Orleans (USA), Christchurch (New Zealand), Innisfail (Australia), and Koidu (Sierra Leone). Population recovery and change is a crude indicator, but it drives important indicators and outcomes of recovery, such as economy, social capital, community restoration and rebuilding, and reestablishment of community function. The specific uncontrolled outcomes that are explored through these case studies are slow population recovery or actual population decline on the one hand, and rapid expansion of growth and new opportunities in recovery on the other.

Population Loss

During reconstruction and recovery, change in demography affects service and infrastructural needs. An altered population and demography after a period of recovery generate needs for new land-use zoning, changes to planning schemes, land-use change, and additional infrastructural needs. Most cities and their plans are predicated on growth. Population growth translates to economic growth, even though this may not necessarily be sustainable in the long term, and drives further expansion of population, jobs, and infrastructure—which in turn attract more jobs and migrants. The incentive for city plans is continual growth. Consequently, recovery from disasters is conceived to be a return to a level of economy and population growth implicit in many city planning schemes. A return to normalcy assumes a continued status quo of steady expansion once infrastructure has been restored, and the necessity for this is that it should occur as rapidly as possible.

Extended decline and loss of population hampers economic recovery and may prompt a different outcome that necessitates plans for long-term decline in the community. Hollander et al. (2009) suggested an alternative planning approach that works toward population shrinkage—a planned decline in population and inevitably in infrastructure and services. Population decline or shrinkage may be an intended outcome after a disaster. Where a location or region may already have been in long-term population decline, the disaster may bring about a speeding up of the process.

We examined the experiences of three locations (King and Gurtner, 2020) that saw strong population decline after a natural disaster, where outcomes of their actual recovery experience were contrary to conventional processes of recovery and restoration taking communities back to that pre-disaster state. Two of these case studies, New Orleans and Innisfail, had been experiencing a long-term decline in population, economy, and regional strategic importance. New Orleans is physically vulnerable to flooding and storm surges consequent upon hurricane impact, compounded by its extremely low elevation below sea level. However, the city had experienced gradual decline in importance for well over a century before Hurricane Katrina brought devastation and loss of life in 2005. The blockade during the American Civil War had reduced New Orleans' significance as a major port, bringing about poverty and socio-economic disadvantage.

Innisfail, located in North Queensland, Australia, had similarly declined in size and importance over many decades. It is a rural service center for a region that produces sugar, bananas, and tropical fruits. In common with many country towns, the mechanization of agriculture along with strong outmigration trends to metropolitan urbanization have led to a population that is aging as its younger residents have moved away from both the town and its surrounding farms. Rural decline and depopulation are a reality for many small country towns and regions.

Both New Orleans and Innisfail have also suffered a succession of disaster impacts. Each disaster recovery has involved a relocation of resources away from longer-term development to the needs of immediate reconstruction. New Orleans has been impacted by a severe hurricane on average every 11 years, although between 1994 and 2019 eight hurricanes impacted Orleans Parish, along with a further 15 tropical storms (NOAA, 2023).

Innisfail experienced destructive cyclones causing loss of life as well as extensive damage to infrastructure and homes in 1918 and 1986, while overall there were seven cyclones between the 1880s and 2011 (BoM, 2023). Two cyclones hit in 2005 and 2011 (Larry and Yasi) such that the region had barely recovered from the earlier cyclone before being impacted by the next. There was no loss of life, but insured damage in the region of impact was estimated to be $500 million in 2005 and $3.4 billion from Yasi in 2011. The population of the Cassowary Coast local government area (of which Innisfail is the main town) at the 2001 census had been 35,408 with 9,719 in Innisfail town. By the 2006 census, both the local government area and the town saw a loss of population and a further loss at the time of the 2011 census after each of cyclones Larry and Yasi. By the 2021 census—and without further cyclones—the population of the local government area had fallen further to 29,157 and the town of Innisfail to 9,257. The decline was absolute but additionally saw a greater loss of younger economically active individuals and families (Boon et al., 2012; King et al., 2014).

The population of greater New Orleans metropolitan area before Hurricane Katrina was estimated to be 1.34 million. New Orleans had peaked at 627,525 in 1960 after which there had been a gradual loss into the metropolitan area that contains the suburbs (Plyer, 2011). During this outmigration the inner-city had become increasingly impoverished. By 2022 (17 years after Hurricane Katrina), New Orleans had regained some population, reaching 369,749 (USCB, 2023)—up from only 158,353 immediately after the hurricane but significantly less than the city population of 455,000 immediately prior to Katrina (Kates et al., 2006). The whole metropolitan area had also absolutely declined from 1.34 million to only 1,005,000 by 2022 (USCB, 2023).

In neither case was recovery based on a planned shrinkage or reduction, but the outcome of the disaster exacerbated a longer-term trend toward population decline. In the case of Christchurch in New Zealand, the population was growing before the earthquakes as it is a major city serving an extensive agricultural region. The earthquakes of 2010/2011 were estimated to have generated a loss of 70,000 people who relocated (King and Gurtner, 2020). Within the greater Christchurch area, an estimated 460,000 people were affected by the earthquakes with 150,000 homes damaged to various extents (King and Gurtner, 2020). The main outflux of people occurred after the 2011 earthquake. By then the city population had reached 367,720 but subsequently recovered slowly to 396,200 by 2023. The greater Christchurch region however reduced from 457,420 at the time of the 2011 earthquake to only 405,000 in 2022 (Christchurch City Council, 2023).

Christchurch before the earthquakes had steadily grown but the population was aging, as was the population of Innisfail and its region. An aging population alone brings different growth planning strategies, but all three of these case study places were, and continue to be, dependent upon tourism for a significant component of their economies. A disaster severely restricts tourism in the short term and requires investment in the restoration of tourist infrastructure before a pre-disaster level of tourist visitation may be anticipated. Global tourism is a robust industry. When a single destination is knocked out, tourists transfer to other markets and may not return in the same numbers even when the industry has been rebuilt in the disaster-affected location. The whole global tourism industry was severely damaged by COVID-19 between 2020 and 2022. Places that were slowly recovering from a disaster, such as a hurricane, cyclone, or earthquake, were then damaged by this catastrophic pandemic (Gurtner and King, 2021). A slowdown of any economic sector constrains recovery of the whole economy and the city.

These were outcomes of disaster recovery that developed separately from conventional strategies of recovery and restoration. Places that are already vulnerable to decline and shrinkage are further pushed along that trajectory by disaster (King and Gurtner, 2020), but conventional recovery models are predicated on restoration to at least pre-disaster levels rather than being oriented toward planned decline or shrinkage strategies. The outcomes for New Orleans and Innisfail are not unexpected in the context of their long-term trajectories of decline, but it is not a desirable outcome of recovery. Significant medium- to long-term loss of population may be suggested as a lack of recovery or loss of resilience. The outcome has not been that which would have been desired. Recovery outcomes diverged from conventional models of recovery strategies and processes. These case studies are examples of places that have experienced a net loss of population following a major natural disaster. They also illustrate the complexity of other factors, trends, and compounding disasters. Many places in the past have declined or disappeared after severe disasters. However, places that were strategically or centrally well located in the first place may return to growth and vitality if their local advantages are reestablished.

There are consequently many examples of places that have exceeded recovery expectations, ultimately progressing to a rebuilt and regenerated community. From a pre-disaster research experience (King, 1979), a case study of Koidu–New Sembehun, or Koidu City, in Sierra Leone, is presented as an example of a city that spontaneously recovered after the complex humanitarian disaster of over ten years of civil war. Koidu–New Sembehun developed as an illicit diamond rush city on top of one of the world's richest deposits of gem diamonds. A government center had been established adjacent to the mining town at Sefadu during the colonial period, such that the city formally carried that name on many maps. Only in the early 1970s was the illicit settlement, by then containing over 68,000 residents (King, 1979) formalized into a city council area. The two chiefdom centers of Koidu and New Sembehun were ultimately amalgamated during the last decade into a single city, Koidu City. Most of the city's history was unplanned and uncontrolled and many of its residents were not counted in earlier censuses. Furthermore, it is relatively remote in a country where colonial governance and postcolonial governance were very poor. The city's growth came from its diamond wealth, although most of its workforce in the mid-1970s was non-mining, providing all of the retail and support services of a city in its central location within Kono District. It was an untidy, overcrowded, dynamic, and lawless city at the height of the diamond rush.

During the civil war from 1991 to 2002, mining continued under various forms of control and contributed toward the conflict but was not the cause of it. The city of Koidu–New Sembehun was razed during the war. Larger buildings were burned and subsequently made unstable, houses were extensively unroofed and wrecked, and most of the population left for safety in villages, forests, and distant locations. At the declaration of peace in early 2002, following a large United Nations intervention, the city was almost totally in ruins. Yet into this wrecked settlement people were reestablishing businesses, food and retail, and mining within days and weeks of the end of the conflict (authors' fieldwork in 2002). Every day more people were moving back in, setting up shelters among the ruins, and beginning to rebuild houses. A spontaneous in-migration had started early in 2002 and has continued.

The Sierra Leone government had only recently returned from exile in neighboring Guinea and had no resources or much of a functioning public service infrastructure to lead recovery, despite initiating strong visions for reconstruction and reconciliation. These strategies had to occur spontaneously with the government initially pursuing an NGO policy (King, 2005) to encourage a range of NGOs to come in and drive recovery and rebuilding, albeit in a very piecemeal manner. There was no centrally operable recovery strategy. As in many lesser-developed countries, recovery was led by the community engaging directly with NGOs and businesses for whatever they were able to offer.

From an almost destroyed city in early 2002 (personal fieldwork observations and key informant interviews with communities and United Nations forces), the city has been rebuilt, improved, and planned. There are still many ruined buildings evident, especially of the multi-story buildings which were unsafe to renovate; however, it is a new and transformed city. By the 2021 census (Statistics Sierra Leone, 2022), the population of Koidu City (formerly Koidu–New Sembehun) had grown to 196,480.

While artisanal diamond mining has become less productive over recent decades, mining by both artisan miners and the mining company continues to support the economy. However, it is no longer a contributor to the conflict, which was driven by poor governance and only ever exacerbated by the resource curse of the diamonds (Wilson, 2013). Koidu City was rebuilt and reestablished despite a lack of guided recovery because it is a natural central place. Its economy is as diversified as during the 1970s diamond rush, but it offers a range of services and functions to surrounding rural districts containing many villages and small towns. This outcome of a high level of sustained population growth is unexpected in terms of recovery where there was no concerted infrastructural restoration plan, but this is otherwise a normal trend on a rapidly urbanizing continent. The devastation of war dispersed urban populations, especially from Koidu, and left wholesale destruction, but the city was able to recover as soon as peace was enabled. Recovery was initiated and driven by peace.

The point illustrated by these three case studies is that actual outcomes of recovery may diverge from intended recovery strategies. There are trends and priorities existing in places prior to a disaster's impact. Recovery strategies cannot apply a standard model to reconstruction in order to take a place back to a perceived pre-disaster state. Recovery strategies must imagine the realistic outcomes of the place or community and tailor the approach to each of those specific communities.

Outcome-Oriented Recovery

Disaster recovery outcomes refer to the measures of success or achievement in the process of restoring a community, infrastructure, and economy after a disaster, with the aim of improving community wellbeing and empowering communities to sustainably manage and drive their own recovery (Government of South Australia, 2022). However, 'disaster recovery programs (and their monitoring and evaluation) have often focused on processes and outputs rather than outcomes' (Government of South Australia, 2022, 3). When these processes are guided by outside agencies, a lack of understanding of the dynamics and characteristics of a community may drive inappropriate recovery responses or may ignore the reality of long-term trends and societal change. Restoring basic services and infrastructure is good and essential, but the post-disaster situation may change priorities, especially if many people leave the place for refuge and ultimately find short- or longer-term opportunities elsewhere. Monitoring and evaluation of recovery may be distorted by mismatched expectations of community responses. Recovery processes may be evaluated as a failure where expectations diverge from outcomes. Such evaluations and monitoring may address quite different facets of recovery and restoration—physical, economic, social, and environmental. Physical restoration of infrastructure is a relatively straightforward measure of progress, but as demonstrated through the case studies, economic recovery is very much dependent upon population movements as well as infrastructure. Social recovery and environmental recovery are especially intangible and difficult to evaluate and quantify such that they may be neglected.

The specific outcomes of disaster recovery will vary depending on the type of disaster, severity of the disaster, and resources available to the community. However, all disaster recovery efforts aim to achieve the following goals:

- Restore the community to its pre-disaster state or functionality as quickly as possible.
- Reduce the risk of future disasters.
- Build a more resilient community.

These outcomes can be categorized into short term, medium term, and long term. Short-term recovery outcomes typically focus on addressing the immediate needs of disaster-affected populations, such as providing emergency shelter, food, water, and medical care. Short-term outcomes may also involve restoring basic services, such as electricity, water, and transportation, and clearing debris from roads and other infrastructure (emergency response).

Medium-term recovery outcomes aim to restore critical infrastructure, such as hospitals, schools, and other public facilities, and to support the recovery of local businesses and industries. This may involve providing financial assistance to affected businesses, supporting the resumption of economic activities, and restoring social services and community infrastructure.

Long-term recovery outcomes involve building resilience and sustainability in the affected communities and addressing the root causes of vulnerability to disasters. This may involve implementing disaster risk reduction measures, such as building codes, land-use planning, and early warning systems, and promoting community-based disaster preparedness and response. The actual outcome after restoration of infrastructure and services may diverge from the planned recovery process. Community members respond in their own best interests in ways that may result in a reduced economy and population, or development and expansion in directions that were not envisaged by the recovery plan. A return to a pre-disaster state may not eventuate.

Types of Disaster Recovery Outcomes

There are many types of disaster recovery outcomes, reflecting diverse priorities and stakeholder interests. The primary aim and outcome of any disaster recovery effort should always be safety and the protection of life. Consistent with the established value of a structurally functional community post-event, the secondary focus is generally on the protection of homes, businesses, and critical infrastructure. Associated objectives are to minimize the economic impact of the disaster with timely restoration of both industry and other infrastructure. Reconstruction and rehabilitation of the built and natural environment has tangible and intangible aesthetic, amenity, and resource value. Ancillary to each of these goals is appreciating the broader social context: the desire to build and enhance community resilience establishing longer-term, sustainable recovery.

The capacity to effectively achieve any, or all, of these recovery outcomes can be affected by several factors. First and foremost is the severity of the disaster event—the more severe in terms of size, scale, impact, and/or damage, the more difficult it will be for a community to recover. Access to resources, such as food, water, and shelter, are central to both survival and sustained efforts. Availability of more complex resources are often influenced by the livelihoods, wealth or poverty, and socio-economic characteristics of the population. While it is not possible to mitigate certain hazard elements, a well-prepared community will recover more quickly and effectively. Similarly, recovery is facilitated by good governance and efficient coordination.

Overall, the success of disaster recovery outcomes can be measured by the extent to which the affected communities and infrastructure are restored to the functionality of pre-disaster conditions, as well as the level of resilience and sustainability achieved in the long term, or specific outcomes that are tailored to the experience of the place and community. Successful disaster recovery outcomes require effective collaboration and coordination among various stakeholders, including government agencies, community-based organizations, private sector, and affected populations.

There are a number of identified barriers to effective disaster recovery outcomes that can impede the successful restoration of communities and infrastructure after a disaster. Logistically, there may be physical and environmental challenges. Rugged terrain and severe weather conditions can make it difficult to access and support the needs of disaster-affected populations.

Outlying and remote communities may remain isolated for extended periods before outside help is able to intervene. Inadequate planning and preparedness can also make it difficult to respond and implement effective recovery programs. While food, water, and shelter are recognized as critical resources, adequate funding, personnel, and equipment are also necessary. Resource issues can further compound the capacity of the affected community to rebuild economy and livelihoods.

In terms of recovery management and organization, political and bureaucratic challenges can create delays in decision-making and implementation of efforts, which can prolong the recovery process and reduce the effectiveness of recovery outcomes. Complex regulatory and legal frameworks can create barriers by limiting the ability of recovery programs to implement necessary measures, such as building codes and land-use planning. Just as poor governance and management can be a concern, limited community participation and engagement reduce the ability of recovery programs to address the needs and priorities of affected populations. Without an established process for communication and ready access to information, a gap may develop between community and recovery institutions.

Addressing these barriers requires a coordinated and collaborative approach that involves multiple stakeholders, including government agencies, community-based organizations, private sector, and affected populations. Successful recovery outcomes depend on effective planning, resource allocation, community participation, and collaboration among stakeholders.

Strategies for Improving Disaster Recovery Outcomes

There are several strategies that can be used to foster disaster recovery outcomes based on a proactive disaster risk reduction/risk management model:

- **Prevention**: The best way to minimize the impact of disasters is to prevent them from happening in the first place. This can be done by building disaster-resistant infrastructure and implementing disaster preparedness plans.
- **Preparedness**: Communities are prepared for disasters by developing education and awareness campaigns that feed into effective disaster preparedness plans that facilitate the training of residents on how to respond to emergencies.
- **Response**: Governments and organizations should have plans in place to respond to disasters quickly and effectively. This includes deploying emergency personnel and resources to affected areas.
- **Recovery**: Disaster recovery efforts should be well-coordinated and focused on the most critical needs of the community. Emergency managers have a role of facilitating responses and recovery but do not have the resources to restore infrastructure themselves. This process begins in pre-disaster preparedness that involves all potential stakeholders and institutions and flows into the response and recovery phases where coordination provides leadership.
- **Mitigation**: Hazard mitigation involves built or engineered structures as well as education and awareness raising. Once a disaster has occurred, the effectiveness of mitigation strategies must be evaluated to take steps to improve mitigation in the face of future disasters.

The Government of South Australia (2022) provides a useful approach to an outcomes-oriented approach in which they define four 'steps towards developing community-centered outcomes: 1. Define the community/communities; 2. Develop and organize tailored outcome(s); 3. Communicate the intended outcomes; 4. Review and adjust outcomes' (Government of South Australia, 2022, 3). This approach extends beyond a recovery model or process in that it requires a continual evaluation and modification of the outcomes of recovery throughout the period in which it unfolds. The community, its needs, and its dynamics comprise the start and finish of recovery, rather than the purely mechanical restoration of services and infrastructure. When the outcome begins to diverge from the recovery model, it should be redefined and refocused, with the community that is being assisted always at the forefront.

An analogy of a process and outcome strategy may be drawn from the allied discipline of urban and regional planning, land-use planning, which builds on extensive planning theory. As a dynamic process of change from one state to another more developed landscape, the planning process is guided and structured by strategic planning supported by legislation. This stipulates a development assessment process that is similar in its intent to recovery guidelines and strategies. It is a staged process that aims to protect communities as it guides them through change over time. An outcome in planning communities is envisaged by master planning. A master plan imagines and defines a community that is to be developed. While it defines what that community may comprise, it also adapts to trends and changes during the building of the community. In getting to an ultimate completion, the master plan follows the stages of development assessment toward a defined end point, making modifications as time passes within the overall vision of the plan. The outcome is not an alternative approach to such processes as recovery or development but is a desirable and clearly defined end point that is based upon knowledge and change in the community.

Conclusion

Recovery models and processes are devised and planned before a disaster occurs and may be generic and standardized to meet the needs of the emergency managers and infrastructure providers who are required to respond to a range of disasters in different locations. While a community disaster plan may adapt and modify the recovery process to fit local conditions, it remains a starting point until a disaster occurs. The recovery plan is at the beginning, but the recovery outcome is the end point of the process. The very nature of a disaster is that it is a human experience and construct. From the moment a disaster occurs, divergence from expectations, plans, and preparations divert or thwart the recovery process. Imagining a desired outcome for the end of the process can drive a more realistic recovery plan, but throughout the recovery period, often extending over years, the people and communities will continually reappraise priorities, options, and opportunities. We hope that a community will bounce forward, but most places will not return to an identical pre-disaster state: the place and life will be different. Anticipating what that scenario may be and modifying outcome plans alongside the community will maximize opportunities and use limited resources most effectively. That is the purpose and significance of a community outcomes approach to recovery.

References

Aguirre, B. *On the Concept of Resilience*. Disaster Research Center, University of Delaware, Delaware. 2009.

Australia-New Zealand Emergency Management Committee. Australian Disaster Recovery Framework. Version 3.0 Endorsed by the Australia-New Zealand Emergency Management Committee. 2022. http://www.recovery.sa.gov.au/_data/assets/pdf_file/0005/933998/ADR-Framework-October-2022.pdf

BoM (Bureau of Meteorology). Database of Past Cyclone Tracks. 2023. http://www.bom.gov.au/cyclone/tropical-cyclone-knowledge-centre/databases/

Boon, H.J., Cottrell, A., King, D., Stevenson, R.B. and Millar, J. Bronfenbrenner's bioecological theory for modelling community resilience to natural disasters. *Natural Hazards*, 60 (2). pp. 381–408. 2012a.

Boon, H.J., Millar, J., Lake, D., Cottrell, A. and King, D. *Recovery from Disaster: Resilience, Adaptability and Perceptions of Climate Change*. National Climate Change Adaptation Research Facility, Gold Coast. 2012b.

Boon, H.J., Cottrell, A. and King, D. *Disasters and Social Resilience: A Bioecological Approach*. Routledge Explorations in Environmental Studies. Taylor & Francis, London. 2016

Buckle, P., Marsh, G. and Smale, S. New approaches to assessing vulnerability and resilience. *Australian Journal of Emergency Management*, 15 (2). pp. 8–14. 2000.

Christchurch City Council. *Facts, Stats and Figures*. Christchurch City Council. 2023. https://ccc.govt.nz/culture-and-community/statistics-and-facts/facts-stats-and-figures#:~:text=Christchurch%20city's%20most%20recent%20population%20estimate%20was%20396%2C200%2C%20at%20June%202023

COAG. *National Strategy for Disaster Resilience: Building the Resilience of Our Nation to Disasters*. Commonwealth of Australia, Canberra. 2011.

Cottrell, A. Communities and bushfire hazard in Australia: More questions than answers. *Environmental Hazards*, 6. pp. 109–114. 2005. ISSN: 1747 7891.

Davis, I. and Alexander, D. *Recovery from Disaster*. Routledge Studies in Hazards, Disaster Risk and Climate Change. Routledge, London and New York. 2016.

Government of South Australia. *State Emergency Management Plan: Guideline C Recovery Outcomes. State Disaster Recovery Coordination Framework*. The Department of the Premier and Cabinet, Adelaide. 2022. https://www.recovery.sa.gov.au/about-recovery/resources-for-recovery-coordinators/resource-documents/C_SDRCF-2022-Guideline-C-Recovery-Outcomes.pdf

Gurtner, Y. and King, D. Traveling safely in an unsafe world – A shared responsibility. In: Wilks, J., Pendergast, D., Leggat, P.A., and Morgan, D. (eds.) *Tourist Health, Safety and Wellbeing in the New Normal*. Springer, New York. pp. 167–187. 2021.

Hollander, J.B., Pallagst, K., Schwarz, T. and Popper, F.J. Planning shrinking cities. *Progress in Planning*, 72 (4). pp. 223–232. 2009.

Kates, R.W., Colten, C.E., Laska, S. and Leatherman, S.P. Reconstruction of New Orleans after Hurricane Katrina: A research perspective. *PNAS*, 103 (40). pp. 14653–14660. October 3 2006.

King, D. Diamond Mining Settlements in Central Kono District, Sierra Leone. PhD thesis, University of London. 1979

King, D.C. Parallel routes to recovery: Community priorities and NGO policy in the post civil war reconstruction of Sierra Leone. *International Journal of Emergency Management*, 2 (3). pp. 149–162. 2005.

King, D., Bird, D., Haynes, K., Boon, H., Cottrell, A., Millar, J., Okada, T., Box, P., Keough, D. and Thomas, M. Voluntary relocation as an adaptation strategy to extreme weather events. *International Journal of Disaster Risk Reduction*, 8. pp. 83–90. 2014.

King, D. and Gurtner, Y. Planning for demographic change and population loss: Land use planning for demographic change after disasters in New Orleans, Christchurch and Innisfail. In: Karacsonyi, D., Taylor, A., and Bird, D. (eds.) *Demography of Disasters: Impacts for Population and Place*. Springer, Cham, pp. 101–123. 2020.

Kirmayer, K.L., Whitley, R., Dandeneau, S.F. and Isaac, C. Community resilience: Models, metaphors and measures. *Journal of Aboriginal Health*. pp. 62–117. November 2009.

McEntire, D.A. *Disaster Response and Recovery: Strategies and Tactics for Resilience*. Wiley, New York. 2009.

Miles, S.B., Henry, V., Burton, H.V. and Kang, H. Community of practice for modeling disaster recovery. *Natural Hazards Review*, 20 (1). p. 04018023. 2019.

Miles, S.B. and Chang, S.E. *Urban Disaster Recovery: A Framework and Simulation Model*. Technical Report MCEER-03-0005, Multidisciplinary Center for Earthquake Engineering Research, University at Buffalo, State University of New York. 2003

Mileti, D.S. *Disasters by Design: A Reassessment of Natural Hazards in the United States*. Joseph Henry Press, Washington, DC. 1999.

National Oceanic and Atmospheric Administration. National Centers for Environmental Information. 2023. https://www.ncei.noaa.gov/

Paton D. and Johnston D. *Disaster Resilience: An Integrated Approach*. Charles C. Thomas Publisher: Springfield IL. 2006.

Patterson, O., Weil, F. and Patel, K. The role of community in disaster response: Conceptual models. *Population Research and Policy Review*, 29. pp. 127–141. 2010.
Phillips, B.D. *Disaster Recovery*. CRC Press, Taylor and Francis, New York. 2009.
Plyer, A. Population Loss and Vacant Housing in New Orleans Neighborhoods. *The Data Center*. 2011. https://www.datacenterresearch.org/reports_analysis/population-loss-and-vacant-housing/. Accessed 7/12/23.
Statistics Sierra Leone. *Preliminary Report of the Final Results of the First Digital Census in Sierra Leone*. Government of Sierra Leone, Freetown. 2022.
Stehlik, D. Community as palimpsest: The example of Perth. *Academy of Social Sciences Dialogue*, 25 (1). pp. 41–49. 2006.
USCB (United States Census Bureau). Quick Facts. New Orleans City, Louisiana. 2023. https://www.census.gov/quickfacts/fact/table/neworleanscitylouisiana/PST045222
Wilson, S.A. Diamond exploitation in Sierra Leone 1930 to 2010: A resource curse? *GeoJournal*, 78. pp. 997–1012. 2013.
Zhou, H., Wang, J., Wan, J. and Jia, H. Resilience to natural hazards: A geographic perspective. *Natural Hazards*, 53. pp. 21–41. 2010.

21
SUSTAINABLE DISASTER RECOVERY

Mitchel Stimers

Introduction

Since the World Commission on Environment and Development's (WCED's) Brundtland Report, 'sustainable development' has acquired significant recognition as a framework for guiding policy and decision-making at the national and international levels (World Commission on Environment and Development (WCED), 1987). The report defined sustainable development as "development that meets the needs of the present without compromising the ability of future generations to meet their own needs" (ch. 2, para. 1). The WCED argued that economic development, social development, and environmental protection are interdependent and mutually reinforcing and that all three must integrate into development planning and decision-making. Sustainable development aims to create a society where natural resources enjoy preservation and maintenance while allowing people to live high-quality lives. Restoring crucial infrastructure after a natural disaster defines 'disaster recovery' (Mileti, 1999). Sustainable development requires developing a comprehensive plan to promptly resume regular community functioning following a disruptive incident to ensure continuity, lessen the disaster's impact on a community, and decrease its impact. Merging the concepts of 'sustainable development' and 'disaster recovery' create a new concept: sustainable disaster recovery (SDR) (Rodríguez et al., 2007).

Sustainable Development

Sustainable development acknowledges the long-term need to balance environmental conservation, social well-being, and economic progress. The concept of sustainable development emerged in the 1980s as a response to growing concerns about the negative impacts of economic growth on the environment and society (WCED, 1987). It emphasizes the need to abandon a business-as-usual approach (i.e., prioritizing short-term economic gains over long-term sustainability). Sustainable development is a comprehensive strategy aiming to ensure the consideration of environmental, social, and economic factors in decision-making processes (Bogati & Gautam, 2021). Environmental sustainability seeks to preserve and protect natural resources and ecosystems for future generations to promote sustainable land use practices, reducing pollution and waste, and mitigating and adapting to the impacts of climate change. Social sustainability entails promoting

DOI: 10.4324/9781003414834-29

social equity, justice, and inclusion, ensuring all individuals have access to basic human needs, such as food, water, shelter, and healthcare, supporting human rights, and reducing inequality. Finally, economic sustainability includes promoting equitable and inclusive economic growth while minimizing negative environmental impacts, accomplished through production and consumption patterns, investing in renewable energy, and creating green jobs.

The awareness of the interconnection of its multidimensionality is one of the fundamental tenets of SD (Shmelev & Shmeleva, 2018; Zhao et al., 2021). For instance, environmental degradation and resource depletion will eventually result from economic expansion at the environment's expense (Eisenmenger et al., 2020). Various policies and practices must be adopted to encourage sustainable use of natural resources, lower greenhouse gas emissions, and increase access to clean water, energy, and healthcare. Adopting a policy may involve several strategies, including supporting sustainable agriculture, implementing green construction standards, and investing in renewable energy sources. The objective is to build a society where environmental preservation, social advancement, and economic growth reinforce one another and collectively ensure a better future for everyone (Silvestre & Ţîrcă, 2019).

Merging Sustainability with Disaster Recovery

The process of reconstructing and repairing communities after a natural or technological disaster that encompasses immediate and long-term needs and encourages sustainability and resilience is known as SDR (Rodríguez et al., 2007; Smith & Wenger, 2007). SDR necessitates using ecologically friendly building materials and methods, such as supporting green infrastructure and renewable energy sources (Islam & Wahab, 2020; Ghanbarzadeh et al., 2021). SDR also unites community engagement and input into the recovery process to ensure that regional needs and objectives motivate recovery activities. Communities can rebuild and recover from a disaster's immediate effects by taking a sustainable approach to disaster recovery and enhancing their long-term possibilities for resilience and prosperity (e.g., enhancing the built environment's sustainability, lowering vulnerability to future calamities, and fostering social and economic development). Collaboration and cooperation across numerous stakeholders, including governmental agencies, nonprofit organizations, and the corporate sector, are necessary to implement SDR.

Communities may create a more sustainable and resilient future via cooperative efforts, making them better equipped to endure and recover from future disasters, what Dale and Newman (2010) termed a 'reconciliation of three imperatives'—namely the ecological/environmental, social, and economic needs of a community—to afford 'equitable access to resources' (p. 2). SDR gives communities a chance to build back better than before the disaster struck and do so in a way that helps the community and the environment. Planning is paramount in SDR, and if done poorly or not at all, it leads to substandard construction, enhanced or newly introduced exposure to hazards, and ignorance of the most vulnerable population's needs (Smith & Wenger, 2007). However, good planning allows communities to seize the 'window of opportunity' (Brundiers & Eakin, 2018).

Sustainable Disaster Recovery

Environmental Considerations

Environmental considerations are paramount in recovering from a disaster, including encouraging environmentally friendly land use techniques, protecting wetlands and green spaces, and ensuring that disaster recovery operations safeguard the environment (Bhattarai et al., 2020).

Disasters can seriously harm the ecosystem, including deforestation, soil erosion, and biodiversity loss. Since disasters can have significant and enduring consequences on ecosystems, wildlife, and natural resources, environmental concerns are crucial in disaster recovery efforts to support sustainable growth and reduce the likelihood of new disasters (Mabon, 2019). Disaster recovery initiatives must restore and safeguard ecosystems, wildlife, and natural resources to address these effects, possibly entailing planting trees, restoring wetlands and other habitats, and safeguarding species. If disaster recovery efforts are not adequately planned and managed, they could also accelerate environmental damage (Budiman et al., 2020). While using large machinery during recovery operations may result in soil erosion and other environmental effects, the reconstruction of buildings and infrastructure may result in deforestation and the extinction of ecosystems (Kirchhoff et al., 2017). Involving locals in SDR efforts is essential in ensuring operations are environmentally sustainable. It is possible to guarantee that recovery activities align with local interests and requirements and support sustainable development by interacting with communities and incorporating them in their design and implementation. Additionally, it may encourage a sense of responsibility and ownership among community members, promoting increased engagement and cooperation in disaster recovery (Berke et al., 2014a).

Community Involvement

Communities frequently have requirements and viewpoints to consider, which planners and policymakers must recognize to meet residents' needs. The community can participate in healing and advancement processes following a disaster through meetings, focus groups, and other types of interaction with stakeholders (Berke et al., 2014b). Practical solutions for disaster risk reduction (DRR) must include SDR, branching off disaster recovery, a stage in the disaster management cycle. Moving sideways or backward via the business-as-usual model (i.e., making no substantive changes to infrastructure, policy, or governance in the wake of a disaster) is a near guarantee that future disasters will produce similar impacts as previous events (Jha, 2010). Infrastructure, houses, and communities must emerge to withstand future disasters, necessitating a sizable investment in resources, including finance, staff, and knowledge for sustaining the affected communities and ecosystems. SDR aids in creating stronger, more resilient communities and better equipped to handle future crises and may encourage a sense of responsibility and ownership among community members, promoting increased engagement and cooperation in SDR and planning strategies (Hamideh, 2020). Involving local communities in SDR is also important for fostering economic growth and sustainability during catastrophe recovery operations.

Economic Considerations

SDR must be economically sustainable, help local economies and small enterprises, and foster long-term economic growth and stability in impacted areas. Disasters disrupt economic activity, harm businesses, and other economic assets, and limit community members' living opportunities (Mannakkara & Wilkinson, 2014). SDR efforts must consider the economic effects of disasters and aid in reconstructing enterprises and other economic assets, including helping affected businesses and people financially and assisting in the revival of critical economic sectors such as agriculture, manufacturing, service-based businesses, and tourism. Since disasters can have a significant and enduring influence on economies, economic sustainability is essential to disaster recovery. As with social and community involvement in SDR, activities must fulfill the requirements of locals to encourage economic sustainability and growth and buy-in from the population (Rouhanizadeh et al., 2020).

Training and support, particularly if substantial damage occurred to economic assets, may also help affected people and communities shift to new economic prospects (Sawalha, 2021). Assuming effective planning and management, recovery activities after a disaster could also help the economy expand and thrive. For instance, the reconstruction of infrastructure and other assets may present investment opportunities, and the emergence of new companies and economic ventures might assist the expansion of regional economies. Additionally, it may encourage a sense of responsibility and ownership among community members, promoting increased engagement and cooperation in SDR processes (Rouhanizadeh et al., 2020). Making sure that SDR operations move forward in a way that encourages economic growth and development is imperative for maximizing opportunities.

The Integration of Technology

Incorporating technology is a crucial component of effective SDR and involves using data and information technologies to aid decision-making and create resourceful responses to disaster problems (Sakurai & Murayama, 2019; Zamanifar & Hartmann, 2020). Technology integration hastens the recovery process, increases its efficacy, and dramatically improves disaster recovery efforts. Integrations include early warning systems that send timely signals of possible disasters to geographic information systems (GISs) and assessment tools that help prioritize recovery activities. Evaluating and mapping disaster effects presents a prime opportunity to incorporate technology into SDR processes. It is feasible to quickly determine the level of damage and designate priority regions for recovery efforts by employing cutting-edge technologies (Abid et al., 2021). Decision-makers may prioritize recovery efforts and allocate resources more effectively using such systems.

Restoring infrastructure and services is another area where technology aids in disaster recovery. Drones and satellite photos can quickly determine the degree of infrastructure damage and speed up the restoration of essential services like power and communications, aiding rehabilitation, and lessening the disaster's effects on the impacted populations (Ahmed et al., 2022). Technology enhances the administration and coordination of disaster recovery activities, fostering well-coordinated recovery efforts and effective and efficient resource use by employing communication systems and resource coordination networks. Integrations can significantly increase the efficiency and speed of recovery operations after a disaster and support effective SDR strategies (Pradhan, 2019).

Sustainable Disaster Recovery and Partnerships

Strong collaborations and cooperation between government institutions, the commercial sector, and community-based organizations (CBOs) are necessary for practical sustainability in disaster recovery. Collaboration among various stakeholders, including government agencies, the corporate sector, and CBOs, is necessary for effective sustainability in disaster recovery (Curnin & O'Hara, 2019). Together, these stakeholders can ensure that disaster recovery activities are well-coordinated, effective, and efficient. Each of these stakeholders has different viewpoints, resources, and capabilities.

Government Agencies

Governmental organizations are essential to disaster recovery since they oversee the organization and managing of the process. They are also in charge of supplying the funding and resources required to aid recovery efforts. Governmental organizations seek to determine the damage following a disaster, create a recovery plan, and allocate funds to help those efforts. Damage assessments frequently mean collaboration with other parties, such as the commercial sector and CBOs, to ensure recovery efforts are well-coordinated and successful (Dzigbede et al., 2020). Governmental organizations oversee adopting various policies and programs to lower the risk of future disasters and enhance the resilience of impacted areas, in addition to coordinating and managing recovery operations. Agencies invest in infrastructure, such as early warning systems and building requirements, and establish zoning and land-use planning regulations that include the risk of future disasters (Wolf-Fordham, 2020). Adopting a comprehensive approach to disaster recovery, including immediate recovery efforts and long-term risk reduction measures, helps government organizations guarantee that affected communities are better prepared for future disasters and can recover more quickly and effectively (Finucane et al., 2020).

Private Sector

The private sector is essential to disaster recovery because it frequently contributes significantly to the local economy and can offer vital resources and experience to aid recovery efforts. Enterprises and firms in the private sector can support disaster recovery efforts financially and technically, as well as by helping to rebuild local businesses and other economic assets (Abe et al., 2019). For instance, they might provide technical know-how by providing items or services like machinery or infrastructure-rebuilding skills. Private sector organizations can contribute to SDR, ensuring that affected communities recover from disasters and are more resilient to future catastrophes by investing in the local economy and assisting in reconstructing companies and other economic assets (Rouhanizadeh et al., 2020). Additionally, the private sector can help advance DRR by promoting laws and initiatives that lessen the likelihood of disasters in the future, such as construction rules and early warning systems. Private sector entities can significantly contribute to catastrophe recovery efforts and long-term sustainability using their financial and technological resources (Wang & van de Lindt, 2021).

Community-Based Organizations

In disaster recovery, CBOs are essential since they are frequently the first to respond to the impacted area's needs. These groups assist afflicted people and communities, covering necessities like food, housing, child and elderly care, and medical attention (Sarabia et al., 2020). CBOs also play a role in encouraging community engagement and participation in the disaster recovery process, ensuring that recovery activities align with the needs and objectives of impacted people and communities by collaborating with and incorporating them in those efforts (Drennan & Morrissey, 2019). Building the capacity of impacted communities and encouraging local ownership of recovery activities are CBO contributions to the sustainability of disaster recovery initiatives. For instance, they can offer resources to assist communities in creating their plans for DRR and promoting readiness for emergencies.

Additionally, CBOs may collaborate with localities to encourage sustainable growth and strengthen economic resilience, which can aid in lowering the likelihood of future disasters and promoting long-term sustainability in impacted areas (Oh & Lee, 2020). CBOs can significantly promote efficient and long-lasting disaster recovery operations by collaborating closely with impacted communities and encouraging community participation and involvement. Effective sustainability in disaster recovery necessitates a collaborative strategy involving public and commercial sector entities and community-based ones (Ali et al., 2019).

Planning, Assessment, and Risk Reduction Strategies to Maintain Sustainability

DRR plan creation is a crucial component assisting communities in preparing for and responding to disasters (Twigg, 2015). DRR connects the sustainably built (or rebuilt) community to the framework of avoiding future disasters whenever possible. Through careful, considered, and calculated DRR processes, communities have a far better chance of lessening the impact of disasters while maintaining current sustainability levels (Raikes et al., 2019). Proper planning involves completing risk assessments, early warning systems, and implementing evacuation plans (Twigg, 2015). A thorough DRR strategy can be built based on risk analysis, including risk mitigation, readiness, reaction, and recovery measures. Mitigation strategies aim to lessen the likelihood or effects of hazards, for instance, by enhancing current infrastructure and building new disaster-resistant and sustainable structures (Albris et al., 2020). Preparedness measures attempt to assist communities and ecosystems in becoming better prepared for the effects of disasters, including response measures, such as the provision of emergency aid, which aim to assist communities in coping with the immediate effects of catastrophes. Upon execution, incorporating resilient infrastructure is a crucial component of efficient sustainability in disaster recovery, including designing structures, transportation networks, and other systems that can withstand disasters to the greatest extent possible (Rouhanizadeh et al., 2020, The United Nations, 2021).

Strategies for DRR are essential for reducing how much natural and technological hazards (and subsequent disasters) affect populations and ecosystems (Marshall, 2020). The subsequent stage of creating a DRR plan is risk analysis. By evaluating the likelihood and effects of risks ascertains the potential risk levels, it is possible to determine the priority risk regions and demographics for targeting in the DRR plan (Twigg, 2015). Developing DRR strategies requires numerous processes, starting with hazard assessment. Hazard assessments identify the types, frequency, potential effects of disasters, and the vulnerability of ecosystems and communities (Anderson & Renaud, 2021). Based on this information, the most vulnerable areas and populations should emerge as a priority. Following a disaster, recovery strategies assist people and ecosystems in returning to normalcy, for instance, by offering long-term assistance. The process' next stage puts DRR strategies into practice, investing in the community, infrastructure, and training and educational initiatives. With these investments, communities and ecosystems will be more resilient to disasters and better able to resist their effects (Murti & Mathez-Stiefel, 2019).

Coordination between various decision-making, planning, and development levels—local, national, and international—is also necessary for effective DRR plan execution (Basher, 2013). Monitoring and evaluating DRR tactics to ensure they continue to be applicable and successful is critical, and that resource use is effective and efficient while revealing up opportunities for improvement; this information can update and alter as necessary to ensure that Quality DRR initiatives lessen the effects of disasters and guarantee a longer-lasting future for ecosystems and communities (Lestari et al., 2020).

Challenges in Sustainable Disaster Recovery

A significant obstacle in SDR is the need for political will to address environmental issues (Berke et al., 1993; Rodríguez et al., 2007; Raju & Van Niekerk, 2013). Many policymakers prioritize short-term economic gains over the planet's long-term health, making it difficult to implement meaningful sustainability policies that benefit disaster recovery efforts (Filho et al., 2019). Resistance to change is yet another substantial impediment (Juran, 2015). People and organizations may frequently avoid adopting new, more environmentally friendly practices and technology because they believe they are excessively expensive or cumbersome (Kioupi & Voulvoulis, 2019). Such avoidance is also true of SDR following a disaster, where conventional techniques and supplies may reduce efficiency.

Another significant challenge in SDR is a need for more resources, particularly financing, and labor. Securing sufficient financing and resources advances sustainability significantly. However, many communities, especially those in developing nations, need more financial and technical resources to switch to more environmentally friendly methods and tools (Barua, 2020; Ahmad et al., 2021; Ogunsanya et al., 2022). Natural disasters can also exacerbate economic and social inequalities, with vulnerable groups frequently bearing the brunt of the effects (Tierney, 2006; Reid, 2013). Furthermore, implementing long-term catastrophe recovery initiatives may become more difficult because of existing inequalities and possibly political instability, which may be more apparent in lesser developed nations (Tierney, 2006).

The speed of technological advancement further hampers SDR (Holmberg & Sandbrook, 2019). Rapid change can make it challenging to keep up with the latest innovations and ensure sustainable use. New technologies frequently emerge without considering their long-term environmental impact (Silvestre & Ţîrcă, 2019; Annosi et al., 2020). A lack of coordination and collaboration among several parties may slow SDR progress. Numerous participants may result in resource waste, competing priorities, and duplication of effort. Inadequate long-term planning and a concentration on short-term fixes deter the efficiency of SDR. Adopting a comprehensive and long-term strategy that includes resilience and sustainability is crucial to lowering the likelihood of future disasters and creating more sustainable communities (Rodríguez et al., 2007; Juran et al., 2022).

There needs to be more information and public awareness regarding the significance of sustainability and the part that people and organizations can play in promoting it (Rieckmann, 2018). A general lack of education and public knowledge on the significance of catastrophe recovery and sustainability highlights another example of SDR issues. People need help to grasp the severity of the situation and take meaningful action to support long-term disaster recovery efforts. The slow rate of SDR may also present issues. Years may pass throughout the reconstruction process due to myriad reasons, for example: (a) lack of funding, (b) depopulation leading to a reduced labor force and/or a severely misbalanced age-gender pyramid for that community, (c) material shortages, or (d) poor or no planning. Any one or combination of these factors could lead to a stalled recovery. In some situations, a community may only partially recover from the harm a disaster has inflicted should recovery lag behind accepted models of recovery timelines (Kates & Pijawka, 1977; Alexander, 2002). Affected communities may experience long-term economic and social hardship because of this.

The frequency and intensity of natural disasters are increasing due to climate change (Benevolenza & DeRigne, 2019), increasing disaster recovery's complexity and cost (Coronese et al., 2019). Frequently, multiple disasters impact the same areas at once, making rebuilding and recovering sustainably difficult. The Bruntland Report (WCED, 1987) identified climate change as one of three significant issues linked to our common survival. It stands to reason that if climate change continues adding to disasters' impacts, climate change and sustainability are also inextricably linked. How many developed countries have addressed (or at least failed to address adequately) future climate change scenarios cascading into sustainable development? This is not to state that progress in the past four decades is absent; some certainly have (e.g., The Kyoto Protocol, The Paris Agreement, UN IPCC Reports). Nevertheless, the anemic pace of climate change action and mitigation adds an astounding layer of complexity to achieving sustainability in communities worldwide. So lethargic is the pace that an effective SDR model has yet to emerge due in part to the uncertainty of the intersection of climate change knowledge and climate change legislation and action (Smith et al., 2018).

Sustainable Disaster Recovery Scenarios

After the 2004 Indian Ocean tsunami, the Indonesian government and foreign groups collaborated to reconstruct the area using sustainable development and DRR. They greatly emphasized resilient and sustainable reconstruction, contributing to the city's excellent recovery efforts. Reconstruction included the development of sustainable farming and fishing methods and the building of new structures and infrastructure that are more resilient to disasters in the future. Indonesia also invested in constructing new healthcare and educational facilities made from sustainable and ecologically friendly materials (Satake, 2014). The United Nations Development Programme (UNDP, 2023) reported that Indonesia's restoration efforts had reduced the likelihood of future disasters and improved the long-term sustainability of communities in the region.

The reconstruction work in New Orleans, Louisiana, following Hurricane Katrina in 2005, is one illustration of a successful SDR project (Fussell, 2015). After the hurricane's devastation, New Orleans tried to rebuild sustainably, utilizing resilient design concepts and green infrastructure. For instance, new structures were built to higher standards to resist storms in the future better, and green spaces soaked up rain and lessened the chance of flooding. The city also implemented a comprehensive sustainability strategy that included steps to increase energy efficiency, cut waste, and encourage green transportation options. Efforts to rebuild New Orleans sustainably have improved the city's resilience and reduced the risk of future disasters (Kates et al., 2006; National Institute of Standards and Technology (NIST), 2006).

Reconstruction activities in Tohoku, Japan, following the 2011 earthquake and tsunami, illustrate successful SDR (Takeuchi et al., 2014). The Japanese government collaborated with neighborhood groups to reconstruct in a way that would lessen the likelihood of future disasters and promote the area's long-term economic and social advancement (Dimmer, 2017). Creating renewable energy sources, including wind and solar energy, adopting green building guidelines, and deploying sustainable transportation networks facilitated the Japanese government's plans (Schreurs, 2021). Additionally, the government promoted the growth of the regional fishing and agriculture industries, which aided in job growth and the nation's economic recovery. These initiatives have aided Tohoku in its recovery from the disaster as well as in constructing a more stable and resilient future. The Tohoku restoration efforts were successful in creating a future for the region that is more robust and sustainable (Kumagai, 2016).

Hurricane Sandy made landfall in New York City in 2012, seriously damaging the city's infrastructure and property (Meixler, 2017). The city's recovery efforts were nevertheless commendable since they placed a heavy emphasis on long-term disaster rehabilitation (Harris & Doerfel, 2017). New York City created a thorough plan that prioritized reconstruction while enhancing the city's storm resilience. The city utilized green infrastructure, such as building additional parks and wetlands, to lower the risk of flooding and enhance the quality of life for people. In addition, the city passed new building regulations that required storm-resistant structures to minimize damage from upcoming storms (Cohen, 2021). The New York City Office of Recovery and Resiliency claims that this method of disaster recovery allowed the city to avoid paying $4 billion in additional expenses for storm damage in the future (The City of New York, 2013).

The recovery efforts after the 2015 earthquakes in Nepal are a further successful illustration of long-term disaster recovery. Following the tragedy, the Government of Nepal and foreign groups collaborated to rebuild while sustainably embracing DRR and sustainable development principles

(Platt et al., 2020; Spoon et al., 2020). Developing earthquake-resistant buildings, encouraging environmentally friendly farming and forestry methods, and creating early warning and DRR systems supported sustainable recovery (Comfort & Joshi, 2017). Nepal's rehabilitation efforts have improved local communities' long-term sustainability and decreased the likelihood of recurrent disasters (Epstein et al., 2018).

Building resilient communities and lowering the risk of future disasters depend on effective, SDR operations. Communities can recover from disasters more sustainably and resiliently by implementing sustainable development methods and resilient design concepts, fostering long-term economic and social development. Hurricane Katrina in New Orleans, the earthquake and tsunami in Tohoku, Japan, and the Indian Ocean Tsunami, among other instances, show the possibility for effective SDR initiatives to aid in community reconstruction and the creation of a more resilient future (Olsen et al., 2005).

Conclusion

Sustainable development after a disaster involves integrating disaster recovery efforts with long-term development planning and risk reduction strategies. It aims to rebuild what was lost and build back better and more resilient communities. This approach prioritizes the involvement of affected communities and collaboration between government agencies, the private sector, and CBOs. The goal of sustainable development after a disaster is to ensure that affected communities are better prepared for future disasters, have greater resilience, and can recover more quickly and effectively. Key elements of sustainable development after a disaster include investing in DRR measures, promoting community engagement and involvement, promoting sustainable economic development, and integrating technology into recovery efforts. By taking a comprehensive and sustainable approach to disaster recovery, affected communities may find support in rebuilding and becoming more resilient to future disasters.

References

Abe, Y., Zodrow, I., Johnson, D. A., & Silerio, L. (2019). Risk informed and resilient development: Engaging the private sector in the era of the Sendai Framework. *Progress in Disaster Science*, *2*, Article 100020. https://doi.org/10.1016/j.pdisas.2019.100020

Abid, S. K., Sulaiman, N., Chan, S. W., Nazir, U., Abid, M., Han, H., Ariza-Montes, A., & Vega-Muñoz, A. (2021). Toward an integrated disaster management approach: How artificial intelligence can boost disaster management. *Sustainability*, *13*(22), Article 12560. https://doi.org/10.3390/su132212560

Ahmad, M., Muslija, A., & Satrovic, E. (2021). Does economic prosperity lead to environmental sustainability in developing economies? Environmental Kuznets curve theory. *Environmental Science and Pollution Research*, *28*(18), 22588–22601. https://doi.org/10.1007/s11356-020-12276-9

Ahmed, H., Bakr, M., Talib, M. A., Abbas, S., & Nasir, Q. (2022, February). *Unmanned Aerial Vehicles (UAVs) and Artificial Intelligence (AI) in fire related disaster recovery: Analytical survey study*. In 2022 International Conference on Business Analytics for Technology and Security (ICBATS) (pp. 1–6). IEEE. https://doi.org/10.1109/ICBATS54253.2022.9759056

Albris, K., Lauta, K. C., & Raju, E. (2020). Disaster knowledge gaps: Exploring the interface between science and policy for disaster risk reduction in Europe. *International Journal of Disaster Risk Science*, *11*, 1–12. https://doi.org/10.1007/s13753-020-00250-5

Alexander, D. (2002). *Principles of Emergency Planning and Management*. Oxford University Press.

Ali, M. S. S., Arsyad, M., Kamaluddin, A., Busthanul, N., & Dirpan, A. (2019). Community based disaster management: Indonesian experience. *Earth and Environmental Science*, *235*(1), Article 012012. https://doi.org/10.1088/1755-1315/235/1/012012

Anderson, C. C., & Renaud, F. G. (2021). A review of public acceptance of nature-based solutions: The "why," "when," and "how" of success for disaster risk reduction measures. *Ambio*, *50*(8), 1552–1573. https://doi.org/10.1007/s13280-021-01502-4

Annosi, C. M., Brunetta, F., Capo, F., & Heideveld, L. (2020). Digitalization in the agri-food industry: The relationship between technology and sustainable development. *Management Decision, 58*(8), 1737–1757. https://doi.org/10.1108/MD-09-2019-1328

Barua, S. (2020). Financing sustainable development goals: A review of challenges and mitigation strategies. *Business Strategy & Development, 3*(3), 277–293. https://doi.org/10.1002/bsd2.94

Basher, R. (2013). *Science and Technology for Disaster Risk Reduction: A Review of Application and Coordination Needs*. United Nations Office for Disaster Risk Reduction. https://caneus.org/undrr/resources/3.Science-and-Technology-for-Disaster-Risk-Reduction.pdf

Benevolenza, M. A., & DeRigne, L. (2019). The impact of climate change and natural disasters on vulnerable populations: A systematic review of the literature. *Journal of Human Behavior in the Social Environment, 29*(2), 266–281. https://doi.org/10.1080/10911359.2018.1527739

Berke, P. R., Kartez, J., & Wenger, D. (1993). Recovery after disaster: Achieving sustainable development, mitigation and equity. *Disasters, 17*(2), 93–109. https://doi.org/10.1111/j.1467-7717.1993.tb01137.x

Berke, P. R., Cooper, J., Aminto, M., Grabich, S., & Horney, J. (2014a). Adaptive planning for disaster recovery and resiliency: An evaluation of 87 local recovery plans in eight states. *Journal of the American Planning Association, 80*(4), 310–323. https://doi.org/10.1080/01944363.2014.976585

Berke, P. R., Lyles, W., & Smith, G. (2014b). Impacts of federal and state hazard mitigation policies on local land use policy. *Journal of Planning Education and Research, 34*(1), 60–76. https://doi.org/10.1177/0739456X13517004

Bhattarai, S., Maycock, B., Alfonso, H., & Reid, A. (2020). Development of an integrated pathways model of factors influencing the progress of recovery after a disaster. *Asia Pacific Journal of Public Health, 32*(5), 226–234. https://doi.org/10.1177/1010539520935386

Bogati, R., & Gautam, M. S. (2021). "Disaster Recovery toward Attaining Sustainable Development Goals." In W. L. Filho, A. L. Salvia, & P. G. Ozuyar (Eds.), *No Poverty* (pp. 211–220). Springer International Publishing.

Brundiers, K., & Eakin, H. C. (2018). Leveraging post-disaster windows of opportunities for change towards sustainability: A framework. *Sustainability, 10*(5), Article 1390. https://doi.org/10.3390/su10051390

Budiman, K., Arini, F. Y., & Sugiharti, E. (2020, June). Disaster recovery planning with distributed replicated block devices in synchronized API systems. *Journal of Physics: Conference Series, 1567*(3), Article 032023. https://doi.org/10.1088/1742-6596/1567/3/032023

Cohen, D. A. (2021). New York City as "fortress of solitude" after Hurricane Sandy: A relational sociology of extreme weather's relationship to climate politics. *Environmental Politics, 30*(5), 687–707. https://doi.org/10.1080/09644016.2020.1816380

Comfort, L. K., & Joshi, J. (2017). Scalability and sustainability in uncertain environments: Transition to recovery from the 2015 Gorkha, Nepal, earthquakes. *Earthquake Spectra, 33*(1_suppl), 385–401. https://doi.org/10.1193/113016eqs217m

Coronese, M., Lamperti, F., Keller, K., Chiaromonte, F., & Roventini, A. (2019). Evidence for sharp increase in the economic damages of extreme natural disasters. *Proceedings of the National Academy of Sciences, 116*(43), 21450–21455. https://doi.org/10.1073/pnas.1907826116

Curnin, S., & O'Hara, D. (2019). Nonprofit and public sector interorganizational collaboration in disaster recovery: Lessons from the field. *Nonprofit Management and Leadership, 30*(2), 277–297. https://doi.org/10.1002/nml.21389

Dale, A., & Newman, L. (2010). Social capital: A necessary and sufficient condition for sustainable community development? *Community Development Journal, 45*(1), 5–21. https://doi.org/10.1093/cdj/bsn028

Dimmer, C. (2017). "Japan after March 11th, 2011: Between Swift Reconstruction and Sustainable Restructuring." In W. Yan & W. Galloway (Eds.), *Rethinking Resilience, Adaptation and Transformation in a Time of Change* (pp. 23–40). https://doi.org/10.1007/978-3-319-50171-0_2

Drennan, L., & Morrissey, L. (2019). Resilience policy in practice—Surveying the role of community based organisations in local disaster management. *Local Government Studies, 45*(3), 328–349. https://doi.org/10.1080/03003930.2018.1541795

Dzigbede, K. D., Gehl, S. B., & Willoughby, K. (2020). Disaster resiliency of US local governments: Insights to strengthen local response and recovery from the COVID-19 pandemic. *Public Administration Review, 80*(4), 634–643. https://doi.org/10.1111/puar.13249

Eisenmenger, N., Pichler, M., Krenmayr, N., Noll, D., Plank, B., Schalmann, E., Wandl, M., & Gingrich, S. (2020). The sustainable development goals prioritize economic growth over sustainable resource use: A critical reflection on the SDGs from a socio-ecological perspective. *Sustainability Science*, *15*, 1101–1110. https://doi.org/10.1007/s11625-020-00813-x

Epstein, K., DiCarlo, J., Marsh, R., Adhikari, B., Paudel, D., Ray, I., & Måren, I. E. (2018). Recovery and adaptation after the 2015 Nepal earthquakes. *Ecology and Society*, *23*(1). https://doi.org/10.5751/es-09909-230129

Filho, L. W., Tripathi, S. K., Andrade Guerra, J. B. S. O. D., Giné-Garriga, R., Orlovic Lovren, V., & Willats, J. (2019). Using the sustainable development goals towards a better understanding of sustainability challenges. *International Journal of Sustainable Development & World Ecology*, *26*(2), 179–190. https://doi.org/10.1080/13504509.2018.1505674

Finucane, M. L., Acosta, J., Wicker, A., & Whipkey, K. (2020). Short-term solutions to a long-term challenge: Rethinking disaster recovery planning to reduce vulnerabilities and inequities. *International Journal of Environmental Research and Public Health*, *17*(2), Article 482. https://doi.org/10.3390/ijerph17020482

Fussell, E. (2015). The long-term recovery of New Orleans' population after Hurricane Katrina. *American Behavioral Scientist*, *59*(10), 1231–1245. https://doi.org/10.1177/0002764215591181

Ghanbarzadeh, G. S., Wedawatta, G., Ginige, K., & Ingirige, B. (2021). Living-transforming disaster relief shelter: A conceptual approach for sustainable post-disaster housing. *Built Environment Project and Asset Management*, *11*(4), 687–704. https://doi.org/10.1108/BEPAM-04-2020-0076

Hamideh, S. (2020). Opportunities and challenges of public participation in post-disaster recovery planning: Lessons from Galveston, TX. *Natural Hazards Review*, *21*(4), Article 05020009. https://doi.org/10.1061/(ASCE)NH.1527-6996.0000399

Harris, J. L., & Doerfel, M. L. (2017). "Interorganizational Resilience: Networked Collaborations in Communities after Superstorm Sandy." In E. C. Jones & A. J. Faas (Eds.), *Social Network Analysis of Disaster Response, Recovery, and Adaptation* (pp. 75–91). Butterworth-Heinemann.

Holmberg, J., & Sandbrook, R. (2019). "Sustainable Development: What Is to Be Done?" In J. Holmberg (Ed.), *Policies for a small planet* (pp. 19–38). Routledge.

Islam, E., & Wahab, H. B. A. (2020). The impact of a cyclonic disaster on coastal communities in Bangladesh: Possible community-led interventions towards sustainable disaster recovery. *Global Social Welfare*, *7*, 339–352. https://doi.org/10.1007/s40609-020-00181-5

Jha, A. K. (2010). *Safer Homes, Stronger Communities: A Handbook for Reconstructing after Natural Disasters*. World Bank Publications.

Juran, L. (2015). A framework for integrating water projects in disaster reconstruction processes in the international context. *Public Works Management & Policy*, *20*(4), 305–321. https://doi.org/10.1177/1087724X15572145

Juran, L., Oliver, R., & Read, D. C. (2022). "Water Sector Reconstruction for Post-Disaster Housing Settlements: A Tale of Two Governance Models." In T. Younos, J. Lee, T. Parece (Eds.), *Resilient Water Management Strategies in Urban Settings: Innovations in Decentralized Water Infrastructure Systems* (pp. 227–243). Springer Water.

Kates, R. W., Colten, C. E., & Leatherman, S. P. (2006). Reconstruction of New Orleans after Hurricane Katrina: A research perspective. *Proceedings of the National Academy of Science*, *103*(40), 14653–14660. https://doi.org/10.1073/pnas.0605726103

Kates, R. W., & Pijawka, D. (1977). "From Rubble to Monument: The Pace of Reconstruction." In J. E. Haas, R. W. Kates, & M. J. Bowden (Eds.), *Disaster* (1–23). MIT Press.

Kioupi, V., & Voulvoulis, N. (2019). Education for sustainable development: A systemic framework for connecting the SDGs to educational outcomes. *Sustainability*, *11*(21), Article 6104. https://doi.org/10.3390/su11216104

Kirchhoff, M., Comino, J. R., Seeger, M., & Ries, J. B. (2017). Soil erosion in sloping vineyards under conventional and organic land use managements (Saar-Mosel Valley, Germany). *Cuadernos de investigación geográfica: Geographical Research Letters*, *43*, 119–140. https://doi.org/10.18172/cig.3161

Kumagai, Y. (2016). "Applying a Sustainability Assessment Framework to Recovery Plans in Eastern Tohoku, Japan." In R. Gibson (Ed.), *Sustainability Assessment* (pp. 208–238). Routledge.

Lestari, P., Ritonga, R., Ruliana, P., & Br Barus, C. C. (2020). Disaster communication uses field training exercise simulation as an important aspect of disaster risk reduction. *Malaysian Journal of Communication*, *36*(1), 166–186. https://ejournal.ukm.my/mjc/article/view/37501/10377

Mabon, L. (2019). Enhancing post-disaster resilience by 'building back greener': Evaluating the contribution of nature-based solutions to recovery planning in Futaba County, Fukushima Prefecture, Japan. *Landscape and Urban Planning*, *187*, 105–118. https://doi.org/10.1016/j.landurbplan.2019.03.013

Mannakkara, S., & Wilkinson, S. (2014). Re-conceptualising "building back better" to improve post-disaster recovery. *International Journal of Managing Projects in Business*, *7*(3), 327–341. https://doi.org/10.1108/IJMPB-10-2013-0054

Marshall, T. M. (2020). Risk perception and safety culture: Tools for improving the implementation of disaster risk reduction strategies. *International Journal of Disaster Risk Reduction*, *47*, Article 101557. https://doi.org/10.1016/j.ijdrr.2020.101557

Meixler, M. S. (2017). Assessment of Hurricane Sandy damage and resulting loss in ecosystem services in a coastal-urban setting. *Ecosystem Services*, *24*, 28–46. https://doi.org/10.1016/j.ecoser.2016.12.009

Mileti, D. (1999). *Disasters by Design: A Reassessment of Natural Hazards in the United States*. Joseph Henry Press.

Murti, R., & Mathez-Stiefel, S. L. (2019). Social learning approaches for ecosystem-based disaster risk reduction. *International Journal of Disaster Risk Reduction*, *33*, 433–440. https://doi.org/10.1016/j.ijdrr.2018.09.018

National Institute of Standards and Technology (NIST). (2006). *Q&As on the National Institute of Standards and Technology (NIST) reconnaissance and assessment of damage from Hurricanes Katrina and Rita*. NIST. https://www.nist.gov/news-events/news/2006/06/qas-national-institute-standards-and-technology-nist-reconnaissance-and

Ogunsanya, O. A., Aigbavboa, C. O., Thwala, D. W., & Edwards, D. J. (2022). Barriers to sustainable procurement in the Nigerian construction industry: An exploratory factor analysis. *International Journal of Construction Management*, *22*(5), 861–872. https://doi.org/10.1080/15623599.2019.1658697

Oh, N., & Lee, J. (2020). Changing landscape of emergency management research: A systematic review with bibliometric analysis. *International Journal of Disaster Risk Reduction*, *49*, Article 101658. https://doi.org/10.1016/j.ijdrr.2020.101658

Olsen, S. B., Matuszeski, W., Padma, T. V., & Wickremeratne, H. M. (2005). Rebuilding after the tsunami: Getting it right. *AMBIO: A Journal of the Human Environment*, *34*(8), 611–614. https://doi.org/10.1579/0044-7447-34.8.611

Platt, S., Gautam, D., & Rupakhety, R. (2020). Speed and quality of recovery after the Gorkha earthquake 2015 Nepal. *International Journal of Disaster Risk Reduction*, *50*, Article 101689. https://doi.org/10.1016/j.ijdrr.2020.101689

Pradhan, M. (2019, April). *Interoperability for disaster relief operations in smart city environments*. In 2019 IEEE 5th World Forum on Internet of Things (WF-IoT) (pp. 711–714). IEEE. https://doi.org/10.1109/WF-IoT.2019.8767169

Raikes, J., Smith, T. F., Jacobson, C., & Baldwin, C. (2019). Pre-disaster planning and preparedness for floods and droughts: A systematic review. *International Journal of Disaster Risk Reduction*, *38*, Article 101207. https://doi.org/10.1016/j.ijdrr.2019.101207

Raju, E., & Van Niekerk, D. (2013). Intra-governmental coordination for sustainable disaster recovery: A case-study of the Eden District Municipality, South Africa. *International Journal of Disaster Risk Reduction*, *4*, 92–99. https://doi.org/10.1016/j.ijdrr.2013.03.001

Reid, M. (2013). Disasters and social inequalities. *Sociology Compass*, *7*(11), 984–997. https://doi.org/10.1111/soc4.12080

Rieckmann, M. (2018). "Learning to Transform the World: Key Competencies in Education for Sustainable Development." In A. Leicht, J. Heiss, & W. J. Byun (Eds.), *Issues and Trends in Education for Sustainable Development* (pp. 39–59). UNESCO Publishing.

Rodríguez, H., Quarantelli, E. L., Dynes, R. R., Smith, G. P., & Wenger, D. (2007). "Sustainable Disaster Recovery: Operationalizing an Existing Agenda" in *Handbook of Disaster Research* (pp. 234–257). https://doi.org/10.1007/978-0-387-32353-4_14

Rouhanizadeh, B., Kermanshachi, S., & Nipa, T. J. (2020). Exploratory analysis of barriers to effective post-disaster recovery. *International Journal of Disaster Risk Reduction*, *50*, Article 101735. https://doi.org/10.1016/j.ijdrr.2020.101735

Sakurai, M., & Murayama, Y. (2019). Information technologies and disaster management–Benefits and issues. *Progress in Disaster Science*, *2*, Article 100012. https://doi.org/10.1016/j.pdisas.2019.100012

Sarabia, M. M., Kägi, A., Davison, A. C., Banwell, N., Montes, C., Aebischer, C., & Hostettler, S. (2020). The challenges of impact evaluation: Attempting to measure the effectiveness of community-based disaster risk management. *International Journal of Disaster Risk Reduction*, *49*, Article 101732. https://doi.org/10.1016/j.ijdrr.2020.101732

Satake, K. (2014). Advances in earthquake and tsunami sciences and disaster risk reduction since the 2004 Indian Ocean tsunami. *Geoscience Letters*, *1*(1), 1–13. https://doi.org/10.1186/s40562-014-0015-7

Sawalha, I. H. (2021). Views on business continuity and disaster recovery. *International Journal of Emergency Services*, *10*(3), 351–365. https://doi.org/10.1108/IJES-12-2020-0074

Schreurs, M. A. (2021). Reconstruction and revitalization in Fukushima a decade after the "triple disaster" struck: Striving for sustainability and a new future vision. *International Journal of Disaster Risk Reduction*, *53*, Article 102006. https://doi.org/10.1016/j.ijdrr.2020.102006

Shmelev, S. E., & Shmeleva, I. A. (2018). Global urban sustainability assessment: A multidimensional approach. *Sustainable Development*, *26*(6), 904–920. https://doi.org/10.4324/9780429287725

Silvestre, B. S., & Ţîrcă, D. M. (2019). Innovations for sustainable development: Moving toward a sustainable future. *Journal of Cleaner Production*, *208*, 325–332. https://doi.org/10.1016/j.jclepro.2018.09.244

Smith, G. P., Martin, A., & Wenger, D. E. (2018). "Disaster Recovery in an Era of Climate Change: The Unrealized Promise of Institutional Resilience." In H. Rodríguez, W. Donner, & Trainor, J. E. (Eds.), *Handbook of Disaster Research* (pp. 595–619). Springer Link.

Smith, G. P., & Wenger, D. (2007). "Sustainable Disaster Recovery: Operationalizing an Existing Agenda." In H. Rodríguez, E. L. Quarantelli, & R. R. Dynes (Eds.), *Handbook of Disaster Research* (pp. 234–257). Springer. https://doi.org/10.1007/978-0-387-32353-4

Spoon, J., Gerkey, D., Chhetri, R. B., Rai, A., & Basnet, U. (2020). Navigating multidimensional household recoveries following the 2015 Nepal earthquakes. *World Development*, *135*, Article 105041. https://doi.org/10.1016/j.worlddev.2020.105041

Takeuchi, K., Elmqvist, T., Hatakeyama, M., Kauffman, J., Turner, N., & Zhou, D. (2014). Using sustainability science to analyse social–Ecological restoration in NE Japan after the great earthquake and tsunami of 2011. *Sustainability Science*, *9*, 513–526. https://doi.org/10.1007/s11625-014-0257-5

The City of New York. (2013). *A stronger, more resilient New York*. NYC Mayor's Office of Climate & Environmental Justic. ESCAP Publications. https://toolkit.climate.gov/reports/stronger-more-resilient-new-york

The United Nations. (2021). *Climate and disaster resilient infrastructure: Building resilience to future uncertainties and shocks*. https://hdl.handle.net/20.500.12870/4543

Tierney, K. (2006). "Social Inequality, Hazards, and Disasters." In R. J. Daniels, D. F. Kettl, & H. Kunreuther (Eds.), *On Risk and Disaster: Lessons from Hurricane Katrina* (pp. 109–128). University of Pennsylvania Press.

Twigg, J. (2015). *Disaster Risk Reduction*. Overseas Development Institute.

Wang, W. L., & van de Lindt, J. W. (2021). Quantitative modeling of residential building disaster recovery and effects of pre-and post-event policies. *International Journal of Disaster Risk Reduction*, *59*, Article 102259. https://doi.org/10.1016/j.ijdrr.2021.102259

Wolf-Fordham, S. (2020). Integrating government silos: Local emergency management and public health department collaboration for emergency planning and response. *The American Review of Public Administration*, *50*(6–7), 560–567. https://doi.org/10.1177/0275074020943706

World Commission on Environment and Development (WCED). (1987). *Report of the world commission on environment and development: Our common future*. UN Documents. Gathering a body of Global Agreements. https://sustainabledevelopment.un.org/content/documents/5987our-common-future.pdf

Zamanifar, M., & Hartmann, T. (2020). Optimization-based decision-making models for disaster recovery and reconstruction planning of transportation networks. *Natural Hazards*, *104*, 1–25. https://doi.org/10.1007/s11069-020-04192-5

Zhao, F., Zhang, S., Du, Q., Ding, J., Luan, G., & Xie, Z. (2021). Assessment of the sustainable development of rural minority settlements based on multidimensional data and geographical detector method: A case study in Dehong, China. *Socio-Economic Planning Sciences*, *78*, Article 101066. https://doi.org/10.1016/j.seps.2021.101066

Section VI

Disaster Recovery Theories and Approaches

22
DISASTER RECOVERY MODELS

Elizabeth Wagemann and Magdalena Gil

Introduction

A model is a human construct, a representation of reality that provides a simple understanding of complex, real-world systems. Models can help define, analyze, synthesize, and communicate a set of concepts using detailed information as inputs and expected results as the output. In public policy, models play a fundamental role in advising decision-makers based on possible scenarios and by providing simplified frameworks to help organize and design actions. However, policy-makers face increasing difficulties using models to address contemporary social issues because the scale, heterogeneity, and complexity of today's challenges make modeling a difficult and uncertain task (Gil and Undurraga, 2022). Also, there is always the risk of oversimplifying our abstractions of reality, which can lead to problems in decision-making and the results emanating from those decisions (Davis and Alexander, 2015). Yet, despite their limitations, models can be valuable if used correctly as a framework for decision-making.

In this chapter, we provide an overview of the most relevant models that have been proposed to understand and manage a very complex but increasingly common challenge: post-disaster recovery. As anthropologists Susanna Hoffman and Anthony Oliver-Smith (1999) stated long ago, disasters should be understood as all-encompassing occurrences that affect nearly all dimensions of human life: social, economic, cultural, spiritual, biological, physical, psychological, and political, among others. Consequently, during recovery and reconstruction processes, many sources and forms of expertise need to be combined and integrated, with different paradigms and values emerging from different disciplines and varied actors. It should not surprise us, then, that several experts consider recovery to be the most complex phase of disaster management (see Chang and Miles, 2004; Smith and Wenger, 2007; Yang and Jahan, 2018). On the other hand, new emerging frameworks are informing decision-making in specific challenges, such as the management of heritage sites affected by disaster (ICOMOS-ICCROM, 2023), the recovery of tourist areas, or sensitive water arrangements (Juran, 2015). However, a common model for understanding, analyzing, and—hopefully—managing recovery has been difficult to achieve.

In this chapter, we explore various disciplinary reviews, empirical cases, and public policy guidelines on recovery in order to extract frameworks and models used for augmented thinking and

DOI: 10.4324/9781003414834-31

decision-making in the aftermath of disaster. Our analysis shows that the development of models over time reflects the increasing complexity of the impacts of disasters on human societies. The scale and multifaceted nature of disaster impacts have required an evolution from simple and isolated disciplinary models to holistic and interdisciplinary models that incorporate numerous intricacies of disaster recovery along with the spatial, temporal, and social dynamics that govern its pace.

All Models Are Wrong, but Some Are Useful

Models have several important limitations, the first being that social life is complicated—sometimes uncertain or even chaotic—and thus can never be fully modeled. Simplifying reality is a condition for models since their power relies in clarifying intricate relationships between different elements or domains. To do so, models rely on a set of assumptions, and the usefulness of a model is contingent on the validity and accuracy of these assumptions. If the assumptions are unrealistic or out of line with real-world conditions, or if the model is biased and omits an important part of the reality it intends to describe, this will affect the reliability of the output for decision-making and policy design. Also, if the model is excessively simple or artificial, it can communicate an incomplete understanding of reality (e.g., by under- or over-exaggerating expectations), thus negatively affecting decision-making and, consequently, those affected by disasters.

There will always be a gap between reality and the models used to inform or guide decision-making. However, this does not mean that models should be seen as a mere theoretical exercise. On the contrary, it is in practice that models find their value. As George Box (1976) once stated, while it is true that all models are wrong, some are useful. Even the most sophisticated and well-informed models are incomplete characterizations of reality, but they can still provide valuable frameworks for understanding, predicting, and making decisions about real-world challenges. Mikael Krogerus and Roman Tschäppeler (2011) provide some criteria that decision-based models should follow to be useful: they need to simplify reality, be pragmatic, summarize complex relationships, be visual to convey concepts that are difficult to explain, provide structure, and provide a method for better understanding reality.

Models act as maps to help people navigate and negotiate through dynamic and increasingly complex processes and relationships. To be useful, though, maps cannot portray a complete picture of the land. Rather, maps selectively represent certain elements of the territory (e.g., roads, rivers, human settlements, and landmarks), just as models selectively represent and account for certain aspects of a system or phenomenon, focusing on the variables and relationships that are most relevant to the specific question or problem at hand. This means that the purpose of models—as of maps—is crucial for evaluating how they are constructed and ultimately their usefulness. A map for electoral redistricting, for example, may not tell us much about the topography of the territory, but instead highlight roads, population centers, and administrative divisions. A topographic map, on the other hand, is necessary for trekking and exploring a territory and thus will preclude irrelevant features that would only serve to cause unnecessary noise. Similarly, reconstruction models aimed at repairing the physical infrastructure in the shortest possible time will be different from reconstruction models that focus on the well-being of disaster-affected families. Finally, we must acknowledge that models and maps are subject to change as reality changes. As our understanding of the challenge evolves and as new data or insights are procured, models require updates and refinements.

When working with models, we should always be aware of their limitations and never afford them excessive authority to predict the future or utilize them as instruments to automate decision-making. However, if we view them as maps, we can better appreciate their potential theoretical and practical value for disaster planning recovery. Unfortunately, an analysis by Ian Davis and David Alexander

(2015) found that most recovery models are designed to be compelling and logical, but also superficial and idealistic, tending to be optimistic. They fail due to lack of resources, skills, knowledge, and political capital, which limits the ability of any government to achieve the goals of the model.

In this chapter, our aim is not to evaluate models but to understand whether models are being used for research and decision-making in the recovery process, or not. We first provide evidence of their use in academic research before discussing their role in public policy through a review of recovery guidelines and manuals. Finally, we suggest a way forward toward a more comprehensive model that has potential to be meaningful for public policy.

Methodology

The analysis presented here is based on several sets of documents gathered in the context of a research project funded by the Fondecyt ANID/INICIACIÓN program. First, we collected and analyzed 11 comprehensive reviews on disaster research from different disciplines. We then complemented this initial analysis with a review of 209 empirical studies on disaster recovery published in the Web of Science or Scopus databases between 2010 and 2020. We procured articles using the words 'disaster' and 'recovery' selecting only articles that: (1) focused on disasters triggered by natural hazards, (2) focused on some aspect of recovery, and (3) provided empirical case studies. The 209 articles used for this entry represent a strategic selection of the most cited case studies from the initially procured collection of 1,104 articles. In each of these case studies, we focused on understanding the role of models and theoretical frameworks for understanding recovery. Finally, we collected manuals and handbooks from international offices, such as United Nations International Strategy for Disaster Reduction (UNISDR) and Economic Commission for Latin America and the Caribbean (ECLAC), that offer practical advice for reconstruction to countries and agencies. We ask the same questions to all this corpus: are models being used and, more importantly, are they useful?

Recovery Models in Academic Research

Our analysis of 11 comprehensive reviews on disaster research came from disciplines labeled as disaster studies (Tierney, 2007), sociology (Arcaya, Raker, and Waters, 2020), anthropology (Barrios, 2017), economics (Kellenberg and Mobarak, 2011), political science (Comfort, 2005), geography (Donovan, 2016), urban planning (Brown et al., 2014; March and Kornakova, 2017), psychology (Masten and Narayan, 2012; Goldmann and Galea, 2014), and public health (Wulff, Donato, and Lurie, 2015). A first conclusion is that the focus of all disciplines has been on understanding impact and not recovery. This is not surprising considering that, due to its potential application to wartime situations, initial disaster research focused on people's responses during events (reactivity), rather than on prevention, mitigation, and recovery (proactivity). This lens changed around the 1990s when sociologists and other social scientists sought to emancipate disasters from their focus on physical events to instead emphasize their social origins (Hewitt, 1983; Wisner et al., 1994; Mileti, 1999; Tierney, 2007). This shift meant a new focus on the pre-onset conditions and the study of the social, economic, and political dynamics and mechanisms that coalesce to co-produce disasters (Tierney, 2007). While this focus has been fruitful in improving prevention, it left recovery and reconstruction somewhat siloed within individual disciplines and often not labeled as 'disaster studies.'

Still, regardless of the discipline, recovery means returning to a pre-disaster state or at least one that is equally acceptable. However, the aspects or variables that are expected to return to an acceptable state vary greatly from discipline to discipline. For example, economics focuses on the recovery of production and consumption, while social capital is a key variable for sociologists and spatial inequities are of import to geographers. Also, articles across the different disciplines show that—even when communities are mourning a pre-disaster state—recovery and reconstruction are equally motivated by the possibilities of social change and, hopefully, the improvement of certain pre-existing conditions. This motivation can go beyond the mantra of 'building back better' to include a profound desire to create something completely different.

We complemented the previous analyses by examining a set of 209 empirical studies of post-disaster recovery. The first finding is that models are rarely explicitly used in these works. Authors do refer to goals such as 'build-back better' or 'sustainable recovery' as models. However, these terms are used as a heuristic; in other words, as guiding or evaluative principles rather than as systematic tools for case studies or to guide decision-making. Still, these heuristics seem to link ideas and concepts across the disciplines, which can be fruitful. The different disciplines share a common language around the concepts of risk, vulnerability, and resilience. For example, individuals, communities, buildings, and political and economic systems can all be considered vulnerable or resilient. Another term common to several disciplines is risk, but there are considerable differences between disciplines that focus on calculating risk and those that perceive it as a social construct.

We also found that case studies are highly disciplinary, with loose links between disciplines. While some of the case studies were published in non-disciplinary disaster-based journals, such as *Disasters* or *International Journal of Disaster Risk Reduction*, most came from disciplinary journals in urban studies, sociology, economics, anthropology, and demography, among others. The concern with recovery and reconstruction appears most prominent in the areas of human geography and urban planning. In this literature, disasters are posited as disruptions to human settlements and recovery as a process aimed at restoring certain equilibria that ensures human and social functions are reestablished. Crucial to this view is the territorially based administrative management of reconstruction processes, which highlights the different ways in which the physical world is partitioned, organized, and politically managed (March and Kornakova, 2017).

While there is no single concept or model that is harnessed across all disciplines, Susan Cutter's (2008) disaster resilience of place (DROP) model stands out as a reference in the literature (cited well over 5,000 times according to Google Scholar). This framework considers ecological, social, economic, institutional, infrastructure, and community dimensions—but its focus is on resilience rather than recovery and its original version fails to explore the links between the different aspects involved. Other notable models are Robert Kates and David Pijawka's (1977) Timetable of Reconstruction and Ian Davis and David Alexander's (2015) sectoral model of recovery (cited 187 and 207 times, respectively). While none of these models have achieved multidisciplinarity, they have all been useful for several disciplines and case studies.

Overall, the empirical articles studied establish that a transdisciplinary model of recovery—one that includes social and physical aspects involved in reconstruction—is a common goal, yet remains a challenge to actually produce. The concepts of sustainable recovery (Smith and Wenger, 2007) or integral recovery (Yang and Jahan, 2018) are present in the literature, but mostly as a goal. In general, empirical studies focus on only one or two aspects of the recovery process, looking to provide knowledge on how buildings, communities, families, organizations, or individuals recover. However, we know very little about how these different processes intersect, interact, and possibly reinforce each other. Unfortunately, the lack of a common model for analyzing recovery shows that transforming this area of expertise into a truly transdisciplinary field remains a challenge (Fortun and Frickel, 2012; Barrios, 2017).

Recovery Models and Public Policy

As mentioned in the introductory section, models find their true value when they allow us to render more informed decisions surrounding real-world challenges. This is why, although models have not been prominent in academic studies, we have found that they are very much present in documents meant to inform public policy and provide recommendations for recovery. These models are derived from different views of the policy process that seek to guide the analysis and overall interpretation of reality for policy-makers. These frameworks represent a set of beliefs, assumptions, goals, and practices that determine if a model is considered useful for decision-making. Here we describe some of the most relevant frameworks, which are also conceptualized in Figure 22.1.

Linear/Rational Framework

The linear/rational framework describes policy-making as a rational, balanced, objective, and analytical problem-solving process that operates in an orderly sequence (Simon, 1947). It assumes that decisions are made in a series of logical, sequential stages, starting with the identification of a problem or issue, its causes, the potential options, and ending with a set of activities to solve or manage the problem or issue (Thomas and Grindle, 1990). The couching of the policy process as a sequence of stages, as described by Harold D. Lasswell (1956), became the starting point for a number of influential typologies (see Anderson, 1975; Jenkins, 1978; Brewer and deLeon, 1983; May and Wildavsky, 1978). The problem is that the linear/rational model was quickly found to be far from reality, thus evolving later into a cyclical process that emphasizes feedback processes of policy-making (Bridgman and Davis, 2003). One of the most widely used post-disaster recovery models—the Robert W. Kates and David Pijawka (1977) Timetable of Reconstruction—follows this framework when describing urban recovery processes as a set of four stages: (1) emergency response, (2) restoration, (3) reconstruction of destroyed assets for functional replacement, and (4) reconstruction for commemoration, betterment, and development. Since then, these phases have become a foundation for many recovery models, even though new phases have been added to the original list, such as mitigation activities and social recovery (Sobhaninia and Buckman, 2022). Still, other theories of how public policy should be conducted have been highly influential as well. Some of these theories reject the notion of a linear/rational decision-making process and instead see policy-makers acting in situations marked by uncertainty and ambiguity (Anderson, 1975; Stone, 1988).

Incremental Framework

The incremental framework conceptualizes public policy as a process of small adjustments to existing policies based on the difficulty of making large policy changes and that small, successive advances may be more realistic and likely to succeed. The political scientist Charles E. Lindblom (1959, 1979) developed this theory in the 1950s as a rational, long-term, goal-oriented policy. This approach has proven useful for existing or ongoing policies and programs, but not for new, critical events that arise such as disasters or war-like situations. Thus, it is more common when attempting to achieve sustainability or resilience, rather than to evaluate or guide recovery processes after disaster. A strategy that includes elements of the linear/rational and incremental approaches has been called a 'mixed-scanning framework' (Etzioni, 1967). This hybrid view considers both possible options looking at past experiences and looks more closely at those that require more in-depth consideration (exhaustive rational approach).

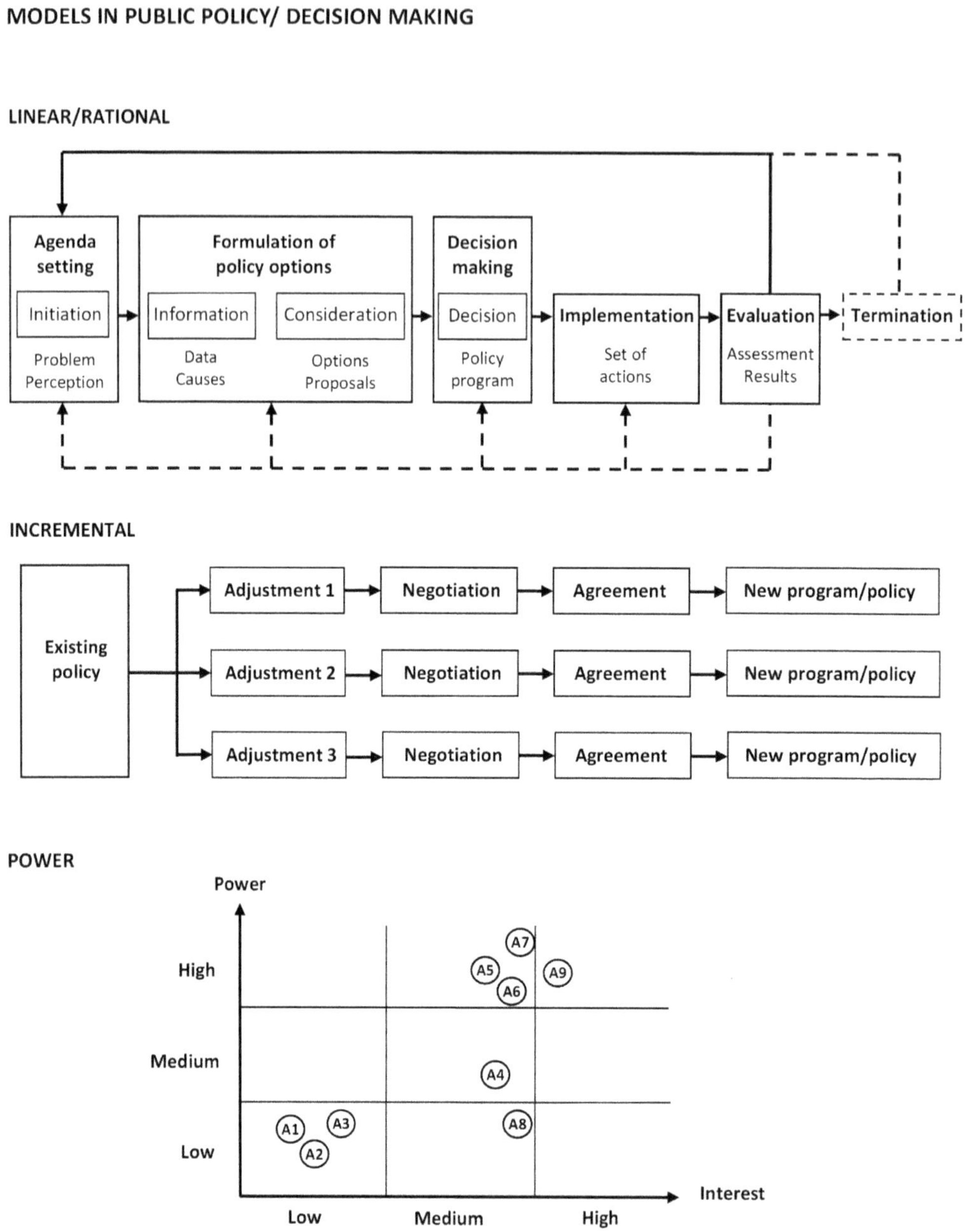

Figure 22.1 Conceptual Diagrams of Relevant Models in Public Policy and Decision-Making: Linear/ Rational Framework (adapted from Lasswell, 1956; Jenkins, 1978; Howlett and Ramesh, 2003); Incremental Framework (adapted from Lindblom, 1959, 1979); and Power Framework (adapted from Mendelow, 1991)

Source: Adopted from Lasswell (1956), Jenkins, (1978), Howlett and Ramesh (2003), Lindblom, (1959), Lindblom (1979), and Mendelow (1991).

Power Framework

This framework is based on the idea that power and the influence of various interest groups serve to shape public policy. Power as a decision-making process focuses on the actions of individuals, groups, and institutions that influence policy decisions (Bachrach and Baratz, 1962). Earlier versions of this framework took a completely top-down perspective in which decisions are shaped by the values and beliefs of policy-makers to examine the extent their legally mandated objectives are achieved over time and why (Sabatier, 1986). In the late 1970s, a bottom-up approach emerged as a response to the perceived weaknesses of the top-down perspective to describe reality more accurately. Rather than starting with a policy decision, it assumes that a multitude of actors interact to shape a particular problem or issue (Sabatier, 1986; Stone, 1988). Aubrey L. Mendelow's Power-Interest Matrix (1991), which considers stakeholders' power to make decisions and act along with their expectations and interests, has been used to define the potential influence of various stakeholder groups. Since the absence of certain groups—particularly (marginalized) communities—in recovery decision-making and planning is believed to induce failure, this stakeholder approach to public policy has been highly influential in disaster policies that mandate community participation.

Guidelines and Manuals for Public Policy

International organizations focused on disaster risk reduction and development have proposed guidelines and manuals to better guide the recovery process, and although they do not describe how the models should be used, they offer an overview of advice that decision-makers should heed when confronted with disaster. Furthermore, these 'gray literature' documents have been produced with the aims of improving transparency and accountability related to international donations, to support evaluations, and to provide empirical evidence of the effectiveness, efficiency, and impact of interventions.

One of the most salient characteristics of these international policy-oriented documents is the sectoral approach. That is, they break down recovery into manageable segments. This allows for organizing and coordinating recovery efforts according to specific sectors, development areas, or industries affected by the disaster. This is very different from what we found in most academic case studies, where links between different aspects of recovery are scarce. However, we know that disasters are events that affect all or most dimensions of human life. For this reason, many sources and forms of expertise should be recognized and integrated during recovery processes, with different paradigms and values emerging from different actors. Manuals and guidelines seemed to recognize this more explicitly than academic articles when discussing post-disaster situations.

For example, a model for disaster analysis proposed by ECLAC (2014) in the Handbook for Disaster Assessment considers economic, social, cultural, environmental, and infrastructural elements. The model is designed to assess impact rather than recovery, which is different from the objectives of this chapter. Nevertheless, it provides important insights into the numerous dimensions that need to be considered. Another example is the Disaster Recovery Framework Guide by the Global Facility for Disaster Reduction and Recovery (2020) that focuses on multi- and inter-sectoral recovery planning aspects, presenting a checklist for each module with correlative primary steps to be followed. As a practical endeavor, it also provides tools for policy-makers and managers of recovery institutions, financial institutions, monitoring and evaluation experts, and governmental implementing agencies to make more astute decisions.

Alternatively, the Handbook on Recovery Institutions (2021) of the United Nations Development Programme (UNDP) focuses on the institutional support needed to deliver effective multidimensional recovery outcomes. Institutional arrangements for recovery refer to the organizational structures, policies, and procedures that political territories put in place to plan, manage, and implement disaster recovery operations. It provides the Tool to Assess Recovery Preparedness at Country Level and sample Terms of Reference for key institutional staff. It also describes institutional options, successful characteristics, and management lessons based on grounded experiences with disaster recovery operations from Chile, India, Indonesia, New Zealand, Serbia, and Mozambique. It describes the critical components of the institutional arrangements that govern the recovery process: capacity, resources, assistance, structures, management, administration, monitoring and evaluation, communication, and collaboration. It also presents a checklist of key considerations.

These guides are designed to be useful to government officials and other types of practitioners who are developing, supporting, or implementing institutional recovery plans. As such, they play an important role in the recovery plans that exist in many countries and jurisdictions. Significantly, most are based on a top-down approach and follow a linear/rational model. In the case of the United States, for example, Philip Berke et al. (2014) and Jennifer Horney et al. (2017) reviewed a list of indicators present in 87 recovery plans showing that these plans include economics, health, housing, and community communication metrics, but do not provide a comprehensive approach or shed light on links among the many aspects evaluated. Several guides present examples of best practices and lessons learned from various recovery processes throughout the globe—but not because the guidelines have been applied in these countries, but rather as examples used in the development of the frameworks. This means that it is very possible that the guides analyzed do not describe the actual use of such models or frameworks but rather the desire for policy-makers to do so.

Finally, it is relevant to reiterate that the guidelines tend to present a top-down and a linear/rational approach based on prioritizing actions and providing checklists of (often discrete) steps to follow. However, several of the best practices highlight the need for stakeholder participation and community engagement to make sure the processes are successful, which is a much-needed step forward.

Conclusion: Toward a More Comprehensive and Useful Model for Recovery

Theoretical models can be useful but they are not always present in disaster research and decision-making. This may be because of the models themselves, or because of the uncertain and rapidly evolving situations that tend to follow disasters, pushing for the use of heuristics instead. Even academic studies of recovery processes tend to rely on frameworks that are more diffuse than any particular model when describing, analyzing, or evaluating specific cases. Moreover, the models that were identified are, for the most part, highly disciplinary and specific to particular aspects of recovery. In contrast, public policy manuals and guides are more explicit in their use of models and emphasize their role in decision-making at the top. These models also tend to be more comprehensive by incorporating several areas of recovery (e.g., infrastructure, social, economic, cultural, and other components), even if it is not clear how these different aspects of recovery interact and potentially influence each other.

Our analysis also shows that there is a perceived need in the literature to move toward models that are more comprehensive yet, simultaneously, more flexible, and nimble than classical models. In other words, models that incorporate multiple aspects of recovery but also allow for adaptation and contextualization to different physical geographies, cultures, situations, and goals. These models should understand recovery as a process that connects many elements and activities—not only the repair and reconstruction of damaged infrastructure but also the economic, cultural, social, emotional, and spiritual revitalization of affected communities. Furthermore, they should also include the recovery of the health and well-being of affected populations, including their sense of community, social and kinship networks, and the costs and benefits of living in a particular natural environment that may have been destroyed. Finally, sustainable recovery necessarily means that there should be disaster risk reduction and better anticipation of future events (Smith and Wenger, 2007; Yang and Jahan, 2018).

How do we ensure that models are actually useful in guiding decision-making and, ultimately, disaster risk reduction? As noted in the introduction, models need to go beyond simplifying reality and summarizing complex relationships: they also need to be practical and visual to explain difficult concepts and allow stakeholders and institutions to choose between different scenarios. In the case of recovery models, this means that models must not only communicate the complexity of multidimensional dynamic processes but more importantly they must be able to define strategies for numerous actors. As a proposal for such a model, we present an adaptation of Ian Davis and David Alexander's (2015) sectoral model of recovery (Figure 22.2). Our version takes into

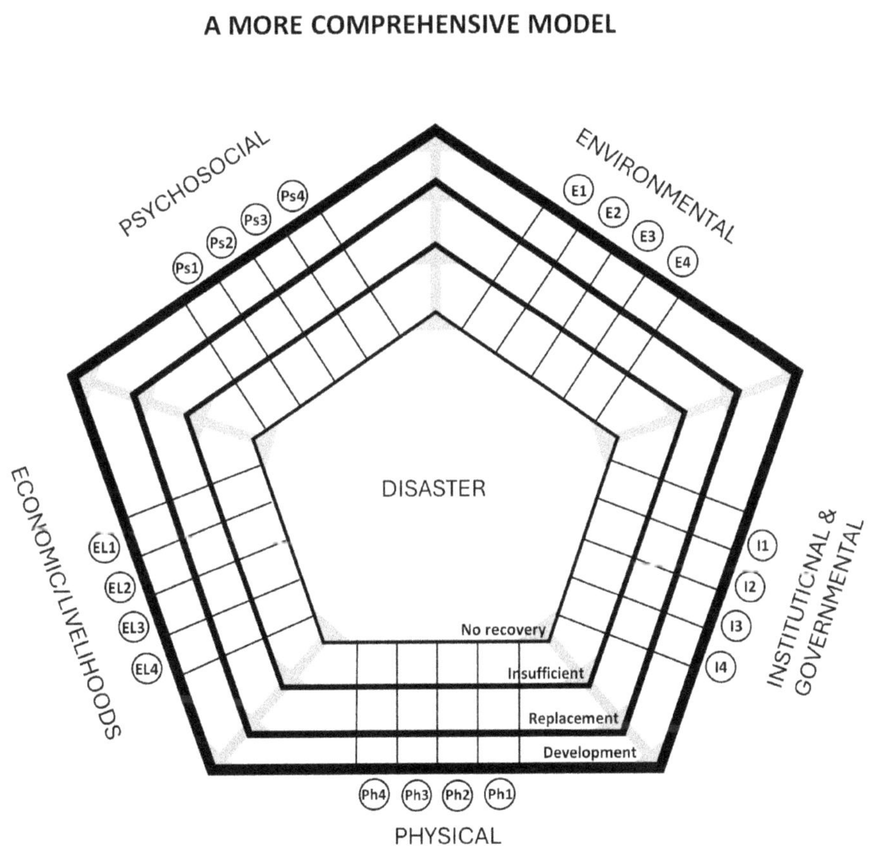

Figure 22.2 A More Comprehensive Model That Incorporates the Sectors Involved in Disaster Recovery and Integrates Indicators for Instrumental Use

Source: Adapted from Davis and Alexander (2015).

the account the gaps identified by the analysis presented in this chapter. We propose a sectoral model because it can provide a comprehensive view of recovery, encompassing economic, social, natural, and built environment recovery. Also, a sectoral approach is beneficial because public policies tend to operate in sectors, and a model that can incorporate this reality has a better chance of being useful. However, it should always be remembered that sectors themselves are social constructs, or simplifications of the complex organization of human settlements and public policy. In reality, each of these sectors is embedded in broader social forces and linked to the other sectors by multiple threads.

The model highlights the different aspects of disaster impacts that need to be included for sustainable recovery and also allows for linking and articulating goals while accounting for each dimension. Goals can include activities usually defined as relief, restoration, recovery, and mitigation by phase models, but in this case, the emphasis is on how the different sectors can advance in each goal at different rhythms depending on their role, the complexity of the tasks, and the capacities at hand. Organizing recovery in such a way has great potential to be used as a monitoring and accountability tool. Furthermore, this type of model enables the incorporation of a stakeholder approach. To do so, we must include the various actors involved in the process of recovery by identifying their sectoral influence, considering their varied interests, and allowing for different levels and sources of power depending on the sector or area of intervention.

Acknowledgments

The research for this chapter was partially funded by the Chilean National Research and Development Agency (ANID) under projects Fondecyt INICIACION.11220562 'Recovery and Reconstruction After Socionatural Disasters: A Model for Interdisciplinary Analysis and Public Policy Intervention' and Fondecyt INICIACION N.11220869 'Relaciones entre espacio doméstico y espacio de trabajo. Actividades económico-productivas en viviendas de emergencia después del evento hidrometeorológico de 2015 en Atacama.'

References

Anderson J. 1975. *Public Policy-Making*. New York, NY: Praeger.
Arcaya M., Raker E.J., and Waters M.C. 2020. "The Social Consequences of Disasters: Individual and Community Change." *Annual Review of Sociology* 46(1). https://doi.org/10.1146/annurev-soc-121919-054827.
Bachrach P., and Baratz M.S. 1962. "Two Faces of Power." *American Political Science Review* 56(4): 947–52. https://doi.org/10.2307/1952796.
Barrios R.E. 2017. "What Does Catastrophe Reveal for Whom? The Anthropology of Crises and Disasters at the Onset of the Anthropocene." *Annual Review of Anthropology* 46(1): 151–66. https://doi.org/10.1146/annurev-anthro-102116-041635.
Berke P., Cooper J., Aminto M., Grabich S., and Horney J. 2014. "Adaptive Planning for Disaster Recovery and Resiliency: An Evaluation of 87 Local Recovery Plans in Eight States." *Journal of the American Planning Association* 80(4): 310–23. https://doi.org/10.1080/01944363.2014.976585.
Bridgman P., and Davis G. 2003. "What Use Is a Policy Cycle? Plenty, If the Aim Is Clear." *Australian Journal of Public Administration* 62(3): 98–102. https://doi.org/10.1046/j.1467-8500.2003.00342.x.
Brewer G.D., and deLeon P. 1983. *The Foundations of Policy Analysis*. Homewood, IL: The Dorsey Press.
Brown D., McGranahan G., Dodman D., and International Institute for Environment and Development. 2014. *Urban Informality and Building a More Inclusive, Resilient and Green Economy*. London: IIED.
Chang S.E., and Miles S.B. 2004. "The Dynamics of Recovery: A Framework." In *Modeling Spatial and Economic Impacts of Disasters*, edited by Okuyama, Y., and Chang, S.E., 181–204. Advances in Spatial Science. Berlin, Heidelberg: Springer. https://doi.org/10.1007/978-3-540-24787-6_10.

Comfort L.K. 2005. "Risk, Security, and Disaster Management." *Annual Review of Political Science* 8(1): 335–56. https://doi.org/10.1146/annurev.polisci.8.081404.075608.
Cutter S.L., Barnes L., Berry M., Burton C., Evans E., Tate E., and Webb J. 2008. "A Place-Based Model for Understanding Community Resilience to Natural Disasters." *Global Environmental Change* 18(4): 598–606. https://doi.org/10.1016/j.gloenvcha.2008.07.013.
Davis I., and Alexander D. 2015. *Recovery from Disaster*. First. Routledge Studies on Hazards, Disaster Risk and Climate Change. London: Routledge.
Donovan A. 2016. "Geopower: Reflections on the Critical Geography of Disasters." *Progress in Human Geography*, January. https://doi.org/10.1177/0309132515627020.
Economic Commission for Latin America and the Caribbean (ECLAC). 2014. *Handbook for Disaster Assessment*. https://www.cepal.org/en/publications/36823-handbook-disaster-assessment.
Etzioni A. 1967. "Mixed-Scanning: A 'Third' Approach to Decision-Making." *Public Administration Review* 27(5): 385. https://doi.org/10.2307/973394.
Fortun K., and Frickel S. 2012. "Making a Case for Disaster Science and Technology Studies". In *STS forum on Fukushima*, 11–12.
Gil M., and Undurraga E.A. 2022. "Living the Modern Dream: Quantification and Modelling during the COVID-19 pandemic in Chile". In *Covid-19 and the Sociology of Risk and Uncertainty: Studies of Social Phenomena and Social Theory across 6 Continents*, edited by Zinn, J., and Brown, P. London: Palgrave-McMillan.
Goldmann E., and Galea S. 2014. "Mental Health Consequences of Disasters." *Annual Review of Public Health* 35: 169–83.
Hewitt K. 1983. *Interpretations of Calamity: From the Viewpoint of Human Ecology*. Boston, MA: Allen & Unwin.
Horney J., Caroline D., Meghan A., Berke P., and Smith G. 2017. "Developing Indicators to Measure Post-Disaster Community Recovery in the United States." *Disasters* 41(1): 124–49. https://doi.org/10.1111/disa.12190.
Howlett M., and Ramesh M. 2003. *Studying Public Policy: Policy Cycles and Policy Subsystems*. Oxford: Oxford University Press.
ICOMOS and ICCROM. 2023. *Guidance on Post-Disaster and Post-Conflict Recovery and Reconstruction for Heritage Places of Cultural Significance and World Heritage Cultural Properties*. Sharjah: ICCROM and ICOMOS. ISBN 978-92-9077-342-9
Jenkins W. 1978. *Policy Analysis: A Political and Organizational Perspective*. London: Martin Robertsen.
Juran L. 2015. "A Framework for Integrating Water Projects in Disaster Reconstruction Processes in the International Context." *Public Works Management & Policy* 20(4): 305–21.
Kates R.W., and Pijawka D. 1977. "From Rubble to Monument: The Pace of Reconstruction." In *Reconstruction Following Disaster*, edited by Haas, John E., Kates, Robert W., and Bowden, Martyn J. 1–23. Cambridge, MA: M.I.T. Press.
Kellenberg D., and Mobarak A.M. 2011. "The Economics of Natural Disasters." *Annual Review of Resource Economics* 3(1): 297–312. https://doi.org/10.1146/annurev-resource-073009-104211.
Krogerus M., and Tschäppeler R. 2011. *The Decision Book: Fifty Models for Strategic Thinking*. London: Profile Books.
Lasswell H.D. 1956. *The Decision Process: Seven Categories of Functional Analysis*. College Park, MD: University of Maryland.
Lindblom C.E. 1959. "The Science of 'Muddling Through." *Public Administration Review* 19(2): 79–88. https://doi.org/10.2307/973677.
Lindblom C.E. 1979. "Still Muddling, Not Yet Through." *Public Administration Review* 39(6): 517. https://doi.org/10.2307/976178.
March A., and Kornakova M. (Eds.). 2017. *Urban Planning for Disaster Recovery*. Oxford: Butterworth-Heinemann, An imprint of Elsevier.
Masten A.S., and Narayan A.J. 2012. "Child Development in the Context of Disaster, War, and Terrorism: Pathways of Risk and Resilience." *Annual Review of Psychology* 63(1): 227–57. https://doi.org/10.1146/annurev-psych-120710-100356.
May J.V., and Wildavsky A. (Eds.) 1978. *The Policy Cycle*, 279–300. Beverly Hills, CA: Sag
Mendelow A.L. 1991. "Environmental Scanning: The Impact of the Stakeholder Concept." *Proceedings from the Second International Conference on Information Systems*, 407–18. Cambridge, MA.

Mileti D.S. 1999. *Disasters by Design: A Reassessment of Natural Hazards in the United States*. Washington, DC: Joseph Henry Press.

Oliver-Smith A., and Hoffman S. 1999. "Anthropology and the Angry Earth: An Overview." In *The Angry Earth: Disaster in Anthropological Perspective*, edited by Oliver-Smith, A., and Hoffman, S. New York, NY: Routledge.

Sabatier P.A. 1986. "Top-Down and Bottom-Up Approaches to Implementation Research: A Critical Analysis and Suggested Synthesis." *Journal of Public Policy* 6(1): 21–48. https://doi.org/10.1017/S0143814X00003846.

Simon H.H. 1947. *Administrative Behavior: A Study of Decision-Making Processes in Administrative Organizations*, vol. 112. New York, NY: Macmillan.

Smith G.P., and Wenger D. 2007. "Sustainable Disaster Recovery: Operationalizing an Existing Agenda." In *Handbook of Disaster Research*, edited by Rodríguez, H., Quarantelli, E.L., and Dynes, R.R., 234–57. Handbooks of Sociology and Social Research. New York, NY: Springer. https://doi.org/10.1007/978-0-387-32353-4_14.

Sobhaninia S., and Buckman S.T. 2022. "Revisiting and Adapting the Kates-Pijawka Disaster Recovery Model: A Reconfigured Emphasis on Anticipation, Equity, and Resilience." *International Journal of Disaster Risk Reduction* 69(February): 102738. https://doi.org/10.1016/j.ijdrr.2021.102738.

Stone D.A. 1988. *Policy Paradox: The Art of Political Decision Making*. New York, NY: Norton.

Thomas J.W., and Grindle M.S. 1990. "After the Decision: Implementing Policy Reforms in Developing Countries." *World Development* 18(8): 1163–81. https://doi.org/10.1016/0305-750X(90)90096-G.

Tierney K.J. 2007. "From the Margins to the Mainstream? Disaster Research at the Crossroads." *Annual Review of Sociology* 33(1): 503–25. https://doi.org/10.1146/annurev.soc.33.040406.131743.

United Nations Development Program (UNDP). 2021. *Handbook on Recovery Institutions. A Guidebook for Recovery Leaders and Practitioners*. New York, NY: United Nations Development Programme. https://www.undp.org/publications/handbook-recovery-institutions-guidebook-recovery-leaders-and-practitioners.

Wisner B., Blaikie P., Cannon T., and Davis I. 1994. *At Risk: Natural Hazards, People's Vulnerability and Disasters*. New York, NY: Routledge.

Wulff K., Donato D., and Lurie N. 2015. "What Is Health Resilience and How Can We Build It?." *Annual Review of Public Health* 36(1): 361–74. https://doi.org/10.1146/annurev-publhealth-031914-122829.

Yang B., and Jahan I. 2018. "Comprehensive Assessment for Post-Disaster Recovery Process in a Tourist Town." *Sustainability* 10(6): 1842. https://doi.org/10.3390/su10061842.

23

RECONFIGURING THE KATES-PIJAWKA RECOVERY MODEL FOR A NEW ERA

Stephen Thomas Buckman and Saeideh Sobhaninia

Introduction

Environmental hazards, such as hurricanes, flooding, drought, and fires, are becoming the norm for many communities in the 21st century. With the uptick in climatic hazards, communities are trying to better understand how to recovery from these events. Over the last few decades, the resiliency notion of 'three As' (anticipate, absorb, and adapt) has led the charge in this context. While resiliency has proven to be worthwhile in many aspects, it tends to leave out post-disaster recovery. Knowledge of the post-disaster recovery process is imperative for understanding how to anticipate and prepare for future shocks. This understanding allows communities to better plan for and anticipate how quickly they can return to equilibrium. Without a strong foundation in the workings of the post-disaster recovery, communities cannot plan for the future and adapt to future shocks.

Recovery models take on many forms and structures. One of the most prominent and widely used post-disaster recovery models over the last 50 years has been the Kates-Pijawka model. Since its introduction in 1977, the Kates-Pijawka model has widely been used to describe the post-disaster recovery process in four phases, reflecting an easy-to-grasp process applicable to multiple disaster scenarios. This simplicity has served the model well and helps explain its tremendous use and longevity. However, evolving social and changing environmental threats that were not on the radar in the mid-1970s require revisiting the model.

Literature Review

Recovery Phases

Generally, researchers accept that recovery includes sequential and simultaneous activities (Mohd et al., 2017) with phases such as relief, restoration, and recovery (Pickett et al., 2013). While these phases have become the foundation to work from, others go beyond the three-phase model. For instance, Burby et al. (1999) defined four phases—preparedness, response, recovery, and mitigation.

Expanding on Burby et al. (1999), further studies have shown that recovery should not only include physical but also social restorations after a disaster. In this sense, post-disaster recovery

DOI: 10.4324/9781003414834-32

involves not just the built environment but also psychological and economic recovery, each with distinct temporal paces (Aguirre & Lane, 2019). Therefore, it is essential to understand a system's capacity to bounce back and learn from the past, but also to achieve revitalization, to take advantage of new opportunities, bounce forward after a disaster, and become more able to mitigate the impacts of future shocks (Newman et al., 2017). Through a comprehensive literature review, we found four key categories of theories in disaster recovery studies: disaster-, people-, place-, and government-related theories.

Disaster-Related Theories

Theories in this category are dependent on the disaster itself (Smith & Birkland, 2012). Put simply, the more significant the damage, the longer the recovery. Wu and Lindell (2004) also emphasized the direct impact of disaster damages and the availability of undamaged housing on the recovery process. Moreover, as Kates and Pijawka (1977) noted, the recovery speed is related to the intensity of the damage. While Kates and Pijawka (1977) found that more significant damage makes the recovery process longer, Aldrich (2012) claimed that more severe damage attracts faster technological investments. On the other hand, Rupakhety and Olafsson (2019) found little evidence that recovery speed is related to the size of the disaster or its magnitude. Their study shows no relationship between the speed of recovery, demographic and economic indicators, or the disaster's magnitude. However, they emphasize the importance of endogenous factors, such as leadership, decision-making, planning, finance, and science.

People-Related Theories

Lindell's (2013) disaster recovery model emphasized community recovery in post-disaster assessment. He declared that neighborhoods within the same community have different recovery rates. Even if the community as a whole recovers, some neighborhoods, households, economic sectors, or individual businesses might not recover. Therefore, anticipating the most vulnerable population segments and economic sectors is essential (Lindell, 2013).

Social Capital and Vulnerability

Social capital is as crucial in recovery as other external factors, such as damage levels and geographical isolation. Roque et al. (2020, 204) found that "social capital facilitated recovery efforts and enhanced resiliency through shared values, network expansion, new partnerships, and a desire to make their communities more robust and less vulnerable to upcoming environmental disturbances." However, Hill and Keys-Mathews (2005) noted that different recovery rates depend on the community's prior vulnerability, pre- and post-disaster resource allocations, and socioeconomic and sociopolitical contexts.

Social Cohesion

The degree of social cohesion is another critical factor facilitating a community's ability to recover post-disaster (Sobhaninia, 2023; Alexander et al., 2007). Social cohesion builds strong ties and assistance between group members influenced by culture, religion, ethnicity, and identity and connected with a common goal. It can lead to having more collective activities in societies and, therefore, shorter recovery time (Nakagawa & Shaw, 2004).

Culture

Many of the failures in post-disaster recovery have resulted from the misunderstanding of cultural nuances of the impacted population (Oliver-Smith, 1991). Changing or preserving the culture after a disaster thus is vital for sustainable reconstruction (Alexander et al., 2007). It is essential then to understand that creating resilient cities is not possible merely with technological advances. It needs to involve testing human capacities through a challenge by changing the culture, the economy, and people's lifestyles (Newman et al., 2017).

Participation

Rumbach (2016, 2054) also emphasized the role of social participation, stating: "public participation in planning and policymaking helps to improve governance by communicating local preferences to decision-makers, building trust in planning processes, improving the quality and implementation of plans, and increasing accountability of organizations and public officials to their constituents." Participatory approaches during post-disaster recovery help guarantee that recovery meets people's needs and is culturally and socially adequate (Felix et al., 2013). Berke and Campanella (2006) argued that since residents are aware of their area's significant weaknesses, they should be involved in the decision-making and planning. Early and ongoing resident participation in the recovery process also impacts better outcomes and equitable recovery (Berke et al., 2002). Thus, a key to post-disaster planning is empowering community members to participate (Townshend et al., 2015).

Place-Based Theories

Place matters, and therefore, the nature of a place can affect the recovery process in different ways. One way has to do with the resources that they offer to residents. Another is placemaking, which is a result of the human-built environment.

Geographical Resources

Geographic vulnerabilities are crucial in the recovery process, particularly those resulting from pre-existing vulnerabilities (Hill & Keys-Mathews, 2005). In addition, recovery time depends on resource availability and limitations. Isumi et al. (1985) presented a resource-constrained recovery model focused on restoring utility infrastructure that estimated the physical infrastructure and system's functionality. This model was based on consumption and restriction of resources by computing post-earthquake behavior related to gas, water, and electricity lifeline systems (Miles et al., 2019). Other place-based factors include energy self-sufficiency, water resources, and proximity to food that can help disaster recovery, as Newman et al. (2017) outlined. Local land use and building practices can also impact the recovery time (Wu & Lindell, 2004).

Placemaking

One of the most catastrophic impacts of disasters is individuals' feeling of 'loss of place' (Diaz & Dayal, 2008). In order to counterbalance the feeling of loss during the recovery process, residents try to reshape their individual and shared identities that are both dependent on the environmental and social characteristics of the place where they live (Cox & Perry, 2011). Sociologists have often emphasized the importance of places and their capabilities to affect community identity.

In addition, human experience can turn a space into a place by ascribing meaning (Sobhaninia et al., 2023). Meanings in a place are influenced by culture and experience, and therefore there exists a positive relationship between a perception of a place and disaster recovery (Sobhaninia et al., 2023). Place and its embedded meaning motivate residents to return to the area when they have been permanently or temporarily dislocated and their houses and possessions are lost (Chamlee-Wright & Storr, 2009).

Government-Based Theories

Where people live, the structures, accessible resources, and the capacity of local governance institutions significantly influence the speed and duration of the short-term recovery process (Rumbach, 2016). Comerio (2013) considered the government a primary factor for classifying housing recovery. Governments can affect the recovery process in different ways, such as the funding they provide and the policies they make.

Government Funding

Factors such as grants-in-aid, available financing, training, information exchange, and planning also impact the speed and quality of recovery outcomes (Wu & Lindell, 2004). One of the direct impacts of resource distribution is that many households turn their temporary housing into permanent housing due to a lack of access to resources since financial assistance often lags in reaching the communities that are most in need (Smith & Birkland, 2012). Yet, government intervention via cash remittance often keeps citizens on the brink of going over the edge.

Governance, Geopolitics, and Political Effects

The speed and success of post-disaster reconstruction also depend on the political and geographical connectedness to the decision-making centers and authorities (Alexander et al., 2007), the disaster management system, economic conditions (Wu & Lindell, 2004), recovery planning (Smith & Birkland, 2012), and cooperative partnerships among involved units (Smith & Birkland, 2012). Mileti (1999) considered several characteristics of a successful government recovery plan, including decent information about the affected area (e.g., vulnerable locations, resources, and population), appropriate cooperation of different groups and organizations, action-oriented procedures, precise damage evaluation, and linking particular needs to existing funding sources.

The Kates-Pijawka Recovery Model

The 1977 Kates-Pijawka model has become the cornerstone of how we think about and deal with urban post-disaster reconstruction. In the model, Kates and Pijawka used four case studies to understand the reconstruction problem: three case studies related to earthquakes (San Francisco, CA in 1906, Anchorage, AK in 1964, and Managua, Nicaragua in 1972) and one to a flooding event (Rapid City, SD in 1972). Drawing upon these cases, they addressed post-disaster reconstruction by proposing a model based on the predictability of the reconstruction process resulting in complete recovery. In their words, “the reconstruction process is ordered, knowable and predictable” (Kates & Pijawka, 1977, 262).

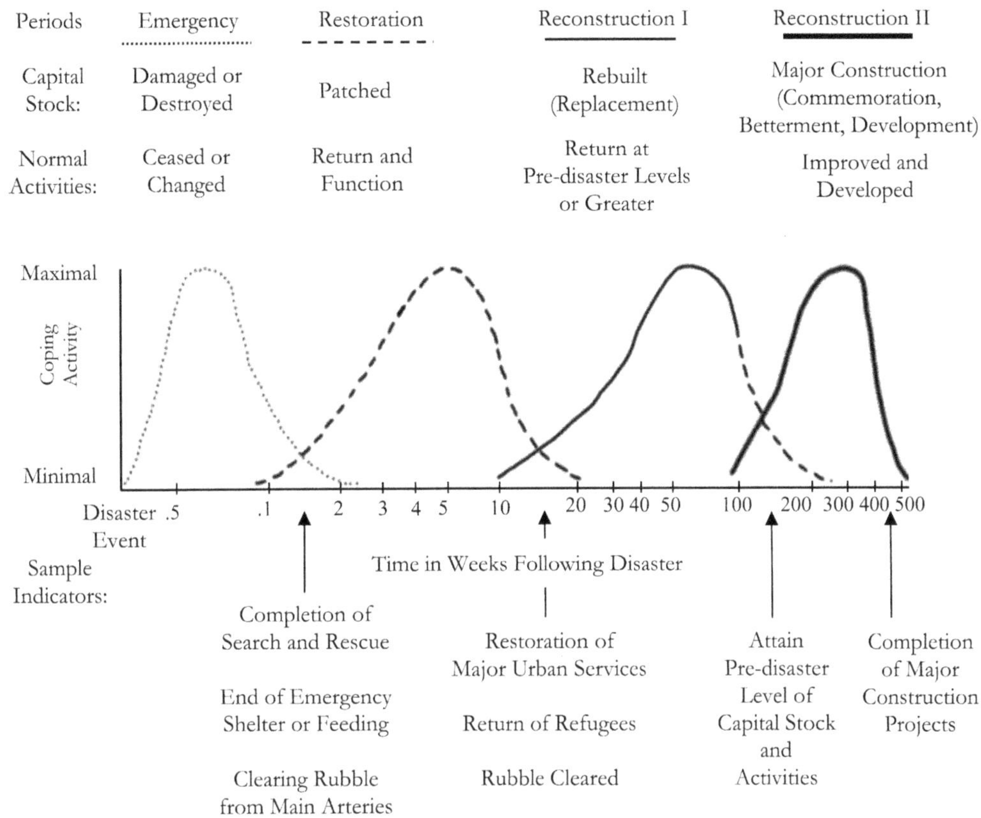

Figure 23.1 Original 1977 Kates-Pijawka Model
Source: Adopted from Kates and Pijawka (1977).

The basis of the Kates-Pijawka model as touched upon is composed of four sequential but overlapping phases, as described below and portrayed in Figure 23.1:

1 *Emergency Response*: The emergency response phase is immediately after the initial shock to the system and can take days to a few weeks. The key during this time is to get back to the essentials and basic triage operations. Search and rescue efforts, emergency feeding and housing for those in need, and debris clearance are primary activities.
2 *Restoration*: The restoration phase takes a few months and constitutes the restoration of utilities, housing, commercial and industrial infrastructure, and getting primary urban services up and running.
3 *Reconstruction (I)*: There are two reconstruction phases. The first phase takes place over many months to years and aims to rebuild to the pre-disaster level (structurally), restore the population to pre-disaster levels, and create a properly functioning society.
4 *Reconstruction (II)*: The second reconstruction phase, which occurs over the years, entails the memorialization or commemoration of the disaster, resulting in monuments and other forms and activities that commemorate the event (Kates & Pijawka, 1977).

Strengths of the Model

The Kates-Pijawka model provides a broad and easy-to-grasp conceptual understanding of the aspects of post-disaster recovery (Smith & Birkland, 2012). It does so by illustrating the phases (emergency response, restoration, recovery, reconstruction I, and reconstruction II) and the timing of the process. The model focuses on the outcomes of numerous recovery cases with overlapping phases. This overlap suggests that recovery phases may start earlier or later for many reasons and that there are factors that can slow the recovery or speed it up. It is important to note that it is just a model, a generalization based on various hazards and real-world case studies, using detailed recovery metrics such as building permits.

A critical finding was that phase two might take years, particularly when recovering cities include replanning. Replanning enables them to improve pre-disaster conditions and go beyond simply reconstruction or replacement of destroyed areas (Alexander et al., 2007). The term 'betterment,' which Kates and Pijawka referred to, can also imply developing new adaptations for safety, relocating parts of the recovering city, or bouncing forward for greater future resiliency.

Moreover, the model uses economic rationalism for decision-making. This approach considers different recovery rates for various economic classes (Alexander et al., 2007). Finally, the model emphasizes city status at the beginning of disaster recovery. It finds that cities with greater access to capital and financial institutions, greater concentrations of wealth, and relatively better economic status recover the fastest (Alexander, 2004).

Weaknesses of the Model

One of the most repeated criticisms of the Kates-Pijawka model is the simplicity of ordering the phases, which ironically is also one of its strengths. But as Hogg (1980) declared, the recovery process is not in a definite order. Alexander et al. (2007) argued that recovery phases are more complicated because of factors, such as vulnerability, political and geographical connectedness, and evolving modern approaches. Similarly, other researchers considered the recovery process as a complex array of overlapping activities, pre-disaster planning results, access to pre- and post-disaster resources and power, institutional deficiencies, and dependence on adaptive post-disaster actions (Alesch et al., 2001; Berke et al., 1993; Bolin & Bolton, 1983).

Another criticism of the model is that the study examined only post-disaster situations (Alexander et al., 2007). However, previous disaster experiences have shown that the recovery process is also highly dependent on pre-disaster vulnerabilities, weaknesses, and strengths and can accelerate past trends. Yet, the past does not necessarily dictate the future (Kates et al., 2006).

Minimizing the role of the community and the effects of different community members and their assistance represent additional criticisms posited by Smith and Birkland (2012). For instance, each community's recovery time and activities are unique since people are not a homogenous group with similar interests and needs (Quarantelli, 1989). Aguirre and Lane (2019) argued that the Kates-Pijawka model does not include the emotional and psychological recovery of disaster-affected populations. Moreover, it does not consider political influences that accelerate or slow the recovery process (Alexander et al., 2007).

Yet another criticism is that Kates and Pijawka base their model primarily on earthquakes as case studies, which are sudden-impact disasters. Although the case studies in the original model were chosen carefully from a wide variety of settings, the model was not tested with other hazards (e.g., slow-onsets such as drought), impacts, or conditions and is, therefore, restricted in scope and applicability. Table 23.1 summarizes both the pros and cons of the model.

Table 23.1 Pros and Cons of the Kates-Pijawka Recovery Model

Pros	*Cons*
A broad concept of post-disaster recovery	Oversimplicity in the ordering of the phases
Illustrates the phases and timing of the recovery process	Overshadows the role of community
Considers cases of multiple types of hazards	Fails to include the emotional and psychological recovery of communities
Recognizes overlapping phases	Does not consider political influences
Provides an opportunity to improve the pre-disaster condition	Does not include pre-disaster vulnerabilities
Emphasizes the status of cities at the beginning of a disaster recovery	Does not include the cyclic impacts of resilience on phases in future shocks
Relies on economic rationalism	

Source: Table Created by authors.

A New Kates-Pijawka Model for a New Era

Through no fault of Kates and Pijawka, the very makeup of society and climate conditions have changed. These changes have resulted in the need for theoretical and practical adaptation. However, the fact that the model continues to play a prominent role in the disaster recovery discussion shows how important and robust it has been from the beginning. Its sustained strength warrants a critical review of how to reconfigure it for today's increasingly dynamic and diverse multi-hazard urban society.

As illustrated, there is a need to address some main weaknesses of the Kates-Pijawka model. The first is the lack of emphasis on the post-disaster context and the absence of the role of community vulnerabilities pre-disaster in the model. A second hindrance is the need for more attention to emotional healing. There is a need to address a community's psychological and sociological impacts in addition to its infrastructure and economics, as a community needs to be strong psychologically and socially (e.g., social networks and social capital) to return to stability. Third, the model has issues concerning equity. It does not consider the need to spread equity throughout the community, especially ensuring that marginalized community sectors have an equal share in the recovery process. Lastly, there is a need to include the concept of resilience in the model. The resiliency that is fostered after experiencing a disaster can lead to quicker recovery in cases of future shocks to an area.

Based on the strengths and weaknesses of the model, including the gaps its weaknesses exposed, we propose four additions. First, there must be an emphasis on the community's pre-disaster makeup. Second, the model needs to address the community's sociological and psychological recovery. Third, and most importantly, there needs to be an emphasis on pre- and post-disaster equity, with the notion being that all community members must benefit from the recovery process. Last, there must be an emphasis on the role of resiliency after a disaster and how it impacts pre-disaster communities' status for future shocks. We address these four changes separately and subsequently fold them into an updated model.

Model Update I: Emphasis on Pre-Disaster Status

Recovery must create a balanced society across social, economic, environmental, and governance aspects (Buckman & Rakhimova, 2020). Communities suffer from different types and levels of vulnerabilities pre-disaster that make the effects of the same disaster harsher for them compared to comparatively less vulnerable communities. Some communities or households are more socially vulnerable (e.g., class structure and gender issues), some are more politically vulnerable (e.g., the form/type of government), and some are more economically vulnerable (e.g., income level and government debt), which can lead to disparate impacts on different households (Kelman et al., 2011).

Model Update II: Emotional Recovery and Issues of Psychological and Sociological Stability

Since the 1970s, there has been a greater emphasis on the psychological well-being of communities post-disaster (Newnham et al., 2020). This concept should be more integrated into hazard analyses since, without it, resilient recovery cannot happen (Amirzadeh et al., 2022). Disasters almost always lead to trauma and psychological distress among survivors, and this emotional distress is often more as a result of the hardship of the recovery process and rebuilding than merely the initial impacts of the event (Hutton, 2001). Thus, the disaster recovery process must consider addressing anxiety and stress, mood difficulties, somatic complaints, and behavioral changes among survivors (Newnham et al., 2020).

It is essential to understand that although disasters impact individuals, they occur at a community level and thus demand collective neighborhood-level actions to address them (Hutton, 2001). Governments can facilitate the emotional recovery of affected communities and individuals in different ways. Some of the methods to faster psychological recovery mentioned in previous studies are improved government and social support (Dhar et al., 2020), ensuring the inclusion of all segments of society (Hutton, 2001), psychological support (Akat & Karataş, 2020), maintaining social relationships and connectedness (Thakur & Jain, 2020), a proper and precise recovery plan for restoring housing (Hutton, 2001), timely and adequate information (Tee et al., 2020), and reconstructing recreational buildings (Kono & Shinew, 2015).

Model Update III: Pre- and Post-Disaster Equity

It should be understood that a just society meets everyone's basic needs. Recovery then will not succeed unless both wealthy and disadvantaged neighborhoods receive equal financial and resource aid. A significant intensifier of disaster risk is the unfair distribution of resources pre- and post-disaster that leads to disproportional impacts of the same disaster across sub-groups and different societies (Kates et al., 2006). As a result, communities do not achieve disaster recovery equally, nor does recovery follow a clear and specific process across all cities (Sullivan, 2003).

As Kelman et al. (2011) noted, inequity comprises poor access to resources such as public services, political networks, and social networks. In addition, lack of awareness, proper management, poor adaptiveness (Rodin, 2014), social weaknesses (Roque et al., 2020), and community support (Pickett et al., 2013) also serve to increase community vulnerability and inequity.

Model Update IV: Resiliency

The first and last stage of the revamped model is the resiliency stage. This stage, in theory, would come before and continue through and after the commemorative stage, which is the fourth stage of the previous model. This stage considers previous disaster impacts and measures used to combat them. It emphasizes how well the community readjusts and reconfigures their society for each base (i.e., social, economic, environmental, and governance). We see the pre-development resiliency phase as the most important in the newly reconfigured model. It lays the foundation for how well a community will rebound from a similar shock to the system.

The New Model

Considering the literature, the proposed model considers four new phases in addition to the four original phases: pre-disaster status, equity, psychological recovery, and resiliency. These four phases are discussed in the following paragraphs.

In the new model, the four original overlapping phases introduced in the Kates-Pijawka Model—emergency, restoration, basic reconstruction, and major construction—remain largely the same in terms of critical tasks and goals. The emergency phase includes having primary urban services: getting back to essentials, search and rescue efforts, emergency feeding and housing of those in need, and debris clearance. The restoration phase includes restoring utilities, housing, commercial, and industrial infrastructure. The basic reconstruction phase includes structurally rebuilding to a pre-disaster level, replacing the population, and building a properly functioning society. Finally, the major construction or reconstruction II phase includes major constructions and commemorations that tend to occur later in the process of recovery. However, to revise the model, it is necessary to see if disaster recovery follows sequential overlapping phases based on 21st-century needs. First, it is essential to consider whether these phases are sequential for all communities. Second, it is vital to consider differences in the duration of overlaps for different communities.

Pre-disaster vulnerability and the inequitable distribution of opportunities, resources, and agency slow down the recovery process in some communities (Sobhaninia, 2023). Pre-disaster status includes the various socioeconomic qualities of a community before the disaster. These attributes, such as level of technology, affluence, and political capital, affect all or some of the recovery phases in communities according to the type and scale of their vulnerability. Furthermore, equity consists of the accessibility of various social, political, geographical, and economic resources for different people within a community or different communities compared to each other. The distribution of and access to these resources is one of the most critical determinants of a community rebounding after a shock (i.e., disaster recovery).

In the updated model, emotional recovery is another determining factor that can facilitate or impinge disaster recovery. This new model proposes that the recovery process includes two overlapping parts that affect each other—physical and emotional recovery. To reduce the stress caused by a disaster and the feeling of uncertainty due to the forced relocation, the recovery process should include the emotional recovery time needed for different individuals and communities as an inseparable component (Newnham et al., 2020) in addition to the four physical recovery phases in the Kate-Pijawka model.

Physical recovery, as mentioned, consists of four similar overlapping phases as introduced in the Kates-Pijawka model: emergency, restoration, basic reconstruction, and major constructions—which are different among each community. Similarly, the psychological impacts of a disaster are rarely evenly distributed among impacted populations. Emotional recovery might slow down the process for some households or even some communities within an affected area. Psychological recovery accounts for the time that it takes for affected people to counterbalance the feelings associated with personal trauma and loss of place, identity, community, assets, and capital. This recovery process depends on the individuals and communities and may speed up or slow the recovery process depending upon idiosyncratic community characteristics. Nevertheless, such recovery is essential since people need time to relieve themselves from the sudden shock and possible loss of friends, family, place, and possessions.

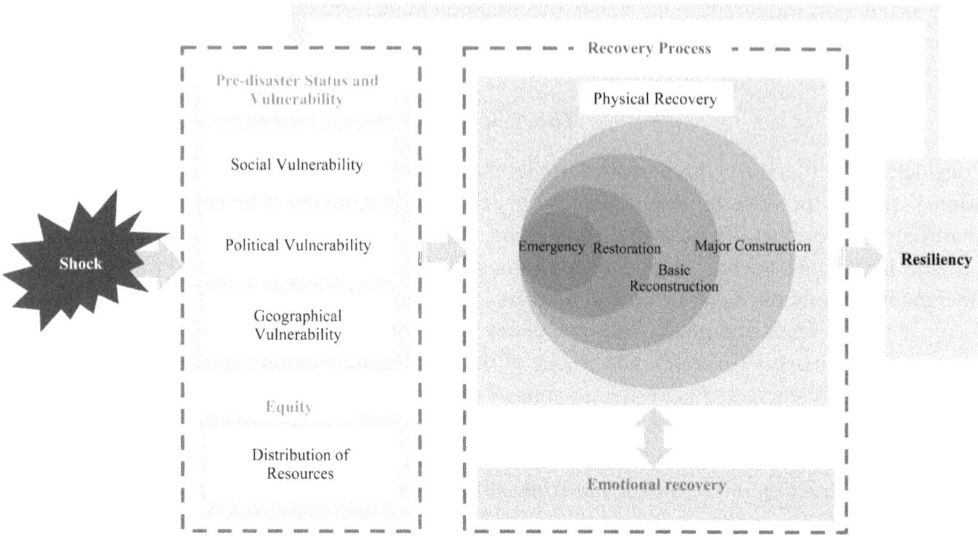

Figure 23.2 The Reconfigured Model of Disaster Recovery

Source: Figure by authors/no sources to cite.

Even if successful disaster recovery happens, communities and cities might become more resilient to future shocks. Therefore, resilience is both the starting and ending point of the newly proposed model. This framing is because previous recovery processes often shape community resiliency and promote faster recovery from future shocks. A successful recovery process can not only help communities and cities recover but also help them become more resilient. Thus, by going through a disaster and learning and recovering from it, communities can adapt to possible future shocks and shorten future recovery time by increasing their resilience.

Having been fully assembled, Figure 23.2 presents the reconfigured Kates-Pijawka model in more detail. It includes all the recovery components involved in disaster recovery based on the literature, including the effects of pre-disaster vulnerabilities (e.g., social, political, and geographical) and equity (e.g., distribution of opportunities and resources) that affect the recovery process—as well as physical and emotional recovery. Physical recovery includes the overlapping four phases (i.e., emergency, restoration, basic reconstruction, and major construction), followed by the resiliency component.

Conclusion

In 1977, an onus was placed on returning to equilibrium as quickly as possible. This equilibrium focused on physical infrastructure and economic recovery. Unfortunately, these well-intentioned goals have become insufficient in the 21st century, and evidence shows returns overwhelmingly favored the status quo and the 'haves' over the 'have nots.' In the four-plus decades since the development of the Kates-Pijawka model, the way we view disaster recovery, urban planning, coupled socio-ecological relationships, societal dynamics, and the climate pressures that befall many urban areas has changed. These changes warrant a new way forward.

Recovery no longer uses a top-down approach concerned with merely infrastructure redevelopment and economic stability. Instead, concern has shifted to a recovery that benefits all segments of community. Decision-makers now recognize the need to address equity, social and emotional recovery, and pre-disaster planning. This chapter folded these issues into an updated Kates-Pijawka model that is current, relevant, and deployable in today's policy and practitioner settings. In doing so, we highlighted the importance of pre-disaster resiliency measures that acknowledge and include both infrastructure and equity issues. Furthermore, we have emphasized social and emotional recovery after an event, as a society can only recover physically if its people have the emotional strength to recover.

Rather than throwing the model out, we have revamped it to reflect present-day and emerging concerns. We believe that this revised model is better able to address current and future disaster recovery efforts.

Acknowledgment

The authors would like to give special thanks to Dr. David Pijawka, one of the two original authors of the Kates-Pijawka model. Through our discussions with Dr. Pijawka, we gained valuable insights into the original model that helped us adapt it as laid out in this chapter.

References

Aguirre, B. E., and Lane, D. (2019). "Fraud in disaster: Rethinking the phases." *International Journal of Disaster Risk Reduction*, 39. DOI: 10.1016/j.ijdrr.2019.101232

Akat, M., and Karataş, K. (2020). "Psychological effects of COVID-19 pandemic on society and its reflections on education." *Electronic Turkish Studies*, 15(4). DOI: 10-7827/TurkishStudies 44336

Aldrich, D.P. (2012). *Building Resilience: Social Capital in Post-Disaster Recovery*". Chicago, IL: University of Chicago Press.

Alesch, D. J., Holly, J. N., Mittler, E., and Nagy, R. (2001). *Organizations at Risk: What Happens When Small Businesses and Non-for-Profits Encounter Natural Disasters.* Fairfax, VA: Public Entity Risk Institute.

Alexander, D. (2004). "Planning for post-disaster reconstruction," In *I-Rec 2004 International Conference 'Improving Post-Disaster Reconstruction in Developing Countries'* (pp. 1–12).

Alexander, D., Davidson, C. H., Fox, A., Johnson, C., and Lizzaralde, G. (2007). *Post-Disaster Reconstruction, Meeting Stakeholder Interests.* Florence: Firenze University Press.

Amirzadeh, M., Sobhaninia, S., and Sharifi, A. (2022). "Urban resilience: A vague or an evolutionary concept?" *Sustainable Cities and Societies*, 81. DOI: 10.1016/j.scs.2022.103853

Berke, P., and Campanella, T. J. (2006). "Planning for post-disaster resiliency." *The Annals of the American Academy of Political and Science*, 604(1): 192–207. DOI: 10.1177/0002716205285533

Berke, P., Ericksen, N., Crawford, J., and Dixon, J. (2002). "Planning and indigenous people: Human rights and environmental protection." *New Zealand Journal of Planning Education and Research*, 22(2): 115–134.

Berke, P. R., Kartez, J. D., and Wenger, D. E. (1993). "Recovery after disaster: Achieving sustainable development, mitigation, and equity." *Disasters*, 17(2): 93–109.

Bolin, R. C., and Bolton, P. A. (1983). "Recovery in Nicaragua and the U.S.A." *International Journal of Mass Emergencies and Disasters*, 1(1): 125–144.

Buckman, S., and Rakhimova, N. (2020). "The resilient City: Combating uncertainty within an urban context." In B. Hagen and D. Pijawka (eds.), *Sustainability in the 21st Century: Pathways, Programs, and Policies* (Third Edition). Dubuque, IA: Kendall Hunt Publishing, pp. 129–148.

Burby, R. J., Beatley, T., Berke, P. R., Deyle, R. E., French, S. P., Godschalk, D. R., Kaiser, E. J., Kartez, J. D., May, P. J., Olshansky, R., Paterson, R. G., and Platt, R. H. (1999). "Unleashing the power of planning to create disaster-resistant communities." *Journal of the American Planning Association*, 65(3): 247–258.

Chamlee-Wright, E., and Storr, V. H. (2009). ""There's No Place like New Orleans": Sense of place and community recovery in the ninth ward after Hurricane Katrina." *Journal of Urban Affairs*, 31(5): 615–634, DOI: 10.1111/j.1467-9906.2009.00479.x

Comerio, M. C. (2013). "Housing Recovery in Chile: A Qualitative Mid-Program Review" Department of Architecture, Pacific Earthquake Engineering Research Center Report. Berkeley: University of California.

Cox, R. S., and Perry, K. M. E. (2011). "Like a fish out of water: Reconsidering disaster recovery and the role of place and social capital in community disaster resilience." *American Journal of Community Psychology*, 48(3): 395–411. DOI: 10.1007/s10464-011-9427-0

Dhar, B. K., Ayittey, F. K., and Sarkar, S. M. (2020). "Impact of COVID-19 on psychology among the university students." *Global Challenges*, 4(11): 2000038.

Diaz, J., and Dayal, A. (2008). "Sense of place: A model for community based psychosocial support programs." *The Australasian Journal of Disaster and Trauma Studies*, 63(8): 820–827.

Felix, D., Branco, J. M., and Feio, A. (2013). "Temporary housing after disasters: A state of the art survey." *Habitat International*, 40: 136–141.

Hill, A. A., and Keys-Mathews, L. D. (2005). "Geography, disaster recovery and remote sensing," *3rd International Workshop for Remote Sensing for Post Disaster Response*.

Hogg, S. J. (1980). "Reconstruction following seismic disaster in Venzone, Friuli." *Disasters*, 4(2): 173–185.

Hutton, D. (2001). *Psychosocial Aspects of Disaster Recovery*. Toronto: Institute for Catastrophic Loss Reduction.

Isumi, M., Nomura, N., and Shibuya, T. (1985). "Simulation of post-earthquake restoration of lifeline systems." *International Journal of Mass Emergencies & Disasters*, 3(1): 87–105.

Kates, R. W. and Pijawka, D. (1977). "From rubble to monument: the pace of reconstruction," In Haas, J. E., Kates, R. W., and Bowden, M. J. (eds.), *Reconstruction Following Disaster*. Cambridge, MA: MIT Press, pp. 1–23.

Kates, R. W., Colten, C. E., Laska, S., and Leatherman, S. P. (2006). "Reconstruction of New Orleans after Hurricane Katrina: A research perspective." *Proceedings of the National Academy of Science*, 103(40): 14653–14660.

Kelman, I., Lewis, J., Gaillard, J. C., and Mercer, J. (2011). "Participatory action research for dealing with disasters on Islands." *Island Studies Journal*, 6(1): 59–86.

Kono, S., & Shinew, K. J. (2015). "Roles of leisure in the post-disaster psychological recovery after the Great East Japan Earthquake and tsunami." *Leisure Sciences*, 37(1): 1–19.

Lindell, M. K. (2013). "Disaster studies." *Current Sociology*, 61(5–6): 797–825.

Miles, S. B., Burton, H. V., and Kang, H. (2019). "Community of practice for modeling disaster recovery." *Natural Hazards Review*, 20(1). DOI: 10.1061/(ASCE)NH.1527-6996.0000313

Mileti, D. (1999). *Disasters by Design: A Reassessment of Natural Hazards in the United States*. Washington, DC: Joseph Henry Press.

Mohd, R., Abu Bakar, N., Hassan, S., and Hussain, A. H. (2017). "Sustainable post-disaster recovery plan for flood victims in Gua Musang and Kuala Krai, Kelantan." *Pertanika Journal of Social Sciences and Humanities*, 25: 1–11.

Nakagawa, Y., and Shaw, R. (2004). "Social capital: A missing link to disaster recovery." *International Journal of Mass Emergencies and Disasters*, 22(1): 5–34.

Newman, P., Beatley, T., and Boyer, H. (2017). *Resilient Cities: Overcoming Fossil Fuel Dependence*. Washington, DC: Island Press.

Newnham, E. A., Gao, X., Tearne, J., Guragain, B., Jiao, F., Ghimire, L., and Chan, E. Y. (2020). "Adolescents' perspectives on the psychological effects of natural disasters in China and Nepal." *SAGE Journals*, 57(1): 197–211.

Oliver-Smith, A. (1991). "Successes and failures in post-disaster resettlement." *Disasters*, 15(1): 12–23.

Pickett, S. T. A., Cadenasso, M. L., and McGrath, B. (2013). *Resilience in Ecology and Urban Design: Linking Theory and Practice for Sustainable Cities*. London, UK: Springer, Volume 3.

Quarantelli, E. L. (1989). "Conceptualizing disasters from a sociological perspective." *International Journal of Mass Emergencies and Disasters*, 7: 243–251.

Rodin, J. (2014). "The Resilience Dividend," *Public Affairs*, New York. ISBN: 978-1-61039-470-3

Roque, A. D., Pijawka, D., and Wutich, A. (2020). "The role of social capital in resiliency: Disaster recovery in Puerto Rico." *Risk, Hazards, & Crisis in Public Policy*, 11(2): 204–235.

Rumbach, A. (2016). "Decentralization and small cities: Towards more effective urban disaster governance?" *Habitat International*, 52: 35–42.

Rupakhety, R., and Olafsson, S. (2019). *Geotechnical, Geological and Earthquake Engineering*. London, UK: Springer International Publishing. ISBN 978-3-319-62099-2

Smith, G., and Birkland, T. A. (2012). "Building a theory of recovery: Institutional dimensions." *International Journal of Mass Emergencies and Disasters*, 30(2): 147–170.

Sobhaninia, S. (2023). "Does social cohesion accelerate the recovery rate in communities impacted by environmental disaster in Puerto Rico? An analysis of a community survey." *Environmental Advances*, 13. DOI: 10.1016/j.envadv.2023.100400

Sobhaninia, S., Amirzadeh, M., Lauria, M., and Sharifi, A. (2023). "The relationship between place identity and community resilience: Evidence from local communities in Isfahan, Iran." *International Journal of Disaster Risk Reduction*, 90, Article 103675. DOI: 10.1016/j.ijdrr.2023.103675

Sobhaninia, S., Buckman, S. T., and Schupbach, J. (2023). "The need for creative placemaking in new development areas: An analysis of the real estate developments of Qom, Iran." *International Journal of Real Estate Studies*, 17(1): 24–39.

Sullivan, M. (2003). "Integrated recovery management: A new way of looking at a delicate process." *The Australian Journal of Emergency Management*, 18(2): 4–27.

Tee, M. L., Tee, C. A., Anlacan, J. P., Aligam, K. J. G., Reyes, P. W. C., Kuruchittham, V., and Ho, R. C. (2020). "Psychological impact of COVID-19 pandemic in the Philippines." *Journal of Affective Disorders*, 277: 379–391.

Thakur, V., and Jain, A. (2020). "COVID 2019-suicides: A global psychological pandemic." *Brain, Behavior, and Immunity*, 88: 952.

Townshend, I., Awosoga, O., Kulig, J., and Fan, H. (2015). "Social cohesion and resilience across communities that have experienced a disaster." *Natural Hazards*, 76(2): 913–938. DOI: 10.1007/s11069-014-1526-4

Wu, J. Y., and Lindell, M. K. (2004). "Housing reconstruction after two major earthquakes: The 1994 Northridge earthquake in the United States and the 1999 Chi-Chi earthquake in Taiwan." *Disasters*, 28(1): 63–81.

24

EVOLUTION OF RECOVERY STUDIES AND RECOVERY DURATION

Saadia Majeed

Introduction

Disaster recovery has been interpreted and defined differently across various disciplines. To comprehend the evolution of recovery studies, it is critical to acknowledge the semantic subtleties related to the term 'recovery' and its application in disaster operations and literature. An investigation into the vocabulary surrounding the term recovery reveals that multitudes of words, including restoration, reconstruction, and reform, have Latin origins and convey the meaning of 'again' or 'back to the starting point' (Quarantelli, 1998). In these cases, the prefix 're-' indicates a return to the original state or normalcy. However, scholars such as Davis and Alexander (2016) have suggested that the prefix bears negative connotations as well, such as, removal or reversal. Nevertheless, the prevailing discourse in disaster literature and operations has predominately embraced the term recovery in a positive light, mainly as a means of attempting to revert to its initial state.

The discourse of recovery is further enriched by a range of technical terms, including relief, resilience, restoration, reconstruction, reform, 'build back better,' and sustainability; all of which aim to preserve, restore, or even surpass a perceived state of normalcy. Various international organizations have formulated their own operational definitions of recovery to suit their respective agendas, and they have played critical roles in shaping and directing disaster recovery concepts. Context-specific and multifaceted definitions of recovery serve as working definitions for their institutional purposes. For instance, the United Nations Office for Disaster Risk Reduction (UNDRR) (United Nations Office for Disaster Risk Reduction (UNDRR), 2017) developed a definition of recovery that aligns with their specific objectives:

> The restoring or improving of livelihoods and health, as well as economic, physical, social, cultural and environmental assets, systems and activities, of a disaster-affected community or society, aligning with the principles of sustainable development and 'build back better', to avoid or reduce future disaster risk.
>
> *(23)*

Global platforms, such as the International Recovery Platform (IRP) and the Global Facility for Disaster Reduction and Recovery (GFDRR) of the World Bank, have developed comprehensive

DOI: 10.4324/9781003414834-33

Disaster Recovery Framework Guides (GFDRR, 2020; IRP, 2012) to emphasize their operational activities on the restoration of livelihoods and living conditions of affected communities and, where feasible, reconstruction and improvement of facilities through the build back better approach.

Similarly, the International Federation of Red Cross (IFRC) and Red Crescent Societies conceptualized recovery to include both rehabilitation and reconstruction, affirming that reconstruction is an element of recovery (IFRC, 2023). Their comprehensive approach aligns with the broader goal of resilient and sustainable recovery efforts after disasters. While the UNDRR definition does not explicitly mention the term reconstruction, it advocates for implementing disaster risk reduction measures and adhering to the build back better principle.

As evidenced by the discourse, the operational aspect of the recovery process encompasses rehabilitation and reconstruction efforts that commence immediately after the emergency phase ends. The guidelines provided by IRP, GFDRR, and IFRC serve as invaluable resources for understanding the operational aspects of effective disaster recovery and reconstruction. Similar recovery guidelines and frameworks establish clear institutional responsibilities, the implementation of pre-existing strategies, and engage the public in recovery efforts to ensure a coordinated and effective process.

A content analysis of mainstream disaster recovery organizations and their respective guidelines and frameworks was conducted. The analysis reveals several critical activities associated with the recovery process. A list of documents used for the content analysis is provided below, and Table 24.1 presents the salient activities and components of recovery that were identified.

- UNDRR Sendai Framework for Disaster Risk Reduction (SFDRR) 2015–2030,
- IFRC Disaster Law Programme and Framework for Community Resilience,
- Asian Disaster Preparedness Center (ADPC) Guidelines for Recovery,
- GRDRR Global Risk Reduction and Resilience Framework 2011–2020,
- World Bank Guidance Notes for Damage and Reconstruction Needs Assessment (DRNA),

Table 24.1 Salient Activities and Components Identified in Major Frameworks for Disaster Recovery

Resilience	*Risk reduction*	*Capacity building*	*Coordination*	*Environmental sustainability*
Infrastructure rehabilitation	Build back better	Humanitarian assistance	Livelihood restoration	Preparedness
Social protection	Early warning systems	Monitoring and evaluation	Community engagement	Inclusiveness
Reconstruction	Legal and institutional frameworks	Rehabilitation	Vulnerability	Disaster preparedness
Recovery planning	Adaptation	Mitigation	Emergency shelter	Sustainable development
Risk management	Equity	Hazard mapping	Public awareness	PDNA
Asset protection	Economic recovery	Community resilience	Risk assessment	Community-based recovery
Public health response	Education support	Relief operations	Infrastructure resilience	Reconstruction planning
Emergency response	Psychological support	Social cohesion	Supply chain management	Stakeholder engagement

Source: Table created by author.

- European Civil Protection and Humanitarian Aid Operations (ECHO) Guidelines for Post Disaster Needs Assessment (PDNA),
- IRP Guidance note on recovery, and
- Sphere Handbook Humanitarian Charter and Minimum Standards for Disaster Response.

Together, the activities and components in Table 24.1 provide a general understanding of the recovery process and encompass a range of terminologies linked to the concept of recovery. The interplay of these terms affects policy, legal matters, and power dynamics, which is evident in recovery process. As experts engage in discussions, debates, and redefinitions highlighting the interconnectedness of various sectors with differing operational responsibilities, the concept continuously evolves.

Nevertheless, it is possible to classify these activities into two primary categories: (1) operational needs, and (2) survivor needs. Davis and Alexander (2016), as well as Quarantelli (1998), have emphasized this binary viewpoint in their foundational works on disaster recovery. They argue that the needs of survivors often diverge from the priorities of political leaders, international financial institutions, non-governmental organizations (NGOs), or construction contractors. Consequently, the meaning of disaster recovery can differ substantially when viewed from the vantage points of those directing recovery efforts and those on the receiving end (Juran, 2015).

From an operational perspective, disaster recovery involves the restoration and enhancement of the affected community's living conditions and environment in the aftermath of a disaster. Governmental and NGOs work tirelessly to implement improved safety measures across all sectors, fortify weakened and damaged structures, and rebuild crucial local institutions with a particular emphasis on strengthening local governments (Alexander, 2007). While external aid may be necessary, it is essential for governments to maintain their authority and avoid inappropriate assistance that undermines their local capacity. To achieve successful disaster recovery, governments must possess sufficient resources, support, a well-thought-out plan, effective management, strong leadership, and a clear future vision. Democratic participation, transparency, and a good human rights record are also vital for creating a trustworthy relationship between the government and the population (Comfort et al., 1999; Diefendorf, 1993; IFRC, 2023; Quarantelli, 1998).

From the survivors' perspective, the primary goal of disaster recovery is to reestablish a sense of familial normalcy in the aftermath of a tragedy, taking into account the loss of life, physical injuries, disabilities, and displacement from homes, jobs, and personal identity. It is crucial for survivors to have a voice in decisions that affect their recovery, and the rebuilding process should incorporate livelihoods (Alexander, 1993; Horney, 2016). The resumption of livelihoods is critical to the success of survivors' recovery efforts. They desire not simply houses but actual homes, protection from future hazards, and the rehabilitation or reconstruction of cultural treasures. Families require safe schools, sufficient healthcare facilities, and essential infrastructure. Whenever feasible, rebuilding should transpire in the vicinity of the original location unless there are significant safety concerns (IFRC, 2023; Juran, 2015; Quarantelli, 1998; Silver and Grek-Martin, 2015).

Evolution of Recovery Studies in the Disaster Literature

The previous discussion highlighted the multifaceted and evolving nature of the recovery process. What was initially perceived as a straightforward process primarily aimed at reconstructing physical infrastructure underwent substantial transformations over time. The current discourse on recovery reflects a deeper understanding of socio-cultural attitudes (Blaikie et al., 1994; Davis, 1978). Although much research has been done on various aspects of recovery, only Davis and Alexander

(2016), in their recent seminal publication on recovery, documented the historical development of recovery in the disaster literature. Building upon their work, we adopted a timeline approach to offer a thorough understanding of the evolution of recovery in the disaster literature.

532–588 AD: (Re)Construction of Hagia Sophia

Hagia Sophia, the historic building in Istanbul, Turkey, was built between 532 and 537 AD. Unfortunately, the building's dome collapsed in 586 AD due to an earthquake. However, the dome was reconstructed in 588 AD utilizing an innovative technique with an iron ring beam. This technique was designed to better withstand the weight of the central dome and the surrounding loading pressure. Later, this technique was employed in cathedral architecture across Europe and can be considered a component of resilient recovery (Davis, 1983).

1666: Great Fire of London

The Great Fire of London in 1666 marked a pivotal moment in urban recovery and building regulation, resulting in the implementation of the London Building Acts, which continue to have a significant impact today. The Acts mandated fire-resistant construction plans, safety legislations, and hazard-resistant urban planning, necessitating advancements in building materials and techniques. In response to the devastating effects of fire, regulations were put in place requiring buildings to be constructed with fire-resistant materials such as brick and stone, along with wider streets and open spaces to inhibit the rapid spread of fires. Additionally, safety measures were introduced, including regulations for installing firefighting equipment and guidelines for constructing fire-resistant walls, roofs, and floors (Davis, 1983). The Acts also influenced urban planning by incorporating setbacks, fire breaks, and creating fire-resistant zones within the city. Consequently, rebuilding after the Great Fire led to advancements in building techniques and materials, setting the stage for modern building codes and regulations focused on safety, resilience, and fire prevention.

1755: Lisbon Earthquake and Tsunami

The 1755 Lisbon earthquake was undoubtedly a pivotal moment in the history of urban planning and reconstruction. Sebastião José de Carvalho e Melo, more commonly known as the First Marquis of Pombal, played a significant role in reconstruction following this catastrophic event. Pombal recognized the necessity for comprehensive safety measures and innovative construction techniques to mitigate the impact of future earthquakes. He enlisted the help of military engineers and planners to aid in the reconstruction process. One of the key innovations introduced was using timber-framed construction techniques specifically designed to enhance earthquake resistance. These measures were part of a broader strategy to rebuild Lisbon and ensure that the city was better prepared for potential future disasters. Pombal's efforts laid the groundwork for Lisbon to emerge as a leader in urban reconstruction and resilience, setting an example for other cities facing similar challenges (Araujo, 2021).

1871: Great Chicago Fire

The Great Chicago Fire of 1871 prompted significant advancements in architecture and engineering, especially in fire-resistant building design. American architect and engineer William Le Baron Jenney played a pivotal role in this transformation. Following the catastrophic damage of an estimated

17,000 structures caused by the fire, Jenney recognized the need for buildings that could withstand such disasters. Utilizing his engineering expertise, he pioneered the use of steel-frame construction in the design of skyscrapers. By employing steel, Jenney created structures that were not only taller but also more resilient to fires compared to traditional masonry buildings. One of Jenney's most important contributions to this new architectural paradigm was the Home Insurance Building, completed in 1885. This structure is widely regarded as the world's first skyscraper, featuring a steel skeleton frame that supported its weight and provided unparalleled fire resistance (Condit, 1964; Davis, 1983).

20th Century: A Paradigm Shift in the Disaster Recovery Discourse

In the 20th century, a significant transformation occurred in disaster recovery approaches. Prior to this period, emphasis was generally placed on restoring physical infrastructure and economic activity as soon as possible, often at the expense of marginalized communities. This approach frequently reinforced existing social inequalities and privileges. Thus, following World War II and the emergence of social science-based approaches, there was a growing awareness of the importance of addressing the social and human dimensions of disasters. Scholars and practitioners began advocating for more inclusive and comprehensive recovery strategies that acknowledge the diverse needs and vulnerabilities of affected populations.

This shift led to the development of frameworks and methodologies rooted in disciplines such as sociology, anthropology, geography, and psychology. These approaches emphasized community participation, social justice, and equity in disaster recovery efforts. The transition towards social science-based approaches marked a pivotal moment in disaster recovery, paving the way for more equitable, inclusive, holistic, and sustainable strategies aimed at building resilience and reducing social disparities in the face of disasters (Quarantelli, 1998, 1989).

1917: Halifax Ship Explosion

Samuel Henry Prince was instrumental in the development of the early social sciences of disaster recovery, particularly through his work in ministering survivors of the Halifax munitions ship explosion in 1917. His efforts laid the groundwork for understanding recovery from a social perspective. In his 1920 landmark publication, "The Aftermath of the Halifax Disaster," Prince examined social and psychological dimensions of recovery following the catastrophic explosion. He investigated not only the physical reconstruction efforts but also the emotional and social aspects of healing and rebuilding communities that were devastated by the disaster (Dynes, 2006; Prince, 1920). Prince's research emphasized the importance of providing psychosocial support to the survivors, addressing their emotional trauma, and fostering community resilience. Prince's work highlighted the need for a comprehensive approach that encompasses the well-being and social cohesion of affected populations—well beyond mere physical reconstruction (Dynes, 2006). By pioneering this social perspective on disaster recovery, Prince significantly contributed to the development of the field, paving the way for subsequent research and practice that prioritize the human and social dimensions of disasters.

1933: Long Beach Earthquake

The Long Beach earthquake of 1933 was a sobering reminder of the catastrophic consequences of seismic activity, emphasizing the critical need for proactive measures to reduce risk. The disaster prompted the introduction of the Field Act by Assemblyman Charles Field of California in 1934, which emerged as a catalyst for similar initiatives worldwide. The Field Act inspired global efforts to enhance building codes and improve structural resilience in earthquake-prone regions, ultimately contributing to the development of earthquake-resistant building codes (Olson, 2003).

1945s–1950s: Post-War Reconstruction

The resurgence of German cities after World War II serves as a compelling testament to their capacity for resilience and determination in the face of catastrophe. Contrary to initial projections of a four-decade reconstruction period, these cities had not only surmounted their housing challenges but had also restored their iconic landmarks, thereby reasserting their individual identities by the late 1950s. Jeffry M. Diefendorf's (1993) work 'In the Wake of War' unveils the complexities of this extraordinary recovery effort. He explores the factors driving this remarkable reconstruction, emphasizing the pivotal roles played by individuals, institutions, and strategic decision-making processes. A central tenet highlighted by the German experience is the judicious utilization of resources. By leveraging existing knowledge, expertise, and infrastructure, the rebuilding process was significantly expedited. Moreover, the practical nature of decision-making played a crucial role in the recovery process. By prioritizing essential infrastructure, housing, and cultural monuments, resources were allocated efficiently, enabling rapid progress rebuilding critical components of the cities. This strategy allowed the authorities to address immediate needs while also laying the groundwork for long-term recovery.

After World War II, Otto H. Königsberger, an architect and planner in Britain, advocated dynamic adaptation and community involvement in recovery planning. His concept of 'action planning' emphasized avoiding paternalism and encouraging local communities to participate actively in their recovery (Davis, 1978). Königsberger contributed to rebuilding efforts by designing and planning for the restoration of cities and communities that suffered significant destruction. His approach to post-war reconstruction was grounded in humanitarian principles, aiming to create spaces that addressed not only the physical needs of the population, but also the social and psychological aspects of rebuilding communities that had experienced trauma during the disaster.

1963: Foundation of Disaster Research Center

The Disaster Research Centre (DRC) was founded in 1963 by sociologists E. L. Quarantelli, Russell Dynes, and Eugene Haas at The Ohio State University, a pioneering institution in the field of disaster research. The center was the first in the world to focus on social science dimensions of disasters, and its founders played a crucial role in advancing the understanding of recovery and reconstruction processes. Their work, which was characterized by systematic studies, challenged conventional assumptions by emphasizing the nonlinear progression of recovery processes (Quarantelli, 1977). Collectively, the founders produced several publications that contested established norms within the field. It should be noted that the DRC is still very active in disaster research and is now located at the University of Delaware.

1970: Bhola Cyclone

The 1970 Bhola Cyclone was a catastrophic tropical cyclone that wreaked havoc in Bangladesh and the Indian state of West Bengal. The Bhola Cyclone remains the deadliest tropical cyclone in recorded history. Following this disaster, the Cyclone Preparedness Programme (CPP) was established and enhanced through the collaborative efforts of the Government of Bangladesh, United Nations, International Red Cross, and Bangladesh Red Crescent Society. CPP has been a key component of Bangladesh's Early Warning System, and its community-based approach has been identified as a critical development in building resilience to disasters Haque et al. (2022).

1976: Guatemala Earthquake

In the aftermath of the Guatemala earthquake, Fred Cuny, an urban planner, played a critical role in shaping post-disaster recovery. He advocated for community involvement and adaptive solutions that had a significant impact on disaster recovery and the provision of safe, low-income housing globally (Cuny, 1983). Cuny's ideas and methodologies have been adopted by humanitarian organizations and disaster response agencies worldwide, influencing their practices and extending beyond specific events to contribute to a broader discussion on effective community-based disaster recovery.

1984: Bhopal Gas Tragedy

The disaster at the Union Carbide Pesticide Plant in Bhopal, India, on December 3, 1984, was the result of a gas leak of methyl isocyanate. The Bhopal disaster caused immediate deaths and long-term health and reproductive issues for thousands of people, making it one of the most severe industrial disasters in history. Consequences of the disaster reverberated internationally and endured for an extended period. As a result, Bhopal sparked a new discourse on industrial crises, crisis management, manmade disasters, the relationship between technology and the environment, and the relationship between technology and society. This has led to the initiation of new research studies and educational programs focused on crisis prevention and management.

1986: Chernobyl Disaster

The catastrophic accident at the Chernobyl Nuclear Power Plant in April 1986 in present day Ukraine remains the most severe nuclear power plant disaster to date. The disaster resulted in widespread contamination and significant health impacts owing to the release of substantial amounts of radioactive material. This tragedy not only had profound global implications but also led to far-reaching consequences in the areas of nuclear safety, reactor design, and emergency response protocols, ultimately shaping public opinion and policy on nuclear energy (Haynes and Bojcun, 1988).

This non-exhaustive timeline was not intended to capture every disaster event, and that task is beyond the scope of this chapter. Rather, our intention was to emphasize some of the most critical occurrences that have progressed of the conceptualization and operationalization of recovery. Prominent figures, such as E. L. Quarantelli, Russell Dynes, John F. C. Turner, Nicholas N. Ambraseys, Robert W. Kates, John Eugene Haas, Martyn J. Bowden, Kathleen Tierney, Dennis S. Mileti, Brenda D. Phillips, Ian Davis, David Alexander, Maureen Fordham, Kathleen Tierney, Tricia Wachtendorf, Ben Wisner, and Susan L. Cutter, have played a pivotal role in transforming the field of disaster recovery. Their innovative perspectives and techniques have greatly influenced the understanding and practice of recovery, leaving a long-lasting impact that extends into the 21st century.

The 21st Century: The Paradigm Shift Continues

The 21st century witnessed a remarkable transformation driven by the profound consequences of the 2001 Gujarat earthquake and the 2004 Indian Ocean Tsunami. In the aftermath of these disasters, immediate and coordinated efforts were undertaken at both national and international levels. Rescue and relief teams were promptly deployed in the affected areas to search for survivors, provide medical assistance, and distribute essential supplies. Both events necessitated massive recovery efforts that focused on rebuilding homes, infrastructure, and public facilities. Collaborative efforts have been directed at constructing earthquake-resistant structures to minimize the impact of future seismic events, and coastal planning emerged as a critical endeavor. Moreover, attention has been devoted to addressing the social and economic dimensions of recovery, encompassing support for livelihood restoration, education, and healthcare in the affected communities (Twigg, 2007).

The notion of 'build back better' gained prominence in the aftermath of the Gujarat earthquake as especially as part of recovery efforts following the 2004 Indian Ocean Tsunami (Juran, 2018). According to Khasalamwa (2009), this approach involves utilizing the reconstruction process to enhance a community's physical, social, environmental, and economic conditions, thereby fostering a more resilient community. Resilience, as defined by Twigg (2007), became a key focus, with build back better advocating for comprehensive and holistic recovery aimed at regenerating safer and stronger communities for the future. Later, these principles were integrated into mainstream international disaster risk reduction frameworks (Khasalamwa, 2009).

The devastating Great Hanshin-Awaji Earthquake of 1995, which occurred in Kobe, Hyogo, compelled the Japanese government and its citizens to address the immediate issue of disaster risk reduction. In the aftermath of this catastrophic event, the Hyogo Framework for Action (HFA) emerged as a comprehensive approach for enhancing global disaster preparedness. The HFA encompasses critical areas such as disaster governance, risk identification and assessment, monitoring and early warning systems, knowledge management and education, and preparedness for effective response and recovery. The primary objective of the HFA was to establish strong governance systems for disaster management by developing legislative and institutional frameworks as well as disaster policies aimed at risk reduction in support of the HFA's key priorities. This extensive strategy was formulated after the Second World Conference on Disaster Reduction in 2005.

The HFA does not explicitly endorse the building back better concept, but emphasizes a resilience-based approach to disaster recovery. Following the implementation of the HFA, disaster management efforts were significantly enhanced, resulting in increased comprehension of the consequences of disasters. The HFA also promoted cross-cutting initiatives, including capacity-building, technology transfer, a multi-hazard approach, gender and cultural diversity perspectives, and community and volunteer participation. The post-HFA approach to disaster risk reduction was formulated and proposed at the Third World Conference on Disaster Reduction, held in Sendai, Japan, in 2015. Following extensive evaluations and reviews of the HFA, and taking into account past disaster lessons learned, representatives from various countries came together to develop a sophisticated 15-year plan for global disaster risk reduction, known as the SFDRR.

The SFDRR aims to achieve a substantial reduction in disaster risk and losses in lives, livelihoods, and health in the economic, physical, social, cultural, and environmental assets of persons, businesses, communities, and countries over the next 15 years (Rajabi et al., 2021; UNISDR, 2015). The SFDRR outlines four focus areas: (1) understanding disaster risk; (2) strengthening disaster risk governance to manage disaster risk; (3) investing in disaster risk reduction for resilience, and (4) enhancing disaster preparedness for effective response, and to 'build back better' in recovery, rehabilitation, and reconstruction. The SFDRR also emphasizes community-based recovery and encompasses a range of approaches: from reactive to proactive steps, from single-agency to partnership-based approaches, from science-driven approach to multidisciplinary approaches, from response and recovery management to comprehensive disaster risk management, and from community participation to community-based disaster risk reduction.

A transition towards prioritizing community-led approaches in decision-making and disaster recovery has been observed in recent years, particularly in the aftermath of devastating bushfires in Australia, such as the 2009 and 2019–2020 Black Summer Bushfires. This shift is rooted in the recognition of the unique strengths, knowledge, and social networks within local communities, which has led to initiatives to empower such valuable resources in the recovery process. These initiatives, which focus on traditional ecological knowledge and cultural practices, have gained prominence by harnessing wisdom embedded in community traditions and utilizing them in contemporary recovery strategies, thereby fostering resilience and a sense of joint-ownership. The same principle was applied in the launch of the Resilient Earthquake Recovery for the People of Syria by the United Nations Development Programme (UNDP) as part of their Earthquake Response Programme for 2023–2025. This community-based and inclusive early recovery approach draws on UNDP's extensive experience, establishes relationships in participatory local planning and management, and facilitates end-to-end recovery processes—from assessing damage and impact to implementing community-driven recovery interventions, by working directly with affected communities (UNDP, 2023).

More recently, the concept of nature-based solutions (NbS) was introduced in recovery studies. The World Bank first introduced this concept in the late 2000s and subsequently adopted it by the International Union for Conservation of Nature (IUCN) in 2016 as Resolution WCC-2016-Res-069. NbS are defined as actions that protect, sustainably manage, and restore natural or modified ecosystems to address societal challenges effectively and adaptively while providing benefits to human well-being and biodiversity (IUCN, 2016). In 2020, the UNDRR provided a comprehensive classification of NbS, which included four primary goals: climate change adaptation, climate change mitigation, disaster risk reduction, and environmental management (including biodiversity conservation). NbS for disaster recovery holds important roles for soil, water, biota, and urban design. Examples include landscape design and vegetation management, reshaping riverine landscapes or wetlands for flood protection, the use of mangroves and sand dunes as natural buffers, and coral reef restoration. Sudmeier-Rieux et al. (2021) and Mabon et al. (2022) documented numerous accounts of ecosystem-based approaches to disaster resilience and recovery globally.

Finally, Figure 24.1 provides a summary of the timeline-based discussion of recovery research that has presented thus far.

Before moving on to the next section, it is worth mentioning the event that will long be remembered in the field of disaster studies: the COVID-19 pandemic. Due to its global reach, loss of life, and economic and social disruption, the event was one of the most devastating disasters since World War II. Despite challenges posed by COVID-19, the pandemic will be remembered for the implementation of comprehensive measures and extensive reforms aimed at achieving recovery in the health, social, cultural, and economic sectors (e.g., logistics, supply chain management, and remote work). In light of the numerous challenges posed by the protracted COVID-19 pandemic, recovery initiatives go beyond the simple aim of reverting to pre-crisis development norms to instead concentrate on redefining the developmental framework. In this regard, the World Health Organization (WHO) has revised its Strategic Preparedness and Response Plan for the period 2023–2025, placing emphasis on principles such as equity, resilience, and environmental sustainability.

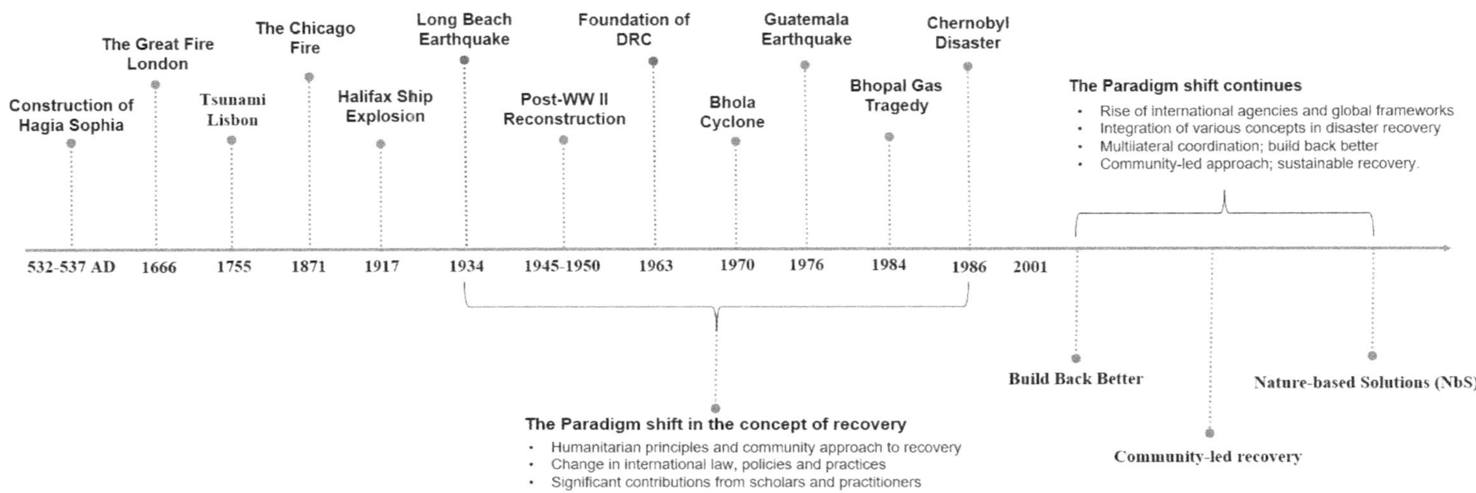

Figure 24.1 Events Shaping the Evolution of Recovery Studies in the Disaster Literature

Source: Figure created by author.

Duration of Recovery: Where to Begin and When to End?

Examining disasters as a means of recovery involves evaluating a diverse array of incidents that can have both immediate and long-lasting effects (Davis and Alexander, 2016; Quarantelli, 1998). According to the EM-DAT database, the world has experienced over 26,000 major disasters since 1900 (EM-DAT, 2023). These disasters can have singular or multiple impacts, resulting in sudden, prolonged, or cascading consequences. In an attempt to explain the recovery process, Robert Kates and David Pijawka published a chapter entitled "From Rubble to Monument: The Pace of Reconstruction" in Reconstruction Following Disaster, edited by Haas et al. (1977). The chapter has been cited numerous times in the academic literature and has become a foundation for both academic and practitioner work. In essence, the model has become the cornerstone of our understanding of the recovery process.

Kates and Pijawka (1977) constructed a post-disaster recovery model to explain how communities recover after a disaster. Using this model (Figure 24.2), Kates and Pijawka (1977) classified the post-disaster recovery process into four stages: (1) emergency response, (2) restoration of the restorable, (3) reconstruction (I) of the destroyed for functional replacement, and (4) reconstruction (II) for commemoration, betterment, and development. The timeline for these four phases spans from a few days to over ten years.

Following the Kates and Pijawka model, numerous scholars and agencies have endeavored to categorize the recovery period into various models and phases over the past five decades. The primary focus has been on optimizing the economic and physical environment, and survivors' preferences in this process have received little attention in the literature (Sobhaninia and Buckman, 2021). Despite efforts made, the duration of the recovery process after a disaster remains largely unanswered. In widely accepted international disaster recovery operations, regardless of type, magnitude, and spatial and temporal scope, the continuum is categorized as (1) short-term or immediate, (2) mid-term or early recovery, and (3) long-term. Immediate relief and response attention are crucial for short-term recovery (Figure 24.3). The first 1–12 months after a disaster

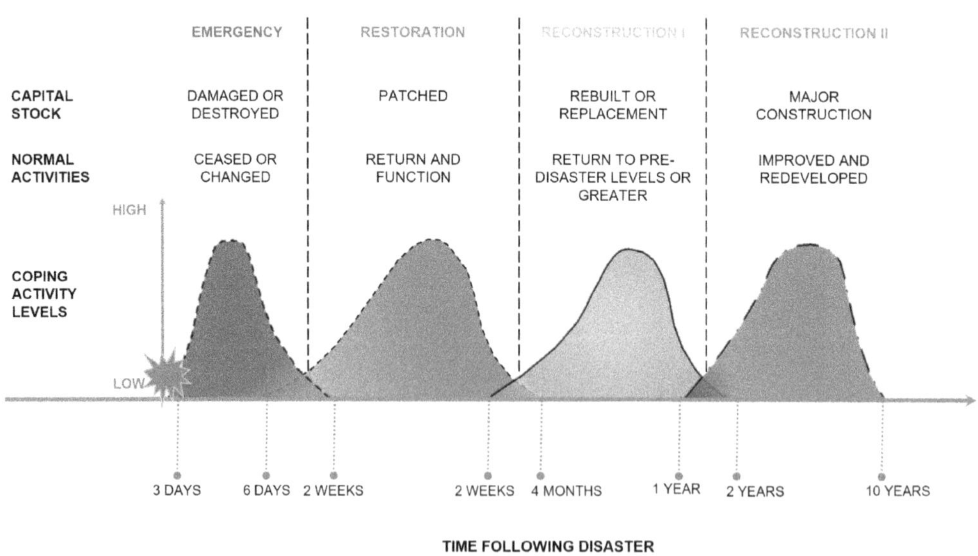

Figure 24.2 Model Illustrating Recovery Activities and the Temporal Phases of a Disaster

Source: Adapted and modified after Kates and Pijawka (1977).

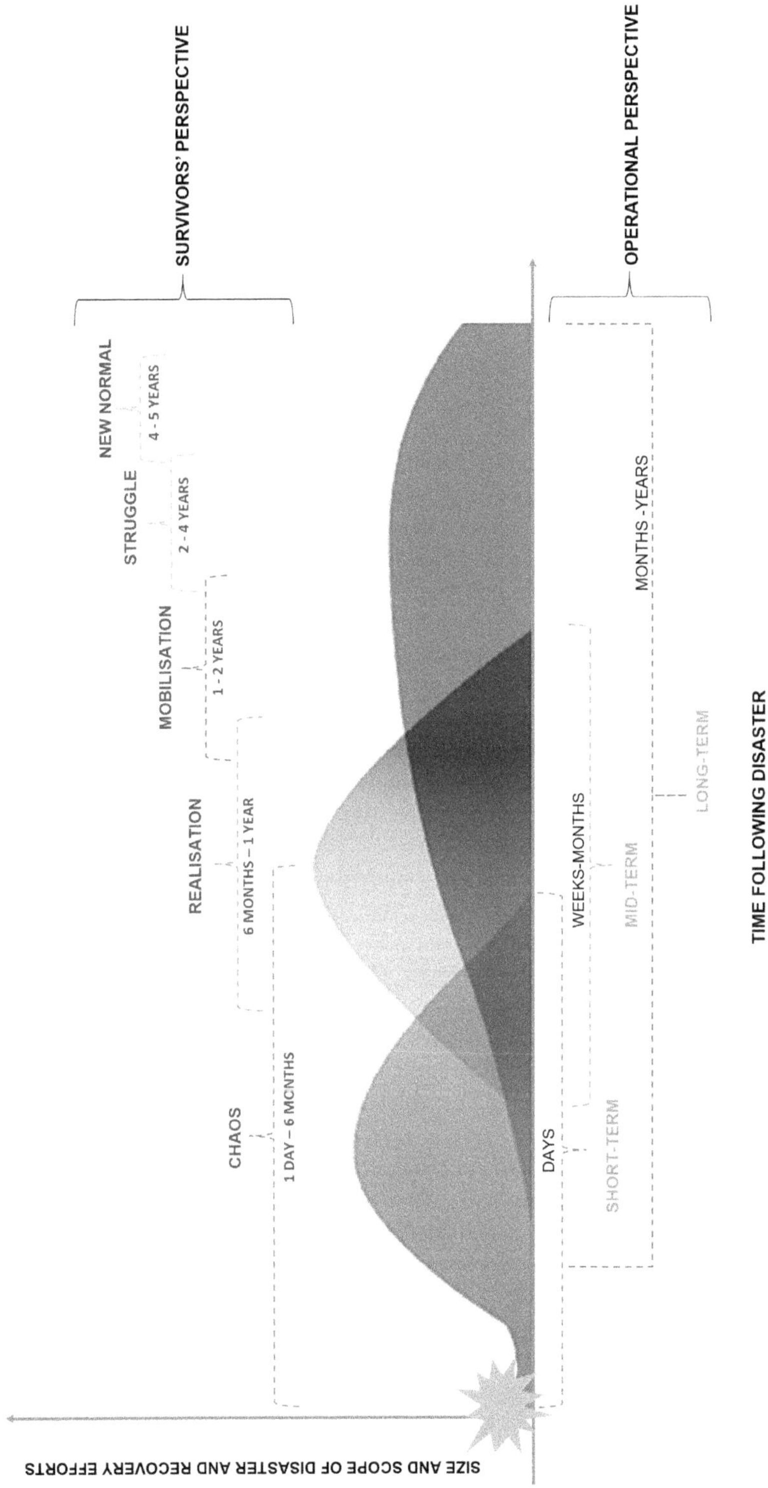

Figure 24.3 Duration of the Recovery Process

Source: Adapted and modified from FEMA (2011); Wilkinson et al. (2014); ADRF (2022).

constitute the early recovery phase, which is focused on assessing impacts and planning recovery activities. Long-term recovery can take years or even decades, during which rebuilding and recovery activities aim to achieve the goal of building back better and ensuring sustainable recovery (Zhou et al., 2022).

The issue of providing immediate or short-term aid versus promoting long-term sustainable recovery presents a significant challenge in the transition from disaster relief to development aid in the recovery timeline. This challenge is further compounded by the diverse operational needs of various agencies and the motivations of numerous stakeholders embedded in the process (Juran, 2015). Moreover, the needs of survivors themselves may not always align with a logical and reasonable recovery process, which can further complicate the efforts to achieve sustainable recovery (Lindell, 2017; Marchezini, 2018; Richardson, 2016). One commonly accepted concept is that events of varying magnitudes are separated by thresholds that correspond to the depth or extent of their impact and the society's ability to cope (Comfort et al., 2011; Richardson Maninger et al., 2016).

Following extensive research conducted for the purpose of this chapter, we identified the recovery phases categorized by Wilkinson et al. (2014) as the appropriate recovery timeline for most disaster events when considering the survivors' perspective. Suzanne and the team at the Resilient Organisations Research Group at the University of Auckland have been studying disaster recovery for a period of 10 years, during which time numerous catastrophes, such as those that occurred in the United States (Hurricane Katrina), Australia (Black Saturday bushfires), China (Wenchuan earthquake), Pakistan (earthquake), and Samoa (tsunami), have been analyzed and recovery patterns established. The stages proposed by Wilkinson et al. (2014) are: (1) Chaos, (2) Realization, (3) Mobilization, (4) Struggle, and (5) the New Normal. Wilkinson et al. explained that after the response stage, where, for instance, the injured and deceased have been taken care of, buildings cordoned off or made temporarily safe, and basic safety and temporary shelter and food have been attended to, real recovery begins. The end of the recovery process is realized when a new normal is in place and largely accepted by society.

As per Wilkinson and others (Mannakkara and Wilkinson, 2014; Wilkinson et al., 2014), the initial stage of disaster recovery comprises a state of chaos that persists for approximately six months. This phase is characterized by confusion and a lack of direction, as evidenced by the prevalent question, 'what do we do?' Assessments of buildings, community information-gathering, multi-stakeholder interactions, information collation and dissemination, and planning arrangements start to form, which are not always well-established yet play a crucial role during this time and in directing the future (Silver and Grek-Martin, 2015). The chaos phase is also marked by waste disposal, demolition, and debris removal as top priorities. The rapid nature of decision-making during this phase may result in the demolition of salvageable buildings despite the need for a more thoughtful approach to land use. For the affected community, this phase often entails temporary accommodation or displacement with minimal or no reconstruction. In terms of operations, the declaration of a dedicated authority or the implementation of a recovery plan following an initial needs assessment typically signifies the end of the chaos phase.

The realization phase typically begins approximately six months after the disaster and is marked by recognition of the extensive recovery effort that will be required. This phase can last anywhere from six months to one year and involves a focus on planning, including rezoning and implementing new building codes designed to enhance resilience. During this phase, resource constraints may become apparent and there may be uncertainty among the population about the future. Businesses are often eager to resume operations, but the reduced or displaced population may be hesitant to return to normalcy. Many individuals may still be living in temporary accommodations, attempting to repair their damaged structures, or have left the affected area altogether (Me-Bar and Valdez, 2004; Oliver-Smith, 1991).

The transition from realization to mobilization represents the subsequent stage, characterized by heightened activity and a resolute commitment to progressing with recovery efforts. This phase commences on the first anniversary of the catastrophe and is marked by an increasingly sunny outlook and sense of accomplishment. The mobilization phase is characterized by the sentiment, 'we're getting on with it,' and lasts from the end of the first year to the end of the second year. During this time, there is a heightened level of decision-making and tangible progress is evident in the form of repairs to public buildings such as schools and community centers. The emergence of new buildings and increased repair activities also draw attention to wider resourcing issues, resulting in rising costs and supply shortages (Mayer, 2019; Smith and Birkland, 2012). However, during this phase the population experiences a decrease in uncertainty as the momentum generated by increased activity fosters optimism. Insurance plays a significant role in driving recovery processes during the mobilization phase, but insurance-related issues have also become more prominent (Kousky, 2019).

The struggle phase, which generally lasts from the conclusion of the second year through the fourth year, follows initial optimism and signifies a period of considerable hardship, as denoted by the term 'struggle.' During this phase, the overarching sentiment is one of the difficulties, and it becomes clear that recovery is not a swift process (Mannakkara and Wilkinson, 2014; Onstad et al., 2012). The struggle phase is marked by a range of activities, including the reconstruction of homes and reestablishment of businesses. Additionally, new business models emerge as the previous ones are unable to adapt to the diminished and/or altered business environment. The struggle phase is also characterized by rising costs, scarcity of materials, increased demand for skilled professionals, reduced housing stock, and issues of affordability. This phase presents particularly challenging circumstances for the community, with many individuals reevaluating their options and contemplating relocation (Me-Bar and Valdez, 2004).

Finally, the new normal phase typically commences around four to five years after a disaster. This phase signifies a period of acceptance and adaptation to the transformed landscape. In the new normal phase, the community comes to terms with the reality that things will never return to the previous state and they learn to navigate and adapt to the novel environment. During this time, businesses and buildings may look different and communities and social facilities may have been relocated or reconfigured (Murao, 2011; Nakagawa and Shaw, 2004). Despite the challenges and changes, people in this phase have learned to accept and adapt to the new normal, marking a significant societal milestone in the long-term recovery process.

Utilizing the insights presented in this section, a diagram was devised to provide a simplified overview of the recovery process was developed (Figure 24.3). The conceptual diagram combines the timelines proposed by Wilkinson et al. (2014), FEMA (2011), and ADRF (2022) to depict the duration of recovery from both survivors' and operational perspectives. Comprising three distinct phases along with essential concepts and timeframes associated with each phase, this framework facilitates the anticipation of obstacles and temporal fuzziness inherent in the recovery process, thereby informing and enhancing future disaster recovery initiatives.

Conclusion

In this chapter, we conducted an extensive yet non-exhaustive examination of the ever-evolving concept of recovery, delving into its historical development and shedding light on the numerous elements that have shaped the process. The overview encompasses significant events and key concepts while simultaneously elucidating the complex factors that influence the duration of recovery.

These factors include but are not limited to economic conditions, policy frameworks, psychological impacts, cultural considerations, and environmental factors.

A crucial insight from our discussion is the indispensable need for operational personnel to clearly articulate their intended meaning when engaging with concepts and strategies related to recovery. This is vital because the term recovery carries the nuanced implication of restoring the post-disaster situation to an acceptable level, which may or may not align with the pre-impact state. Furthermore, our exploration highlights the significance of adopting a comprehensive recovery approach. Beyond the traditional emphasis on rebuilding physical structures and reviving economic activities, modern conceptualizations of recovery recognize the need for a holistic recovery process that extends its reach across various sectors and their numerous social dimensions. By underscoring cross-cutting themes, our chapter advocates an approach that transcends mere restoration, acknowledging the interconnectedness of different elements in the recovery journey. This holistic perspective aligns with contemporary understandings of recovery, which recognize that true recovery rises above reconstruction and involves fostering resilience across diverse structural, nonstructural, cultural, and environmental dimensions.

References

ADRF. *Australian Disaster Recovery Framework*. 2022, https://www.recovery.sa.gov.au/__data/assets/pdf_file/0005/933998/ADR-Framework-October-2022.pdf

Alexander, D. *Natural Disasters*. Routledge, 1993.

Alexander, D. *Planning for Post-Disaster Reconstruction*, University de Montreal, 2007.

Alexander, D. E. "Disaster and Emergency Planning for Preparedness, Response, and Recovery." *Oxford Research Encyclopedia of Natural Hazard Science*. Oxford University Press, 2015, pp. 1–20.

Alexander, D., C. H. Davidson, A. Fox, G. Lizzaralde, and C. Johnson. *Post-Disaster Reconstruction, Meeting Stakeholder Interests*. Oxford University Press, 2007.

Araújo, A. C. "The 1755 Lisbon Earthquake: The Catastrophe and the Reconstruction." *Storicamente*, vol. 17, no. 5, 2021. https://doi.org/10.52056/9788833138732/05

Asian Disaster Preparedness Center (ADPC). "Guidelines for Recovery." ADPC. Vol. 10, no. 5, pp. 408–419. https://r4r.adpc.net/content-page/guidance-and-tools/

Blaikie, P., T. Cannon, I. Davis, and B Wisner. *At Risk: Natural Hazards, People's Vulnerability and Disasters*. 1st ed. Routledge, 1994.

Choi, C., and R. Honda. "Motive and Conflict in the Disaster Recovery Process." *International Journal of Disaster Resilience in Business*, 11 November 2019.

Comfort, L., B. Wisner, S. Cutter, R. Pulwarty, K. Hewitt, A. Oliver-Smith, J. Wiener, M. Fordham, W. Peacock, and F. Krimgold. "Reframing Disaster Policy: The Global Evolution of Vulnerable Communities." *Global Environmental Change Part B: Environmental Hazards*, vol. 1, no. 1, 1999, pp. 39–44. https://www.tandfonline.com/doi/abs/10.3763/ehaz.1999.0105

Condit, C. W. *"The Chicago School of Architecture: A History of Commercial and Public Building in the Chicago Area, 1875–1925*. University of Chicago Press, 1964. https://www.cambridge.org/core/journals/historical-journal/article/abs/1666-and-londons-fire-history-a-reevaluation/BF9FC59C9B9746B624D1628170DBE5E2

Cuny, F. *Disasters and Development*. Oxford University Press, 1983.

Davis, I. "Disasters as Agents of Change, or, Form Follows Failure." *Habitat International*, vol. 7, no. 5–6, 1983, pp. 277–310.

Davis, I. *Development by People for People*. Oxford Brookes University, 1978.

Davis, I. *Learning from Disaster Recovery: Guidance for Decision Makers*. International Recovery Platform, 2006.

Davis, I., and D. Alexander. *Recovery from Disaster*. Routledge, 2016.

Diefendorf, J. M. *In the Wake of War: The Reconstruction of German Cities after World War II*. Cambridge University Press, 1993.

Dynes, R. R. "The Dialogue between Voluntary Organizations and the State." In *Handbook of Disaster Research*, pp. 171–188. Springer, 2006.

ECHO. European Civil Protection and Humanitarian Aid Operations, "Guidelines for Post Disaster Needs Assessment (PDNA)." ECHO, 2013. https://www.undp.org/sites/g/files/zskgke326/files/publications/PDNA%20Volume%20A%20FINAL%2012th%20Review_March%202015.pdf

EM-DAT. *Inventorying Hazards & Disasters Worldwide since 1988*. CRED. Brussels, Belgium. Guide to Developing Disaster Recovery Frameworks. Washington, DC, 2023. https://www.emdat.be/

FEMA. *National Disaster Recovery Framework*. FEMA, 2011. www.fema.gov/media-library-data/20130726-1820-25045-5325/508_ndrf.pdf

GFDRR. 2015a. https://www.gfdrr.org/en/publication/guide-developing-disaster-recovery-frameworks

GFDRR, Global Facility for Disaster Reduction and Recovery. Disaster Recovery Framework. 2020, Washington, D.C.

GRDRR. Global Platform for Disaster Risk Reduction. Guide to Developing Disaster Recovery Frameworks. Washington, DC, 2015b. https://www.gfdrr.org/en/publication/guide-developing-disaster-recovery-frameworks

Haas, J. E., R. W. Kates, and M. J. Bowden. "Reconstruction Following Disaster." *Reconstruction Following Disaster*. The MIT Press, 1977.

Haque, A., D. Haider, M. S. Rahman, L. Kabir, and R. P Lejano. "Building Resilience from the Grassroots: The Cyclone Preparedness Programme at 50." *International Journal of Environmental Research and Public Health*, vol. 19, no. 21, 2022, pp. 14503. https://www.ncbi.nlm.nih.gov/pmc/articles/PMC9657222/

Haynes, V., and M. Bojcun. *The Chernobyl Disaster*. Hogarth, 1988.

Horney, J., M. C. Simon, K. Ricchetti-Masterson, and P Berke. "Resident Perception of Disaster Recovery Planning Priorities." *Business*, 27 October 2016.

IFRC. Laws, Policies and Plans for Disaster Recovery: Multi-Country Synthesis Report. Geneva, Switzerland, 2023. https://disasterlaw.ifrc.org/media/4230.

IFRC. Framework for Community Resilience. Geneva: Switzerland. https://www.ifrc.org/document/ifrc-framework-community-resilience, 2014.

IRP. *Guidance Note on Recovery: Pre-Disaster Recovery Planning*. New York, 2012. https://recovery.preventionweb.net/publication/guidance-note-recovery-pre-disaster-recovery-planning

IUCN. "WCC-2016-Res-069-EN Defining Nature-based Solutions". Gland, Swizerland, 2016. https://portals.iucn.org/library/sites/library/files/resrecfiles/WCC_2016_RES_069_EN.pdf

Juran, L. "A Framework for Integrating Water Projects in Disaster Reconstruction Processes in the International Context." *Public Works Management & Policy*, vol. 20, no. 4, 2015, pp. 305–321.

Juran, L. "Disaster, Development, and Water: the Reconstruction and (re)fabrication of Hazardous Waterscapes in Post-Tsunami India." In *The Asian Tsunami and Post-Disaster Aid*, edited by S. Reddy, pp. 79–99. Springer Nature, 2018.

Kates, R. W., and D. Pijawka. "From Rubble to Monument: the Pace of Reconstruction." *Reconstruction Following Disaster*, vol. 1, 1977, pp. 1–23.

Kennedy, J., J. Ashmore, E. Babister, and I Kelman. "The Meaning of 'build back better': Evidence from Post-Tsunami Aceh and Sri Lanka." *Journal of Contingencies and Crisis Management*, vol. 16, no. 1, 2008, pp. 24–36.

Khasalamwa, S. "Is 'Build Back Better' a Response to Vulnerability? Analysis of the Post-Tsunami Humanitarian Interventions in Sri Lanka." *Norwegian Journal of Geography*, vol. 63, 2009, pp. 73–88.

Kousky, C. "The Role of Natural Disaster Insurance in Recovery and Risk Reduction", 2019. https://www.annualreviews.org/doi/abs/10.1146/annurev-resource-100518-094028

Lindell, Michael. *Recovery and Reconstruction after Disaster*. Economics, 2017.

Mabon, L., L. Barkved, K. de Bruin, and W. Y Shih. "Whose Knowledge Counts in Nature-Based Solutions? Understanding Epistemic Justice for Nature-Based Solutions through a Multi-City Comparison across Europe and Asia." *Environmental Science & Policy*, vol. 136, 2022, pp. 652–664.

Mannakkara, S., and S. Wilkinson. "Re-Conceptualising "Building Back Better" to Improve Post-Disaster Recovery." *International Journal of Managing Projects in Business*, vol. 7, no. 3, 2014, pp. 327–341.

Marchezini, V. "Volume number." *Disaster Prevention and Management*, Vol. 28, no. 1, 5 September 2018, pp. 143–152,

Mayer, B. "A Review of the Literature on Community Resilience and Disaster Recovery." *Current Environmental Health Reports*, vol. 6, no. 3, 2019, pp. 167–173.

Me-Bar, Y., and F. Valdez Jr. "Recovery Time after a Disaster and the Ancient Maya." *Journal of Archaeological Science*, vol. 31, no. 9, 2004, pp. 1311–1324.

Murao, O. *Recovery Curves for Housing Reconstruction from the 2011 Great East Japan Earthquake and Comparison with Other Post-Disaster Recovery Processes*. International Research Institute of Disaster Science, Tohoku University, 2017.

Nakagawa, Y., and R. Shaw. "Social Capital: A Missing Link to Disaster Recovery." *International Journal of Mass Emergencies & Disasters*, vol. 22, no. 1, 2004, pp. 5–34.

Oliver-Smith, A. "Successes and Failures in Post-Disaster Resettlement." *Disasters*, vol. 15, no. 1, 1991, pp. 12–22.

Olson, D. *The California Earthquake of April 18, 1906: Early Research and Noteworthy Studies*. U.S. Geological Survey, 2003.

Onstad, P. A., S. M. Danes, A. M. Hardman, P. D. Olson, M. S. Marczak, R. K. Heins, S. R. Croymans, and K. A. Coffee. *The Road to Recovery from a Natural Disaster: Voices from the Community. Community Development*, vol. 43, no. 5, 2012, pp. 566–580. https://doi.org/10.1080/15575330.2012.699081

Prince, Samuel H. *A Report on the Halifax Disaster: December 6th, 1917*. Oxford University Press, 1920.

Quarantelli, E. L. "Conceptualizing Disasters from a Sociological Perspective." *International Journal of Mass Emergencies and Disasters*, vol. 7, no. 3, 1989, pp. 243–251.

Quarantelli, E. L. "Disaster Recovery: Research Based Observations on What It Means, Success and Failure, Those Assisted and Those Assisting." Preliminary Paper No. 263. Disaster Research Center, University of Delaware, 1998. Available at http://udspace.udel.edu/bitstream/handle/19716/282/PP+263.pdf.txt?sequence=3

Rajabi, E., J. Bazyar, V. Delshad, and H. R Khankeh. "The Evolution of Disaster Risk Management: Historical Approach." *Disaster Medicine and Public Health Preparedness*, vol. 16, no. 4, 2022, pp. 1623–1627.

Richardson, B. K., and L. Maninger. "We Were All in the Same Boat: An Exploratory Study of Communal Coping in Disaster Recovery." *Sociology*, 8 March 2016. Vol. 42, pp. 32–41

Silver, A., and J. Grek-Martin. "Now We Understand What Community Really Means: Reconceptualizing the Role of Sense of Place in the Disaster Recovery Process." *Psychology*, 1 June 2015 Vo. 42, pp. 32–41.

Smith, G., and T. Birkland. "Building a Theory of Recovery: Institutional Dimensions." *International Journal of Mass Emergencies & Disasters*, vol. 30, no. 2, 2012, pp. 147–170.

Sobhaninia, S., and S. T. Buckman. "Revisiting and Adapting the Kates-Pijawka Disaster Recovery Model: A Reconfigured Emphasis on Anticipation, Equity, and Resilience." *Journal of Disaster Research*, vol. 45, no. 2, 2021, pp. 123–140.

Sphere Handbook. "Humanitarian Charter and Minimum Standards for Disaster Response." Sphere, 2018. https://spherestandards.org/handbook/editions/

Sudmeier-Rieux, K., et al. "Scientific Evidence for Ecosystem-Based Disaster Risk Reduction." *Nature Sustainability*, vol. 4, no. 9, 2021, pp. 803–810.https://doi.org/10.1038/s41893-021-00732-4

Twigg, J. *Characteristics of a Disaster-Resilient Community – A Guidance Note*. DFID Disaster Risk Reduction Interagency Coordination Group, 2007.

UNDP. "Supporting a Resilient Earthquake Recovery for the People of Syria – UNDP's Earthquake Response Programme 2023–2025." March 19, 2023. https://www.undp.org/syria/publications/supporting-resilient-earthquake-recovery-people-syria-undps-earthquake-response-programme-2023-2025

United Nations Office for Disaster Risk Reduction (UNDRR). UNISDR Annual Report 2017. New York, 2017. https://www.unisdr.org/we/inform/terminology

UNISDR. *Sendai Framework for Disaster Risk Reduction 2015–2030*. United Nations Office for Disaster Risk Reduction, 2015.

Wilkinson, S., A. Chang-Richards, and J. O. B. Rotimi. "Reconstruction Following Earthquake Disasters." In *Encyclopedia of Earthquake Engineering*, edited by M. Beer, et al., Springer, 2014.

Zhou, B., H. Zhang, and R. Evans. "Build Back Better: A Framework for Sustainable Recovery Assessment." *International Journal of Disaster Risk Reduction*, vol. 76, 2022, pp. 102998. https://doi.org/10.1016/j.ijdrr.2022.102998

25

SOCIAL SCIENCE METHODOLOGIES IN DISASTER RECOVERY RESEARCH

Naxhelli Ruiz-Rivera

Introduction

In the field of disaster studies, there has been a substantial development of methodological issues presented (Donner and Diaz, 2018; King and Gurtner, 2021; Norris, 2006; Rivera, 2022). Many of these works have focused on the emergency phase, exploring the human experience of disaster through different research strategies including qualitative (Phillips, 2014), quantitative (Karatani and Hayashi, 2007), mixed methods (Russo, 2022), as well as geographic information systems and crowdsourced geographic information (Goodchild and Glennon, 2010; Tomaszewski, 2021; Tomaszewski et al., 2015). This chapter addresses methodological issues in recovery as something that disaster researchers share in general, but with a distinctiveness associated with its social science underpinnings, particularly its temporality and focus on social change.

Studying recovery inherently raises methodological issues that have been widely discussed in the context of disaster aftermaths. However, what perhaps distinguishes recovery from other components of disaster or emergency management is that it implies an underlying positive form of social change. This threshold is used to monitor and evaluate recovery outcomes. This is clear in the United Nations Office for Disaster Risk Reduction's definition of recovery as "the restoring or improving of livelihoods and health, as well as economic, physical, social, cultural and environmental assets, systems and activities, of a disaster-affected community or society, aligning with the principles of sustainable development and 'build back better', to avoid or reduce future disaster risk" (United Nations, 2016, 21). Note that the terms 'improving' and 'build back better' are included, which imbue expectations of social change. Additionally, scholars such as Davis and Alexander (2016, 23) have defined recovery as "decisions and actions taken after a disaster in order to restore and improve pre-disaster living conditions and the environment of the affected community."

A core element in these conceptualizations is well-being, which has been identified as "an emergent property of recovery systems" (Blackman et al., 2023, 3). Thus, the idea of recovery generally involves improvement—due to steps taken over time in the recovery process—from a condition in which something was lost to a change in the conditions that led to disaster in the first place. That is, recovery is a normative concept. However, it should be noted that many studies in this field also address recovery not as a 'process towards a good outcome,' but as a phase in which socio-physical and trauma conditions could worsen following the hazard impact, instead of

DOI: 10.4324/9781003414834-34

improving. The recovery phase may even deepen the loss and become a 'disaster after the disaster' (Juran, 2018).

While normative approaches revolve around key factors that evaluate the outcomes of post-disaster interventions (for example, well-being indicators), as Ryan et al. (2016) point out, such evaluations are diverse in terms of scope, depth, variables, and objectives. A variety of indicators such as business activity, labor market, mental health, out-migration, level of public participation, and social connectedness have been used to understand the extent to which recovery is achieved. However, it remains hard to decipher whether disaster-affected people recover, and to what extent such outcomes actually stem from effective recovery policies.

In this chapter, we explore these two broad types of social scientific methodological approaches. First, we investigate normative methods that intend to understand recovery as a socio-political process or describe or analyze aspects such as institutional arrangements, social networks, and cultural meanings of changes after disasters. Next, we summarize frameworks that are typically employed by institutions to evaluate recovery, particularly under their needs for policy design. These frameworks, such as needs, damage, and loss assessments, are often based on guidelines and principles.

Normative Approaches to Analyzing Disaster Recovery

Ethical Considerations

A great deal of attention has been given to ethical matters and the role of a researcher within the disaster context, given the intense vulnerability that disaster-stricken areas experience, and how this vulnerability could be worsened by research activities. Recovery is about vulnerable populations and their losses, which means that ethical matters lie at the core of all research designs. Although this goes beyond the most 'critical' phase of the emergency, ethical responsibilities such as informed consent, avoidance of deception, anonymity, and clear respect for participants' rights are essential (Phillips, 2014). The scope and purposes of the research need to be clarified, so the intellectual merit is parallel to accountability, caring, empathy, and reducing human suffering (Denzin and Lincoln, 2005).

Normative approaches in recovery are based on the ethical ideals of human dignity and the universality of rights—no matter who, where, or when. Key ethical dimensions of such approaches consider the extent to which future vulnerability is avoided in the recovery process and how freedom, opportunities, and human agency are respected (O'Mathúna and Gordjin, 2018). Longitudinal designs, social network analysis, and narrative techniques integrate these ethical criteria into the research design, including the assessment of recovery outcomes and other elements such as individual or collective capacities and the possible reduction of human suffering.

Longitudinal Designs

Longitudinal research designs are relevant given their focus on scrutinizing evolving pathways and unequal recovery outcomes at several points across time (Hamideh et al., 2023). Longitudinal designs allow us to understand how persons, households, and geographical units change over time. Specifically, they assist in identifying factors that enable or constrain recovery (Sou et al., 2021)—factors sometimes referred to as 'recovery predictors' (Tatsuki and Kawami, 2023). Thus, longitudinal metrics can be used as proxies to detect problems with recovery (Horney et al., 2018). Both qualitative and quantitative approaches to longitudinal research provide valuable evidence of the trajectories of disaster-affected populations in terms of relevant outcomes such as livelihood paths, health, income, or displacement.

Quantitative longitudinal designs to understand disaster impacts and recovery factors have been utilized in countries such as Japan and the United States and also as a basis for comparative studies (Bui et al., 2021; Choudhury and Wu, 2023; Tatsuki and Kawami, 2023; Terumoto et al., 2021). They have been used to understand specific dimensions of recovery, such as well-being and life satisfaction (Fujimoto et al., 2022), migration (Gray et al., 2014), economic resilience, business trajectories (Peacock et al., 2014; Schrank et al., 2013; Watson et al., 2020), and mental health (Merdjanoff et al., 2023; van der Velden et al., 2009). These studies generally use regression models as the basis of statistical analysis.

The challenge for longitudinal designs is the ability to sustain the probabilistic integrity of the sample, which needs to guarantee the coherence of the cohort units (e.g., individuals or households) over time (Goff et al., 2023). Cohort follow-up is generally more difficult and expensive in disasters due to significant rates of (forced) residential mobility. However, displaced populations are crucial to understanding the impacts of mobility/displacement as well as the dynamics and perspectives of population recovery more broadly (Morrow-Jones and Morrow-Jones, 1991).

Another version of longitudinal methodologies is qualitative. This approach chooses a point in time in which it is already possible to identify some recovery outcomes and define a retrospective analysis of the explanatory factors (also known as causal configurations) leading to those results (Sou et al., 2021; Tetsuya, 2017). Qualitative longitudinal analyses typically employ interviews and observations but may also include participant self-reporting techniques, such as diary methods (Terzis et al., 2022), and text and photos shared on social media platforms (Liu et al., 2008).

Social Network Analysis

Social network analysis is a commonly used methodological approach in disaster studies. In conceptual terms, social network analysis is frequently used to understand the role that social capital, collective action, and human agency play in disasters (Misra et al., 2017). Various aspects of disasters—such as social vulnerability, coping capacity, emergency response, and recovery—have been studied using social networks as a vehicle to better understand trust, social resources, knowledge, the flow of aid, and other disaster dynamics (Aldrich, 2019; Benedict et al., 2023; Jones and Faas, 2016; Lee et al., 2019). Some studies argue that social networks are, in fact, the basic unit of disaster response (Sanyal and Routray, 2016; Wilkin et al., 2019).

According to Misra et al. (2017), emergencies lead to atypical social networks that develop emergent properties. There are two types of such properties: the structural properties are the dyadic interactions between two nodes (i.e., actors), while the compositional properties define these actors' attributes, such as social characteristics, geographical location, and organization size (Wasserman and Faust, 1994). From these two basic sets of properties, key topics in recovery such as altruism, social support, disaster management capacities, decision-making, and adaptation can be empirically documented and represented.

There are two distinct types of social network analysis, egocentric networks and sociocentric networks. Egocentric networks focus on the more personal relationships of respondents and are used to understand "the local social environment surrounding the ego" (Varda, 2016, 44). A well-known application of egocentric networks in recovery research relates the role of the three classical types of social capital in disasters: bonding (horizontal connections between similar persons), bridging (horizontal connection between dissimilar persons), and linking (vertical connection with actors with political power and authority) (Aldrich, 2019; Hawkins and Maurer, 2010; Marín et al., 2015).

On the other hand, sociocentric networks are increasingly important for analyzing interorganizational systems in which aspects such as governance and policy implementation take place (Kapucu and Demiroz, 2016). Topics such as communication and social media (Kim and Hastak, 2018), the effectiveness of coordination (Hossain and Kuti, 2010; Kapucu and Hu, 2016), and organizational learning and leveraging factors or obstacles to recovery (Islam and Walkerden, 2017) are compatible with this methodological approach. Such approaches can utilize various quantitative and qualitative techniques to gather sociocentric network data, such as surveys, interviews, and participant observation.

Narrative Techniques

Narrative approaches focus on understanding disaster and recovery experiences for both analytical purposes and to contribute to communities' psychosocial healing process from distress, loss, or trauma (Kargillis et al., 2014). Narrative approaches are particularly appropriate for exploring meanings, emotions, and survivors' diverse insights into the multifaceted disaster experience (Fortune, 2018; Jerolleman, 2021; Rahiem et al., 2021). Also, narrative techniques can harness storytelling to produce 'alternative' representations of the disaster aftermath, particularly in cases when such evidence challenges overly optimistic views of the recovery process (Few et al., 2021) or when survivors are engaged in struggles against the official, hegemonic portrayal of the disaster and are seeking recognition, justice, and social repair—or what Gorostiza and Armiero (2021) call 'narrative reparation.'

Memory and risk awareness play a prominent role in disaster recovery, and storytelling—whether oral, written, digital, maps, performative, or through another medium—plays a crucial part in how the narrative of the disaster is constructed within the community that faces it (Graham, 2022; Monteil et al., 2020). This narrative becomes a vital part of the process of social learning and healing, which are vital elements of resilience. Narrative construction takes place in households, at community gatherings and memorials (Hayashi, 2016; Ono et al., 2021; Sakamoto, 2020), in museums and tourist sites as part of the historical representation of the recovery process (Mahn et al., 2021; Tanaka et al., 2021), and ultimately in the memories, rituals, and other cultural expressions that eventually become part of the community's heritage.

Narrative techniques can make use of multiple sources, such as interviews, letters, photography, focus groups, blogs, or diaries (Guest et al., 2013). In terms of sampling, the use of narrative techniques in recovery studies usually employs purposive criteria, in which participants are purposefully selected through specific profiles to capture similarities, and contrasts in the informants' experiences of those events. Snowball sampling and other techniques are often used to arrive at a level of data saturation and triangulation to help buttress the narrative. The experiences that individuals elaborate on and share among themselves and with the researcher are essential for documenting the recovery context and understanding the knowledge that emerges from the experience of recovery.

Frameworks for Analyzing Disaster Recovery

Post-Disaster Needs, Loss, and Recovery Assessments

Methodologies to assess post-disaster impact, needs, loss, and damage are an important part of disaster management in both the response and recovery phases. These methodologies shape many dimensions of recovery, from the availability of financial resources to the priorities and goals that interventions should pursue. The first assessments were developed to evaluate the impacts of the Managua earthquake of December 1972 in Nicaragua (Bello et al., 2015). In this decade, the Economic Commission for Latin America and the Caribbean (ECLAC) developed the first disaster impact assessments, which were the seed for some of the most relevant loss and damage methodologies. ECLAC developed a series of handbooks (see ECLAC, 1991, 2003, 2014), which have been used worldwide.

In 2008, the European Commission, the United Nations Development Group, and the World Bank produced the Joint Declaration on Post-Crisis Assessments and Recovery Planning (European Commission, United Nations Development Group, and the World Bank, 2008), a common platform for action that constitutes the basis of the current needs-based methodologies for recovery. The declaration is a combination of a revised ECLAC Quantitative Damage and Loss Assessment (DaLA), a Disaster Recovery Framework (DRF), and a Human Recovery Needs Assessment (HRNA). Next, the Global Facility for Disaster Risk Reduction (GFDRR) of the World Bank developed multiple documents and guidelines, such as three volumes on the Damage and Loss and Needs Assessment (DaLA) Guidance Notes (see GFDRR, 2010a, 2010b, 2010c) and several revised versions of the methodologies for Post Disaster Needs Assessments (PDNAs) (GFDRR, European Commission, and UNDP, 2018; GFDRR, European Commission, and UNDP 2013a, 2013b). The GFDRR documents and guidelines establish the scope of these methodologies to assess damage for reconstruction needs and losses for recovery needs.

PDNAs are based on four elements: a baseline of preexisting vulnerabilities and capacities; an estimation of disaster effects (i.e., damages to infrastructure and assets); an assessment of disaster impacts (i.e., macro and long-term economic and human development impacts); and a recovery strategy comprising an outline of the recovery needs, guiding principles, intended sectoral results, and implementation arrangements (GFDRR, European Commission, and UNDP 2013b). These elements have contributed to understanding the resources and logistics of large-scale post-disaster processes and how involved institutions can better attain their intended recovery goals. Furthermore, PDNAs provide a standardized framework for recovery planning for practitioners and authorities. They have provided a space for developing recovery indicators, relating them to financial resource allocation, and more effectively achieving milestones of the recovery process.

Non-Economic Loss and Damage Assessments

Not all disaster losses are directly or easily translatable in terms of physical, tangible, or quantifiable measures. To that end, the non-economic loss and damage (NELD) framework is an instrument stemming from the Warsaw International Mechanism for Loss and Damage that was conceived in 2013. The foci of NELD are losses associated with slow-onset events along with limits to adaptation to climatic risks. NLED has been used in recovery analysis as an "assessment lens for recording unavoided climate impacts" (Serdeczny et al., 2016, 1) while considering both material and non-material aspects that cannot be addressed within conventional market-driven domains. Aspects such as cultural heritage, identity, sense of belonging, and local knowledge are often excluded from traditional frameworks based on damage assessment due to incommensurability (Menk et al., 2022). NELD has been a referent for the recovery debates that differentiate 'metrics of value' from the 'metrics of harm' that could serve as direction on how the burden of recovery from 'irreversible and unavoidable loss' should be distributed among ecosystem actors considering their responsibility in disaster risk causes (Boyd et al., 2017; McShane, 2017).

Table 25.1 Types of Incommensurable Losses Addressed by the Non-Economic Loss and Damage (NELD) Framework

	Intrinsic	*Instrumental*
Material	Biodiversity Culturally important landscapes Heritage Life Homes	Productive land Communal sites Ecosystem services
Non-material	Dignity Physical and mental well-being Intrinsic value of biodiversity	Social cohesion Sovereignty Security Identity and sense of belonging Sense of place

Source: Adapted from Serdeczny et al. (2016); Created by author.

Recent advances in the NELD framework propose a matrix of two axes—each with two cells—that constitute the material and non-material and intrinsic and instrumental aspects of disaster recovery (Table 25.1). Such a conceptualization could provide guidance and paradigmatic support in the assessment of the impacts of both sudden and slow-onset events (Serdeczny et al., 2016). The nexus of material and intrinsic aspects encompasses, among other things, biodiversity loss, habitat destruction, and loss of cultural and architectural/vernacular heritage. Non-material intrinsic intersections include values such as dignity and physical and mental well-being. On the other hand, material and instrumental intersections include loss of productive land, destruction of markets and religious sites, and loss of ecosystem services. Finally, non-material instrumental aspects consist of social cohesion, adverse health impacts, disability-adjusted life years (DALYs), loss of identity, and decline of traditional/local (ecological) knowledge. Collectively, this matrix can help to affect policy and decision-making by generating awareness of the seemingly less empirical or tangible components of loss and recovery.

NELD categories have been incorporated as part of the research design in studies on recovery enablers, the transition from non-economic loss to non-economic impacts, and measures to compensate for such impacts (Ayeb-Karlsson et al., 2023; Pill, 2022). Additionally, NELD valuation methodologies have been deployed to estimate damages to cultural heritage in post-disaster scenarios (Romão and Paupério, 2021). NELD approaches can lend themselves to narrative techniques that were discussed earlier in this chapter, which help to comprehensively address the human experience of suffering throughout the recovery process.

Forensic Approaches

Forensic approaches focus on "understanding the factors that contributed to the production and occurrence of a disaster" (Mota-Ferreira et al., 2023, 2). Forensic methodologies have a variety of purposes but are primarily focused on exploring the factors that govern the type and intensity of disaster impacts. Some forensic methods are oriented to demonstrate a nexus between human actions and negative disaster impacts, such as 'incident investigation techniques' (focused on infrastructure failures) or the procedures developed for judicial inquiries and criminal investigations (Mendoza, 2019). Other forensic methods are employed to assess the long-term causes and consequences of a disaster, as well as the effectiveness of interventions.

Each disaster forensic methodology has developed a battery of guidelines, questions, fieldwork techniques, timelines, and outcomes. For example, Forensic Investigation of Disasters (FORIN) focuses on understanding the root causes of disasters through a causal chain (Oliver-Smith et al., 2016). FORIN outlines key questions about the characteristics of the triggering event, the socio-economic profile of the affected communities, land use, governance, and other attributes. FORIN suggests four research strategies: (1) retrospective longitudinal analysis, (2) disaster scenario building, (3) comparative case analysis, and (4) meta-analysis (Oliver-Smith et al., 2016).

A complementary approach to FORIN is Forensic Disaster Analysis (FDA), a set of tools developed at the Center for Disaster Management and Risk Reduction Technology (CEDIM) in Germany. FDA uses tools for near real-time assessments mainly through multiple-source databases and remote sensing to produce indicators that could be used in the disaster aftermath and recovery process (Prizzia, 2016). Further, the Detecting Disaster Root Causes framework, developed by the German Committee for Disaster Reduction (DKKV), is based on case study analyses of the drivers and root causes of disaster risk across different categories such as development, governance, perception, political environment, and physical conditions (Witting, 2012). The framework is primarily based on a qualitative design and seeks to inform intervention options and produce recommendations. The results are framed in Table 25.2, which portrays the moments and constituents of the framework: drivers and root causes of the disaster on the vertical axis, and components on the horizontal axis.

Forensic perspectives consider not just the pre-disaster conditions and direct disaster impacts, but also how inadequate response and deficient recovery interventions amplify the disaster's negative consequences. In particular, the Post Event Review Capability (PERC) framework builds its perspective considering all elements of the traditional disaster management cycle, including recovery capabilities and actions, as part of the forensic analysis of the disaster process (Venkateswaran et al., 2020). The PERC framework's methodological guide examines the extent to which recovery actions reduce long-term impacts, whether such actions enable or constrain recovery, who benefits from the measures implemented, as well as the long-term impacts of recovery measures for the most vulnerable groups in society (Venkateswaran et al., 2015).

Several works have combined forensic perspectives with other incident analysis and industrial incident methodologies for a deeper understanding of disaster drivers and impacts. A prime example is the Integrative Post-Event Impact Assessment Framework, which proposes a systematic

Table 25.2 The German Committee for Disaster Reduction's Detecting Disaster Root Causes Framework

	Before	*After*
Levels of analysis	Vulnerability and disaster risk	Disaster risk management, response, and recovery
Root causes	Underlying patterns and structures beyond individual crises or events: 1 Development 2 Governance 3 Awareness and perception 4 Political environment 5 Physical and environmental conditions	
Drivers	Specific impacts and insecurities such as marginalization, fragile livelihoods, and negligence of disaster preparedness	

Source: Adapted from Witting (2012); Created by author.

analysis of the cascading impacts of volcanic eruptions combining approaches such as FORIN and PERC with root cause analysis and some of its visualization tools (Dominguez et al., 2021). Another example of this combination is the Risk and Root Cause Assessment (RRCA) Framework, a technique developed within the PEARL Project (Preparing for Extreme and Rare Events in coastal regions) (Fraser et al., 2014, 2016, 2020). Both perspectives use mixed methods approaches, each applying qualitative and quantitative techniques such as text analysis, interviews, field surveys, and assessment through the development of indicators to in-depth and/or comparative case studies.

Auditing Methodologies

Auditing is a form of public administration assessment that necessarily involves an evaluation of whether an institution has performed adequately in a public task. Auditing the process of disaster risk reduction or any part of the disaster cycle, such as recovery, is a complex responsibility that requires an analysis of how the public sector faces development processes, organizes with wider sectors of its society, and plans for uncertainties. Auditing methodologies can relate closely to the methods and challenges of forensic approaches, given that they both assess past events vis-à-vis the achievement of specific goals.

Although audits cannot necessarily change recovery contexts or directly improve accountability, an audit supports a long-term timeframe for disaster risk reduction activities. Audits establish normative parameters for institutional reforms and for designing future recovery interventions. The process can contribute to institutional learning, and the implementation of adaptive practices, and guide public actors to gain credibility through long-term interventions (Labadie, 2008).

There are three main types of auditing approaches: financial audits, compliance audits, and performance audits (INTOSAI, 2021). The first two types analyze the extent to which a subject matter—for example, a specific disaster event—adequately responds to previously established financial provisions, regulations, or management procedures, which constitute the referent for the assessment. On the other hand, performance audits investigate "whether interventions, programs, and institutions are performing by the principles of economy, efficiency, and effectiveness and whether there is room for improvement" (INTOSAI, 2021, 11). This type of evaluation is designed to provide recommendations, which means that it must analyze the complexities of disaster risk reduction more broadly, acknowledging international standards and best practices from a comparative perspective. These standards are considered in the risk assessments and problem analysis of the audited event.

The development of auditing methodologies in emergency management and disaster recovery started being considered after the 2004 Indian Ocean tsunami, which impacted 14 countries and mobilized a complex international process of humanitarian aid to the affected countries (Engwirda, 2010). In the face of this massive challenge and the simultaneous lack of methodologies to address the type of context, the International Organization of Supreme Audit Institutions (INTOSAI) developed the Task Force on the Audit and Accountability of Disaster Aid, which subsequently created the Working Group for the Audit of and Accountability for Disaster Aid in 2007 (Anderson, 2018). The Working Group developed guidelines and best practices in a series of International Standards of Supreme Audit Institutions (ISSAIs), which were known as ISSAI 5500 standards (INTOSAI, 2019). The standards include the audit of disaster risk reduction (ISSAI 5510); the audit of disaster-related aid (ISSAI 5520); adapting audit procedures to take account of the increased risk of fraud and corruption in the emergency phase following a disaster (ISSAI 5530); and the use of geospatial information to audit disaster management and disaster-related aid (ISSAI 5540). Each of these standards includes different procedures, guiding questions, data collection and organization strategies, and practical examples of how to conduct the assessment.

ISSAI 5500 standards were withdrawn in 2020 and substituted with the document "Guidance on Auditing Disaster Management," known as GUID 5330 (INTOSAI, 2020). Although the guidelines are not mandatory, they provide direction on how to assess whether the public sector achieves its objectives and priorities in the disaster management cycle. The guidelines also assess the extent to which institutional arrangements are adequate to manage relief and recovery operations. They particularly adopt the G20/OECD methodological framework (OECD, 2012), which recommends a risk assessment focusing on how well financial instruments, incentives, and risk transfer tools deal with disaster costs and losses. The G20/OECD framework recommends the use of various approaches such as impact assessment procedures, risk matrices, and scenarios to better prepare countries to finance recovery. GUID 5330 incorporates this as a general guide to understanding disaster preparedness planning but also develops further guiding questions to understand disaster response procedures and results, governance frameworks, and conducting risk assessments.

Conclusion

In this chapter, we explored two social science methodologies that seek to understand what disaster recovery means: normative approaches that examine the medium- and long-term changes that stem from disaster, and frameworks that evaluate the effectiveness of recovery interventions. Each of these perspectives defines disaster recovery and its components in a different manner, and each explores why and how societies change when confronted with the experience of deep loss, trauma, and uncertainty. In recent decades, recovery research has considered a wide range of matters and methodological approaches, including the deployment of mixed methods and combined quantitative and qualitative techniques, various sampling strategies, and diverse analytical frameworks. These tools have helped scientists to more clearly understand the social experience and to assist in the reduction of suffering in the years after a disaster. Although there are many challenges in post-disaster settings, the methodologies and works reviewed in this chapter might be useful for future researchers to inform, improve, and imagine more innovative and impactful approaches to disaster recovery.

References

Aldrich, Daniel P. *Black Wave: How Networks and Governance Shaped Japan's 3/11 Disasters*. The University of Chicago Press, 2019.

Anderson, Phyllis. "Cooperating for Results: Developing INTOSAI Guidelines to Audit Disaster-Related Aid." *Journal of the European Court of Auditors*, vol. 11, 2018, pp. 49–51.

Ayeb-Karlsson, Sonja, Alvi Chandra, and Karen E. McNamara. "Stories of Loss and Healing: Connecting Non-Economic Loss and Damage, Gender-Based Violence and Wellbeing Erosion in the Asia–Pacific Region." *Climatic Change*, vol. 176, no. 11, Nov. 2023, p. 157. *DOI.org (Crossref)*, https://doi.org/10.1007/s10584-023-03624-y

Bello, Omar, Laura Ortiz, and Joseluis Samaniego. *Assessment of the Effects of Disasters in Latin America and the Caribbean, 1972–2010*. United Nations, Economic Commission for Latin America and the Caribbean (ECLAC), 2015. https://www.cepal.org/sites/default/files/pages/files/s1501201_en.pdf.https://www.cepal.org/sites/default/files/pages/files/s1501201_en.pdf

Benedict, Bailey C., Seungyoon Lee, Caitlyn Jarvis, Laura K. Siebeneck, and Rachel Wolfe. "Utilising Qualitative Data for Social Network Analysis in Disaster Research: Opportunities, Challenges, and an Illustration." *Disasters*, 2023. 12605. *DOI.org (Crossref)*, https://doi.org/10.1111/disa.12605

Blackman, Deborah, Girish Prayang, Hitomi Nakanishi, Jo Chaffer, and Ben Freyens. "Wellbeing in Disaster Recovery: Understanding Where Systems Get Stuck." *International Journal of Disaster Risk Reduction*, vol. 95, Sept. 2023, p. 103839. *DOI.org (Crossref)*, https://doi.org/10.1016/j.ijdrr.2023.103839

Boyd, Emily, Rachel A. James, Richard G. Jones, Hannah R. Young, and Friederike Otto. "A Typology of Loss and Damage Perspectives." *Nature Climate Change*, vol. 7, no. 10, Oct. 2017, pp. 723–29. *DOI.org (Crossref)*, https://doi.org/10.1038/nclimate3389

Bui, Bonnie Khanh Ha, Philip Anglewicz, and Mark J. VanLandingham. "The Impact of Early Social Support on Subsequent Health Recovery after a Major Disaster: A Longitudinal Analysis." *SSM - Population Health*, vol. 14, Jun. 2021, p. 100779. *DOI.org (Crossref)*, https://doi.org/10.1016/j.ssmph.2021.100779

Choudhury, Mahed-Ul-Islam, and Haorui Wu. "Learning Capacity and Diversification, Enabling and Constraining Factors, and External Assistance: A Cross-National Comparative Analysis of Long-Term Livelihood Recovery." *International Journal of Mass Emergencies & Disasters*, vol. 41, no. 1, Mar. 2023, pp. 66–84. *DOI.org (Crossref)*, https://doi.org/10.1177/02807270231171511

Davis, Ian, and David Alexander. *Recovery from Disaster*. Routledge Taylor & Francis Group, 2016.

Denzin, Norman K., and Yvonna S. Lincoln, editors. *The SAGE Handbook of Qualitative Research*. 3rd edition, Sage Publications, 2005.

Dominguez, Lucia, Constanza Bonadonna, Corine Frischknecht, Scira Menoni, and Alexander Garcia. "Integrative Post-Event Impact Assessment Framework for Volcanic Eruptions: A Disaster Forensic Investigation of the 2011–2012 Eruption of the Cordón Caulle Volcano (Chile)." *Frontiers in Earth Science*, vol. 9, Jun. 2021, p. 645945. *DOI.org (Crossref)*, https://doi.org/10.3389/feart.2021.645945

Donner, William, and Walter Diaz. "Methodological Issues in Disaster Research." *Handbook of Disaster Research*, edited by Havidán Rodríguez et al., Springer International Publishing, 2018, pp. 289–309. *DOI.org (Crossref)*, https://doi.org/10.1007/978-3-319-63254-4_15

ECLAC. *Handbook for Disaster Assessment*. Economic Commission for Latin America and the Caribbean, 2014. https://repositorio.cepal.org/server/api/core/bitstreams/ade10e58-089e-44e2-b781-5282e32687e7/content

ECLAC. *Handbook for Estimating the Socio-Economic and Environmental Effects of Disasters*. Economic Commission for Latin America and the Caribbean, 2003. https://reliefweb.int/report/world/handbook-estimating-socio-economic-and-environmental-effects-disasters

ECLAC. *Manual for Estimating the Socio-Economic Effects of Natural Disasters*. Economic Commission for Latin America and the Caribbean, 1991. https://repositorio.cepal.org/server/api/core/bitstreams/21d50cc2-020d-4b73-b8df-c276beed5d57/content

Engwirda, Maarten B. "Can Auditors Make a Difference in Accounting for and Auditing Disaster-Related Aid?" *International Journal of Government Auditing*, vol. 37, no. 2, Apr. 2010, pp. 10–13.

European Commission, The United Nations Development Group, and the World Bank. *Joint Declaration on Post-Crisis Assessments and Recovery Planning*. 2008. https://eird.org/cd/recovery-planning/docs/3-pdna-process/PCNA-PDNA-Joint-Decl-EU-UNDG-WB.pdf

Few, Roger, Hazel Marsh, Garima Jain, Chandni Singh, and Mark Glyn Llewellyn Tebboth. "Representing Recovery: How the Construction and Contestation of Needs and Priorities Can Shape Long-Term Outcomes for Disaster-Affected People." *Progress in Development Studies*, vol. 21, no. 1, Jan. 2021, pp. 7–25. *DOI.org (Crossref)*, https://doi.org/10.1177/1464993420980939

Fortune, Joanne H. "Narrative Analysis of the 3-Year Recovery of Superstorm Sandy Survivors." *ProQuest Dissertations and Theses*, 2018. Walden University, Ph.D., https://www.proquest.com/docview/2092214551. ProQuest One Academic; Publicly Available Content Database, 2092214551.

Fraser, Arabella, Shona Paterson, and Mark Pelling. "Developing Frameworks to Understand Disaster Causation: From Forensic Disaster Investigation to Risk Root Cause Analysis." *Journal of Extreme Events*, vol. 03, no. 02, Jun. 2016, p. 1650008. *DOI.org (Crossref)*, https://doi.org/10.1142/S2345737616500081

Fraser, Arabella, Mark Pelling, Anna Scolobig, and Stavros Mavrogenis. "Relating Root Causes to Local Risk Conditions: A Comparative Study of the Institutional Pathways to Small-Scale Disasters in Three Urban Flood Contexts." *Global Environmental Change*, vol. 63, Jul. 2020, p. 102102. *DOI.org (Crossref)*, https://doi.org/10.1016/j.gloenvcha.2020.102102

Fraser, Arabella, Mark Pelling, Tobias Blätgen, and Jörn Birkmann. *The Risk and Root Cause Analysis (RRCA) Framework. Work Package 1. Deliverable Report 1.1. Grant Agreement N° 603663*. European

Union Seventh Framework Programme, 2014. https://www.academia.edu/34623396/The_PEARL_Risk_and_Root_Cause_Analysis_Framework

Fujimoto, Shinya, Fuminori Kawami, Anna Matsukawa, Shosuke Sato, and Shigeo Tatsuki. "Pre-Existing Inequalities, Post-Event Social Environmental Changes, and Long-Term Trajectories of Life Recovery: Findings from Five-Wave Natori City Life Recovery Panel Survey." *Journal of Disaster Research*, vol. 17, no. 2, Feb. 2022, pp. 207–16. *DOI.org (Crossref)*, https://doi.org/10.20965/jdr.2022.p0207

GFDRR. *Damage, Loss, and Needs Assessment Guidance Notes. Volume 1. Design and Execution of a Damage, Loss, and Needs Assessment Guidance Note for Task Managers*. The World Bank, Global Facility for Disaster Reduction and Recovery (GFDRR), 2010a. https://www.preventionweb.net/media/42962/download?startDownload=true

GFDRR.*Damage, Loss, and Needs Assessment Guidance Notes. Volume 2. Conducting Damage and Loss Assessments after Disasters*. The World Bank, Global Facility for Disaster Reduction and Recovery (GFDRR), 2010b. https://www.preventionweb.net/media/42961/download?startDownload=true

GFDRR. *Damage, Loss, and Needs Assessment Guidance Notes. Volume 3 Estimation of Post-Disaster Needs for Recovery and Reconstruction*. The World Bank, Global Facility for Disaster Reduction and Recovery (GFDRR), 2010c. https://www.preventionweb.net/media/42963/download?startDownload=true

GFDRR. *PDNA Guidelines Volume B. Gender*. Global Facility for Disaster Risk Reduction World Bank, European Commission, United Nations Development Group, 2013a. https://www.preventionweb.net/publication/post-disaster-needs-assessment-guidelines-volume-b-gender?preview=7827c50ee54

GFDRR. *Post-Disaster Needs Assessments Volume A. Guidelines*. Global Facility for Disaster Risk Reduction World Bank, European Commission, United Nations Development Group, 2013b. https://www.gfdrr.org/sites/default/files/publication/pdna-guidelines-vol-a.pdf

GFDRR. *Post-Disaster Needs Assessment PDNA. Lessons from a Decade of Experience*. Global Facility for Disaster Risk Reduction World Bank, European Commission, United Nations Development Group, 2018. https://www.gfdrr.org/sites/default/files/publication/Final_PDNA_Evaluation_Report.pdf

Goff, Amber Burtt, Sarah Friedman, and David Abramson. "Strategies and Costs of Building and Maintaining a Longitudinal Disaster Cohort." *International Journal of Mass Emergencies & Disasters*, vol. 41, no. 1, Mar. 2023, pp. 26–38. *DOI.org (Crossref)*, https://doi.org/10.1177/02807270231171502

Goodchild, Michael F, and J. Alan Glennon. "Crowdsourcing Geographic Information for Disaster Response: A Research Frontier." *International Journal of Digital Earth*, vol. 3, no. 3, Sept. 2010, pp. 231–41. *DOI.org (Crossref)*, https://doi.org/10.1080/17538941003759255

Gorostiza, Santiago, and Marco Armiero. "Repairing as Struggle for Narrative Justice." *Rethinking Post-Disaster Recovery*, edited by Laura Centemeri et al., 1st edition. Routledge, 2021, pp. 87–109. *DOI.org (Crossref)*, https://doi.org/10.4324/9781003184782-7

Graham, Lindsay. "Recovery Projects Haitian Memory, Humanitarian Response and the Affordances of the Digital Disaster Archive." *Negotiating Institutional Heritage and Wellbeing*, edited by Elisabeth Punzi et al., BRILL, 2022, pp. 90–114. *DOI.org (Crossref)*, https://doi.org/10.1163/9789004468900

Gray, Clark, Elizabeth Frankenberg, Thomas Gillespie, Cecep Sumantri, and Duncan Thomas. "Studying Displacement after a Disaster Using Large Scale Survey Methods: Sumatra After the 2004 Tsunami." *Annals of the Association of American Geographers. Association of American Geographers*, vol. 104, no. 3, Jan. 2014, pp. 594–612. *PubMed*, https://doi.org/10.1080/00045608.2014.892351

Guest, Greg, Emily E. Namey, and Marilyn L. Mitchell. *Collecting Qualitative Data: A Field Manual for Applied Research*. SAGE Publications, Ltd, 2013. *DOI.org (Crossref)*, https://doi.org/10.4135/9781506374680

Hamideh, Sara, Sabine Loos, Jason Rivera, Alessandra Jerolleman, Heather Champeau, and Haorui Wu. "IJMED Special Issue: Longitudinal Recovery." *International Journal of Mass Emergencies & Disasters*, vol. 41, no. 1, Mar. 2023, pp. 4–8. *DOI.org (Crossref)*, https://doi.org/10.1177/02807270231184213

Hawkins, Robert L, and Katherine Maurer. "Bonding, Bridging, and Linking: How Social Capital Operated in New Orleans Following Hurricane Katrina." *British Journal of Social Work*, vol. 40, no. 6, Sept. 2010, pp. 1777–93. *DOI.org (Crossref)*, https://doi.org/10.1093/bjsw/bcp087

Hayashi, Isao. "Museums as Hubs for Disaster Recovery and Rebuilding Communities." *New Horizons for Asian Museums and Museology*, edited by Naoko Sonoda, Springer Singapore, 2016, pp. 165–76. *DOI.org (Crossref)*https://doi.org/10.1007/978-981-10-0886-3_12

Hirsch, Thomasch, Sabine Minninger, Sophia Wirsching, Isaiah Toroitich, Elena Cedillo, Michiel Hermans, Endeshaw Kassa, Habib Torikul, and Efaz Ahmed. *Non-Economic Loss and Damage. With Case Examples from Tanzania, Ethiopia, El Salvador, and Bangladesh. Discussion Paper 76*. Act Alliance, Bread for the World, Protestant Agency for Diakonie and Development, 2017. https://reliefweb.int/report/world/non-economic-loss-and-damage-case-examples-tanzania-ethiopia-el-salvador-and-bangladesh

Horney, Jennifer, Caroline Dwyer, Bhagath Chirra, Kerry McCarthy, Jennifer Shafer, and Gavin Smith. "Measuring Successful Disaster Recovery." *International Journal of Mass Emergencies & Disasters*, vol. 36, no. 1, Mar. 2018, pp. 1–22. *DOI.org (Crossref)*, https://doi.org/10.1177/028072701803600101

Hossain, Liaquat, and Matthew Kuti. "Disaster Response Preparedness Coordination through Social Networks." *Disasters*, vol. 34, no. 3, Jul. 2010, pp. 755–86. *DOI.org (Crossref)*, https://doi.org/10.1111/j.1467-7717.2010.01168.x

INTOSA. *Introduction to the 5500 Series of ISSAIs and INTOSAI GOV 9250*. International Organization of Supreme Audit Institutions (INTOSAI), 2019. https://reliefweb.int/sites/reliefweb.int/files/resources/23056785.PDF

INTOSA. *GUID 5330. Guidance on Auditing Disaster Management*. International Organization of Supreme Audit Institutions (INTOSAI), 2020. https://www.issai.org/wp-content/uploads/2020/12/GUID-5330-English.pdf

INTOSA. *ISSAI 100. Fundamental Principles of Public-Sector Auditing*. International Organization of Supreme Audit Institutions (INTOSAI), 2021. https://www.issai.org/wp-content/uploads/2021/02/ISSAI-100-related-to-new-ISSAI-150-FIN-1.pdf

Islam, Rabiul, and Greg Walkerden. "Social Networks and Challenges in Government Disaster Policies: A Case Study from Bangladesh." *International Journal of Disaster Risk Reduction*, vol. 22, Jun. 2017, pp. 325–34. *DOI.org (Crossref)*, https://doi.org/10.1016/j.ijdrr.2017.02.011

Jerolleman, Alessandra. "Storytelling and Narrative Research in Crisis and Disaster Studies." *Oxford Research Encyclopedia of Politics*, by Alessandra Jerolleman, Oxford University Press, 2021. *DOI.org (Crossref)*, https://doi.org/10.1093/acrefore/9780190228637.013.1615

Jones, Eric C., and A. J. Faas, editors. *Social Network Analysis of Disaster Response, Recovery, and Adaptation*. 1st edition, Elsevier, 2016.

Juran, Luke. "Disaster, Development, and Water: The Reconstruction and (Re)Fabrication of Hazardous Waterscapes in Post-Tsunami India." *The Asian Tsunami and Post Disaster Aid*, edited by Sunita Reddy. 1st edition, Springer, 2018, pp. 79–99. *DOI.org (Crossref)*, http://link.springer.com/10.1007/978-981-13-0182-7_5

Kapucu, Naim, and Fatih Demiroz. "Interorganizational Networks in Disaster Management." *Social Network Analysis of Disaster Response, Recovery, and Adaptation*, edited by Eric C. Jones and A. J. Faas, 1st edition, Elsevier, 2016, pp. 25–39

Kapucu, Naim, and Qian Hu. "Understanding Multiplexity of Collaborative Emergency Management Networks." *The American Review of Public Administration*, vol. 46, no. 4, Jul. 2016, pp. 399–417. *DOI.org (Crossref)*, https://doi.org/10.1177/0275074014555645

Karatani, Yuka, and Haruo Hayashi. "Quantitative Evaluation of Recovery Process in Disaster-Stricken Areas Using Statistical Data." *Journal of Disaster Research*, vol. 2, no. 6, 2007, pp. 453–64, https://doi.org/10.20965/jdr.2007.p0453

Kargillis, Christina, Mayumi Kako, and David Gillham. "Disaster Survivors: A Narrative Approach towards Emotional Recovery." *Australian Journal of Emergency Management*, vol. 29, no. 2, 2014, pp. 25–30

Kim, Jooho, and Makarand Hastak. "Social Network Analysis: Characteristics of Online Social Networks after a Disaster." *International Journal of Information Management*, vol. 38, no. 1, Feb. 2018, pp. 86–96. *DOI.org (Crossref)*, https://doi.org/10.1016/j.ijinfomgt.2017.08.003

King, David, and Yetta Gurtner. "Focusing Post-Disaster Research Methodology: Reflecting on 50 Years of Post-Disaster Research." *The Australian Journal of Emergency Management*, vol. 36, no. 4, Oct. 2021, pp. 32–39. agis.20211218058593.

Labadie, John R. "Auditing of Post-Disaster Recovery and Reconstruction Activities." *Disaster Prevention and Management: An International Journal*, vol. 17, no. 5, Nov. 2008, pp. 575–86. *Crossref*, https://doi.org/10.1108/09653560810918612

Lee, Seungyoon, Arif Mohaimin Sadri, Satish V. Ukkusuri, Rosalee A. Clawson, and Justin Seipel. "Network Structure and Substantive Dimensions of Improvised Social Support Ties Surrounding Households during Post-Disaster Recovery." *Natural Hazards Review*, vol. 20, no. 4, Nov. 2019, p. 04019008. *DOI.org (Crossref)*, https://doi.org/10.1061/(ASCE)NH.1527-6996.0000332

Liu, Sophia, Leysia Palen, Jeannette Sutton, Amanda Hughes, and Sarah Vieweg. "In Search of the Bigger Picture: The Emergent Role of On-Line Photo Sharing in Times of Disaster." *Proceedings of the 5th International ISCRAM Conference*, Qatar Computing Research Institute, 2008, p. 11, https://

citeseerx.ist.psu.edu/document?repid=rep1&type=pdf&doi=14ddb00802c004ec94fe1bc84b3539c1f353bec9

Mahn, Churnjeet, Caroline Scarles, Justin Edwards, and John Tribe. "Personalising Disaster: Community Storytelling and Sharing in New Orleans Post-Katrina Tourism." *Tourist Studies*, vol. 21, no. 2, Jun. 2021, pp. 156–77. *DOI.org (Crossref)*, https://doi.org/10.1177/1468797620967597

Marín, Andrés, Örjan Bodin, Stefan Gelcich, and Beatrice Crona. "Social Capital in Post-Disaster Recovery Trajectories: Insights from a Longitudinal Study of Tsunami-Impacted Small-Scale Fisher Organizations in Chile." *Global Environmental Change*, vol. 35, Nov. 2015, pp. 450–62. *DOI.org (Crossref)*, https://doi.org/10.1016/j.gloenvcha.2015.09.020

McShane, Katie. "Values and Harms in Loss and Damage." *Ethics, Policy & Environment*, vol. 20, no. 2, May 2017, pp. 129–42. *DOI.org (Crossref)*, https://doi.org/10.1080/21550085.2017.1342960

Mendoza, Marina Tamara. Forensic Investigation of Post-Flood Damage Data to Support Spatial Planning. 2019. Politecnico di Milano, Ph.D. Thesis in Urban Planning, Design, and Policy, https://www.politesi.polimi.it/handle/10589/145774

Menk, Linda, Thomas Schinko, Veronica Karabaczek, Isabel Hagen, and Stefan Kienberger. "What's at Stake? A Human Well-Being Based Proposal for Assessing Risk of Loss and Damage from Climate Change." *Frontiers in Climate*, vol. 4, Dec. 2022, p. 1032886. *DOI.org (Crossref)*, https://doi.org/10.3389/fclim.2022.1032886

Merdjanoff, Alexis, Sarah Friedman, Rachael Piltch-Loeb, and David Abramson. "The Detrimental Effect of High Mobility on Long-Term Mental Health Trajectories Following Hurricane Katrina." *International Journal of Mass Emergencies & Disasters*, vol. 41, no. 1, Mar. 2023, pp. 121–32. *DOI.org (Crossref)*, https://doi.org/10.1177/02807270231171630

Misra, Sanchayeeta, Rupak Goswami, Tandra Mondal, and Rabindranath Jana. "Social Networks in the Context of Community Response to Disaster: Study of a Cyclone-Affected Community in Coastal West Bengal, India." *International Journal of Disaster Risk Reduction*, vol. 22, Jun. 2017, pp. 281–96. *DOI.org (Crossref)*, https://doi.org/10.1016/j.ijdrr.2017.02.017

Monteil, Charlotte, Jenni Barclay, and Anna Hicks. "Remembering, Forgetting, and Absencing Disasters in the Post-Disaster Recovery Process." *International Journal of Disaster Risk Science*, vol. 11, no. 3, Jun. 2020, pp. 287–99. *DOI.org (Crossref)*, https://doi.org/10.1007/s13753-020-00277-8

Morrow-Jones, Hazel A, and Charles R. Morrow-Jones. "Mobility Due to Natural Disaster: Theoretical Considerations and Preliminary Analyses." *Disasters*, vol. 15, no. 2, Jun. 1991, pp. 126–32. *DOI.org (Crossref)*, https://doi.org/10.1111/j.1467-7717.1991.tb00441.x

Mota-Ferreira, Adriano, Victor Marchezini, Tatiana Sussel Gonçalves, Miguel Angel Trejo-Rangel, and Allan Yu Iwama. "A Systematic Review of Forensic Approaches to Disasters: Gaps and Challenges." *International Journal of Disaster Risk Science*, Oct. 2023, https://doi.org/10.1007/s13753-023-00515-9

Norris, Fran H. "Disaster Research Methods: Past Progress and Future Directions." *Journal of Traumatic Stress*, vol. 19, no. 2, Apr. 2006, pp. 173–84. *DOI.org (Crossref)*, https://doi.org/10.1002/jts.20109

OECD. *Disaster Risk Assessment and Risk Financing. A G20/OECD Methodological Framework.* Organization for Economic Co-operation and Development:. 2012. https://www.oecd.org/finance/insurance/G20disasterriskmanagement.pdf

Oliver-Smith, Anthony, Irasema Alcántara-Ayala, Ian Burton, and Allan Lavell. *Forensic Investigations of Disasters (FORIN): A Conceptual Framework and Guide to Research (IRDR FORIN Publication No.2).* Integrated Research on Disaster Risk. 2016. https://www.researchgate.net/publication/291349173_Forensic_Investigations_of_Disasters_FORIN_a_conceptual_framework_and_guide_to_research

O'Mathúna, Dónal P, and Bert Gordijn. "Conceptualizing and Assessing Disasters: An Introduction". *Disasters: Core Concepts and Ethical Theories*, edited by Dónal P. O'Mathúna, Vilius Dranseika, and Bert Gordijn, 1st edition, SpringerOpen, 2018, pp. 1–9. *DOI.org (Crossref)*, https://doi.org/10.1007/978-3-319-92722-0_1

Ono, Yuichi, Marlene Murray, Makoto Sakamoto, Hiroshi Sato, Pornthum Thumwimol, Vipakorn Thumwimol, and Ratchaneekorn Thongthip. "The Role of Museums in Telling Live Lessons." *Journal of Disaster Research*, vol. 16, no. 2, Feb. 2021, pp. 135–40. *DOI.org (Crossref)*, https://doi.org/10.20965/jdr.2021.p0135

Peacock, Walter Gillis, Shanon Van Zandt, Yang Zhang, and Wesley E. Highfield. "Inequities in Long-Term Housing Recovery after Disasters." *Journal of the American Planning Association*, vol. 80, no. 4, Oct. 2014, pp. 356–71. *DOI.org (Crossref)*, https://doi.org/10.1080/01944363.2014.980440

Pill, Melanie. "Re-Framing Non-Economic Losses to Non-Economic Impacts for Effective Policymaking: Evidence from the Caribbean." *Climate and Development*, vol. 14, no. 8, Sept. 2022, pp. 770–79. *DOI.org (Crossref)*, https://doi.org/10.1080/17565529.2021.1987852

Phillips, Brenda D. *Qualitative Disaster Research*. Oxford University Press. 2014.

Prizzia, Ross. "Climate Change and Disaster Forensics." *Disaster Forensics*, edited by Anthony J. Masys, Springer International Publishing, 2016, pp. 223–47. *DOI.org (Crossref*https://doi.org/10.1007/978-3-319-41849-0_9

Rahiem, Maila D. H, Robien Ersing, Steven E. Krauss, and Husni Rahim. "Narrative Inquiry in Disaster Research: An Examination of the Use of Personal Stories from the Child Survivors of the 2004 Aceh Tsunami." *International Journal of Disaster Risk Reduction*, vol. 65, Nov. 2021, p. 102544. *DOI.org (Crossref)*, https://doi.org/10.1016/j.ijdrr.2021.102544

Rivera, Jason David, editor. *Disaster and Emergency Management Methods: Social Science Approaches in Application*. 1st edition. Routledge Taylor & Francis Group, 2022.

Romão, Xavier, and Esmeralda Paupério. "An Indicator for Post-Disaster Economic Loss Valuation of Impacts on Cultural Heritage." *International Journal of Architectural Heritage*, vol. 15, no. 5, May 2021, pp. 678–97. *DOI.org (Crossref)*, https://doi.org/10.1080/15583058.2019.1643948

Russo, Barbara. "Mixed-Methods Research in Disaster and Emergency Management." *Disaster and Emergency Management Methods: Social Science Approaches in Application*, edited by Jason David Rivera, 1st edition. Routledge Taylor & Francis Group, 2022, pp. 67–84.

Ryan, Roberta, Liana Wortley, and Éidín Ní Shé. "Evaluations of Post-Disaster Recovery: A Review of Practice Material." *Evidence Base: A Journal of Evidence Reviews in Key Policy Areas*, no. 4, Jan. 2016, pp. 1–33. informit.212676752567547.

Sakamoto, Mayumi. "Disaster Memories in Museums and Disaster Recovery. Disaster Reduction Institute and 1995 Hanshin-Awaji Earthquake Recovery." *17th World Conference on Earthquake Engineering, 17WCEE*, vol. 10a–0009, 2020, p. 10.

Sanyal, Saswata, and Jayant K. Routray. "Social Capital for Disaster Risk Reduction and Management with Empirical Evidences from Sundarbans of India." *International Journal of Disaster Risk Reduction*, vol. 19, Oct. 2016, pp. 101–11. *DOI.org (Crossref)*, https://doi.org/10.1016/j.ijdrr.2016.08.010

Schrank, Holly L, Maria I. Marshall, Adrienne Hall-Phillips, Renee F. Wiatt, and Nicole E. Jones. "Small-Business Demise and Recovery after Katrina: Rate of Survival and Demise." *Natural Hazards*, vol. 65, no. 3, Feb. 2013, pp. 2353–74. *DOI.org (Crossref)*, https://doi.org/10.1007/s11069-012-0480-2

Serdeczny, Olivia, Eleanor Waters, and Sander Chan. *Non-Economic Loss and Damage in the Context of Climate Change: Understanding the Challenges*. Deutsches Institut für Entwicklungspolitik. 2016. https://www.idos-research.de/uploads/media/DP_3.2016.pdf

Sou, Gemma, Duncan Shaw, and Felix Aponte-Gonzalez. "A Multidimensional Framework for Disaster Recovery: Longitudinal Qualitative Evidence from Puerto Rican Households." *World Development*, vol. 144, Aug. 2021, p. 105489. *DOI.org (Crossref)*, https://doi.org/10.1016/j.worlddev.2021.105489

Stallings, Robert A., editor. *Methods of Disaster Research*. Xlibris. 2002.

Tanaka, Naoto, Satoru Ikaptra, Mariko Kusano, Kuzuo Yamasaki, and Matsumoto. "Disaster Tourism as a Tool for Disaster Story Telling." *Journal of Disaster Research*, vol. 16, no. 2, Feb. 2021, pp. 157–62. *DOI.org (Crossref)*, https://doi.org/10.20965/jdr.2021.p0157

Tatsuki, Shigeo, and Fuminori Kawami. "Longitudinal Impacts of Pre-Existing Inequalities and Social Environmental Changes on Life Recovery: Results of the 1995 Kobe Earthquake and the 2011 Great East Japan Earthquake Recovery Studies." *International Journal of Mass Emergencies & Disasters*, vol. 41, no. 1, Mar. 2023, pp. 94–120. *DOI.org (Crossref)*, https://doi.org/10.1177/02807270231171504

Terumoto, Kiyomine, Yoriko Tsuchiya, Rie Otagiri, Hironobu Nakabayashi, and Itsuki Nakabayashi. "Individual Disaster Recovery: A Framework in the Long-Term Recovery Process after the Great East Japan Earthquake." *International Journal of Disaster Risk Reduction*, vol. 60, Jun. 2021, p. 102280. *DOI.org (Crossref)*, https://doi.org/10.1016/j.ijdrr.2021.102280

Terzis, Lauren D, Leia Y. Saltzman, Dana A. Logan, Joan M. Blakey, and Tonya C. Hansel. "Utilizing a Matrix Approach to Analyze Qualitative Longitudinal Research: A Case Example During the COVID-19 Pandemic." *International Journal of Qualitative Methods*, vol. 21, Jan. 2022, p. 160940692211237. *DOI.org (Crossref)*, https://doi.org/10.1177/16094069221123723

Tetsuya, Okada. Acknowledging Local Sociality in Disaster Recovery: A Longitudinal, Qualitative Study. 2017. Macquarie University, PhD Thesis in Environmental Science, https://figshare.mq.edu.au/articles/thesis/Acknowledging_local_sociality_in_disaster_recovery_a_longitudinal_qualitative_study/19430861/1

Tomaszewski, Brian, Michael Judex, Joerg Szarzynski, Christine Radestock, and Lars Wirkus. "Geographic Information Systems for Disaster Response: A Review." *Journal of Homeland Security and Emergency Management*, vol. 12, no. 3, Jan. 2015. *DOI.org (Crossref)*, https://doi.org/10.1515/jhsem-2014-0082.
Tomaszewski, Brian, editor. *Geographic Information Systems (GIS) for Disaster Management*. 2nd edition, Routledge, Taylor & Francis Group, 2021.
United Nations. *Report of the Open-Ended Intergovernmental Expert Working Group on Indicators and Terminology Relating to Disaster Risk Reduction. A/71/644*. United Nations General Assembly. 2016. https://www.preventionweb.net/files/50683_oiewgreportenglish.pdf
van der Velden, Peter G, Joris Yzermans, and Linda Grievink. "Enschede Fireworks Disaster." *Mental Health and Disasters*, edited by Yuval Neria, Sandro Galea and Fran H. Norris, Cambridge University Press, 2009, pp. 473–96.
Varda, Danielle. "Strategies for Researching Social Networks in Disaster, Response, Recovery and Mitigation." *Social Network Analysis of Disaster Response, Recovery, and Adaptation*, edited by Eric C. Jones and A. J. Faas, 1st edition, Elsevier, 2016, pp. 41–56.
Venkateswaran, Kanmani, Karen MacClune, Adriana Keating, and Michael Szönyi. *The PERC Manual. Learning from Disasters to Build Resilience: A Simple Guide to Conduct a Post Event Review*. ISET-International, IIASAZurich Insurance Group Ltd. 2015. https://www.i-s-e-t.org/_files/ugd/558f8a_1c59df95b2d84086a89b23a92b3ffa77.pdf
Venkateswaran, Kanmani, Karen MacClune, Adriana Keating, and Michael Szönyi. *The PERC Manual. Learning from Disasters to Build Resilience: A Guide to Conducting a Post-Event Review – (2020)*. ISET-International, IIASAZurich Insurance Group Ltd. 2020. https://www.i-s-e-t.org/_files/ugd/558f8a_b56c2c34ae8d4ff0b0637c6af7c7aacd.pdf
Wasserman, Stanley, and Katherine Faust. *Social Network Analysis: Methods and Applications*. Cambridge University Press. 1994.
Watson, Maria, Yu Xiao, Jennifer Helgeson, and Maria Dillard. "Importance of Households in Business Disaster Recovery." *Natural Hazards Review*, vol. 21, no. 4, Nov. 2020, p. 05020008. *DOI.org (Crossref)*, https://doi.org/10.1061/(ASCE)NH.1527-6996.0000393
Wilkin, Joanna, Eloise Biggs, and Andrew Tatem. "Measurement of Social Networks for Innovation within Community Disaster Resilience." *Sustainability*, vol. 11, no. 7, Apr. 2019, p. 1943. *DOI.org (Crossref)*, https://doi.org/10.3390/su11071943
Witting, Maximilian. *Detecting Disaster Root Causes a Framework and an Analytic Tool for Practitioners*. German Committee for Disaster Reduction (DKKV). 2012. https://www.preventionweb.net/files/globalplatform/entry_bg_paper~studydetectingdisasterrootcausesweb.pdf

Section VII

Disaster Recovery Contexts

26

THE ROLE OF HOUSEHOLDS AND COMMUNITIES IN DISASTER RESPONSE AND RECOVERY

The Case of the Pacific and Small Island Developing States

Anand Chand and Luke Juran

Introduction

Small island developing states (SIDS) are exposed to multiple hazard risks, particularly the increasing intensity of cyclones and sea-level rise due to anthropogenic climate change (Corlew, 2012; Nunn & Kumar, 2018; Weber et al., 2015). The most common disasters in the Pacific region are cyclones, floods, landslides, tidal waves ('king tide'), droughts, forest fires, earthquakes, and tsunamis (IPCC, 2012; Yila et al., 2014). The Intergovernmental Panel on Climate Change (IPCC, 2012) defines disaster response as 'actions taken directly before, during or immediately after a disaster to save lives, reduce health impacts, ensure public safety and meet the basic subsistence needs of the affected people … it is predominantly focused on immediate and short-term needs and is sometimes called disaster relief' (231). Disaster recovery is a long-term process to return normalcy to the community. Recovery activities may begin while emergency response actions are in progress ('early recovery'), and decisions made during the response phase directly influence recovery.

Both state and non-state actors are involved in disaster response and recovery in Pacific island countries (Chand & Taupo, 2020). At the national/state level, most Pacific island countries, with the support of international donor agencies, have developed National Disaster Management Plans, established National Disaster Management Councils (NDMCs), and formed various committees to focus on disaster management. For example, Kiribati developed its National Disaster Risk Management Plan (NDRMP) in 2022. International disaster relief organizations have realized that top-down, state-level response has not been effective in the region and now recognize that utilizing bottom-up, community-based approaches (CBAs) is more successful and garners greater buy-in (Nunn & Kumar, 2018; Weber et al., 2015; Yila et al., 2014). In the Pacific, CBA is a small-scale, grassroots approach that employs and enhances existing social capital resources such as kinship, social networks, communal systems, traditional governance and leadership, reciprocity, collective labor, and exchanging food and services as critical response and recovery mechanisms. Ratuva (2014) does an excellent job illustrating these social protection

DOI: 10.4324/9781003414834-36

mechanisms in Fiji, Samoa, Kiribati, and Vanuatu. A major advantage of CBA is that it harnesses communities' existing knowledge, capacities, needs, and priorities to inform disaster response and recovery (Ensor & Berger, 2009; Glauberman et al., 2024; Kirkby et al., 2017). This chapter explores the role of households and communities in disaster response and recovery in Oceania with parallels for SIDS and lesser-developed states that share similar issues of remoteness, fragmentation, scarcity of built infrastructure, and multi-hazard risk.

Literature Review

There is a robust literature on the role of state and non-state actors across the continuum of disaster management. For example, scholars have focused on the role of the national government (Djalante & Thomalla, 2011), police functions (Blaustein et al., 2023), international disaster relief operations (Paul, 2019), early warning systems, disaster preparedness (Cox et al., 2020), multi-hazard modeling (Aksha et al., 2020), the role of the church (Holmgaard, 2019), community resilience (Weber et al., 2015), and the disproportional gendered impacts of disasters (Juran, 2012). This chapter focuses on non-state actors—specifically households and communities—in disaster response and recovery in the Pacific. International disaster relief organizations have recognized that state-level disaster response is ineffective in Pacific island countries (Nunn & Kumar, 2018; Weber et al., 2015; Yila et al., 2014). There is growing scholarship on the role of non-state actors. Much of the literature argues that the household and community play a vital role in disasters and that we should thus focus on such actors in disaster management—particularly in relatively poor and lesser-developed countries (Mann & Williams, 2021). The means and modalities of household and community response vary across countries. For instance, literature situated in Vanuatu and elsewhere elucidates the importance of social capital resources such as kinship, reciprocity, and communal relationships as principal social protection mechanisms (Pelling & High, 2005; Ratuva, 2010; Sadri et al., 2018).

Moreover, studies have highlighted the role of informal and semi-formal indigenous institutions (e.g., chiefs, village leaders, 'big men,' youth groups, and religious congregations) in mobilizing communities both pre- and post-disaster (Cox et al., 2020; Fernandez & Shaw, 2015; Schuermann & Lauer, 2016). This may involve providing community leaders and institutions with authority and resources as crucial emergency relief partners (Rumbach & Foley, 2014). Other informal actors also participate in disaster response and recovery. For instance, research has explored the role of religious institutions and churches in supporting their members through various caretaking initiatives (Ha, 2015; Sheikhi et al., 2021). As demonstrated by Holmgaard (2019) in Samoa, religious institutions are not only a central fabric of society, but they also have resources (e.g., buildings for shelter, safety, and storage as well as kitchens and everyday items) that are extremely valuable during disasters. Furthermore, the literature has established the importance of women's groups in disaster response and recovery, particularly the value of female (and other genders) experiences, knowledge, and participation (i.e., gender mainstreaming) (Chowdhury et al., 2022; Juran & Trivedi, 2015).

In light of the current literature, this chapter seeks to fill existing research gaps by examining the importance of both recognizing and leveraging household and community-based disaster risk reduction strategies in Pacific island countries. Successful disaster response and recovery must acknowledge that households and communities exhibit a broad range of needs and priorities. Therefore, leveraging such units in disaster processes is essential to responding to the grounded realities and fostering community cohesion and engagement.

Coping Strategies in the Pacific Islands

There are several disaster-related coping strategies in the South Pacific. Prominent among these are kinship and the cultural practices of reciprocity and communal labor, each of which is discussed in this section. First, kinship is one of the most fundamental strategies. Ratuva (2014) contends that 'one of the most resilient mechanisms for social protection in the Pacific is kinship because it provides the basis for collective support for the community's social, economic, and psychological needs in times of crisis' (46). Nunn and Kumar (2018) argue that kinship in the Pacific islands encompasses a range of unique resilience characteristics that help them cope with climate change, including high degrees of social coherence, closeness to nature, and knowledge of the sensitivity of coastal fringes to climate-environmental changes. In this sense, kinship embodies Bourdieu's (2018) conceptualization of kinship as cultural capital that people can draw upon during times of social and economic uncertainty. In the South Pacific, kinship and extended family relationships are expressed in local languages such as *veiwekani* in Fiji, *au ainga* in Samoa, *utu* in Kiribati, and *wantok* in Vanuatu (Ratuva, 2014). Kinship relations predominantly focus on people's daily lives; thus, such networks are practical, and functional, and provide a safety net during 'normal' times and especially in times of disaster. For example, during disasters in the Pacific, kin members are expected to distribute goods and services and failure to do so is frowned upon by other community members. When national and local government disaster response fails, communities in Pacific island countries are forced to rely on social support mechanisms such as kinship for both immediate survival and long-term recovery. Ratuva (2014) argues that community-based protection systems are deeply entrenched in Pacific cultures. These cultural values, norms, customs, and practices—of which kinship is central—comprise a set of vital coping mechanisms that communities in the South Pacific activate during disasters.

Reciprocity is another key social protection mechanism in the Pacific. Reciprocity refers to the exchange of goods and services among community members. This helps safeguard the community and is especially common and effective when people cannot purchase things due to financial and/or crises. In Hau'ofa's formative essay 'Our Sea of Islands' (1994), reciprocity is couched as 'not dependence but *interdependence* [emphasis added]' (157). Reciprocity extends dignity to others in the community and often occurs with goods produced by the individuals themselves, thereby strengthening bonds between both humans and the land. Ratuva (2014) further stresses that reciprocity 'reinforces kinship links and community identity and defines the boundaries and the types of goods to be produced in a community's exchange economy' (46). For instance, the practice of *kerekere* (requesting something such as food or household items from another individual) is a crucial part of the exchange economy and thus culture in the case of Fiji (Ravuvu, 1987). In Samoa, this practice is known as *o au nei o oe taeao*, roughly translated as 'you support me and I will support you next time' (Macpherson, 1999). Among indigenous Hawaiians, the practice is known as *kokua*, literally meaning to 'relieve' or 'assist,' and it is recognized as a form of empathetic exchange when someone needs help (Kanahele, 1986; McCubbin et al., 2008). In Tuvalu, the concept of reciprocity is known as *bubuti* (Taupo et al., 2018), and in the Easter Island, the first exchange between two individuals is known as *hōo* with subsequent acts of reciprocity referred to as *hakahoki* (McCall, 1980). While examples of reciprocity exist in all places, its role and cultural significance are distinct in the Pacific, particularly during disasters.

Third is communal labor, which is an extension of kinship obligation. Communal labor brings together a group of people to achieve a certain goal through the utilization of public skills to develop society. In the Pacific islands, if a household needs something to be done—such as building a house, creating a new vegetable garden, making a canoe, or preparing for a marriage or funeral ceremony—the community will assist without remuneration (Ratuva, 2014; Ravuvu, 1987). This is different from individual paid labor as practiced in the West and departs from the contemporary global economy (McCall, 1980), as aptly demonstrated by Pendry (2022) in Tonga. For instance, in the Pacific islands, entire villages take turns building a house or creating a new vegetable garden for families in the community, and the process keeps perpetuating. This practice is known as *ala ile pule ole tautua* (the path to leadership is service) in the Samoan language, it is referred to as *solesolevaki* (to support each other) in the Fijian language, and it is termed *karekare* (to take turns) in Kiribati (Ravuvu, 1987). Collectively, cultures forged and maintained by indelible traits of kinship, reciprocity, and communal labor help to sustain Pacific island communities during times of disaster.

Household and Community Response and Recovery in the Pacific

Given limited state resources and capacity, the predominance of poorly built infrastructure, and the isolated and fragmented nature of small island nations, households and communities in the Pacific play an outsized role in disaster response and recovery. In other words, households and communities cannot depend on formal and state actors to assist them in times of crisis. Rather, they must prepare for disasters and initiate recovery by themselves. Thus, in Pacific island countries, disaster response and recovery are often best achieved when affected households and communities do it themselves with external assistance seeking to galvanize and complement such efforts (i.e., a high degree of self-determination). This section will discuss these issues.

Disaster Response by Households

As first responders, households play a primary role in disaster response as they cannot rely on state agencies (Lindell et al., 2015). Before a disaster, whether it be a cyclone, storm, or flood, households make preparations. First, families stock up on food and clean water to ensure adequate supplies are available during and post-disaster. Previous studies demonstrate that food and water are the most stored items by households in the Pacific islands (Elliott et al., 2017; Thaman et al., 2002).

Safe drinking water and access to multiple sources are essential as water cuts, water pollution, and the spread of waterborne diseases are common during and in the aftermath of disasters (Elliott et al., 2017; McNamara et al., 2020). Households store enough food supplies to last for the anticipated duration of the disaster, and families supplement their food supply with home gardens and help from kin and social networks. In agricultural and rural communities, households harvest and store root crops in cases of heavy rain and flooding as such sub-surface crops tend to rot and get damaged. The storage of food is needed as most rural communities in the Pacific are engaged in subsistence agriculture. Thus, crops represent a significant source of survival and a shock to the agricultural base could have far-reaching consequences. Households also protect crops that are not mature enough to harvest by shielding them with the shredded leaves of other plants to reduce the impacts of wind and rain. This is a traditional food security measure that farmers have learned over generations orally from their ancestors.

In addition to securing water and food supplies before disaster, households in the Pacific store batteries to power radios and torches/flashlights. This enables households to listen to warning notices and provides visibility during electricity cuts. It is important to emphasize the importance of radios in the Pacific as the internet and televisions are relatively uncommon compared to high- and middle-income countries, and cellular service is nonexistent in many places (especially during disasters). Households also prepare differently for different types of disasters, indicating that they know to respond accordingly. For example, during the COVID-19 pandemic, households stored medicine, face masks, hand sanitizers, and food (Chand et al., 2022).

Finally, before disasters, community members, neighbors, and villagers assist each other in securing weak houses. This is done by tying ropes and wires to prevent the roofs from being blown away during cyclones and strong winds. Shutters are also placed over windows to prevent them from being broken by wind and debris. Furthermore, households perform tasks such as cutting old and dead tree branches to avoid property damage and providing shelter to family members and neighbors with structurally weaker houses.

After disasters, household recovery activities begin while the emergency response is still in progress, and this transition from response to recovery has been termed 'early recovery' (ADRF, 2022). As disasters subside, households begin cleaning houses and clearing debris from their yards and community compounds. The community helps each other clean up the debris, rebuild homes, and shelter families with damaged homes (Fletcher et al., 2013). Household characteristics and relationships also affect disaster recovery in the Pacific. For example, a household's close relationships with neighboring households, extended family, and the broader community fundamentally shape their recovery processes, channels, and ultimately outcomes.

Role of the Community in Disasters

Given the prevalence of poor infrastructure and limited physical and financial resources, governments in the Pacific islands rely heavily on local communities and non-state actors to mitigate the effects of disasters. Pacific island societies are communal—everything revolves around the community and extended family, and this tradition goes back to ancient times (McCall, 1980). As such, research in the Pacific islands has highlighted the importance of community-based social protection systems (CBSPS) (Hau'ofa, 1994; Ratuva, 2014).

The role of CBSPS and a prevailing spirit of communitas are integral to disaster response and recovery in the Pacific, where the community anchors everyday life. Szreter (2009) describes CBSPS as an 'endowment of mutually respecting and trusting relationships' (290) and notes that both how CBSPS manifests and its depth are a function of the quality of the set of relationships of a social group. A strong bond exists between the community and individual members of the community group in the form of a wider, non-genealogical kinship (Ratuva, 2014). For example, the cultural practice of reciprocity, as well as communal labor, allows individuals to borrow items (from community members) such as knives, axes, and chainsaws to cut old and dead trees that are likely to fall in a cyclone, in addition to other tools to support disaster preparedness and recovery. In Tuvalu, the cultural practice of *bubuti*—roughly translated as 'request' in English—involves asking for goods and labor services from relatives and neighbors (Taupo et al., 2018). *Bubuti* is common and assumes an elemental role in times of disaster.

In the Fiji Islands, prior studies have found that communities deploy social capital to augment disaster response. This comes in the form of sharing information, participating in search and rescue (SAR) operations, mutual exchange of labor, and revival of small businesses (Chand & Taupo, 2020; Fletcher et al., 2013; Nanau, 2011; Weber et al., 2015). These organically derived techniques help to foster resilience and recovery at the household and community scales (Yila et al., 2014). Dupont (2004) refers to social bonds among community members as manifestations of social capital, and this is particularly relevant in Pacific island countries where communities know that they need to protect themselves. Thus, self-protection and relying on the broader community—as opposed to relying on government and formal actors—is the predominant channel for coping with disaster-related issues. In fact, in the Pacific, it could be argued that the state relies heavily on local communities to protect themselves during disasters and to initiate the recovery process, with government and formal actors merely supplementing ongoing local systems.

Additionally, in the Pacific, community-based traditional ecological knowledge (TEK) both observed and passed down orally over generations has assisted societies in preparing for, managing, and resolving problems caused by natural disasters (Fletcher et al., 2013). The advantage of utilizing CBA in the Pacific is the ability to leverage TEK, local capabilities, the spirit of communitas, and lived experiences in a way that accounts for the uniqueness of the place and its numerous dynamic systems. Over generations, households and communities have become resilient as a function of their daily experiences and their trials and errors (Nunn, 2000; Weber et al., 2015). These accumulated lessons coalesce over time to create localized disaster knowledge. The literature highlights that disaster recovery is best achieved when families and communities exercise a high degree of self-determination, which better allows them to navigate within the bounds of their sociocultural, economic, and ecological limitations and capacities (Comfort, 1995; McNamara et al., 2020). In a way, the constant ebbs and flows of the natural environment—which require Pacific islanders to respond, adapt, and innovate—mirror the gives and takes (i.e., interdependencies) that Pacific islanders exercise with each other.

Role of Village Leaders in Disasters

Most countries in the South Pacific retain socially stratified societies with hierarchical rungs headed by chiefs, elders, and 'bigmen.' This hierarchy is a dominant and culturally accepted feature of the traditional governance system that has persisted for generations (McNamara et al., 2022; Chand & Taupo, 2020). During disasters, these informal leaders are turned to and perform a critical role in both organizing and mobilizing resources, including boats, labor, services, human capital, and other needed instruments in the Fiji Islands (Weber et al., 2015; Yila et al., 2014). It is simpler and swifter, compared to relying on state actors, for village leaders to direct and adjudicate these tasks as they possess legitimacy and a trust ecology with their community counterparts. In times of disaster, the elevated status of village leaders allows them to instigate community support activities and guide response and recovery in a way that is grounded both geographically and culturally.

Role of Religious Organizations in Disasters

Religion is an institution upon which Pacific islanders seek spiritual and practical mooring. This is equally true in the sphere of disasters (McGeehan & Baker, 2017). Pastors (e.g., of the Protestant or Catholic denomination) and churches play a significant role in the diffusion of social, economic, and political mores in Pacific island societies (Cox et al., 2020; Macpherson & Macpherson, 2011). Churches do not merely bring together society in times of desperation, rather they offer space for shelter, psychosocial and emotional support, sanitary facilities, food preparation, and other valuable functions.

Monetary and nonmonetary contributions stemming from the Christian faith—both locally/domestically and through international church networks and faith-based humanitarian organizations—help support household and community needs during disasters. As an example, churches and faith-based schools and institutions are critical transmitters of disaster/emergency warnings, and they regularly prepare communities for preparedness and evacuation through their schools, congregations, outreach activities, and overall community presence. When there is a disaster warning in the Pacific, communities are typically instructed to 'go to the church.' Thus, while religion is certainly spiritual, it is also functional as it embodies a platform for providing evacuation space, shelter, water, and food packages to disaster-affected populations. The church is also a vehicle for delivering counseling and psychosocial, emotional, and trauma-based therapy to community members, as shown by Gillard and Paton (1999) after Cyclone Nigel in the Fiji Islands. Finally, Holmgaard's (2019) exploration of how Samoans from two tsunami-affected villages made sense of the disaster and understood its causes and impacts based on different Christian rationalizations—as a punishment or a 'second coming'—further establishes the church as a forum for community recovery.

Discussion

In the Pacific, households and communities must develop social capital, rely on social networks, and utilize cultural assets such as reciprocity and informal institutions to manage disasters; hence, the significance of community resilience. Community resilience is defined as 'the ability of a community to recover from a disaster or persist sustainably in the face of a new, ongoing hardship' (Blaustein et al., 2023, 2). In Pacific island countries, local communities activate everyday social protection mechanisms for purposes of disaster resilience, which is essential since government and formal assistance are slow, often lack context, and are less effective. As aptly stated by Ratuva (2014), 'the metropolitan ideologies of formal and institution-based social protection policies do not consider the culturally-based indigenous social protection systems to build resilient communities' (4). These social protection systems involve mobilizing community leaders and informal institutions with authority and resources as key emergency management partners. This suggests that emergency management mentalities, methods, and capabilities in Oceania and SIDS must be theorized and conceptualized as an extension of everyday community relations. In this regard, the household-community-village leader-faith nexus is a valuable model for disaster response and recovery.

Given the relatively weak built and economic infrastructures coupled with a lack of state resources, this chapter argues that the impacts of formal, state-centric responses to disasters are limited without the critical roles played by households and communities. International donor agencies and local governments have acknowledged this and are now concentrating on developing community-based disaster mitigation and resilience. The literature indicates that disaster response and recovery are more effective when collaborations exist among the community, state, and non-state actors—particularly when the partnership is strong (Blaustein et al., 2023; Mann & Williams, 2021; Mutongwizo et al., 2021). This chapter recommends such partnerships as an efficient way forward vis-à-vis constraints in built infrastructure, state and formal resources, and the pending forces of climate change. In the Pacific, anthropogenic climate change seriously threatens fisheries, subsistence agriculture, infrastructure, water quality, natural resources, biodiversity, the formal and informal economy, and culture and daily life writ large. Thus, collaborating with the communities who are confronting these unprecedented challenges (at a level of citizen control) is urgent, and this joint approach also allows for the use of TEK and indigenous solutions to localized problems.

Ultimately, this chapter seeks to redirect our focus on the importance of social capital and social networks among households and communities in the disaster context. Characteristics of these social institutions—kinship, reciprocity, and communal labor, for example—must be recognized and leveraged as integral components of disaster management in the Pacific. Households and communities must be factored into responding to disasters, initiating recovery and reconstruction processes, and developing long-term resilience in a way that meshes with their local cultures and environments. The case of Pacific island countries, and SIDS in general, is a testament to this assertion.

Conclusion

This chapter examined disaster response and recovery by households and communities in Pacific island countries and SIDS more broadly. Given poor built infrastructure and a lack of state resources, household and community resilience are needed. This chapter highlighted that social capital and networks—particularly partnerships among kin, neighbors, and community—are crucial in lesser-developed countries. Thus, in terms of resilience, the role of governments and policymakers should be to build upon existing informal institutions and capital to augment community-based social protection mechanisms. Communities recognize that they cannot rely on governments and formal actors for everything, but supporting cultural capital, social networks, and enhanced nature-society relationships is a welcome path forward.

This chapter underscored the significant contributions of traditional village leadership, churches, and other non-state actors in disaster response and recovery. Since disasters perturb the complex, informal systems of Pacific households and communities, these systems and units must be directly involved in disaster management. To that end, employing CBAs and further developing the nexus of household, community, village leader, and faith represents a pertinent strategy. We argue that households and communities must be brought to the center stage of any discussion or analysis of disaster response and recovery to fully understand what is happening in contexts such as the Pacific. Finally, more empirical research is needed on the roles, functions, and capacities of households and communities to elicit lessons learned, and best practices, and ultimately improve resilience for all.

References

Aksha, S.K., Resler, L.M., Juran, L., & Carstensen, L.W. (2020). A geospatial analysis of multi-hazard risk in Dharan, Nepal. *Geomatics, Natural Hazards & Risk*, *11*(1), 88–111.

Australian Disaster Recovery Framework (ADRF). (2022). *The Australia-New Zealand Emergency Management Committee*. 3rd edition. Sydney: ADRF.

Blaustein, J., Miccelli, M., Hendy, R., & Burns, K.H. (2023). Resilience policing and disaster management during Australia's Black Summer bushfire crisis. *International Journal of Disaster Risk Reduction*, *95*(103848), 1–14.

Bourdieu, P. (2018). The forms of capital. In *The Sociology of Economic Life*, pp. 78–92. London: Routledge.

Chand, A., Karan, M.F., Goundar, P., & Reddy, N. (2022). Policing the COVID-19 pandemic: Police and public perceptions of enforcement of health protocols in the Fiji Islands, *Island Studies Journal*. Published online, Early access December 2022. https://doi.org/10.24043/isj.410

Chand, A., & Taupo, T. (2020). Impact of natural disasters and climate change on national security in the Pacific: Kiribati and Tuvalu Case studies. In *Mapping Security in the Pacific*, Sara Amin, Dnielle Watson, and Christian Girard. (eds.), pp. 59–72. London: Routledge.

Chowdhury, T.J., Kako, M., Arbon, P., Muller, R., Gebbie, K., & Steenkamp, M. (2022). Understanding the experiences of women in disasters: Lessons for emergency management planning. *The Australian Journal of Emergency Management*, *37*(1), 72–77.

Comfort, L.K. (1995). *Self-organization in disaster response and recovery. The Great Hanshin, Japan earthquake*. Available from: https://www.connectivetx.org/ [Accessed, 15 February 2023].

Corlew, L.K. (2012). *The Cultural Impacts of Climate Change: Sense of Place and Sense of Community in Tuvalu, a Country Threatened by Sea Level Rise* (Doctoral dissertation, Honolulu: University of Hawaii at Manoa, May 2012).

Cox, J., Varea, R., Finau, G., Tarai, J., Kant, R., Titifanue, J., & Neef, A. (2020). Disaster preparedness and the abeyance of agency: Christian responses to tropical cyclone Winston in Fiji. *Anthropological Forum*, *30*(1), 125–140.

Djalante, R., & Thomalla, F. (2011). Community resilience to natural hazards and climate change: A review of definitions and operational frameworks. *Asian Journal of Environment and Disaster Management*, *3*(3), 339–355.
Dupont, B. (2004). Security in the age of networks. *Policing and Society*, *14*(1), 76–91.
Elliott, M., MacDonald, M.C., Chan, T., Kearton, A., Shields, K.F., Bartram, J.K., & Hadwen, W.L. (2017). Multiple household water sources and their use in remote communities with evidence from Pacific Island countries. *Water Resources Research*, *53*(11), 9106–9117.
Ensor, J., & Berger, R. (2009). *Understanding Climate Change Adaptation: Lessons from Community-Based Approaches*. Rugby, UK. Practical Action.
Fernandez, G., & Shaw, R. (2015). Youth participation in disaster risk reduction through science clubs in the Philippines. *Disasters*, *39*(2), 279–294.
Fletcher, S.M., Thiessen, J., Gero, A., Rumsey, M., Kuruppu, N., & Willetts, J. (2013). Traditional coping strategies and disaster response: Examples from the South Pacific region. *Journal of Environmental and Public Health*, *Vol. 1*, 1–9.
Gillard, M., & Paton, D. (1999). Disaster stress following a hurricane: The role of religious differences in the Fijian Islands. *The Australasian Journal of Disaster and Trauma Studies*, *2*(24).
Glauberman, G., Mendoza Kabua, P., Camba, M., Dela Cruz, M., & Fontenot, H.B. (2024). Perspectives on emergency preparedness among indigenous pacific people in Hawaii: A qualitative study. *Journal of Community Health Nursing*, *41*(3), 189–202. https://doi.org/10.1080/07370016.2024.2309375
Ha, K.M. (2015). The role of religious beliefs and institutions in disaster management: A case study. *Religions*, *6*(4), 1314–1329.
Hau'ofa, E. (1994). Our sea of islands. *The Contemporary Pacific*, *6*(1), 147–116.
Holmgaard, S.B. (2019). The role of religion in local perceptions of disasters: The case of post-tsunami religious and social change in Samoa. *Environmental Hazards*, *18*(4), 311–325.
Intergovernmental Panel on Climate Change (IPCC). (2012). *Managing the Risks of Extreme Events and Disasters to Advance Climate Change Adaptation*. A Special Report of Working Groups I and II of the Intergovernmental Panel on Climate Change (Cambridge and New York: Cambridge University Press).
Juran, L. (2012). The gendered nature of disasters: Women survivors in post-tsunami Tamil Nadu. *Indian Journal of Gender Studies*, *19*(1), 1–29.
Juran, L., & Trivedi, J. (2015). Women, gender norms, and natural disasters in Bangladesh. *Geographical Review*, *105*(4), 601–611.
Kanahele, G.H.S. (1986). *Ku kanaka: Stand Tall*. Honolulu: University of Hawaii Press.
Kiribati National Disaster Risk Management Plan (NDRMP). (2022). https://www.climate-laws.org/geographies/kiribati/policies/national-disaster-risk-management-plan-d3e8d4fc-2eef-481f-9460-2426ce424c63.
Kirkby, P., Williams, C., & Huq, S. (2017). Community-based adaptation (CBA): Adding conceptual clarity to the approach and establishing its principles and challenges. *Climate and Development*, *10*, 577–589.
Lindell, M.K., Prater, C.S., Gregg, C.E., Apatu, E.J., Huang, S.K., & Wu, H.C. (2015). Households' immediate responses to the 2009 American Samoa Earthquake and Tsunami. *International Journal of Disaster Risk Reduction*, *12*(2015), 328–340.
Macpherson, C. (1999). Changing contours of kinship: The impacts of social and economic development on kinship organization in the South Pacific. *Journal of Pacific Studies*, *22*(2), 71–95.
Macpherson, C., & Macpherson, L. (2011). Churches and the economy of Sāmoa. *The Contemporary Pacific*, *23*(2), 303–338.
Mann, J.P., & Williams, B.D. (2021). Policing in the Eye of the Storm. *Journal of Police and Criminal Psychology*, *36*(2), 183–192.
McCall, G. (1980). Kinship and association in Rapanui reciprocity. *Pacific Studies*, *3*(2), 1–21.
McCubbin, L.D., Ishikawa, M.E., & McCubbin, H.I. (2008). The *Kanaka Maoli*: Native Hawaiians and their testimony of trauma and resilience. In *Ethnocultural Perspectives on Disaster and Trauma. International and Cultural Psychology*, A.J. Marsella, J.L. Johnson, P. Watson, & J. Gryczynski (eds.), New York: Springer. https://doi.org/10.1007/978-0-387-73285-5_9
McGeehan, K.M., & Baker, C.K. (2017). Religious narratives and their implications for disaster risk reduction. *Disasters*, *41*(2), 258–281.
McNamara, K.E., Clissold, R., Westoby, R., Piggott-McKellar, A.E., Kumar, R., Clarke, T., & Nunn, P.D. (2020). An assessment of community-based adaptation initiatives in the Pacific Islands. *Nature Climate Change*, *10*(7), 628–639.

Mutongwizo, T., Holley, C., Shearing, C.D., & Simpson, N.P. (2021). Resilience policing: An emerging response to shifting harm landscapes and reshaping community policing. *Policing: A Journal of Policy and Practice*, *15*(1), 606–621.
Nanau, G. (2011). The wantok system as a socio-economic and political network in Melanesia. *The Journal of Multicultural Society*, *2*(1), 31–55.
Nunn, P., & Kumar, R. (2018). Understanding climate-human interactions in Small Island Developing States (SIDS) implications for future livelihood sustainability. *International Journal of Climate Change Strategies and Management*, *10*(2), 245–271.
Paul, B.K. (2019). Channeling disaster aid: Process and problems. In *Disaster Relief Aid: Changes and Challenges*. Berlin, Germany: Palgrave Macmillan, Springer Nature.
Paul, B.K. (2006). Disaster relief efforts: An update. *Progress in Development Studies*, *6*(3), 211–223.
Pelling, M., & High, C. (2005). Understanding adaptation: What can social capital offer assessments of adaptive capacity? *Global Environmental Change*, *15*(4), 308–319.
Pendry, D.A. (2022). Communal labor and sharing systems. In *The Routledge Handbook of Indigenous Development*, K. Ruckstuhl, I.A. Velásquez Nimatuj, J.-A. McNeish & N. Postero (eds.), pp. 370–378. London: Routledge.
Ratuva, S. (2010). Back to basics: Towards integrated social protection for vulnerable groups in Vanuatu. *Pacific Economic Bulletin*, *25*(3), 40–63.
Ratuva, S. (2014). Failed or resilient subaltern communities? Pacific indigenous social protection systems in a neoliberal world. *Pacific Journalism Review*, *20*(2), 40–58.
Ravuvu, A. (1987). *Fijian Ethos: The Fijian Way of Life*. Suva: Institute of Pacific Studies.
Rumbach, A., & Foley, D. (2014). Indigenous institutions and their role in disaster risk reduction and resilience: Evidence from the 2009 tsunami in American Samoa. *Ecology and Society*, *19*(1), 9–17.
Sadri, A.M., Ukkusuri, S.V., Lee, S., Clawson, R., Aldrich, D., Nelson, M.S., Seipel, J., & Kelly, D. (2018). The role of social capital, personal networks, and emergency responders in post-disaster recovery and resilience: A study of rural communities in Indiana. *Natural Hazards*, *90*, 1377–1406.
Schuermann, S., & Lauer, M. (2016). Disaster recovery in the western Pacific: Scale, vulnerability, and traditional exchange practices. *Natural Hazards*, *84*, 1287–1306.
Sheikhi, R.A., Seyedin, H., Qanizadeh, G., & Jahangiri, K. (2021). Role of religious institutions in disaster risk management: A systematic review. *Disaster Medicine and Public Health Preparedness*, *15*(2), 239–254.
Szreter, S. (2009). The new political economy: The importance of social capital. In *The Global Financial Crisis and Its Impact on the Work of the Global Financial System*, J. Somavia (ed.), Geneva: UN System Chief Executives Board for Coordination.
Taupo, T., Cuffe, H., & Noy, I. (2018). Household vulnerability on the frontline of climate change: The Pacific atoll nation of Tuvalu. *Environmental Economics and Policy Studies*, *20*(4), 705–739.
Thaman, R.R., Meleisea, M., & Makasiale, J. (2002). Agricultural diversity and traditional knowledge as insurance against natural disasters. *Pacific Health Dialog*, *9*(1), 76–85.
Watson, D., Amin, S.N., Wallace, W., Akinlabi, O., & Ruiz-Vásquez, J.C. (eds.) (2022). *Policing the Global South: Colonial Legacies, Pluralities, Partnerships, and Reform*. London: Routledge.
Weber, E., Yila, O.M., & Neef, A. (2015). *Community Resilience against Natural Hazards: Case Studies from the Pacific Islands*. United Nations Office for Disaster Risk Reduction (UNISDR).
Yila, O., Weber, E., & Neef, A. (2014). The role of social capital in post-flood response and recovery among downstream communities of the Ba River, Western Viti Levu, Fiji Islands. In *Risks and Conflicts: Local Responses to Natural Disasters*, Vol. 14, pp.79–107. Emerald Group Publishing Limited.

27

OPPORTUNITIES THAT ARISE DURING A CRISIS

The Case of Ho Chi Minh City, Vietnam

Mokbul Morshed Ahmad and Tran Thi Minh Hien

The "Invisible Threat": Identifying Challenges and Opportunities from the COVID-19 Pandemic

Declaring the COVID-19 pandemic a global disaster is a recognition of its multidimensional economic, social, and psychological severity and an opportunity to focus resources and a worldwide effort to confront it. While measures to control the COVID-19 pandemic have brought positive results, the reality is that we still face significant long-term challenges both economically and socially.

In the economic sector, businesses and industry face uncertainty and slow recovery. Businesses, especially those in the tourism and entertainment sector, face economic recession and bankruptcy risk. Global supply chains are disrupted, causing problems in the production and distribution of goods (Butt, 2021). Unemployment and loss of income continue to be complex problems, affecting the quality of life of many people and severely impacting families and communities with poor economic conditions. As a result, governments were forced to deploy financial stimulus measures to support businesses and workers.

On the social front, health and education systems also face major challenges in adjustment and recovery. The upheaval from the pandemic has created psychological and social challenges, including increased stress and fear (Stogner et al., 2020). Communities face the possibility of social loss and isolation (Marston et al., 2020), with vulnerable groups such as the elderly and the poor facing more significant difficulties as their resources are limited. Social divisions may increase without effective political and social adaptations (Marston et al., 2020).

The education system also has difficulty converting to online instruction to manage disruptions to the school schedule. This poses many challenges regarding educational rights when studying online, highlighting the inequity between students in different areas (Czerniewicz et al., 2020). Students in low-resource settings often face difficulties accessing online education due to a lack of devices and unstable Internet connections. In addition, isolation and lack of face-to-face communication can affect the psychology of students and teachers, creating stress and isolation (Mizani et al., 2022).

However, amidst the challenges of the pandemic, significant opportunities emerge. The need for digital transformation is emphasized, encouraging innovation and flexibility in society and the

DOI: 10.4324/9781003414834-37

economy. The COVID-19 pandemic, often referred to as the "invisible threat," serves as a stark warning and a unanimous call for action from the international community. Cooperation and connectivity are also important keys to rebuilding a stronger society and economy after these difficult challenges. Facing the above challenges and taking advantage of the opportunities, we can rebuild a strong economy and society that is more flexible and fair while working toward a future in which everyone can be proud and proactively contribute.

Innovation in Crisis: Opportunities for Rapid Development and Application of Innovative Solutions

Transition to and Integration of Remote Work

Embracing the shift to remote work and virtual collaboration is an important step in building a new work platform, where flexibility and resilience are not just trends but represent the power to shape the future of work and careers. The widespread adoption of remote working amid the challenges posed by the COVID-19 pandemic have opened up countless development opportunities. This shift has accelerated the evolution of workplace norms that emphasize flexibility and digital connectivity (de Lucas Ancillo et al., 2023). Businesses have the chance to access the global talent pool, promoting diversity and innovation (Haque, 2023). The rise of remote work has also driven technological advances in collaboration tools, cybersecurity, and virtual communication platforms, creating a more seamless and productive remote work experience (Haque, 2023). Organizations redefining their work structures can prioritize employee well-being, promoting a healthier work-life balance (Feeney & Stritch, 2019). Remote working has become a catalyst for reimagining traditional office models, providing the opportunity to cultivate a more adaptable and resilient workforce. From creating flexible working policies to providing training and emotional support, we can build a foundation for workforce resilience and prosperity. Companies can take advantage of these changes to create favorable work environments that encourage creativity and boost performance. Seizing these opportunities can address the pandemic's immediate challenges and pave the way for a future where work is inclusive, more flexible, and suitable for the growing needs of individuals and organizations.

Flexible Supply Chain

The upheavals brought about by the COVID-19 pandemic have underscored the imperative for a paradigm shift in supply chain dynamics. This transformative moment has illuminated the vulnerabilities inherent in traditional supply chain models, thereby accentuating the necessity for diversified sourcing, agile logistics, and adaptive manufacturing processes (Bastas & Garza-Reyes, 2022). The disruptive forces unleashed by the pandemic have become a crucible for innovation, pushing businesses and industries to reconceptualize and reconfigure their supply chains for heightened flexibility.

In this transformation background, a notable opportunity emerges that extends beyond recovery and envisions a resilient and responsive supply chain architecture. This shift extends beyond immediate concerns and encompasses a broader spectrum of potential disruptions, encompassing health crises, environmental challenges, and geopolitical shifts.

The pandemic serves as a catalytic agent for reimagining supply chain management strategies, prompting the exploration of innovative approaches that prioritize adaptability and resilience. Technological solutions, such as blockchain, present themselves as formidable tools to augment transparency and traceability within supply networks (Centobelli et al., 2022). These technological interventions not only address immediate challenges but also herald a new era of digital transformation, promising to reshape industries, foster global cooperation, and propel a more agile and interconnected future.

As we navigate the path to recovery, the flexible supply chain emerges as not merely a response to a crisis but a strategic evolution that positions businesses and industries to thrive in uncertainty. It is an opportune moment for stakeholders to leverage technological advancements, embrace innovative methodologies, and collaboratively shape a future where supply chains are resilient, dynamic, and responsive to the ever-changing contours of the global landscape.

Opportunities to Improve the Healthcare System

The challenges posed by the COVID-19 pandemic have shed light on important opportunities to strengthen healthcare systems worldwide. The opportunity lies in reassessing and strengthening healthcare capacity, ensuring adequate resources, and promoting research and development. The urgency of the crisis has driven investments in healthcare infrastructure, workforce training, and technology integration. The opportunity to leverage artificial intelligence, data analytics, and digital communications tools not only facilitates more effective pandemic response strategies but also lays the foundation for a future technology savvy. The combination of contact tracing applications, automated notification systems, and remote diagnostic technology has provided accurate and rapid information on the epidemic situation (Asadzadeh et al., 2020; Costa & Peixoto, 2020; Ye et al., 2022). In addition, artificial intelligence and machine learning are used to predict epidemic trends and provide more detailed information on how to respond (Van der Schaar et al., 2021).

By seizing these opportunities, countries can strengthen their ability to respond to immediate health crises and build resilient healthcare systems that are patient-centered and prioritize equity, access, and long-term sustainability. The pandemic has now become a catalyst for transformative changes that have the potential to leave a lasting, positive impact on global healthcare. This crisis also shines a light on the importance of preventive and primary care, mental health support, and community engagement.

Digital Transformation in Education

The COVID-19 pandemic has created meaningful opportunities for educational development by accelerating the integration of technology into learning environments. As traditional education models face disruption, adopting online learning platforms, virtual classrooms, and digital resources becomes imperative. This presents an opportunity to reimagine and enhance educational practices through technology. Schools and organizations can leverage interactive and personalized learning experiences that reach students regardless of geographic limitations. Additionally, integrating EdTech (Educational Technology) tools can also accommodate diverse learning styles and disabilities (e.g., hard of hearing through automatic captions), promoting a more inclusive and accessible education system (Aithal & Aithal, 2023). This transition period presents an opportunity to close the digital divide, ensuring that students of all backgrounds have equitable access to quality education. By seizing this opportunity, educators and policymakers can lay the foundations for a flexible education system and adopt new teaching methods, including the use of technology to personalize learning and develop flexibility in teaching to prepare students to meet the demands of an increasingly digital and connected world.

Opportunity to Rebuild Infrastructure to Cope with Epidemics

Rebuilding infrastructure in the aftermath of epidemics offers a unique opportunity to redefine the way we approach housing, workplaces, and public spaces, placing a renewed emphasis on residents' well-being, adaptability, and environmental sustainability. The transformation of housing into multifunctional spaces becomes a central focus, integrating living and remote working activities seamlessly. This approach not only addresses the need for social distancing but can also promote a sustainable and holistic lifestyle. Incorporating recreational spaces and green gardens into housing designs contributes to a healthier living environment, with proven benefits for mental health and overall well-being. By prioritizing both remote work and recreational activities in housing structures, communities can better prepare for future challenges while nurturing the health and happiness of their residents.

Transforming workplaces into resilient and adaptable environments is another critical aspect of post-epidemic infrastructure redevelopment. Renovating workplaces to include isolation systems, stringent cleaning protocols, and advanced technologies enhances workplace safety and flexibility. This shift addresses immediate health concerns and establishes a more agile and technology-driven work culture that is better equipped to handle future challenges and promote a healthy work-life balance.

Public spaces like schools and stadiums can be redesigned with flexible functions, allowing quick conversion into vaccination sites or disease isolation areas during health crises. This adaptive approach ensures that critical infrastructure can be repurposed efficiently, optimizing the use of existing facilities to address immediate public health needs. Integrating flexibility into urban planning and infrastructure design provides communities with a rapid response mechanism, fostering resilience and preparedness for future epidemics or health emergencies.

Strengthening emergency medical infrastructure stands as a critical component of post-epidemic reconstruction. Investing in additional medical facilities, enhancing the capacity of existing healthcare facilities, and stockpiling essential medical supplies and equipment ensure a more resilient response to future health crises. Integrating advanced technologies, data analytics, and telemedicine solutions facilitates timely and efficient emergency medical services. Prioritizing and bolstering backup infrastructure significantly enhances communities' capacity to cope with epidemics and other crises, creating a more prepared and responsive healthcare system.

Incorporating comprehensive planning strategies that account for natural disasters and complex land use functions becomes a cornerstone of reimagined infrastructure. Smart land use planning helps create buffer zones, emergency evacuation routes, and strategically located medical facilities. This holistic approach ensures communities are well-prepared for the interconnected challenges of epidemics and natural disasters, fostering long-term resilience and sustainability.

Toward Sustainable Development: Opportunities for Community Solidarity, Global Cooperation, and Social Justice

Opportunity to Increase Health Awareness in the Community

The COVID-19 pandemic has opened up a rare opportunity to increase public health awareness and education, especially in mental health. Given the challenges posed by the pandemic, the chance to make positive changes in how we understand and care for community psychology has become apparent.

Education plays a vital role in creating awareness and understanding of public health. Governments and organizations can invest in education campaigns to address the immediate issues related to the pandemic and focus on building a wise and equipped community for the future. Organizations and communities can collaborate to create emotional support strategies, providing resources and helpful information so people can face the mental challenges of post-pandemic recovery.

Raising awareness about preventive measures and the benefits of vaccination brings immediate results and creates the foundation for a responsible and proactive community. Through that, we can build a healthy community that can withstand pandemics and all health challenges. Furthermore, the digital age ushers in opportunities for continuous innovation in increasing health awareness in the community. Using social media and technology to communicate health messages can increase impact and reach. Online campaigns and mobile applications not only help provide complete and accurate information but also facilitate interaction and support within the community. Collectively, by seizing these opportunities, we are not only facing the pandemic positively but also building the foundation for a strong, wise, and supportive community that helps people overcome difficulties and enjoy healthier lives.

Opportunity to Engage the Community

The COVID-19 pandemic represents both a challenge and an opportunity for the community to create cohesion and solidarity. In difficult times, communities across the globe have demonstrated incredible resilience and solidarity. Developing a vaccine in record time is a medical achievement and a testament to the power of global collaboration and commitment to innovation in challenging environments (Kim et al., 2020).

Shared experiences from challenges have created a solid foundation for long-term connections and mutual support in the community. Development opportunities lie not only in encouraging creativity at the grassroots level but also in promoting positive responses from community leaders and expanding social connections. Collaborative efforts to address local needs, from food distribution to mental health support, have become powerful examples of the power of collective action.

The emergence of community bonds is not only a special financial tool but also a symbol of support and commitment to the community (Campbell-Scherer et al., 2021). In the difficult circumstances of the pandemic, community bonds can play an important role in mobilizing resources to support those most in need. Corporate and socially responsible organizations can issue bonds to fund relief projects, provide medical materials, and even build the infrastructure needed to deal with the pandemic. Communities can also organize donations, charity events, and other support activities to help those in need during times of emergency. Community bonds not only provide financial resources but are also a symbol of unity and emotional support.

Community solidarity is not only a natural reaction to the crisis but also a pioneering force in building a strong and vibrant future and an opportunity for us to enjoy community values and recognize that we can achieve more when we are together. An emphasis on community engagement also provides chances to stimulate diverse voices and ensure that the most vulnerable members receive the care and support they need. As we navigate the recovery, the spirit of solidarity forged during the pandemic can serve as a driving force for sustainable development in which communities actively shape their futures through collaborative efforts and a renewed sense of connection.

Opportunity for Global Cooperation in the Digital Era

The pandemic knows no borders, and we need a limitless approach to face it. The COVID-19 pandemic has highlighted the critical importance of global cooperation, providing a unique opportunity for countries and international organizations to join forces to address common challenges. Collaboration between governments, international organizations, and the scientific community has created a model for sharing information, experience, and resources. The development and distribution of vaccines have demonstrated the power of collaborative efforts between countries, pharmaceutical companies, and research institutions working together to accelerate the delivery of life-saving solutions. These joint efforts helped to control the pandemic and were important steps toward global understanding and consensus, laying the foundation for future collaborative initiatives. Opportunities for increased collaboration extend beyond health care to areas such as climate change, economic recovery, and technology governance.

Leveraging digital platforms to share and collaborate information in real-time has opened up a new world of collaborative potential. Artificial intelligence and big data technology help track and predict infections and create opportunities for research groups worldwide to work together to find solutions (Awotunde et al., 2022). Online platforms and mobile applications have connected people with necessary information and facilitated consensus in implementing preventive measures (Kusumastuti et al., 2022). The impact of the pandemic is not just a challenge but also an opportunity to demonstrate the capabilities of humans and technology when they stand together and collaborate to respond to the invisible threat flexibly and creatively. While these advances bring many benefits, we must also consider the challenges that overreliance on technology can bring. Fears of job loss due to automation, privacy concerns, and social disconnection are all aspects to consider when making decisions about technological developments in society and economy.

By fostering a spirit of mutual support and information sharing, the world can harness collective wisdom and resources, paving the way for a more interconnected and resilient future. The lessons learned from this crisis provide a blueprint for strengthening international partnerships, reinforcing the view that, with solidarity, the global community can navigate and overcome challenges shared.

Equity and Inclusion in Recovery

During recovery from the COVID-19 pandemic, many opportunities arose, opening up space to create a more equitable and inclusive society. Faced with the challenges and pain of the pandemic, we are not only witnesses of loss but also those responsible for shaping the future. Focusing on targeted and equitable policy development is important to support disproportionately affected demographic groups, such as those with low incomes, unstable workers, and disadvantaged local communities. In some cases, people from middle-class backgrounds may be financially devastated and less likely to receive support from the government (Megasari et al., 2021). This may partly reflect social injustice, as these people do not receive the help because of their social status. Protecting workers' rights, ensuring social security, and increasing financial support for those in difficulty, regardless of circumstances, occupations, etc., can help build a society where everyone has equal opportunities and rights.

In addition, we also have the opportunity to build a more resilient and equitable future. We can create a responsible and stable community by creating a business and social environment that welcomes everyone, regardless of economic status or class. Investing in renewable energy, promoting green industry, and supporting socially responsible businesses are important steps in shaping a sustainable future.

Ultimately, this opportunity is also a clear guide to creating a society that leaves no one behind. Amid challenges and upheaval, we can work together to build a diverse and inclusive community where every individual is valued and respected. By creating opportunities for the participation and contribution of all, we can shape a future where no one is left behind.

Reality in Ho Chi Minh City: Opportunity to Alleviate Poverty During the COVID-19 Pandemic

A study on the impact of the COVID-19 pandemic on the poor in Ho Chi Minh City was conducted by the authors from September to October 2023. The study area was the five districts with the highest poverty rate, namely Cu Chi, Binh Chanh, Hoc Mon, Nha Be, and Can Gio districts. Questionnaires (n=400) with randomly selected households living in poverty and 20 in-depth interviews with purposefully selected local officials and social organizations were conducted. The results demonstrate the opportunities that the pandemic brings in reducing poverty in the study area.

Opportunities to Promote Proactive Spirit

For the poor, the pandemic is not only a challenge but also opens up new opportunities, promoting a spirit of initiative and creativity in overcoming difficulties and building a stronger future. Questionnaires with poor households show that despite significant support from the government, organizations, and officials (even "meals with more meat and vegetables than before the pandemic" or cash support that almost doubles or triples their average income), more than 70% of households express a desire to end social distancing measures soon. When given the choice between (1) continue to receive aid in the context of a prolonged pandemic, and (2) quickly end the pandemic to earn their income independently, they prefer the latter, aiming to resume work and increase their income. Surveyed individuals believe that the stability of finances that comes from their work is more secure than the support from others. Local officials shared that as soon as the social distancing order ended, some workers immediately continued to maintain their old jobs. Others, especially those who lost their jobs due to the pandemic, actively sought job opportunities through acquaintances or local job platforms and job fairs. Local authorities play an important role in supporting communities by providing employment courses and job placement. This not only helps create a favorable environment for people to access job opportunities but also promotes positive interaction between the government and the community.

However, there are limitations: the jobs that workers in poor households choose are often jobs with immediate income that are not sustainable. Jobs such as washing dishes, serving meals, doing housework by the hour, or jobs "doing whatever someone hires" have become popular. Although local authorities have tried to encourage, create conditions, and organize more stable vocational classes (e.g., bartending, salon, orchid growing, sewing), people often choose jobs that they can find themselves or are introduced to through personal relationships because they are "used to the old job" and "don't know how to do anything else."

When asked about improving people's skills, local officials said that poor households rarely participate in vocational training classes because they "spend that time earning money to buy food during the day" or "just stay home because they don't want to go to work." For these cases, local officials have recorded in writing and can propose to remove from the list of poor households with working-age members who are capable of working but refuse to work. This policy also encourages poor households to participate in labor to increase their income without depending much on support from the government.

Remote Working Opportunities

In the current context, with the spread of the pandemic, remote work has become more important than ever. For poor households in Ho Chi Minh City, this situation is not only a challenge but also an opportunity to change and improve their daily lives. Most laborers from poor households work in the informal sector, requiring work at a specific location such as trading at markets, selling lottery tickets, and driving motorbike taxis. However, in the future, there is a need to adapt and expand the scope of remote working, creating new opportunities such as instructions for combining online sales, moving from traditional motorbike taxis to motorbike taxi with the support of technology (e.g., Grab is one of the big apps of the technology motorbike taxi), and vocational training to transition to jobs with more stability and adaptability (e.g., learning to sew and taking on contracts to rework garments at home). Although many poor households currently have difficulty accessing technology and equipment, this is also an opportunity for society and the government to invest in training and providing resources. This helps the poor to adapt to the new working environment and enhances their skills and career opportunities.

In the manufacturing industry in five districts of Ho Chi Minh City (e.g., textiles, mechanics, processing), although currently there are only a small number of contracted employees working remotely from home, with the adjustment and recovery, an increase in the use of remote workers can be expected. This helps to reduce the pressure facing supply disruptions and creates new opportunities for workers.

Some people from other poor households, although they do not have experience working remotely, shared that this policy can be suitable and convenient for their circumstances. They are people who have difficulty moving, taking care of the sick, or have reached the age of overwork but are still healthy. However, they "want to work to earn extra income" and are "willing to take on more jobs at home." This can motivate them to participate in the remote work revolution, bringing equal benefits and career opportunities to all classes of society.

Opportunities from Flexible Supply Chains

In the current economic and social situation, Ho Chi Minh City faces a series of simultaneous challenges from the post-COVID-19 crisis and global upheavals, such as the Ukraine-Russia war and the Israel-Palestine war. These impacts have posed many problems for businesses and communities while simultaneously opening up opportunities for flexibility and innovation in the supply chain.

The Ukraine-Russia War has caused significant fluctuations in the energy market, especially gasoline prices, directly affecting maritime activities and fishing in Thanh An commune of Can Gio district. This not only places pressure on transportation costs but also affects the stability of the city's fuel supply. In addition, the Israel-Palestine war has strongly impacted industries such as textiles, mechanics, machining, and food processing, all of which are important industries to the local economy. Businesses in Cu Chi, Hoc Mon, and Binh Chanh are facing a decline in the export of their products, which may lead to a reduction in job supply and income for workers in the industry.

Poor communities, especially those working in the grocery and fruit and vegetable trade, are facing difficulties due to the reduced purchasing power of consumers. Despite receiving financial support, this problem can only be solved in the short term, and sustainable solutions are needed to ensure a stable source of living. In this context, supply chain opportunities arise from businesses' and communities' flexibility and creativity in facing simultaneous fluctuations and challenges from many sources, including the post-COVID-19 crisis and other challenges and global disturbances such as the Ukraine-Russia war and the Israel-Palestine war.

A significant opportunity is the ability to find alternative sources of supply for raw materials and products affected by global events. Businesses can research and develop new supply sources from other countries or regions, minimizing risks and increasing supply chain stability. Innovation in products and services is also an important aspect of this opportunity. Businesses can focus on developing new products that suit emerging market needs, thereby creating new sources of income and ensuring diversity in revenue. Cooperation with the international community is another strategy that can be applied to overcome challenges. Cooperation with international partners not only opens the door to new markets but also strengthens the vibrance and competitiveness of businesses.

Opportunities to Improve Healthcare

The opportunity to improve the healthcare system for poor households in Ho Chi Minh City after the pandemic opens up important prospects for reshaping the public healthcare picture and creating a more effective care environment, reflecting a proactive and humane spirit. One of the particularly positive points is the popularity of COVID-19 testing and vaccination in poor communities. Implementing this method helps prevent the spread of the virus and enhances community resistance. Making these services free reduces the financial burden on poor households and encourages them to actively participate in epidemic prevention measures.

During difficult times, authorities and humanitarian organizations have introduced social distancing measures, and special attention has been paid to poor households. The home care model has reduced travel pressure and increased access to medical services. Being proactive in notifying and contacting local management when there are signs of disease is an opportunity to strengthen the ability to detect and control disease outbreaks early.

However, improvements in the medical examination and treatment process are needed to take full advantage of this opportunity. Reducing hassles and optimizing processes helps create a convenient and transparent environment for people using medical services. At the same time, synchronizing these processes on a national scale will help reduce difficulties and increase equity in access to healthcare. In summary, the opportunity to improve the healthcare system after the pandemic period is an important step to ensure sustainability and proactiveness in maintaining public health. This is especially important in facing challenges from epidemics, pandemics, and global fluctuations.

Opportunities for Digital Transformation in Education

Digitally transforming the education sector of poor communities in Ho Chi Minh City after the epidemic period is opening up new opportunities and, at the same time, posing complex methods that need to be resolved. Although most people have proven flexible and accepting of the possibility of online learning, many have found difficulty accessing educational technology. Poor households often do not own computers and smartphones, which reduces their access to online learning sessions. Children sometimes have to share learning devices with their parents, while some do not have access to online learning due to a lack of finances.

To improve opportunities in education after the pandemic, comprehensive solutions are needed. First, governments and non-profit organizations must focus on providing network connection devices to poor households, especially families with children. In addition, it is necessary to increase technology education so the community can understand and use technology effectively. At the same time, innovative solutions such as mobile education application development can help improve children's access to education. The government can cooperate with businesses and social organizations to offer support policies, discounts, or equipment sponsorship for poor communities. In short, digital transformation in education for poor communities in Ho Chi Minh City is an opportunity to eliminate inequality in access to education and create a fair and sustainable learning environment after the pandemic period.

Opportunity to Rebuild Infrastructure to Cope with Epidemics

Ho Chi Minh City's infrastructure redevelopment in response to epidemics is characterized by a comprehensive and inclusive approach, extending support to the entire population. A key focus is the maintenance and support of the network system (e.g., FPT, Viettel). This proactive measure ensures the stability of communication networks and aims to reduce costs for residents. This initiative is crucial for facilitating smooth remote working and learning activities, particularly during epidemic seasons and periods of social distancing.

In addition to the city's concerted efforts, remote medical consultation service systems, encompassing psychological support and epidemic prevention instructions, have been implemented and progressively refined since the onset of the COVID-19 pandemic. This initiative underscores the city's commitment to leveraging technology for enhanced healthcare accessibility.

Concurrently, workshops focused on enhancing building functionality and land use to adapt and integrate epidemic responses are organized. These workshops serve as platforms to raise awareness and facilitate the implementation of strategies that promote a more resilient and prepared urban environment. In summary, Ho Chi Minh City's multifaceted approach addresses immediate challenges and positions the city as a model for inclusive, technologically driven, and forward-thinking urban development in the context of public health and infrastructure.

Opportunity to Increase Public Health Awareness

The pandemic creates opportunities to raise public awareness and promote health knowledge in the community. People have become more aware of the importance of health and placed it at the top of their list of personal priorities. Under the impact of the pandemic, there is a shift in the priority needs of poor households when health insurance becomes the main concern instead of income and employment support. This reflects a deep awareness of the importance of personal health protection since the COVID-19 pandemic. Choosing health insurance is not only about ensuring peace of mind in medical treatment but also about investing in the ability to work and create a stable income in the future. Raising health awareness and choosing health insurance brings personal benefits and contributes to the community's sustainable development. This can be considered one of the crucial achievements in recovery and growth after the pandemic.

In addition, health concerns are shown through poor households receiving information and strictly complying with COVID-19 pandemic prevention solutions. Even though the pandemic is under control, wearing masks and washing hands are still maintained by poor households as a new habit in life.

Although the pandemic has brought various challenges, it has also created vital opportunities for poor communities in Ho Chi Minh City to recover and develop. Combining personal motivation, community solidarity, improvement opportunities from local authorities, and opportunities to raise public health awareness create a positive picture where everyone can look forward to a more optimistic future.

Opportunity to Engage the Community

The spirit of solidarity in the community has played an important role in supporting the poor. Sponsors inside and outside of the region have actively shared resources and supported the poor, creating united communities. Poor families are guaranteed to "eat full," and there are cases where they even "eat more vegetables and meat than usual," demonstrating a situation where the community supports each other.

Community engagement was facilitated by numerous entities. The government and various civil society organizations contributed to this effort: religious establishments helped their congregations by providing essential necessities; the Fatherland Front Committee played a central and guiding role in all strategies to support the poor; farmers' associations provided seedlings, fishing nets, clam rakes, food trucks, etc.; the Women's Association took care of female workers and used association funds to support low- or no-interest business loans and exceptional care for reproductive health and safety measures; the Elderly Association supported living expenses for any member aged 60 or older (allowance was 540,000 Vietnam dongs/month from 60 to 80 years and 750,000 Vietnam dongs/month for over age 80); and the Veterans Association participated by providing scholarships to children of poor households to help them have the opportunity to go to school and improve their career capacity. Solidarity especially shone through in Binh Chanh commune, Binh Chanh district, through the one-on-one support policy. This means that each business or association is assigned to a poor household, supporting them until they escape poverty. This is an important step toward creating strong bonds between community members and helping those in need overcome challenges.

Equity and Inclusion in Recovery

Opportunities for equity and inclusion in the recovery of poor households in Ho Chi Minh City after the epidemic pose an important context for equitable development and inclusivity in the recovery process. In particular, prioritizing support for poor households before, during, and after the pandemic is a positive sign.

The local government has applied a fair support distribution method by focusing on the district party committee and slowly spreading it to the commune, village, and neighboring areas, finally reaching poor households. Poor households are prioritized and guaranteed to receive the first support, even when funding is limited.

Local people and officials find it important to ensure fairness in distributing support. The fact that poor households compare and reflect on the results of this process themselves plays a vital role in evaluating the performance and fairness of local government. Village heads/deputies and self-management groups also play an important role in assessing the current situation of poor households in the area, helping to determine the quality of life, support status, and ability to escape poverty. In short, the opportunity for equity and inclusion is not just a matter of distributing support but also a chance to build a more equitable and inclusive society in the post-pandemic recovery.

Opportunity to Review and Improve the Effectiveness of Supporting Poor Households

The opportunity to review and improve the support effectiveness of Ho Chi Minh City's poor community after the epidemic is an essential aspect of the recovery and community rebuilding process. The COVID-19 pandemic is not only a challenge but also a valuable lesson about local governments' management and ability to respond. Under pressure from the crisis, local authorities have introduced effective prevention and demographic support measures. However, opportunities open up when they can evaluate their performance and consider what was successful and where improvement is needed. This review process not only helps optimize resources but also provides an opportunity better to define the specific priorities and needs of the community. Furthermore, civil society organizations and associations also have opportunities to strengthen their roles and influence. By actively participating in the review process, they can offer opinions, suggest solutions, and share their unique local knowledge. This strengthens the relationship between the community and these organizations while also creating opportunities for consensus and initiative in confronting future challenges. In general, the opportunity to review and improve support performance is not only about optimizing resources but also seeking cooperation and creativity to build a stronger community after the epidemic period.

Conclusions and Lessons from the Pandemic: Building a More Robust Future

Lessons from the COVID-19 pandemic represent both an emotional outpouring of loss and a profound reflection on the opportunities and teachings we can draw from the pandemic itself. It is not just about cultivating prevention; it is also about building the foundation for a more robust and secure future.

One of the key points is to recognize the importance of emergency response capabilities. The pandemic has highlighted the strengths and weaknesses in how we respond to the challenge of healthcare and global economic and social disruptions. In addition, the pandemic opens up opportunities for us to research and apply improvement measures. From investing in public health research, digital transformation in education, remote working, restructuring supply chains, and rebuilding infrastructure to developing flexible emergency response plans, we can transform the lessons from disasters into positive actions.

It is also impossible to miss the opportunity to consider how our society has faced difficulties. From community solidarity to individual autonomy, the pandemic has made us see the actual value of solidarity and willingness to help. Ultimately, the lesson from the pandemic is not just about looking back but also an opportunity to look forward. Our drive is to shape a more robust future where we can face every challenge confidently to build a more united and tolerant society.

References

Aithal, P., & Aithal, S. (2023). Predictive analysis on future impact of ubiquitous education technology in higher education and research. *International Journal of Applied Engineering Management Letters*, 7(3), 88–108.

Asadzadeh, A., Pakkhoo, S., Saeidabad, M. M., Khezri, H., & Ferdousi, R. (2020). Information technology in emergency management of COVID-19 outbreak. *Informatics in Medicine Unlocked, 21*, 100475.

Awotunde, J. B., Oluwabukonla, S., Chakraborty, C., Bhoi, A. K., & Ajamu, G. J. (2022). "Application of Artificial Intelligence and Big Data for Fighting COVID-19 Pandemic" In Said Ali Hassan & Ali Wagdy Mohamed & Khalid Abdulaziz Alnowibet (ed.), *Decision Sciences for COVID-19: Learning through Case Studies* (pp. 3–26). Springer.

Bastas, A., & Garza-Reyes, J. A. (2022). Impact of the COVID-19 pandemic on manufacturing operations and supply chain resilience: Effects and response strategies. *Journal of Manufacturing Technology Management, 33*(5), 962–985.

Butt, A. S. (2021). Strategies to mitigate the impact of COVID-19 on supply chain disruptions: A multiple case analysis of buyers and distributors. *The International Journal of Logistics Management, 2*. https://doi.org/10.1108/IJLM-11-2020-0455

Campbell-Scherer, D., Chiu, Y., Ofosu, N. N., Luig, T., Hunter, K. H., Jabbour, B., …, & Awasis, F. (2021). Illuminating and mitigating the evolving impacts of COVID-19 on ethnocultural communities: A participatory action mixed-methods study. *Canadian Medical Association Journal, 193*(31), E1203–E1212.

Centobelli, P., Cerchione, R., Del Vecchio, P., Oropallo, E., & Secundo, G. (2022). Blockchain technology for bridging trust, traceability and transparency in circular supply chain. *Information Management, 59*(7), 103508.

Costa, D. G., & Peixoto, J. P. J. (2020). COVID-19 pandemic: A review of smart cities initiatives to face new outbreaks. *IET Smart Cities, 2*(2), 64–73.

Czerniewicz, L.et al. (2020). A wake-up call: Equity, inequality and COVID-19 emergency remote teaching and learning. *Postdigital Science and Education, 2*(3), 946–967. https://doi.org/10.1007/s42438-020-00187-4

de Lucas Ancillo, A., Gavrila, S. G., & del Val Núñez, M. T. (2023). Workplace change within the COVID-19 context: The new (next) normal. *Technological Forecasting Social Change, 194*, 122673.

Feeney, M. K., & Stritch, J. M. (2019). Family-friendly policies, gender, and work–life balance in the public sector. *Review of Public Personnel Administration, 39*(3), 422–448.

Haque, S. M. S. (2023). The impact of remote work on HR practices: Navigating challenges, embracing opportunities. *European Journal of Human Resource Management Studies, 7*(1), 55–83.

Kim, Y. C., Dema, B., & Reyes-Sandoval, A. (2020). COVID-19 vaccines: Breaking record times to first-in-human trials. *Npj Vaccines, 5*(1), 34.

Kusumastuti, R. D., Nurmala, N., Rouli, J., & Herdiansyah, H. (2022). Analyzing the factors that influence the seeking and sharing of information on the smart city digital platform: Empirical evidence from Indonesia. *Technology in Society, 68*, 101876.

Marston, H. R., Musselwhite, C., & Hadley, R. (2020). COVID-19 vs social isolation: The impact technology can have on communities, social connections and citizens. *The British Society of Gerontology*.

Megasari, L. A., Mas' udah, S., & Febrianto, P. T. (2021). The impact of the COVID-19 pandemic and work from home policy on poor communities and women in rural area. *Society, 9*(2), 522–538.

Mizani, H., Cahyadi, A., Hendryadi, H., Salamah, S., & Retno Sari, S. (2022). Loneliness, student engagement, and academic achievement during emergency remote teaching during COVID-19: The role of the God locus of control. *Humanities and Social Sciences Communications, 9*(1), 1–9.

Stogner, J., Miller, B. L., & McLean, K. (2020). Police stress, mental health, and resiliency during the COVID-19 pandemic. *American Journal of Criminal Justice, 45*, 718–730.

Van der Schaar, M., Alaa, A. M., Floto, A., Gimson, A., Scholtes, S., Wood, A., …, & Ercole, A. (2021). How artificial intelligence and machine learning can help healthcare systems respond to COVID-19. *Machine Learning, 110*, 1–14.

Ye, F., Majumder, S., Jiang, W., Li, X., Balakrishnan, N., Zhang, Y., & Deen, M. J. (2022). A framework for infectious disease monitoring with automated contact tracing—A case study of COVID-19. *IEEE Internet of Things Journal, 10*(1), 144–165.

28

SOURCES AND DISTRIBUTION OF FUNDING

The Case of Housing Reconstruction after the 2015 Nepal Earthquake

Barsha Manandhar

Introduction

The 7.8 magnitude earthquake in April 2015 severely affected Nepal. The earthquake impacted the lives, livelihoods, and property of many people in 32 of 77 districts of Nepal. Many buildings collapsed in urban and rural areas. The earthquake killed 8,790 people and displaced about 1 million people. In the Kathmandu Valley alone, earthquakes damaged or destroyed 114,398 private houses (Housing Recovery and Reconstruction Platform, 2020). Subsequently, the Government of Nepal (GON), with international support, established a new governance arrangement for housing reconstruction in affected districts. It also provided USD 3.94 billion in financial aid for recovery and reconstruction. Recent reports from government and donor organizations claimed that 92% have received third financial installment and reconstructed 704,341 housing units (Gyewali, 2021; Shrestha, 2021). Yet, the housing reconstruction data on urban Kathmandu Metropolitan City (KMC) revealed that only 18% of affected households received all financial assistance for housing reconstruction (Housing Recovery and Reconstruction Platform, 2020; Manandhar, 2021).These data points suggest inequity issues in the distribution of funding for post-disaster housing reconstruction, which is one of the central issues in post-disaster housing recovery and reconstruction (Peacock et al., 2014; Ganapati & Mukherji, 2019; Emrich et al., 2022).

Using the case of the 2015 Nepal earthquake, this chapter examines the causes and consequences of the discrepancy in the distribution of financial assistance for housing reconstruction in the urban Kathmandu city. In particular, the chapter examines who makes and implements policies and decisions related to financial assistance for post-disaster housing reconstruction. The question will contribute to the growing literature on disaster governance and the social equity issue in disaster and emergency management. The study conducted and analyzed semi-structured interviews (n=81) from July 2019 to September 2019, with respondents from government, non-government, and local organizations that were involved in post-disaster recovery initiatives after the 2015 Nepal earthquakes. The analysis of interviews and documents revealed that powerful stakeholders, such as development agencies and government bureaucracies, significantly influenced policy and decision-making, leading to restricted and overlooked support for the allocation of funding for urban housing reconstruction. This negligence stemmed from a focus on simpler, quantifiable reconstruction projects in rural areas. The next section provides

DOI: 10.4324/9781003414834-38

background information and an overview of the literature on the management and mobilization of financial assistance for post-disaster housing recovery and reconstruction. Following that, the chapter presents research findings and discussions. Finally, the chapter concludes by offering brief policy implications.

Research Background

Nepal was struck by two severe earthquakes, measuring 7.8 and 7.3 on the Richter scale, in late April and mid-May 2015, affecting 32 of its 77 districts. These natural disasters resulted in approximately 8,790 deaths and displaced 8 million people. This happened during a politically and economically fraught period of Nepal following the decade of Maoist war[1] (Hutt, 2020). The Post Disaster Needs Assessment (PDNA), conducted by the Nepali government and donor agencies following the earthquakes, estimated damages at USD 7 billion, with 498,852 houses completely destroyed and 256,697 partially damaged (Shrestha, 2021). In response, the Government of Nepal organized an international donor conference, leading to the establishment of the Post Disaster Recovery Framework (PDRF), which initially proposed a budget of approximately USD 9.38 billion for 23 reconstruction programs. The National Reconstruction Authority (NRA) was then established to oversee the reconstruction processes, supported by committees including the Advisory Council and a Steering Committee chaired by the prime minister, as well as an Executive Committee led by the CEO. The NRA coordinated the efforts of various ministries, which implemented projects through their respective Central and District Level Project Implementation Units (Nepal Disaster Report, 2017).

The post-earthquake reconstruction efforts in Nepal were undertaken by substantial financial support from a wide range of international development organizations, including multilateral and bilateral agencies, as well as international non-governmental organizations (INGOs). Prominent multilateral and financial bodies such as the Asian Development Bank (ADB), the World Bank, and the International Monetary Fund, along with bilateral partners from countries including Japan, India, Saudi Arabia, the European Union, the UK, the US, China, and Switzerland, played key roles in the recovery effort (Lamsal, 2021). For example, the World Bank's International Development Association (IDA) provided a significant loan of around USD 700 million between 2015 and 2019 for the construction of housing in rural districts, while the ADB signed an agreement to provide a loan of USD 322 million. The Japanese government contributed approximately USD 360 million, primarily through loans, aiding in the construction of thousands of homes. In addition, substantial loans and grants were provided by the governments of India and the Saudi Development Fund, with other countries such as China, the US, and the UK contributing over USD 100 million each in grants. This funding was channeled through various mechanisms, including the government's budgetary processes and off-budget programs, and was instrumental in reconstructing private housing, hospitals, schools, public buildings, and cultural sites (Gyewali, 2021; Lamsal, 2021).

The distribution and management of funds for post-earthquake reconstruction in Nepal after the devastating 2015 earthquakes involved multiple entities and a complex allocation process. The Reconstruction Act mandated the creation of a "reconstruction fund" under the NRA, yet the Ministry of Finance of Nepal was tasked with the management and mobilization of most funds via the government's budgetary processes. International support, primarily through non-budgetary projects, played a pivotal role in the reconstruction efforts. The World Bank also established the Multi-Donor Trust Funds (MDTF) for housing reconstruction in rural districts, with contributions from the US, the UK, Switzerland, and Canada. Moreover, INGOs such as Save the Children, World Vision

International, Oxfam-GB, and the UN Mission to Nepal made significant contributions, with INGO spending amounting to USD 92 million (Lamsal, 2021).

The mechanism for distributing reconstruction funds was multifaceted and varied. The NRA played a key role by providing USD 3,000 in cash assistance for house reconstruction, disbursed in three tranches through banking channels. However, this system faced challenges, particularly in rural areas, due to its complexity and the resulting delays. Additionally, the Government of Nepal (GoN) and the NRA facilitated low-interest housing loans for a limited time to affected individuals in both urban and rural settings. International development organizations, including humanitarian organizations also directly provided financial and technical assistance through their affiliated local organizations and consultants (Manandhar, 2021). Notably, in urban Kathmandu, only 18% (as of 21 October 2019) of the affected households had received complete financial assistance from government for housing reconstruction (Housing Recovery and Reconstruction Platform, 2020) (Figure 28.1).

Urban Kathmandu, Kathmandu Metropolitan City (KMC), is in the Kathmandu Valley, Nepal. The urban core of KMC has old housing structures, heavily impacted by the earthquake. KMC is the biggest urban city of Nepal and consists of 32 wards (smallest administrative unit). Interviews were conducted with local government officials, civil society organizations, and social mobilizers using semi-structured interviews to collect primary data from interlocutors. In particular interviews were conducted with public managers of the NRA (n = 35), representatives

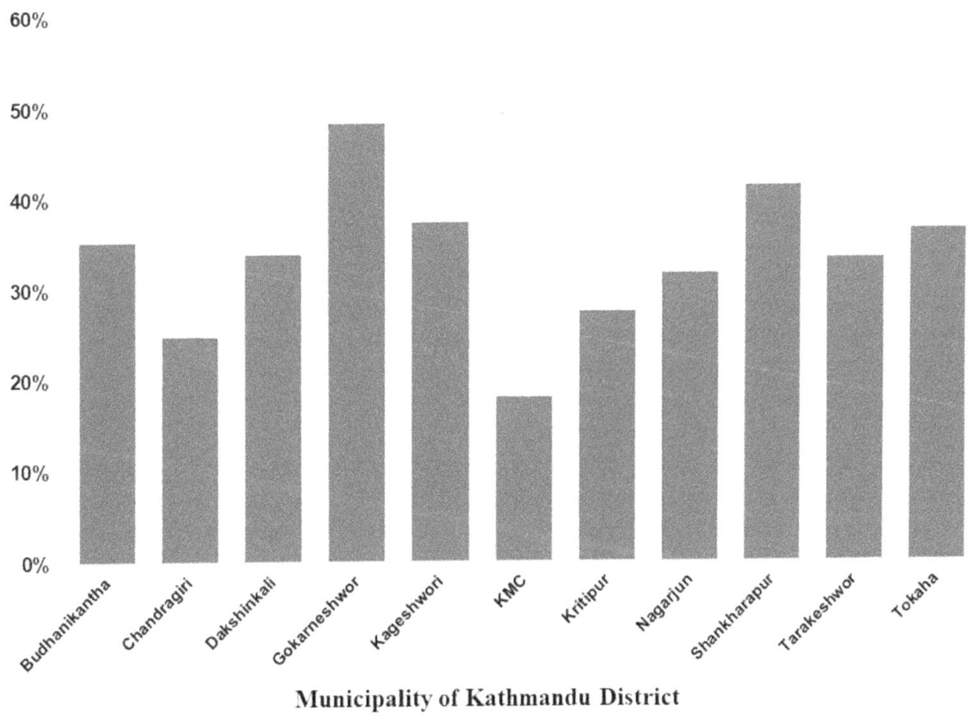

Figure 28.1 Percentage of Households Receiving the Third Cash Assistance in Municipalities Located in Kathmandu District

Source: Housing Recovery and Reconstruction Platform (2019); Figure created by author.

from INGOs (n = 15), and ward leaders (n = 31) of KMC. Participants were selected using purposive and snowball sampling techniques. The interviews were audio recorded with the verbal consent of the participants with each interview lasting 30–45 minutes.

Literature Review

Following a disaster, housing reconstruction emerges as a critical yet complex undertaking, requiring substantial financial resources. However, governments of developing countries often lack the financial capacity and management expertise to undertake such large-scale reconstruction. In these contexts, international financial institutions, multilateral and bilateral organizations, and humanitarian agencies play a crucial role (Johnson & Olshansky, 2016; Mukherji, 2019; Juran et al., 2022). They provide essential resources, including grants and loans, to support reconstruction efforts. For instance, the World Bank frequently offers financial assistance to governments for post-disaster infrastructure rebuilding. International Financial, World Bank, and bilateral organizations are increasing mobilizing large funding through MDTF (Wolfgang et al., 2008; Ganapati & Mukherji, 2014; Mukherji, 2019). Additionally, international non-governmental organizations, bilateral organizations, and humanitarian agencies contribute through grants and donations, further bolstering reconstruction efforts in developing countries. Even domestic sources, such as government reserve funds, insurance programs, and personal savings, can play a role in financing reconstruction, as demonstrated by the Wenchuan earthquake in China, where the central government ensured accountability and availability of funding for housing reconstruction (Xiao et al., 2020).

While the availability of financial resources is crucial for post-disaster recovery, simply having funds does not guarantee their equitable distribution among different people, places, and sectors. A significant body of post-disaster recovery and reconstruction research highlights the persistent issue of inequity in the distribution of post-disaster financial assistance for housing reconstruction (Green et al., 2007; Fussell et al., 2010; Taheri Tafti & Tomlinson, 2013; Ganapati & Mukherji, 2014; Peacock et al., 2014; Mukherji, 2015; Emrich et al., 2022). Despite the mobilization of substantial resources, their disbursement often fails to adequately address the needs of all affected communities. These studies consistently find that marginalized groups, such as women, renters, the elderly, and those living in poverty, often receive inadequate support. Furthermore, in areas with existing socioeconomic disparities, wealthier households and neighborhoods tend to have access to better resources, faster aid, and more substantial reconstruction efforts (Finch et al., 2010). This pattern not only exacerbates existing inequalities but also leaves vulnerable populations more susceptible to the impacts of future disasters.

Several studies paint a stark picture of inequitable post-disaster financial distribution and housing reconstruction. Nakashima and Okada (2018) expose the limitations of the Japanese government's reconstruction policies, which overlooks regional disparities in cost and socio-economic factors, placing a financial burden on victims and hindering their long-term recovery. Similarly, Mukherji (2015) and Taheri Tafti and Tomlinson (2013) reveal how donors' reconstruction policies in Bam, Iran, Bhuj, India, and Gujrat, India, prioritized the majority while failing to provide timely and appropriate support for lower-income groups, tenants, and informal settlers, leading to inequitable outcomes. These studies illustrate how seemingly neutral policies can disproportionately impact certain groups due to pre-existing vulnerabilities, such as lack of land ownership, economic disadvantage, or marginalized housing status. As highlighted in above paragraphs, the role of powerful actors in shaping inequitable distribution is further explored by studies like (Jones et al., 2016; Ruszczyk, 2019; Shrestha et al., 2019). They highlight how powerful international organizations and financial institutions, like the United Nations and World Bank, influenced policy decisions in disaster-affected regions like Nepal, privileging certain agendas and knowledge systems over local needs and priorities.

The impact of inequitable distribution extends beyond restrictive policy of powerful organizations. Studies like Finch et al. (2010) and Kamel and Loukaitou-Sideris (2004) demonstrate how factors like race, ethnicity, and class shape access to resources and recovery. In New Orleans after Hurricane Katrina, the slow recovery of populations in the middle range of social vulnerability was linked to their ineligibility for outright assistance and lack of resources for self-recovery. Similarly, Green et al. (2007) expose how the African American population in Katrina's path faced systemic barriers due to their concentration in an area excluded from FEMA flood zones and lacking access to resources for rebuilding. These studies emphasize that recovery and reconstruction is not solely determined by policy but also by the identities and vulnerabilities of affected communities. These studies demonstrate the need for further research exploring the link between diverse funding sources and the stakeholders involved, and their impact on the distribution of resources. Examining how different stakeholders contribute to inequitable funding distribution is crucial for developing effective interventions and ensuring equitable reconstruction outcomes.

Findings

Collectively, the findings detail various governance challenges and their multiple influences on the distribution of financial assistance in urban KMC.

Governance Structure

Nine months after this devastating event, the Nepali government, backed by international donor agencies, crafted a governance structure for the design and implementation of reconstruction tasks and resources in the affected districts. This structure, grounded in the newly promulgated Reconstruction Act, consisted of five high-level policy-making committees. The prime minister, leading national politicians, senior government bureaucrats from various ministries, officials from governmental development and enforcement agencies, ex-bureaucrats, and government-appointed experts chiefly occupied or dominated these committees. Furthermore, a new Nepal Reconstruction Authority (NRA) was established in December 2015 for five-year time for the management and coordination of reconstruction resources and services. It consisted of several administrative and implementation bodies that were filled with public officers from government ministries and temporary human resources such as consultants and engineers, who had distinct backgrounds and loyalties. Its hierarchy ranged from a Chief Executive Officer (CEO) to senior bureaucrats, executives, and career-public managers. Despite the diversity of its staff, traditional bureaucracies, career-public managers, and administrators from government ministries and departments dominated decision-making and implementation bodies.

Furthermore, the limited participation of local governments, independent NGOs, and communities hinted the dominance of government bureaucrats and administrators in post-disaster governance. Initially, attempts were made to implement local-level governance structures, such as "resource centers," but these efforts were short-lived and limited in scope. These local structures were never fully expanded and were quickly dismantled. An officer from a project implementation unit expressed this concern, noting, "NRA's structure looks good from outside, but we have nothing at the ward level [of local government]." Reflecting on the potential benefits of the resource centers, the Deputy Director of the Central Level Project Implementation Unit (CLPIU) for Education said, "If we had resources centers, we could have mobilized resources or coordinated ground level staffs and engineers, and we could have done better quality work and completed work faster." Prior to its restructuring in July 2018, the NRA's local-level post-disaster governance structure was embodied by District Coordination Committees (DCCs). These committees, established in 22 highly affected districts, were the only bodies that held the power and authority to coordinate, monitor, and appraise NRA's activities. However, in a move that further emphasized the centralization of governance, the GoN formally disbanded all DCCs in 2018. The responsibilities of these committees were transferred to NRA's District Project Implementation Unit (DLPIU), which resulted in a lack of representation for local governments and communities within any governing structure. Moreover, while political leaders from the earthquake-impacted districts found representation in some of these policy-making committees, the decision-making and implementation bodies lacked participation from independent international NGOs.

Consequently, an interviewee from an NGO critiqued the less inclusive governing bodies and the disproportionate influence of the government bureaucracies, stating, "Unlike [collaborative] recovery processes in Gujarat's earthquake, the [Nepali] government has adopted one-door policy. The government determines housing designs, regulates budget allocation, and makes decisions on [reconstruction] policies." Despite the Reconstruction Act's provision for NRA's financial autonomy, including the power to establish, mobilize, and access the Reconstruction Fund, this provision has remained largely unimplemented. Many NRA officials expressed concerns about their organization's limited financial resources and authority. During a public forum designed to evaluate the NRA's accomplishments, a senior NRA official emphasized the organization's limited financial resources, and the Ministry of Finance was repeatedly mentioned as exercising strict control over disaster finances. For example, a representative from the CLPIU-GMALI (Grant Management and Local Infrastructure) said, "The NRA lacks the ability to manage even a small budget and relies heavily on the Ministry of Finance." In addition, others reported insufficient and untimely support from the Ministry, which has often been accused of freezing or delaying the financial process. This overarching control not only signals to limited decentralization of rights and resources to the newly established reconstruction authority but also to local government and communities.

For instance, even after the promulgation of the new constitution and implementation of the new federal system and local governments representatives which all guaranteed the local governments and communities rights, local governments had limited rights to make decisions and manage financial resources related to private housing reconstruction—"*People say that 'Singha durbar ko adhikar ghar ghar m*a' [It refers to the decentralization of rights to local government and communities], *but our [local] rights are no different than the Panchayat system,*[2] A local ward leader said." Moreover, other ward leaders mentioned that local government rights, in the case of post-earthquake reconstruction, do not even include local attesting (certification) rights because the NRA even did not approve local governments' list of beneficiaries of post-earthquake cash assistance which they prepared following the earthquake. Rather than more rights that were promised by the new federal governance system, local leaders mentioned more local rights and responsibilities were recentralized to the government after federalism was implemented in Nepal.

Coordination and Communication

The emphasis on collaboration and communication in policy documents for effective and efficient management of reconstruction fundings and activities were consistently undermined by a series of challenges within NRA and between stakeholders. Within NRA these obstacles primarily revolve around power dynamics, personal egos, and political affiliations among employees representing different ministries. One of the primary obstacles is the varying degrees of authority and power possessed by the NRA employees. Drawn from different government ministries, the discrepancy in their statuses frequently triggers conflicts over hierarchy and tensions among managers and staff. This stands in stark contrast to the intended collaborative activities. The influence of personal ego is exemplified in the daily joint meetings of the NRA. An engineer based at the NRA headquarters said,

> We have different departments and there are four to five undersecretaries in different departments. Undersecretaries do not attend meetings if other undersecretaries call a meeting because of personal ego and issue of who is superior among undersecretaries,

These power dynamics and egos also extend to affect communication and information sharing among NRA officers. One officer pointed out the need for ego management and unbiased recognition of staff from different departments. Further emphasizing this point, a high-ranking CLPIU officer highlighted the detriments of maintaining personal hierarchy to the loss of effective communication and information sharing. A high-level CLPIU officer said, "if we only compete to maintain our own hierarchy, then there will be a problem of communication and information sharing." The challenges are exacerbated by the political affiliations of NRA officers. Several interviewees indicated that such affiliations often determine the transparency and distribution of information, leading to an environment of mistrust among officers. This lack of transparency was criticized by one engineer who mentioned, "We have a politically appointed CEO and if someone in the lower rank is politically active and has political connections, information directly goes from the lower to CEO without sharing of information with others" (Public Manager, Engineer Main office, 21 August 2019).

At the community level, local leaders, who consider NRA as just another government body, consistently reported limited engagement, coordination, and communication with the NRA—the Nepali government. The primary concern voiced by local leaders was that their opinions and problems were not being considered. The NRA's modus operandi was criticized for its top-down approach, where NRA junior staff would visit once a year for a perfunctory survey, then quickly shift responsibility onto the locals. An interviewee noted, "Only junior staff come here once a year. An engineer came here recently to do re-survey…these officials come for one day and immediately ask us to contact them with the beneficiary" (Ward Leader, 10 July 2019). This approach, criticized as monopolistic, led to local leaders feeling unheard and unvalued, which was symbolized by the ward leader's complaint: "They do not take our opinion" and "listen to our problems." The NRA's reluctance to share information about policy changes added to the frustration. Local ward leaders and social mobilizers reported having to resort to the NRA's website or newspapers for updates, further highlighting the absence of direct communication. Information about changes in cash assistance and bank loan processes for beneficiaries was particularly crucial, yet, as a leader pointed out, the NRA did not communicate about the "recent change in loan processes" (Ward Leader, 14 July 2019).

The nature of collaboration, coordination, and communication between the NRA and partner organizations was both dynamic and complex. Many senior NRA interviewees acknowledged "good working relationships" and "collaboration" with donors due to their direct provision of technical and financial resources (Interviews). However, an undercurrent of distrust was also expressed due to differing conditions and requirements for accessing loans and grants. As one senior career-public officer from NRA main office stated, "When they provide loans and grants, they will present a lot of conditions. They will say to do this and do that, fulfill this, and fulfill that." This officer further elaborated that such conditions could sometimes become barriers in fostering effective collaborations with donors.

The mistrust between the NRA and development organizations was further exacerbated by the practices of some donor agencies. For instance, the lack of transparency in disaster finances from some partner organizations that came through off-budgetary processes created concerns among public officers. A career-public officer from CLPIU highlighted the necessity of transparency in disaster finances that are funneled through off-budgetary processes. Similarly, many public officers expressed distrust in the practices of INGOs and NGOs largely because of their unwillingness to abide by the initial MOU with GoN. For instance, an NRA career-public officer expressed that INGOs and NGOs are reluctant to share information and come under the oversight of the NRA. "We have some problem working with INGOs and NGOs... they do not want to come under our platform, and they want to work outside of our oversight." Another public officer expressed distrust in INGOs because of their unwillingness to share information about reconstruction with the NRA:

> We face difficulty working with I-NGOs ... we have large investment, but very weak communication. They are doing work, but they do not report about what type of work they are doing and with what mission. They do not inform. They get approval for one thing and do something different in field. We have misunderstanding and mistrust.
>
> (Career public manager, NRA main office,25 July 2019)

Urban Complexities and Needs

The findings on governance challenges have directly influenced the design and implementation of reconstruction policies pertaining to the management and allocation of fundings. Specifically, major post-disaster housing reconstruction policies were restrictive, and the implementation often did not adequately consider the complexities of urban issues and the needs of the affected people. For example, stringent land ownership criteria and urban neglect hampered the distribution of financial assistance for housing reconstruction. These decisions are further discussed below in different headings.

Following the earthquake, the NRA collaborated with partner organizations to assess the damage to houses. Their goal was to provide appropriate financial support to those affected. In Kathmandu Metropolitan City (KMC), local representatives of each ward, trained by the NRA and its donors, categorized the houses as either damaged or partially damaged. These social mobilizers then surveyed the properties and compiled a list of potential beneficiaries. Roughly a year after the earthquake, NRA engineers performed supplementary surveys in KMC. Their objective was to identify and remove "fake beneficiaries" from the initial list, thereby creating a revised list. However, these repeated assessments by the NRA did not involve local authorities or delegate tasks to local governments. This approach contributed to delayed decisions concerning financial assistance and resulted in the exclusion of many affected households from the beneficiaries' list, although affected households had opportunities to appeal if excluded. Many questioned the validity of the assessment and expressed that the NRA failed to understand local contexts related to housing damage. A local leader who participated in the initial survey critiqued the NRA engineers' understanding of the housing complexities faced by KMC's urban residents. This leader pointed out, "NRA engineers did not even go inside houses to assess the damage....so how do they know about the shared wall system" (Ward Leader, 4 August 2019). This raised significant questions about the validity and effectiveness of the assessment process implemented by the NRA.

Furthermore, the Post Disaster Recovery Framework (PDRF) proposed 23 programs for urban housing reconstruction (PDRF (Post Disaster Recovery Framework), 2016). These programs, which included initiatives such as "housing reconstruction," "housing credit, repair and retrofit," "material supply, enterprise, livelihood," "sustainable housing services," and "rental housing for poor" were intended to support the recovery of affected households. While most programs conceived in the PDRF were never implemented, the government and donors did launch financial assistance programs for urban housing reconstruction. After the identification of beneficiaries of cash assistance, reconstruction projects adopted a "blanket approach" to provide cash assistance to all earthquake-affected households in 32 districts. The government provided the first installment (USD 500) to all affected people after the agreement. The second and third installments were contingent upon evidence of construction of the plinth and first floor, respectively. In addition to repair and retrofitting subsidies of USD 1,000 for affected houses, the GoN and NRA briefly offered a low-interest housing credit or bank loan of USD 25,000 for urban areas and USD 15,000 for rural areas (Ward Leader, 5 August 2019). Local residents interviewed unanimously reported that constructing "safe houses" in KMC was unfeasible with such limited cash assistance due to the high costs of urban construction. "We need USD 5,000 just to demolish and remove the debris of a house," said one leader. Several NRA engineers remarked that the retrofitting program proposed by donors was disconnected from local realities, as the costly design exceeded the USD 1,000 allotted for retrofitting. Many urban beneficiaries vented frustrations about the government's one-size-fits-all policy, ignoring the complex problems of urban affected households, and claimed that "the policy for the Kathmandu [city] should not be same as the policy for [rural] Humla and Jumla [districts]" (Ward Leader, 9 July 2019).

Furthermore, access to cash assistance and low-interest bank loans was complicated by building permit requirements from the GoN and KMC. Stringent land ownership criteria proved particularly difficult. For example, loan applicants needed to present the land title as personal collateral, with one leader disclosing that eligibility for a USD 25,000 bank loan necessitated a minimum of 855.625 sq ft land (plus an additional 10 feet if adjacent to a road) and a monthly income of USD 600. This led to many affected individuals failing to receive financial assistance or being unable to build houses.

Similarly, the land and housing tenure of affected households heavily influenced access to financial assistance and the progress of housing reconstruction initiatives. Many affected households in KMC lack private land and housing ownership as these are often formally titled to the male head of the family. Fragmentation of houses and land among siblings has resulted in smaller parcels, which fail to meet the land ownership and land size requirements for obtaining financial assistance or bank loans. Such issues seriously affected access to financial assistance and the reconstruction of damaged houses, as affected households could neither transfer nor mortgage these small plots. Additionally, some families living on *Guthi* land (state land granted to institutions) do not possess land titles. Despite their long-term residence, they only have informal usage rights granted by the owning institutions. Many families living on such land or renting it out face similar issues of insecurity and potential eviction if they attempt to reconstruct their damaged houses. These issues apply equally to families in squatter (*sukumbasi)* settlements, where tenure security is non-existent. As a result, these families continue living in damaged houses or rent them out to non-family members.

The government and donors even neglected the urban housing reconstruction problem, initially framing it as a "rural problem." For instance, most peer-reviewed articles, donors reports, and news articles started to name the earthquake that impacted 30 districts as the "Gorkha Earthquake," essentially implying it as the rural problem. Furthermore, the widespread damage and poverty in urban areas like KMC were ignored. As one local leader lamented, "The Nepal government thinks that Kathmandu people are rich, and they have ignored us. They only think that people outside Kathmandu is in need of the grant money" (Ward leader, 9 July 2019). Another leader emphasized the deep-seated poverty in his constituency, stating, "In Kathmandu, many people are extremely poor. People are living in abject poverty" (Ward Leader, 4 August 2019). As a result, many earthquake-affected people are either living in structurally risky housing conditions, constructed non-compliance houses, or displaced from the location.

Instead, donor agencies, INGOs, and NGOs seemed to prioritize specific places, such as rural areas, and specific sectors like infrastructure for reconstruction efforts. An interviewee from an organization, closely associated with the NRA and donor agencies, pointed out, "The World Bank and other donors don't want to go in places where there are [reconstruction] gaps or problems" (INGO officer, 13 September 2019). This interviewee further explained that these entities tend to prioritize areas where they find it "easier to spend money." Similarly, NGOs preferred working in districts where they already had existing projects and familiarity, leading to a gap in attention toward earthquake-affected districts. Moreover, some interviewees mentioned that the preference of donor agencies, INGOs, and NGOs for the "software part" of reconstruction, such as educational and awareness programs, over the "hardware part," like the construction of houses and schools.

Discussion of the Findings

This chapter examines inequities in the distribution of financial assistance for recovery and reconstruction following the 2015 Nepal earthquake. The chapter highlights linkages between disaster governance challenges and the distribution of post-disaster housing reconstruction funding. Based on the findings, the chapter argues that powerful stakeholders play an important role in shaping inequities in the distribution of post-disaster financial assistance through their influence on policies, priorities, and discourses. Following paragraphs further discussed major findings.

The top-down governance structure infers that central government and bureaucracies dominate policy-making and implementing processes and practices, including the management and distribution of funding. Similarly, since international development organizations are the major sources of financial and technical assistance for reconstruction, they influenced all stages of policy design and implementation. Johnson and Olshansky (2016) also noted that the entity controlling disaster funds invariably influences decision-making and implementation. Many disaster studies have also stressed the importance of centralized governance and the central role of national governments and development agencies in post-disaster recovery and reconstruction, including managing large-scale post-disaster reconstruction funding (Ganapati & Ganapati, 2009; Comerio, 2014; Ganapati & Mukherji, 2014; World Bank Group, 2014; Johnson & Olshansky, 2016; Bakema et al., 2017; Daly et al., 2017; Xiao et al., 2020; Juran et al., 2022). Such influence of powerful stakeholders has often led to exclusion of local communities and organizations, a disconnect from local needs, reduced accountability, and delays and inequities in the distribution of funding and services for housing reconstruction. The dominance of Nepali governments and government ministries in shaping disaster governance structure and financial resources could be due to a variety of reasons. First, it may be linked it to concerns over potential financial mismanagement and control by donor and implementing organizations, as observed in Haiti and other countries in the contexts of post-earthquake reconstruction (O'Connor et al., 2014; Hsu & Schuller, 2020). Such concern was clear from my findings, where government, although completely dependent on development organizations, largely mistrust them in efficiently managing large reconstruction aids. This is also a popular perception of Nepali people, also expressed by some respondents, who believe the reconstruction as another "*magi khani bhado*" [used as a prop to ask for help or aid] of international communities. Second, it's essential to view this centralize control of post-disaster finances in Nepal through broader political and economic context. The 2015 earthquake and subsequent construction took place against a backdrop of a decade-long violent Maoist uprising in Nepal, followed by another decade of ethnic contestation and conflicts over Nepal's new constitution and federal structure. The rise of the Maoist party and indigenous groups as potent political forces, coupled with the shift to a federal governance system, challenged traditional political parties and centralized bureaucratic control. These dynamics possibly fueled the recentralization of governance institutions and finances by traditional bureaucracies, which was common in other social, environmental, and political spheres (Shneiderman et al., 2016; Dongol & Neumann, 2021).

Another finding underscores the significant coordination and communication challenges among stakeholders, including government agencies, NGOs, and local communities. As a result, a mutual distrust among stakeholders significantly hindered leveraging support for funding and reconstruction in the urban KMC. Essentially, the preferences and discourses of donors and bureaucracies shaped the reconstruction and funding policies, processes, and priorities, leading to an uneven resource distribution for urban private housing. Past studies on Nepal's urban reconstruction indicate that unclear urban reconstruction policies, limited governance capacity, and the neglect of municipal and ward-level officials hampered progress (Daly et al., 2017; Lam & Kuipers, 2019). Furthermore, Shrestha et al. (2019) linked the limited opportunities for local community participation and inclusion in Nepal's disaster management policies and practices to the nexus of experts from bureaucracy, NGOs, and donor agencies and their underlying "custodial culture and scientific tradition" (Shrestha et al., 2019). Similarly, Ganapati and Mukherji (2014) examine the World Bank-funded housing reconstruction projects in India and Turkey following significant earthquakes in 2001 and 1999, respectively. It highlights a prevailing urgency, driven by the Bank's emergency recovery loan mechanism and success measures (WB's internal culture and routines), which compromised the flexibility required during the project's implementation. This urgency hampered the integration of crucial aspects such as planning, land readjustment, and public participation for housing reconstruction.

In KMC, this article revealed flaws in the assessment of affected houses due to insufficient local government and leader involvement. These locals were well-acquainted with the intricate housing and land tenure systems. The NRA disregarded the local government's beneficiary list, leading to repeated house assessments to identify "fake beneficiaries." This centralized and discriminatory approach delayed evaluations, obscured problem identification, and excluded many households from beneficiary lists. Additionally, the government used a blanket approach to provide cash assistance. This one-size-fits-all cash assistance policy, tied to property ownership, failed to address the complex urban challenges. Many poor individuals, particularly those without formal property titles or those living on "Guthi land" or in "Squatter settlements," were excluded from aid programs. Moreover, despite some residents reconstructing their homes without formal legal processes early on, the government later enforced the national building code in KMC.

Another significant observation relates to framing earthquake damages as a "rural problem," and the characterization of the earthquake as the "Gorkha earthquake." Consequently, these organizations and donors were largely absent in urban areas, sidelining urban issues including the distribution of financial resources and support for housing reconstruction. Instead of complex and structural issues of urban reconstruction, government and donors primarily focused on easier, familiar, and measurable issues. That is, reconstruction was geared toward issues such as the distribution of financial assistance to rural affected household, vulnerable and landless household of rural areas, construction of schools, cultural sites, public buildings, etc.

Conclusion

In conclusion, this chapter shows how powerful stakeholders often overshadow the root or complex causes of urban vulnerabilities, including the equitable distribution of financial assistance for housing reconstruction. If governments and international agencies genuinely aim to build "resilient cities," they must address and understand these urban vulnerabilities. In addition, policymakers should revamp strategies to promote active involvement from local governments and communities in the decision-making process. Granting local governments more authority would be beneficial since they possess a more profound understanding of their regions and residents. However, merely allocating authority and resources is not sufficient. Measures must be put in place to ensure increased local leader accountability and minimize the influence of local power dynamics. Furthermore, it's essential for policymakers to discern between urban and rural settings when crafting policies. These two areas present distinct challenges; hence, a one-size-fits-all solution will not suffice.

Notes

1 Between 1996 and 2006, the Communist Party of Nepal (Maoists) started an armed revolution against the government. After the comprehensive peace process of 2006 between the Nepali government and Maoists, Nepal was declared The Federal Republic of Nepal. The new constitution of the Federal Republic of Nepal was promulgated in 2015, which overlaps with the 2015 Nepal earthquake.

2 Panchayat System is a party-less political system that existed between 1960 and 1990 in Nepal. The King of Nepal directly ruled the government through his appointed ministers. After 1990, Nepal adopted a multiparty democratic governance system.

References

Bakema, M. M., Parra, C., McCann, P., Dalziel, P., & Saunders, C. (2017). Governance in Shaky Societies: Experiences and lessons from Christchurch after the earthquakes. *Environmental Policy and Governance*, *27*(4), 365–377. https://doi.org/10.1002/eet.1764

Comerio, M. C. (2014). Disaster recovery and community renewal: Housing approaches. *Cityscape*, *16*(2), 51–68.

Daly, P., Ninglekhu, S., Hollenbach, P., Duyne Barenstein, J., & Nguyen, D. (2017). Situating local stakeholders within national disaster governance structures: Rebuilding urban neighbourhoods following the 2015 Nepal earthquake. *Environment and Urbanization*, *29*(2), 403–424. https://doi.org/10.1177/0956247817721403

Dongol, Y., & Neumann, R. P. (2021). State making through conservation: The case of post-conflict Nepal. *Political Geography*, *85*, 102327. https://doi.org/10.1016/j.polgeo.2020.102327

Emrich, C. T., Aksha, S. K., & Zhou, Y. (2022). Assessing distributive inequities in FEMA's disaster recovery assistance fund allocation. *International Journal of Disaster Risk Reduction*, *74*, 102855. https://doi.org/10.1016/j.ijdrr.2022.102855

Finch, C., Emrich, C. T., & Cutter, S. L. (2010). Disaster disparities and differential recovery in New Orleans. *Population and Environment*, *31*(4), 179–202. https://doi.org/10.1007/s11111-009-0099-8

Fussell, E., Sastry, N., & VanLandingham, M. (2010). Race, socioeconomic status, and return migration to New Orleans after Hurricane Katrina. *Population and Environment*, *31*(1–3), 20–42. https://doi.org/10.1007/s11111-009-0092-2

Ganapati, N. E., & Ganapati, S. (2009). Enabling participatory planning after disasters. *Journal of the American Planning Association, 75*(1), 41–59. https://doi.org/10.1080/01944360802546254

Ganapati, N. E., & Mukherji, A. (2014). Out of sync: World Bank funding for housing recovery, post disaster planning, and participation. *Natural Hazards Review, 15*(1), 58–73. https://doi.org/10.1061/(ASCE)NH.1527-6996.0000120

Ganapati, N. E., & Mukherji, A. (2019). ""Housekeeping": Managing Post-Disaster Housing Recovery Long Term" in *Housing Recovery after Disasters*. Edited by Frances L. Edwards. Lanham, Maryland: Lexington Books, (p. 167–191).

Green, R., Bates, L. K., & Smyth, A. (2007). Impediments to recovery in New Orleans' Upper and Lower Ninth Ward: one year after Hurricane Katrina. *Disasters, 31*(4), 311–335. https://doi.org/10.1111/j.1467-7717.2007.01011.x

Gyewali, S. (2021). *Exemplary Reconstruction*. Rebuilding Nepal. Kathmandu, Nepal: National Reconstruction Authority, Kathmandu, Nepal

Housing Recovery and Reconstruction Platform (2019), *Monthly District Report*, Kathmandu, Nepal: Housing Recovery and Reconstruction Platform.

Housing Recovery and Reconstruction Platform (2020), *Urban Housing Recovery: Compilation of Case Studies from Nepal and Beyond*. Kathmandu, Nepal: Housing Recovery and Reconstruction Platform.

Hsu, K. J., & Schuller, M. (2020). Humanitarian aid and local power structures: Lessons from Haiti's 'shadow disaster. *Disasters, 44*(4), 641–665. https://doi.org/10.1111/disa.12380

Hutt, M. (2020). Before the dust settled: Is Nepal's 2015 settlement a seismic constitution? *Conflict, Security & Development, 20*(3), 379–400. https://doi.org/10.1080/14678802.2020.1771848

Johnson, L. A., & Olshansky, R. B. (2016). *After Great Disasters: How Six Countries Managed Community Recovery*. Cambridge: Lincoln Institute of Land Policy.

Jones, S., Oven, K. J., & Wisner, B. (2016). A comparison of the governance landscape of earthquake risk reduction in Nepal and the Indian State of Bihar. *International Journal of Disaster Risk Reduction, 15*, 29–42. https://doi.org/10.1016/j.ijdrr.2015.10.011

Juran, L., Oliver, R., & Read, D. C. (2022). "Water Sector Reconstruction for Post-Disaster Housing Settlements: A Tale of Two Governance Models" in *Resilient Water Management Strategies in Urban Settings: Innovations in Decentralized Water Infrastructure Systems*. Edited by Tamin Younos, Juneseok Lee, Tammy E. Parece. Switzerland: Springer Water, pp. 227–243.

Kamel, N. M. O., & Loukaitou-Sideris, A. (2004). Residential assistance and recovery following the Northridge earthquake. *Urban Studies, 41*(3), 533–562. https://doi.org/10.1080/0042098042000178672

Lam, L. M., & Kuipers, R. (2019). Resilience and disaster governance: Some insights from the 2015 Nepal earthquake. *International Journal of Disaster Risk Reduction, 33*(October 2018), 321–331. https://doi.org/10.1016/j.ijdrr.2018.10.017

Lamsal, R. (2021). *Exemplary Reconstruction*. Rebuilding Nepal. Kathmandu, Nepal: National Reconstruction Authority.

Manandhar, B. (2021). *Managing Collaborative Networks in Post-Disaster Recovery: A Case Study of 2015 Nepal Earthquake* [Florida International University]. https://doi.org/10.25148/etd.FIDC010218

Mukherji, A. (2015). From tenants to homeowners: Housing renters after disaster in Bhuj, India. *Housing Studies, 30*(7), 1135–1157. https://doi.org/10.1080/02673037.2015.1008423

Mukherji, A. (2019). Funding Flows: Transboundary Considerations of Disaster Recovery. *Oxford Research Encyclopedia of Natural Hazard Science*. https://doi.org/10.1093/acrefore/9780199389407.013.223

Nakashima, T., & Okada, S. (2018). Financial imbalances in regional disaster recovery following earthquakes—Case study concerning housing-cost expenditures in Japan. *Sustainability, 10*(9), 3225. https://doi.org/10.3390/su10093225

Nepal Disaster Report 2017: The Road to Sendai (2017).Kathmandu, Nepal: Government of Nepal-Ministry of Home Affairs.

O'Connor, D., Brisson-Boivin, K., & Ilcan, S. (2014). Governing failure: Development, aid and audit in Haiti. *Conflict, Security & Development, 14*(3), 309–330. https://doi.org/10.1080/14678802.2014.923150

PDRF (Post Disaster Recovery Framework) (2016). Kathmandu, Nepal: Government of Nepal-National Reconstruction Authority.

Peacock, W. G., Van Zandt, S., Zhang, Y., & Highfield, W. E. (2014). Inequities in long-term housing recovery after disasters. *Journal of the American Planning Association, 80*(4), 356–371. https://doi.org/10.1080/01944363.2014.980440

Ruszczyk, H. A. (2019). Ambivalence towards discourse of disaster resilience. *Disasters*, *43*(4), 818–839. https://doi.org/10.1111/disa.12385

Shneiderman, S., Wagner, L., Rinck, J., Johnson, A. L., & Lord, A. (2016). Nepal's ongoing political transformation: A review of post-2006 literature on conflict, the state, identities, and environments. *Modern Asian Studies*, *50*(6), 2041–2114. https://doi.org/10.1017/S0026749X16000202

Shrestha, N. (2021). *Exemplary Reconstruction*. Rebuilding Nepal. Kathmandu, Nepal: National Reconstruction Authority.

Shrestha, K. K., Bhattarai, B., Ojha, H. R., & Bajracharya, A. (2019). Disaster justice in Nepal's earthquake recovery. *International Journal of Disaster Risk Reduction*, *33*, 207–216. https://doi.org/10.1016/j.ijdrr.2018.10.006

Taheri Tafti, M., & Tomlinson, R. (2013). The role of post-disaster public policy responses in housing recovery of tenants. *Habitat International*, *40*, 218–224. https://doi.org/10.1016/j.habitatint.2013.05.004

Wolfgang, F., Ishan, A., & Kai, K. (2008). *Managing Post-Disaster Reconstruction Finance*. (Vol. 4475). Washington, DC: World Bank Publications.

World Bank Group. *Pakistan Earthquake 2005: The Case of Centralized Recovery Planning and Decentralized Implementation*. (2014). Washinton, DC: World Bank Group.

Xiao, Y., Olshansky, R., Zhang, Y., Johnson, L. A., & Song, Y. (2020). Financing rapid community reconstruction after catastrophic disaster: Lessons from the 2008 Wenchuan earthquake in China. *Natural Hazards*, *104*(S1), 5–30. https://doi.org/10.1007/s11069-019-03789-9

29

RECOVERY AMONG VULNERABLE POPULATIONS

The Case of 'Forgotten People' after the 2015 Nepal Earthquake

Sanam K. Aksha, Prabin Acharya, and Shirish Bhat

Introduction

Nepal experienced a strong earthquake of a moment magnitude 7.8 (M_w) on 25 April 2015. The epicenter was in Gorkha district in central Nepal, 80 km northwest of Kathmandu, the capital of the country. The earthquake claimed approximately 9,000 lives, injured more than 22,300 persons, and over 1 million buildings were completely or partially destroyed (NPC, 2015). The earthquake impacted more than 8 million people, with 32 out of 77 districts categorized as affected (Acharya et al., 2022a; NPC, 2015) (Figure 29.1). According to the disaster risk reduction portal of the Government of Nepal, at least 773,095 private dwellings and 2,687 government buildings were completely destroyed. Also, 298,998 private houses and 3,776 government units were partially damaged (Government of Nepal, n.d.). About 50,000 classrooms in 7,553 educational institutions were also damaged (Acharya et al., 2022a).

The estimated recovery cost was $7 billion USD (NPC, 2015), representing a significant portion of the country's GDP of around $24 billion USD (Worldometer, 2024). The most impacted districts were predominantly rural areas where communities lost their dwellings, livestock, crops and harvests, and agricultural tools. The condition of grazing and crop fields further deteriorated after the earthquake, and agricultural inputs and services (such as irrigation) remained unavailable, damaged, or inaccessible. According to the Post-Disaster Needs Assessment (PDNA) prepared by the National Planning Commission of the Government of Nepal (NPC, 2015), the earthquake impacted the livelihoods of about 2.29 million households. The PDNA estimated that over 17,000 cattle and approximately 40,000 smaller domesticated animals perished. Additionally, the earthquake damaged irrigation facilities and triggered landslides in several locations, rendering nearly 1,000 hectares of land useless, as documented in the PDNA report.

The distribution of impacts across individuals and communities was not even. Data obtained from the disaster risk reduction portal of the Government of Nepal reveal that about half of the mortality (47%) was attributed to age-dependent populations who are 65+ years old or age 5 or less (Government of Nepal, n.d.). Research has found that characteristics such as age, monthly household income, and type of dwelling played a significant role in earthquake-related injuries (Paul & Ramekat, 2018). Most importantly, according to a village-level social vulnerability analysis, the most hard-hit geographic areas also correspond with populations categorized as exhibiting

DOI: 10.4324/9781003414834-39

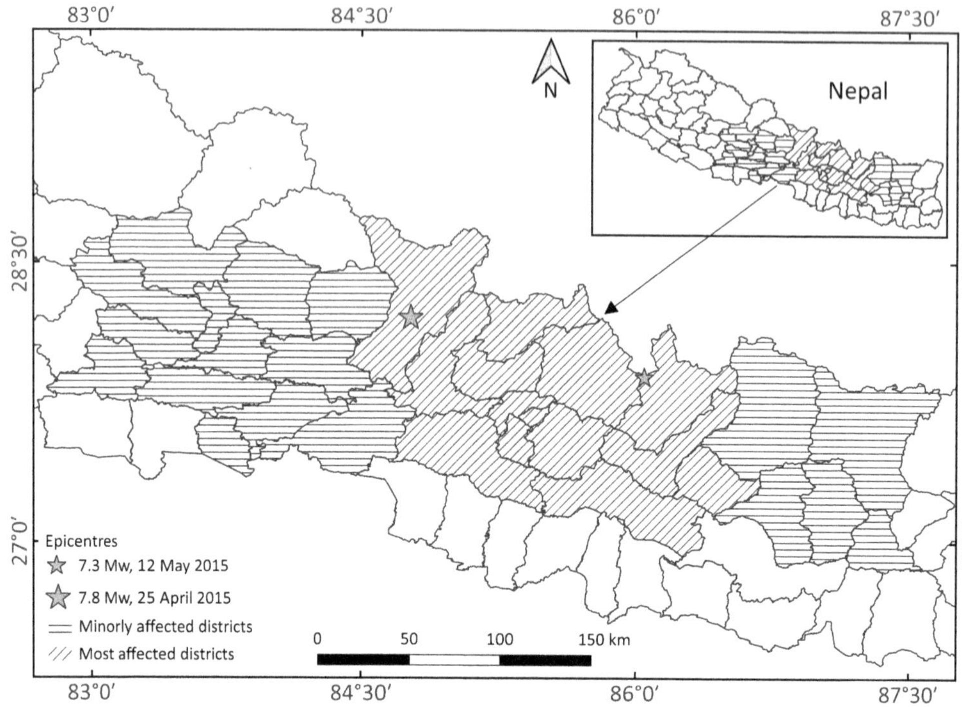

Figure 29.1 The Most Severely and Minorly Affected Districts due to the 2015 Gorkha Earthquake in Nepal
Source: Figure by Authors/Original Figure.

medium to high social vulnerability (Aksha et al., 2019). According to the latest 2021 census, Nepal had 142 castes and ethnic groups, most of which live in the country's rural areas (National Statistics Office, 2023). The earthquake disproportionately affected marginalized communities and lower castes. Furthermore, the politicization of aid during response efforts created conflict within communities (Melis, 2022). The earthquakes were most devastating in rural areas characterized by a higher concentration of ethnic communities, scattered settlements, poor road networks, and traditional mortar-brick homes. Due to the predominance of centralized politics and poor development outside of Kathmandu, the rescue, response, and recovery during and after the earthquakes were challenging, essentially depriving needy communities of much-needed resources (Hülssiep, Thaler, & Fuchs, 2021).

The Disaster Recovery Process

Disasters disrupt social and community functioning. Disasters displace people, close schools and businesses, and damage or interrupt critical infrastructure such as hospitals, power stations, water treatment and supply, wastewater treatment, communication systems, and transportation networks (Tierney, 2019). Consequently, local resources become overwhelmed and affected communities require assistance from other jurisdictions, including state and national governments as well as the international community. Thus, it requires relentless and complex efforts for communities to recover and function normally. Disaster recovery is generally defined as a process of restoring, rebuilding, and reshaping physical infrastructure, repopulation by survivors and new residents, the gradual resumption of regular daily routines, and ultimately returning to a stable, acceptable post-disaster state (Cutter et al., 2014; Smith, Martin, & Wenger, 2018). Disaster recovery is an intricate and holistic process that encompasses multiple, complexly interwoven, and inextricably linked sectors such as private and public infrastructure, economic activities, cultural elements, ecosystem services, and social institutions and organizations (Tierney & Oliver-Smith, 2012).

Disaster recovery is typically considered the last phase of the cyclical disaster management cycle. The recovery process is the longest and most expensive phase (Alcántara-Ayala et al., 2023). Disaster recovery can broadly be divided into three stages: short-, intermediate-, and long-term (Emrich, Aksha, & Zhou, 2022; Frerks et al., 1995; Phillips, 2016). Short-term recovery immediately follows a disaster event. This phase generally lasts for days to weeks and focuses on distributing relief materials such as food, water, and temporary shelter. Intermediate-term recovery typically focuses on rehabilitating disaster-impacted infrastructure, homes, and communities. This stage, lasting weeks to months, aims to restore essential services. Long-term recovery—which lasts months to years to over a decade—seeks to reconstruct affected communities to a level that is 'better' and more resilient to future disasters. Although early disaster recovery research characterized it as a singular, sequential, measurable outcome, recent research has emphasized that recovery is marked by numerous complex, multistage processes that overlap and, in many cases, are not necessarily distinctive (Tierney & Oliver-Smith, 2012). The recovery process is often highly dependent upon initial damages, local resource availability, and recovery program execution timelines, among other factors (Mukherji, 2017). Recent literature emphasizes that recovery is a forward-looking process of restoring, rebuilding, and reshaping the physical, social, economic, and natural environment through pre-event planning and post-event actions (Smith et al., 2018).

Disasters reveal differential impacts on vulnerable and marginalized populations. The social, economic, demographic, cultural, and built environment characteristics of an individual and a community play influential roles in determining the magnitude of disaster damages and losses. These disadvantaged groups, including the elderly, low-income families, individuals with disabilities, and subaltern communities, typically experience heightened difficulties due to pre-existing social, economic, and health-related vulnerabilities. Hence, vulnerable populations face distinct challenges during post-disaster recovery, often enduring protracted, long-term impacts. If executed poorly, the recovery phase can exacerbate pre-existing vulnerabilities and lead to increased dependency on aid. Marginalized communities often have limited access to recovery resources such as financial assistance, housing, healthcare, employment opportunities, and formal banking and financing. The lack of adequate support and tailored recovery strategies can lead to prolonged displacement, persistent health issues, and deepening poverty, thereby hindering the ability of vulnerable populations to rebuild their lives and communities effectively (Alcántara-Ayala et al., 2023; Mukherji, 2017).

Vulnerable Populations in the Disaster Recovery Process

In the disaster literature, much is known conceptually about what renders individuals and communities more vulnerable to disasters, shocks, and stresses (Cutter, Boruff, & Shirley, 2003). However, how these concepts manifest during the recovery process—specifically in terms of the distribution of post-disaster recovery support—is less known and relatively little explored (Emrich et al., 2022). Following disruptive disaster events, residents and communities often desire a rapid return to normalcy (Gall & Friedland, 2020). However, a disaster recovery's pace and trajectory are primarily determined by when and how requisite support is provided to disaster survivors (Finch, Emrich, & Cutter, 2010). In the past four decades, significant investments have been put into research to understand the risks of various hazards, develop early warning tools and techniques, and implement more informed recovery programs and policies. As a result of technological advancements in computing capacity, abundant impact and recovery data have been gathered. Yet, despite these significant advances, society continues to lose more to disasters (UNDRR, 2020; White, Kates, & Burton, 2001). Such losses are often unequally borne by the most marginalized and vulnerable communities who struggle to prepare for, respond to, and rebound from disasters (Peacock et al., 2014; Tierney, 2019).

Previous research has shown that, in general, socioeconomic and demographic variables such as population density, pre-existing vulnerability, access to and receipt of aid packages, and governance—in addition to damages from the actual disaster—are primary determinants of post-disaster recovery (Aldrich, 2012; Drakes et al., 2021; Phillips, 2016). Although disasters produce adverse outcomes for most individuals and communities, they especially impact the poor, marginalized, and disadvantaged (Shrestha et al., 2019). In disasters, poorer households sustain less absolute property damage, but the impacts relative to pre-event financial conditions can be larger than wealthier households. With less capacity for recovery, households in poor economic conditions depend more on external assistance for recovery support. However, research has shown that low-income areas often receive less external assistance relative to damage levels (Emrich et al., 2022; Mukherji, 2017; Peacock et al., 2014; Tierney & Oliver-Smith, 2012). Housing tenure is another important variable in determining the recovery capacity of disaster survivors (Mukherji, 2010). In urban areas, renters and tenants must often leave their dwellings and the affected area as a result of a lack of livable housing (Juran, Adams, & Prajapati, 2019). Furthermore, they may be excluded from receiving (some) disaster recovery support due to a lack of property ownership documents. After a major disaster, rental rates also tend to rise and thus may become unaffordable for renter populations (Mukherji, 2017; Wilson, Tate, & Emrich, 2021).

Moreover, the critical role of vulnerable populations in the recovery process is frequently overlooked (Finch et al., 2010; Jerolleman, 2019). This leads to recovery plans that do not adequately address the specific needs of vulnerable populations or harness their lived experiences and potential contributions. Despite facing significant challenges, disadvantaged communities possess invaluable knowledge and resilience strategies that can enhance the effectiveness and sustainability of recovery efforts. Their active participation and inclusion in planning and implementing recovery processes is crucial for ensuring that recovery efforts are equitable and inclusive (Juran, Oliver, & Read, 2022). Recovery strategies can be more successfully implemented if they are tailored to prioritize the diverse needs of vulnerable populations and facilitate their engagement in decision-making processes. These strategies seek to ensure that no one is left behind in the journey toward recovery and resilience.

Recovery and Reconstruction Approaches after the 2015 Earthquake

After the 2015 earthquake, the Government of Nepal mobilized its resources and also requested assistance from international communities. Emergency rescue and relief materials were provided to survivors through the Prime Minister's Disaster Relief Fund, as guided by the 'Disaster Act, 1983.' Thirty-four countries assisted in high-level rescue activities, and national, international donor, and humanitarian organizations provided immediate relief resources (NPC, 2015). The National Planning Commission (NPC) was tasked along with various ministries of the government and key international organizations such as the Asian Development Bank (ADB), European Union

(EU), United Nations Development Programme (UNDP), World Bank (WB), and Japan International Cooperation Agency (JICA) to assess disaster damages and losses and estimate the funds needed to build back better and create resilient infrastructure and communities. NPC prepared a Post-Disaster Needs Assessment (PDNA) report that gathered data on the four sectors (i.e., social, productive, infrastructure, and cross-cutting issues) and numerous sub-sectors. The PDNA concluded that 'the country will require substantial external assistance to meet the rehabilitation and reconstruction costs, estimated to be at least NPR 669 billion (~US$6.7 billion) over several years, depending on the sector' (NPC, 2015, xx).

Establishment of the National Reconstruction Authority (NRA)

On 25 December 2015, eight months after the devastating event, the National Reconstruction Authority (NRA) was established to oversee and coordinate post-earthquake recovery and reconstruction efforts (Acharya et al., 2022b). Delayed formation of the NRA can be attributed to the unstable political climate within the country. The scale of the disaster, coupled with a substantial recovery project requiring significant financial and human resources, led to political parties competing for prominent roles in order to influence the recovery efforts and take credit for the process. Also, there were frequent changes to the leadership and decision-making positions within and outside the NRA, further delaying formation of the NRA and the recovery process writ large (Mukherji, Ganapati, & Manandhar, 2021).

The NRA faced daunting challenges as it commenced its operations from scratch without any pre-established organizational structure or recovery plans. Initially, the NRA was tasked with developing policies, plans, and programs to reconstruct physical infrastructure, private residences, and community facilities. The Prime Minister oversaw the National Reconstruction Advisory Council (NRAC) and its steering committee, while the executive committee was chaired by the NRA Chief Executive Officer (CEO), who also served as the head of the NRA (Acharya et al., 2022b). The advisory council's role was to guide the steering committee regarding the formulation of reconstruction plans and policies. The steering committee possessed the authority to endorse the organizational framework, authorize financial allocations, sanction certain strategies and directives developed by the executive committee, and guide the NRA to enhance the overall efficiency of the reconstruction endeavor. The executive committee was responsible for overseeing the NRA's operations. Additionally, three Central Level Project Implementation Units (CLPIUs) were established to assist in the reconstruction of housing, schools, and local infrastructure. At the district level, District Level Project Implementation Units (DLPIUs) were established to lead local-level recovery efforts.

Despite its mandate, the NRA faced challenges enforcing major action plans due to its lower perceived position compared to other ministries at the central level. Similarly, the NRA confronted difficulties coordinating with various government tiers, international funding agencies, and public and private development partners. Additionally, the public, media, and political leaders focused primarily on the delivery of physical and numerable outputs, particularly in the reconstruction of private housing. Therefore, more nuanced approaches, such as the establishment of integrated settlements, special provisions for vulnerable populations, and community-based reconstruction efforts, were overlooked in order to address the pressures of a challenging political environment. Also, concepts of zoning, land use planning, and different requirements for minority populations—which could have disrupted pre-disaster inequities and fostered long-term development—were not pursued. It wasn't until January 2018, nearly three years after the earthquake, that procedures were established to identify and track vulnerable beneficiaries (NRA, 2017). These included single women over age 65, all individuals over age 70, minors under age 16, and individuals with disabilities officially recognized or approved by the government.

Recovery Model and Approach

The NRA adopted an 'owner-driven' approach to reconstruction in which the government provided \$3,000 USD in financial support in three installments to each eligible impacted household along with technical assistance to promote earthquake-resistant construction (Acharya et al., 2022a, 2022b). The beneficiaries constructed their houses and technical assistance was provided by a team comprised engineers, sub-engineers, and assistant sub-engineers. The same approach to technical and financial assistance was utilized by the NRA regardless of the socioeconomic condition, geographic location, or capacity of the beneficiaries to rebuild and recover. No additional arrangements were put in place to support the needs of the most vulnerable populations, including marginalized and low-caste communities, women, children, the elderly, and persons with disabilities. While some vulnerable populations, as identified by the NRA, received additional support from international/non-governmental organizations (I/NGOs)—including \$500 USD for housing reconstruction—this support was not mandatory, and not every vulnerable individual received it (NRA, 2017).

Financial support provided through the NRA was insufficient to construct earthquake-resistant buildings that fulfilled family, social, and cultural requirements. Additionally, the technical teams were not adequately equipped to provide timely and necessary technical support. The technical teams were limited in resources, had to cover large, remote areas, and were tasked with completing and submitting forms to DLPIU in order to approve financial support. The approval process included submitting physical and digital forms to district offices along with cross-verification procedures, indicating a lengthy process. Consequently, the technical teams were overly focused on procedural aspects, and the true spirit of technical assistance was neither delivered nor achieved in many instances. For example, engineers and technical staff were often unwilling to work in rugged and remote locations, were frequently absent from the field, and were not adequately monitored by the DLPIUs. Moreover, positions for engineers and technical staff in remote areas were regularly vacant due to frequent resignations and lengthy recruitment processes. Overall, the resultant scenario demonstrates the cumbersome task of managing such a large-scale, owner-driven reconstruction effort. As a consequence, vulnerable populations and those requiring supplemental support did not receive special attention or consideration in the form of additional financial support, meaningful technical assistance, and timeliness of operations. For many households, an additional \$500 USD without further technical or financial assistance from the NRA was still insufficient to fully reconstruct their homes and recover.

Funding Distribution and Accountability

The funding distribution approach of the NRA focused primarily on addressing the housing impacts of the disaster (Acharya et al., 2022b). Specifically, families for which one or multiple homes were destroyed and no other livable structure was available, were eligible to receive a reconstruction grant totaling \$3,000 USD. Housing grants of \$1,000 USD were provided to retrofit partially damaged structures. The funds were allocated in three installments (\$500 USD after signing the contract, \$1,500 USD after completion to the plinth level, and \$1,000 USD after construction to the roof level) facilitated through the banking system to provide a transparent and effective channel. It must be noted that many houses were constructed solely to access or utilize financial support, and the NRA did not prioritize the preservation of vernacular architecture.

Thus, broader societal needs were not met in the housing reconstruction process, particularly among the most vulnerable.

While the NRA did make an effort to categorize vulnerable individuals into four groups (i.e., single women over age 65, all individuals over age 70, minors under age 16, and individuals with disabilities), they overlooked those who were economically disadvantaged, ethnically marginalized, or with limited literacy skills (NRA, 2017). Approximately 18,500 beneficiaries across the 32 earthquake-affected districts were identified as vulnerable (NRA, 2024a) Among them, single women over age 65 and individuals over age 70 constituted approximately 98% of the total, with the remaining 2% consisting of minors and disabled individuals (NRA, 2024a). As mentioned, provisions were made to provide an extra $500 USD to support vulnerable populations through an I/NGO route. However, there was no requirement for I/NGOs to include vulnerable populations in their support efforts. Consequently, most of the vulnerable beneficiaries did not receive additional assistance—and the poor, low caste, illiterate, ethnically marginalized, and other subaltern communities were not even identified as vulnerable.

Subsequently, in May 2020, more than five years after the earthquake, the NRA executive committee decided to provide an additional US $500 to support the identified vulnerable populations (NRA, 2024b). However, it was stipulated that those who already received support from I/NGOs, or those who had received all tranches of support, or who had completed reconstruction were ineligible for additional assistance. Thus, the majority of the identified vulnerable population did not receive the supplemental funds, or if they did, it was provided in an untimely manner more than half a decade after the event. This clearly indicates that even the identified vulnerable individuals (not considering the poor, low caste, and numerous others) were not prioritized in recovery efforts.

The 'Forgotten People' and a Missed Opportunity

Nepal is exposed to a multitude of natural hazard risks due to its location in the seismically active Himalayan region, diverse geological formations, presence of glacial lakes, heavy rainfall during the monsoon season, and haphazard land use practices. Additionally, these physical factors interact with unique social and cultural environments to create a vulnerable geographic region (Acharya, 2017; Aksha et al., 2019). Nepal has witnessed strong earthquakes over the span of every 80–100 years, but the country lacks a documented history of how those earthquakes have impacted people, how they were dealt with, and what was learned from the events and experiences (Rana, 1936). Hence, the country lacked and continues to lack a comprehensive strategy to deal with earthquakes and other natural hazards and disasters. The 2015 earthquake provided a poignant opportunity to address deep-seated vulnerabilities and societal inequities, mitigate risks, and overhaul the disaster governance system to set future standards. However, the opportunity to move toward a more resilient infrastructure and society has been lost.

The response and recovery strategies were characterized by a blanket approach, providing financial assistance primarily based on damage to a house instead of specifically targeting the needs of the people, especially the vulnerable and marginalized (Hülssiep et al., 2021; Melis, 2022). The post-disaster recovery framework (Box 29.1) explicitly specifies the objectives of building more resilient societies, strengthening the capacity of the people, reducing vulnerability, and fostering social cohesion (NRA, 2016). However, the post-disaster recovery process was principally focused on private housing reconstruction. Given that housing was the most affected sector and it represents a basic need, it is predictive and reasonable that housing reconstruction should be emphasized. Through that avenue, there existed a unique opportunity for sustainable hazard mitigation and to build resilient societies, particularly in a manner that addressed the

Box 29.1 Vision and Objectives of Nepal's 'Post Disaster Recovery Framework'

Recovery Vision

Establishment of well-planned, resilient settlements and a prosperous society

Strategic Recovery Objectives

Objective 1: Restore and improve disaster-resilient housing, government buildings, and cultural heritage in rural areas and cities.
Objective 2: Strengthen the capacity of people and communities to reduce their risk and vulnerability and to enhance social cohesion.
Objective 3: Restore and improve access to services and improve environmental resilience.
Objective 4: Develop and restore economic opportunities and livelihoods and re-establish productive sectors.
Objective 5: Strengthen the capacity and effectiveness of the state to respond to the people's needs and to effectively recover from future disasters.

Source: (NRA, 2016)
Box created by authors; Data from NRA (2016)

needs of vulnerable and marginalized populations (Mukherji, 2010; Oliver-Smith, 1990). Yet the process focused entirely on housing reconstruction without addressing the needs of vulnerable populations—even when they were identified and targeted. To these limited identified groups, there was a provision for additional cash support of 50,000 NPR or ~$500 USD toward the end of the reconstruction process. Many beneficiaries did not receive additional financial support, and for the few who did, the support was neither timely nor sufficient. Furthermore, the funds were provided without any clearly stated objectives for how the money was supposed to support them.

Another challenge was associated with the eligibility requirements to receive financial assistance. To receive government housing assistance, affected individuals had to ensure that the impacted house was registered under their name (Rawal et al., 2021). One of the most vulnerable groups in Nepal is the Dalit community, who represent the lowest social hierarchy and lowest economic stratus. Dalits often occupy marginal public lands that are not registered in their names (Biswakarma, 2018). Thus, they were disqualified from receiving housing aid due to not officially owning their homes. Several such reports were reported in mainstream news outlets. For instance, six years after the earthquake, a Dalit community of 25 households in Barpak Gorkha, near the epicenter of the earthquake, were still living in temporary shelters as they were unable to rebuild their houses (Annapurna Post, 2021). Although attempts were made to identify vulnerable people using different categories (e.g., orphan children below 16 years, single women, elderly above 70 years, and persons with a disability), no specific programs were designed to address their immediate needs and long-term recovery requirements (Karki, Matthewman, & Grayman, 2022; Rawal et al., 2021). Furthermore, in the immediate aftermath of the disaster, it was documented that Dalit and low-caste communities were sometimes excluded from receiving relief aid and barred from accessing critical information from local officials (Eichenauer et al., 2020; Lam & Kuipers, 2019).

Survivors of disasters are capable agents and should not be treated as victims (Jerolleman, 2019). However, the top-down, techno-centric, and centralized management approach of reconstruction practices provided minimal opportunities for local communities to actively participate in disaster recovery (Karki, Matthewman, & Grayman, 2023; Shrestha et al., 2019). In fact, as argued by Krüger et al. (2015), the government's approach effectively undermined the contribution of local knowledge and negated their role in sustainable hazard mitigation and building resilient infrastructure. Local cultures, vernacular architecture, and historical approaches were ignored during housing reconstruction design and implementation. Additionally, although gender equity and social inclusion were identified as guiding principles in the PDNA report, female representation in post-earthquake decision-making bodies, including the NRA, was small (Shrestha et al., 2019).

The recovery process in Nepal adopted an owner-driven housing reconstruction approach, which has been cited as a successful model in other parts of the region, particularly after the 2001 Bhuj earthquake in Gujarat, India, and the 2005 Kashmir earthquake in Kashmir, Pakistan (Lam, 2022; NPC, 2015). The autonomous nature of the newly established NRA and its vested power to distribute resources made it attractive because slow and lengthy bureaucratic mechanisms create unnecessary hurdles and delays. However, the political landscape and tussle of Nepali political parties delayed the reconstruction approach and the entire recovery vision. Slow response of the government created stress for disaster-affected populations who were waiting for government support and approval to begin the reconstruction process. The lack of proper guidance on constructing earthquake-resilient buildings, coupled with inadequate communication, forced many households to unnecessarily live in temporary shelters for an extended period of time (Kotani et al., 2020).

As a result of the slow process—and because moving to their homestead and building a permanent were critical needs—households with resources started building their homes without consulting engineers and other technical resources. By the time the government-approved template of earthquake-resilient building plans reached the villages, many households, mostly those with family members working in Arab Gulf states, had already begun or completed constructing their own homes, relying on their own scarce and stressed resources. Meanwhile, households that did not have enough resources to rebuild were forced to live in damaged, unrepaired houses, temporary shelters, or occupy risky areas (Kotani et al., 2020). All of the houses that were constructed out of line with the government template were disqualified from receiving cash support as they lacked specified criteria. To receive the cash support, these families often built another one- or two-room structure adjacent to their already built homes, even though many did not intend to use it as a domicile (Lam, 2022). As a result, earthquake-resilient dwellings were built, albeit on paper, and it could even be argued that it was not resilient infrastructure that was created but, rather, a reification of pre-disaster risks.

Global communities responded to the impacts of the 2015 Gorkha earthquake by providing $282 million USD, a total of only 66.8% of the pledged support (Eichenauer et al., 2020). However, while allocating these funds across numerous projects, the various entities did not take enough consideration of the physical and socioeconomic vulnerabilities of the geographic locations. An analysis of aid allocation established that *Palikas* (higher caste) near Kathmandu and political patronage played a significant role in the distribution of emergency funds with a lack of considerations for the composition of vulnerable and marginalized populations or the geographic areas most in need (Eichenauer et al., 2020). Another missing component of post-earthquake recovery was the consideration of other relevant hazard risks. For example, the impacted regions are also at high risk to landslides and floods. Landslide research has shown that slope instability significantly increases after major seismic activities in mountainous areas (Dahal et al., 2013; Kincey et al., 2021, 2023; Tian et al., 2020). After the earthquake, many households opted to relocate to safer locations rather than rebuilding in the same area due to the imminent threat of seismic activity (He, 2019). However, risk to other hazard types, such as landslides and floods, may not have been considered, and in actuality all should be considered together. Given the scale of impact and perennial multi-hazard risk, housing recovery should have included (seismic-induced) landslides, floods, and other common hazards in their recommendations to vulnerable mountain communities. Unfortunately, the time and policy windows as well as influx of external assistance have passed, and the opportunity has been missed.

Conclusion

It is impossible to achieve truly sustainable disaster recovery without considering and including vulnerable and marginalized communities. In addition to functional and systemic disadvantages, such populations often lack the physical resources and disposable incomes required to fully recover from disasters—let alone to 'build back better' primarily on their own. The disaster recovery and justice literature demonstrates that vulnerable populations require additional support and/or nuanced approaches to adequately complete the recovery processes. Disaster reconstruction and recovery encapsulate many components, including housing, business, cultural, economic, and social and psychological recovery. Nepal's post-earthquake recovery activities focused on housing reconstruction. Since rural areas were the hardest hit and more than 1 million houses and buildings were destroyed, it is no surprise that housing was the central pillar of recovery activities.

For housing recovery, the Government of Nepal chose an owner-driven approach that aimed to facilitate the reconstruction of houses as per the needs of individual households in more swift, less bureaucratic manner. However, through the adoption of a universal, templated approach and treatment of all affected households equally by providing the same amount of recovery resources, the model failed to address the needs of those who required the most assistance. The need for official property ownership documents to receive financial assistance restricted vulnerable populations from accessing available resources, and the delayed disbursement of funds impinged the recovery process for those who did receive financial assistance. Additionally, throughout the process, a consideration of multi-hazard risk arose as a considerable missed opportunity. Recovery programs should consider multi-hazard risk from the outset so that effective mitigation strategies can be implemented and just, resilient communities can be built. Research has warned that equating reconstruction of the physical environment with social recovery leaves certain strata of society, primarily vulnerable and marginalized populations, out of the recovery process. This practice can exert additional pressure and further marginalize such populations. Nepal seems to have fallen into this trap of attempting to recover the built environment regardless of the needs of vulnerable populations.

References

Acharya, M. (2017). The 1950 Indo-Nepal Treaty of Peace and Friendship: Road to Review. In P. Jaiswal (Ed.), *Revisiting Nepal's Foreign Policy in Contemporary Global Power Structure* (pp. 65–78). New Delhi: G. B. Books.

Acharya, P., Sharma, K., Pokharel, G. R., & Adhikari, R. (2022a). Managing postdisaster reconstruction after the 2015 Gorkha, Nepal earthquake and lessons learned. *Natural Hazards Review*, *23*(4), 04022029.

Acharya, P., Sharma, K., Pokharel, G. R., & Adhikari, R. (2022b). Reviewing the progress of reconstruction five years after the 2015 Gorkha earthquake, Nepal. *Building Research & Information*, *50*(6), 595–609.

Aksha, S. K., Juran, L., Resler, L. M., & Zhang, Y. (2019). An analysis of social vulnerability to natural hazards in Nepal using a modified social vulnerability index. *International Journal of Disaster Risk Science*, *10*(1), 103–116. doi:10.1007/s13753-018-0192-7

Alcántara-Ayala, I., Gomez, C., Chmutina, K., van Niekerk, D., Raju, E., Marchezini, V., … Gaillard, J. (2023). *Disaster risk* (First ed.). New York: Routledge.
Aldrich, D. P. (2012). *Building resilience: Social capital in post-disaster recovery*. Chicago: University of Chicago Press.
Annapurna Post. (2021). Bhukampa ko 6 Barsha: Epicenter ka Samudaya Tahara Bhitrai (In Nepali). Retrieved from https://annapurnapost.com/news/178540/
Biswakarma, T. (2018). Citizenship and social security of landless Dalits in Nepal. *Globe: A Journal of Language, Culture and Communication*, *6*, 52–65.
Cutter, S. L., Boruff, B. J., & Shirley, W. L. (2003). Social vulnerability to environmental hazards. *Social Science Quarterly*, *84*(2), 242–261. Retrieved from ≤Go to ISI≥://WOS:000183276700002
Cutter, S. L., Emrich, C. T., Mitchell, J. T., Piegorsch, W. W., Smith, M. M., & Weber, L. (2014). *Hurricane Katrina and the forgotten coast of Mississippi*: Cambridge University Press.
Dahal, R. K., Bhandary, N. P., Timilsina, M., Yatabe, R., & Hasegawa, S. (2013). *Earthquake-induced landslides in the roadside slopes of east Nepal after recent September 18, 2011 earthquake*. Paper presented at the Earthquake-Induced Landslides: Proceedings of the International Symposium on Earthquake-Induced Landslides, Kiryu, Japan, 2012.
Drakes, O., Tate, E., Rainey, J., & Brody, S. (2021). Social vulnerability and short-term disaster assistance in the United States. *International Journal of Disaster Risk Reduction*, *53*, 102010.
Eichenauer, V. Z., Fuchs, A., Kunze, S., & Strobl, E. (2020). Distortions in aid allocation of United Nations flash appeals: Evidence from the 2015 Nepal earthquake. *World Development*, *136*, 105023.
Emrich, C. T., Aksha, S. K., & Zhou, Y. (2022). Assessing distributive inequities in FEMA's Disaster recovery assistance fund allocation. *International Journal of Disaster Risk Reduction*, *74*, 102855.
Finch, C., Emrich, C. T., & Cutter, S. L. (2010). Disaster disparities and differential recovery in New Orleans. *Population and Environment*, *31*(4), 179–202.
Frerks, G. E., Kliest, T., Kirkby, S., Emmel, N. D., O'keefe, P., & Convery, I. (1995). A 'disaster' continuum? *Disasters*, *19*(4), 362–366.
Gall, M., & Friedland, C. J. (2020). If mitigation saves 6 per every 1 spent, then why are we not investing more? A Louisiana perspective on a national issue. *Natural Hazards Review*, *21*(1), 04019013.
Government of Nepal. (n.d.). Nepal Disaster Risk Reduction Portal. Retrieved from http://www.drrportal.gov.np/
He, L. (2019). Identifying local needs for post-disaster recovery in Nepal. *World Development*, *118*, 52–62.
Hülssiep, M., Thaler, T., & Fuchs, S. (2021). The impact of humanitarian assistance on post-disaster social vulnerabilities: some early reflections on the Nepal earthquake in 2015. *Disasters*, *45*(3), 577–603.
Jerolleman, A. (2019). *Disaster recovery through the lens of justice*: Springer.
Juran, L., Adams, E. A., & Prajapati, S. (2019). Purity, pollution, and space: Barriers to latrine adoption in post-disaster India. *Environmental Management*, *64*(4), 456–469.
Juran, L., Oliver, R. D., & Read, D. C. (2022). Water Sector Reconstruction for Post-disaster Housing Settlements: A Tale of Two Governance Models. In *Resilient Water Management Strategies in Urban Settings: Innovations in Decentralized Water Infrastructure Systems* (pp. 227–243): Springer.
Karki, J., Matthewman, S., & Grayman, J. H. (2022). Nayã Ghar (A new house): Examining post-earthquake housing reconstruction issues in Nepal. *International Journal of Disaster Risk Reduction*, *78*, 103116.
Karki, J., Matthewman, S., & Grayman, J. H. (2023). Paranirvar mānis (dependent people)? Rethinking humanitarian dependency syndrome: A Bourdieusian perspective. *Disasters*, *47*(3), 630–650.
Kincey, M. E., Rosser, N. J., Densmore, A. L., Robinson, T. R., Shrestha, R., Singh Pujara, D., … Arrell, K. (2023). Modelling post-earthquake cascading hazards: Changing patterns of landslide runout following the 2015 Gorkha earthquake, Nepal. *Earth Surface Processes and Landforms*, *48*(3), 537–554.
Kincey, M. E., Rosser, N. J., Robinson, T. R., Densmore, A. L., Shrestha, R., Pujara, D. S., … Swirad, Z. M. (2021). Evolution of coseismic and post-seismic landsliding after the 2015 Mw 7.8 Gorkha earthquake, *Nepal. Journal of Geophysical Research: Earth Surface*, *126*(3), e2020JF005803.
Kotani, H., Honda, R., Imoto, S., Shakya, L., & Shrestha, B. K. (2020). Transition of post-disaster housing of rural households: A case study of the 2015 Gorkha earthquake in Nepal. *International Journal of Disaster Risk Reduction*, *44*, 101443.
Krüger, F., Bankoff, G., Cannon, T., Orlowski, B., & Schipper, E. L. F. (2015). *Cultures and disasters: understanding cultural framings in disaster risk reduction*: Routledge.
Lam, L. M. (2022). Against the trend: evaluation of Nepal's owner-driven reconstruction program. *Housing Studies*, 39(8), 1–28.

Lam, L. M., & Kuipers, R. (2019). Resilience and disaster governance: Some insights from the 2015 Nepal earthquake. *International Journal of Disaster Risk Reduction*, *33*, 321–331.
Melis, S. (2022). Post-conflict disaster governance in Nepal: One-door policy, multiple-window practice. *Disasters*, *46*(1), 226–245.
Mukherji, A. (2010). Post-Earthquake Housing Recovery in Bachhau, India: The Homeowner, the Renter, and the Squatter. *Earthquake Spectra*, *26*(4), 1085–1100. doi:10.1193/1.3474646
Mukherji, A. (2017). Post-disaster housing recovery. In *Oxford Research Encyclopedia of Natural Hazard Science*.
Mukherji, A., Ganapati, N. E., & Manandhar, B. (2021). Panacea or problem: New governance structures for disaster recovery. *International Journal of Disaster Risk Reduction*, *52*, 101960.
National Statistics Office. (2023). *National Population and Housing Census 2021 (National Report)*. Kathmandu, Nepal: National Statistics Office
NPC. (2015). *Nepal Earthquake 2015: Post Disaster Needs Assessment, Vol. A: Key Findings*. Kathmandu: Government of Nepal, National Planning Commission (NPC)
NRA. (2016). *Nepal Earthquake 2015: Post Disaster Recovery Framework 2016–2020*. Kathmandu: National Reconstruction Authority (NRA), Government of Nepal
NRA. (2017). *Procedures for Identifying Vulnerable Beneficiaries Affected by the Earthquake*. Kathmandu: National Reconstruction Authority (NRA), Government of Nepal Retrieved from http://www.nra.gov.np/en/content/acts/0
NRA. (2024a). 2015 Earthquake Reconstruction Portal. Retrieved from https://portal.nra.gov.np/infographics/private-building
NRA. (2024b). *The NRA Executive Committee Decisions*. Kathmandu: National Reconstruction Authority (NRA), Government of Nepal Retrieved from http://www.nra.gov.np/en/content/executive_committee_decision/0
Oliver-Smith, A. (1990). Post-disaster housing reconstruction and social inequality: A challenge to policy and practice. *Disasters*, *14*(1), 7–19.
Paul, B. K., & Ramekat, A. (2018). Host characteristics as factors associated with the 2015 earthquake-induced injuries in Nepal: A cross-sectional study. *International Journal of Disaster Risk Reduction*, *27*, 118–126.
Peacock, W. G., Van Zandt, S., Zhang, Y., & Highfield, W. E. (2014). Inequities in long-term housing recovery after disasters. *Journal of the American Planning Association*, *80*(4), 356–371.
Phillips, B. D. (2016). *Disaster Recovery* (Second Edition ed.). Boca Raton, FL: CRC Press, Taylor & Francis Group.
Rana, B. S. (1936). *Nepalko Mahabhukampa (in Nepali)* (Fourth ed.). Kathmandu: Publication Nepa~laya.
Rawal, V., Bothara, J., Pradhan, P., Narasimhan, R., & Singh, V. (2021). Inclusion of the poor and vulnerable: Learning from post-earthquake housing reconstruction in Nepal. *Progress in Disaster Science*, *10*, 100162.
Shrestha, K. K., Bhattarai, B., Ojha, H. R., & Bajracharya, A. (2019). Disaster justice in Nepal's earthquake recovery. *International Journal of Disaster Risk Reduction*, *33*, 207–216.
Smith, G., Martin, A., & Wenger, D. E. (2018). Disaster recovery in an era of climate change: the unrealized promise of institutional resilience. In *Handbook of disaster research* (pp. 595–619): Springer.
Tian, Y., Owen, L. A., Xu, C., Ma, S., Li, K., Xu, X., … Wang, S. (2020). Landslide development within 3 years after the 2015 Mw 7.8 Gorkha earthquake, Nepal. *Landslides*, *17*, 1251–1267.
Tierney, K. (2019). *Disasters: A sociological approach*. Massachusetts: Polity Press.
Tierney, K., & Oliver-Smith, A. (2012). Social dimensions of disaster recovery. *International Journal of Mass Emergencies & Disasters*, *30*(2), 123–146.
UNDRR, C. (2020). *Human cost of natural disasters: An overview of the last 20 years (2000-2019)*. Retrieved from.
White, G. F., Kates, R. W., & Burton, I. (2001). Knowing better and losing even more: The use of knowledge in hazards management. *Global Environmental Change Part B: Environmental Hazards*, *3*(3), 81–92.
Wilson, B., Tate, E., & Emrich, C. T. (2021). Flood recovery outcomes and disaster assistance barriers for vulnerable populations. *Frontiers in Water*, 159.
Worldometer. (2024). *GDP by country*. https://www.worldometers.info/gdp/gdp-by-country/

30

COMPUTER MODELING TO INFORM DISASTER RECOVERY

The Case of Hurricane Matthew in North Carolina

Mohammad Yunus Naseri and Luke Juran

Introduction

Background

Flooding events constitute a significant proportion of global and regional disasters (Dworkin, 1974; CRED, 2015; Aksha et al., 2018). Thus, understanding, assessing, and predicting floods and their impacts have become an important part of disaster risk reduction. For these purposes, flood inundation methods and models have been developed and refined.

Flood inundation methods and models, which can be deployed across a wide range of scientific, practical, and policy arenas, can be used for purposes such as risk mapping, flood damage assessment, real-time flood forecasting, and water resources management (Apel et al., 2006; Bhuiyan and Dutta, 2012; Shastry and Durand, 2019; Aksha et al., 2020). Additionally, models have been used to investigate riverbank erosion, sediment transport, contaminant transport, floodplain ecology, fluvial geomorphology, and catchment hydrology (Marriott, 1992; Panagoulia, 1992; Panagoulia, 2006; Karim et al., 2020). Furthermore, flood models can be applied to examine climate change impacts, risk reduction, and adaptation and resettlement (Västilä et al., 2010; Karim et al., 2020).

There are various ways to classify methods for flood inundation methods, one classification being empirical methods, hydrodynamic methods, and simplified-conceptual methods. This study uses a simplified-conceptual approach known as Height Above Nearest Drainage (HAND). Due to its low computational cost and reduced data requirement, HAND can be utilized to simulate flood inundation extent in large floodplains and to facilitate scenario modeling, probabilistic risk mapping, and environmental impact mapping (Speckhann et al., 2018; Johnson et al., 2019; Carvalho et al., 2020). HAND was first developed by Rennó et al. as a terrain descriptor. This terrain descriptor used a digital elevation model (DEM) to define the height of a cell above a certain cell in the nearest riverbed to which this cell flowed (Garousi-Nejad et al., 2019). In subsequent years, Nobre et al. used HAND to determine soil water potential, Antonio Donato Nobre et al. used it for flood potential, Rahmati et al. used it for groundwater potential, and Johnson et al. integrated it with the National Water Model (NWM).

DOI: 10.4324/9781003414834-40

Objectives

This study delineates flood inundation extent using the HAND simplified-conceptual model. After a non-exhaustive review of inundation delineation models, this study meets the following objectives using a historic flood in North Carolina in the wake of Hurricane Matthew in 2016 as the modeling event. By detailing the construction of our HAND-based model, the study and objectives also satisfy our purpose of providing a detailed framework on how to construct a HAND-based model for future researchers and practitioners.

1 Produce a flood inundation extent using the HAND model;
2 Compare HAND model results to an inundation extent produced using US Geological Survey (USGS) high-water mark (HWM) reference data;
3 Identify and discuss avenues for HAND model improvement based on results of the first two objectives.

Review of Inundation Models

Model Types

There is a large literature on methods for flood inundation mapping. The methods can be classified in numerous ways based on different criteria. For example, methods can be classified as lumped, semi-distributed, and distributed based on spatial representation of the study site, while they can also be classified as empirical, hydrodynamic, and simplified-conceptual based on structure of the model. Here, we classify and illustrate the various methods using the latter classification. Empirical methods include measurements, surveys, and statistical models that are based on observations. Hydrodynamic methods are based on solving equations that model physical laws, such as conservation of mass and momentum. Simplified-conceptual models have gained increasing popularity in recent years. Simplified-conceptual models differ from hydrodynamic models because they are based on more modest physical processes. As a result, simplified-conceptual models have shorter run times that render them more suitable for simulating flood extent and depth in large and data-poor floodplains.

Empirical Models

Empirical models are based on collecting, processing, and analyzing flood-related data called 'observations' (Smith, 1997). These data include on-the-ground measurements, surveys, aerial photography, and remotely sensed imagery. These observations, which reflect the past and thus do not provide predictive measurements, can be used as inputs to other flood inundation methods/models as well as for calibrating and validating hydrodynamic models (Pappenberger et al., 2007; Teng et al., 2017; Smith, 1997).

Empirical models are the oldest of all three methods for flood extent delineation. Empirical methods are generally considered to be accurate. However, their accuracy depends on the techniques used for data collection and processing, with more costly and complex instruments required to collect and process high accuracy data (Didier et al., 2015; Gallien, 2016). A major limitation of empirical models is that they reflect a snapshot of past events and do not necessarily provide information on future events due to continually changing topographic and infrastructural conditions (Madsen and Rosbjerg, 1997; Teng et al., 2017). Coarse spatial and temporal resolution data can also present limitations, but improvements in remote sensing can help mitigate these drawbacks

(Smith, 1997; Teng et al., 2017). Still, limitations in engineering (e.g., design of sensors, telemetry, carriers), environmental data (e.g., clouds, wind, vegetation cover), and processing (e.g., algorithms for mining data) can detract from the accuracy of empirical methods.

Hydrodynamic Models

Hydrodynamic models are mathematical models that simulate the motion of fluids by solving equations that model physical laws; hence, they are also referred to as physically based models. Hydrodynamic models can simulate floodplains in one, two, or three dimensions and are therefore categorized as 1D, 2D, and 3D models. 1D models simulate the flow of water in one dimension along the river channel and can be used to simulate river flows in cases when the flow is significantly one-dimensional or modeling flow in two dimensions is not necessary (e.g., confined channel flow, pipe flow) (Brunner, 2016). 2D models simulate flow in two dimensions assuming that the third dimension (depth) is small or insignificant compared to the other two (Roberts et al., 2015). These models solve the conservation of mass and momentum, or 'shallow water equations', in two dimensions and help determine velocity and depth over defined space and time intervals. Meanwhile, 3D models can be used to model vertical turbulence, vortices, and spiral flow at bends during catastrophic incidents such as floods resulting from dam failure, flash floods, landslides, and tsunamis (Huang and Dai, 2014; Yu et al., 2016; Teng et al., 2017).

Hydrodynamic models are used in flood modeling due to the many advantages they provide. Hydrodynamic models can be linked to hydrologic and river models for flood risk mapping, flood forecasting, and scenario analysis, and the model inputs can be manipulated to examine changes in initial and boundary conditions (Teng et al., 2017). In general, 1D versions of hydrodynamic models are computationally efficient, 2D models can accurately simulate inundation spatially and temporally, and 3D models can represent vertical features of the flow (Monaghan, 1994; Ye and McCorquodale, 1998). Limitations of 1D hydrodynamic models are their inability to model the lateral diffusion of flood waves, discretization of topography into cross-sections rather than a continuous surface, and subjectivity in the location and orientation of these cross-sections (Néelz and Pender, 2013; Teng et al., 2017). 2D and 3D models can be data and computationally intensive and are generally not viable to use in areas over 1000 km^2 (Néelz and Pender, 2013; Teng et al., 2017).

Simplified-Conceptual Models

Simplified-conceptual models, sometimes called 0D or 0-term models, are considered 'simplified' because they do not simulate the physical processes of inundation but are instead based on simple hydraulic concepts. Different simplified-conceptual models are based on different hydraulic concepts, and the models can be used in a predictive way (Néelz and Pender, 2013; Teng et al., 2017; Pender, 2006).

One example of a simplified-conceptual model is the Rapid Flood Spreading Method (RFSM), which divides the floodplain into elementary areas and then spreads the flood by filling the areas using a filling/spilling process (Lhomme et al., 2008). RFSM can produce the approximate inundation extent of a flood up to 1000 times faster than a typical hydrodynamic model (Flood Modelling). Other examples of simplified-conceptual models are models based on the theory of the 'planner method'. These models produce flood inundation extent by dividing a high-resolution DEM into minute intervals and then linking the water stage/volume to the flood extent (Teng et al., 2017). HAND is another simplified-conceptual model that normalizes topography relative to the local riverbed (Nobre et al., 2011). HAND then calculates the inundation extent for an input height by selecting all cells with HAND values (i.e., the relative height of each cell to the local riverbed to which it drains) less than that height.

Simplified-conceptual models are many times faster than 3D, 2D, and even 1D hydrodynamic models, making them suitable for simulating flood inundation across large areas (e.g., the continental US in the case of HAND) while still providing accuracy at finer geographical scales. Where clear flow paths exist across the simulated region, the models can be used effectively to identify inundation extent with accuracy similar to 2D hydrodynamic models (Kuldeep and Garg, 2016; Teng et al., 2017). However, results of simplified-conceptual models diverge from hydrodynamic models when the modeling occurs in complex topographies (i.e., areas where momentum conservation is important). Therefore, applications of these models face limitations where the dynamic effects of flow are more significant in determining the direction of flow.

Model Selection

Model selection for simulating water flow or determining inundation extent depends on multiple factors, such as type of output needed, level of accuracy required, data availability for model construction and validation, and computational demands of the model (Tilford et al., 2007; Di Baldassarre et al., 2009; Kauffeldt et al., 2016). For example, a model with fast run time and real-time data assimilation is necessary for flood forecasting, whereas a numerical model that can accurately simulate supercritical flow and other fluid dynamics is important for flood risk assessment in urban areas. Models that can accurately represent velocity are required in flood damage assessment and erosion studies. For hazard mapping, environmental flow assessment, and water resources planning, models that can simulate maximum inundation extent and depth work well. In the selection of an inundation model, therefore, a balance should be struck among model demands, model complexity, processing speed, and data requirements and availability in addition to other technical and practical considerations (Asselman et al., 2008; Kvočka et al., 2017; Sehgal et al., 2018; Aksha et al., 2020).

Model Selected for This Study

This study uses the HAND method to delineate flood inundation based on the relative advantages of simplified-conceptual models and our goals of identifying areas for model improvement and providing a practical framework for others. The USGS empirical model, which is based on the interpolation of HWMs, is used as a reference. HWMs are field-verified identifications of the highest water levels for a respective flood event. This method includes the production of flood elevation surfaces by interpolation of HWMs and a combination of flood elevation surfaces with high-resolution DEMs (Liu et al., 2014; Musser et al., 2017).

Study Area and Event

The study area is Rockfish Creek at Hope Mills, North Carolina (Figure 30.1), which experienced a devastating flood during Hurricane Matthew on 8–9 October 2016. The extent of destruction is detailed in a USGS report (Musser et al., 2017), and according to the Hurricane Matthew Resilient Redevelopment Plan, the floods resulted in 26 fatalities and damaged more than 100,000 residential, business, and governmental buildings estimated at $4.8 billion USD in North Carolina alone (NCEM, 2017). Additionally, thousands of residents were forced to flee their homes and hundreds of thousands lost access to electricity.

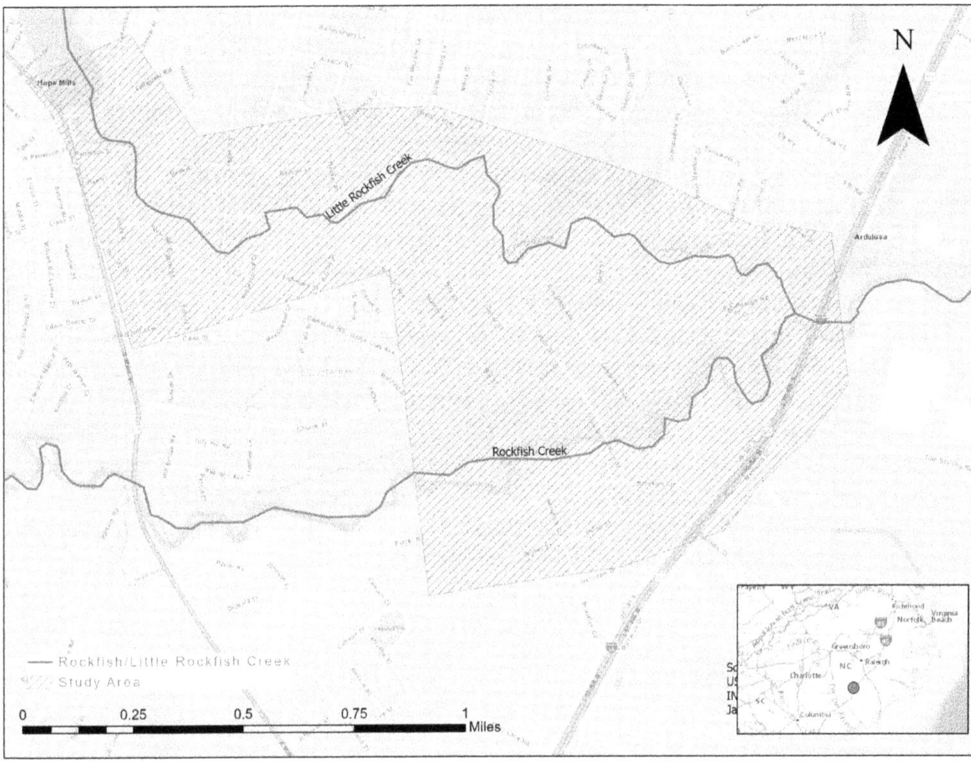

Figure 30.1 The Study Area of Rockfish Creek at Hope Mills, North Carolina

Source: Original figure/figure created by authors.

The study area includes part of the Town of Hope Mills and contains three hydrologic unit code (HUC) watersheds all at the HUC12 level: Little Rockfish Creek, Middle Rockfish Creek, and Lower Rockfish Creek. An HUC is a hierarchical ranking system for watersheds. According to the latest HUC delineation, there are six categories of watersheds ranging from regions (the largest watersheds, with a 2-digit HUC) to sub-catchments (the smallest watersheds, with a 12-digit HUC).

Materials and Methods

Producing the flood inundation extent using HAND consisted of three processes. First, the HAND raster was created. Next, using the HAND raster, hydraulic properties and synthetic rating curves were produced for each DEM-derived river reach in the study area. Finally, the inundation extent for each DEM-derived catchment was calculated using data from the NWM reanalysis discharge dataset. The following subsections describe the delineation process using the HAND method and incorporation of NWM reanalysis discharge data.

Production of HAND Raster

The HAND raster was produced with ArcGIS Pro 2.5.0, which was also used in subsequent processes to derive reach-averaged hydraulic properties and flood inundation maps. Data used to produce the HAND raster were a 3 m DEM from USDA Geospatial Gateway and NHDPlus v2.1 flowlines. All data layers were projected in UTM zone 17 and exported to a geodatabase for better data management. Production of the HAND raster consisted of three steps: (1) rasterizing NHDPlus stream origins; (2) producing a stream raster from the DEM; and (3) production of HAND values.

In the first step, a binary raster layer with values of one on sources of NHD flowlines and zero for the rest was produced. The raster had the same extent (i.e., number of rows and columns and cell size) as the 3 m DEM. In the second step, a stream raster from the DEM was produced that originated from the same points as NHDPlus flowlines but stayed consistent with the DEM flow directions. This was done using the Fill, Flow Direction, Flow Accumulation, and Con functions in ArcGIS Pro. There are some inconsistencies between the DEM-derived streams and NHDPlus streams. The inconsistencies/misalignments are partly because DEM-derived lines are produced using a 3 m DEM, while the NHDPlus lines were mapped at a medium resolution of 1/100,000 scale (30 m DEM) (Merwade et al., 2008; Maidment et al., 2016; Shastry and Durand, 2019). In the final step, the HAND raster where each cell has a HAND value/index (i.e., vertical distance to the local stream network) was produced using the Flow Direction and Flow Distance functions in ArcGIS Pro. Figure 30.2 shows the HAND raster for the study area; the HAND value is zero on the streambed and then increases with increasing distance from the river bed.

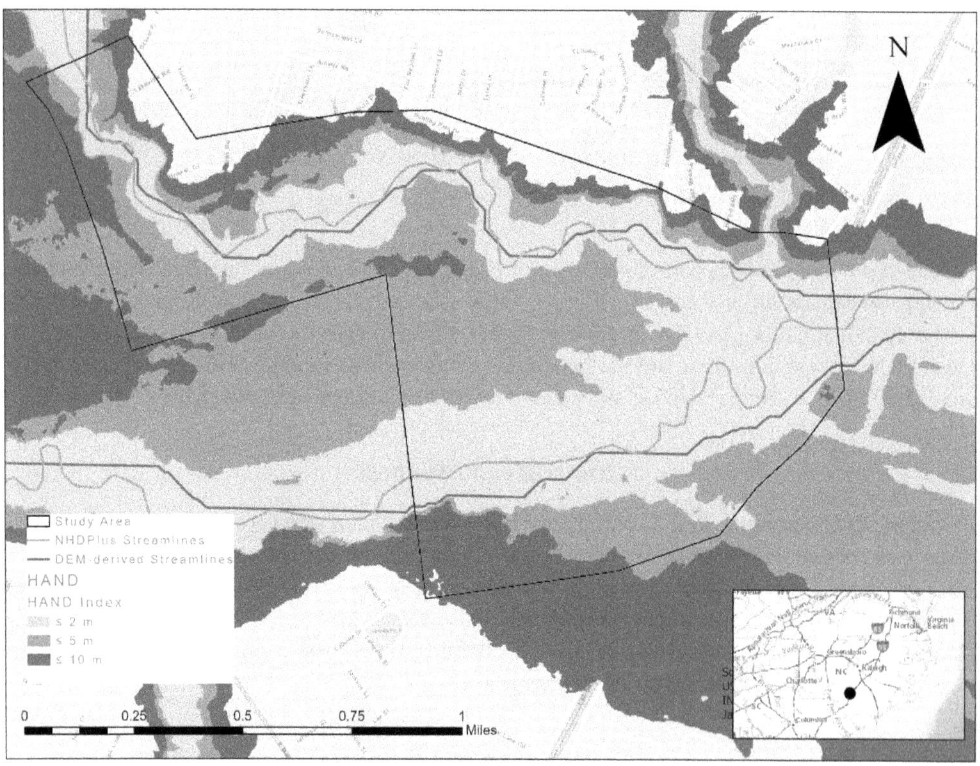

Figure 30.2 HAND Raster for the Study Area

Source: Original figure/figure created by authors.

Calculation of Hydraulic Properties and Synthetic Rating Curves

The hydraulic properties and synthetic rating curves for each DEM-derived river reach were calculated using the HAND raster and supplementary data. First, each river reach in the stream raster produced from the 3 m DEM was assigned a unique value, which enabled the production of catchments for each uniquely identified river reach in ArcGIS Pro. The uniquely identified river reaches and their catchments were then converted to vector format. The conversion to vector is important because, in the calculation of hydraulic properties, each catchment (and their respective river reaches) should be selected separately. As for the DEM-derived stream network, the catchments also exhibited inconsistencies with the NHD catchments. The differences were minor for some catchments and greater for others. The misalignment of DEM-derived and NHD-derived catchments is considered one of the challenges of using the NHDPlus river network with HAND, and this issue is the subject of ongoing research to better partition stream reaches and their associated areas (Zheng et al., 2018).

Next, the reach-averaged hydraulic properties (e.g., slope, wetted perimeter, area) that were eventually used for the derivation of synthetic rating curves were calculated. These properties were calculated based on the assumption that the DEM used to produce the HAND raster adequately represented the channel bed elevation. Hydraulic properties were not calculated based on surveyed cross-sections, which is the most accurate and conventional method in flood studies. In the HAND method for flood inundation delineation, surveyed cross-sections are not necessary. One major advantage of the HAND method in calculating the hydraulic properties of a river reach is that HAND considers the entire river reach as a prism (Zheng et al., 2018; Garousi-Nejad et al., 2019). This means that the cross-section from which geometric properties were calculated is along the length of the reach rather than along a vertical slice, which is at a point or series of points. The hydraulics of a river make themselves felt on all the water, and there is not a critical point along a river reach that controls the hydraulics (i.e., rating curve) of a river. Rather, it takes a certain length of the river for hydraulics to make themselves felt on the water, ultimately resulting in a consistent relationship between depth and flow. Thus, the rating curves of points along a river just a short distance from each other are quite different, and it is only along with a certain distance that the depth-discharge relationship of a river stabilizes (i.e., one rating curve can represent the relationship between depth and discharge of the river) (Zheng et al., 2018).

The process for calculating hydraulic properties, based on Zheng et al., uses iterative applications of Manning's equation at different discharge levels to establish a relationship between discharge and depth. In this process, each DEM-derived catchment has one river reach to which it drains. The steps for calculating hydraulic properties include using various ArcGIS Pro functions as well as the formulas shown in Table 30.1.

Inundation Extent Production Using NWM Discharge Data

The NWM, first implemented in 2016 by the National Oceanic and Atmospheric Administration (NOAA), simulates the water cycle in the continental US through the mathematical representation of hydrologic processes (Maidment et al., 2016; Cohen et al., 2018). The discharge data used in this study are from the NWM reanalysis dataset, which contains output products from NWM historical simulations from 1993 to 2018. The reanalysis simulations use observed precipitation data and other meteorological inputs from weather reanalysis datasets. This dataset is publicly available through big data providers such as Amazon Web Services.

Table 30.1 Formulas for Calculating Hydraulic Properties of Each DEM-Derived River Reach

Target Parameter	*Formula*	*Parameter Description*
Single-cell plane area (A_C)	$dx \times dy$	dx = Width of the cell dy = Length of the cell
Surface area (A_s)	$\sum A_c$	
Wetted bed area (A_b)	$\sum A_c \times \sqrt{1+slp^2}$	slp = Percent slop
Volume (V)	$\sum A_c \times (h - h_d)$	h = Arbitrary inundation height h_d = Inundation depth for the given stage, h
Cross-sectional area (A)	$\frac{V}{L}$	V = Volume L = Length of the river reach
Wetted perimeter (P)	$\frac{A_b}{L}$	
Top width (T)	$\frac{A_s}{L}$	
Hydraulic radius (R)	$\frac{A}{P}$	

Source: Table created by authors/original table.

NWM outputs are stored in NetCDF format in point-type (e.g., streamflow), 250 m gridded (e.g., ponded water depth), and 1 km gridded (e.g., land surface variables). Inundation extents for each DEM-derived catchment were produced using discharge data from 8 to 9 October 2016. Discharge data were introduced to the synthetic rating curves to calculate respective inundation heights. Next, inundation heights for each river reach were introduced to their respective HAND raster layers to find the inundation extent in each DEM-derived catchment. Inundation extents from all sub-catchments were then joined to produce the total inundation extent, to which NHD-HR permanent water pixels were added to create a final inundation map (Figure 30.3). Model results indicate that approximately 1.87 km^2 of the 2.93 km^2 study area (63.7%) was inundated.

Results

A reference flood inundation extent using USGS HWMs was generated to compare to the HAND-derived extent presented in Figure 36.3. The reference extent was produced using the observed HWMs. The USGS documented 267 HWMs after Hurricane Matthew, of which 254 were field-verified (Musser et al., 2017). Seven inundation maps were created for hard-hit communities in central and eastern North Carolina using these HWMs. One of the maps was created for Rockfish Creek at Hope Mills, NC, which was used as a validation basis of this study.

The HAND inundation map was compared to the one produced using HWMs. The following process was used to produce inundation maps using HWMs according to Musser et al. (2017). Before production of the reference extent, accuracy of HWM locations and elevations were checked by comparing field note diagrams and descriptions to aerial photographs and detailed street and parcel maps. Some HWMs were not used in the mapping process due to issues such as low location and elevation accuracy (e.g., locations could not be determined accurately, elevations were substantially different from neighboring HWMs) and localized flooding heights (i.e., not representing water surface elevation of surrounding areas). The

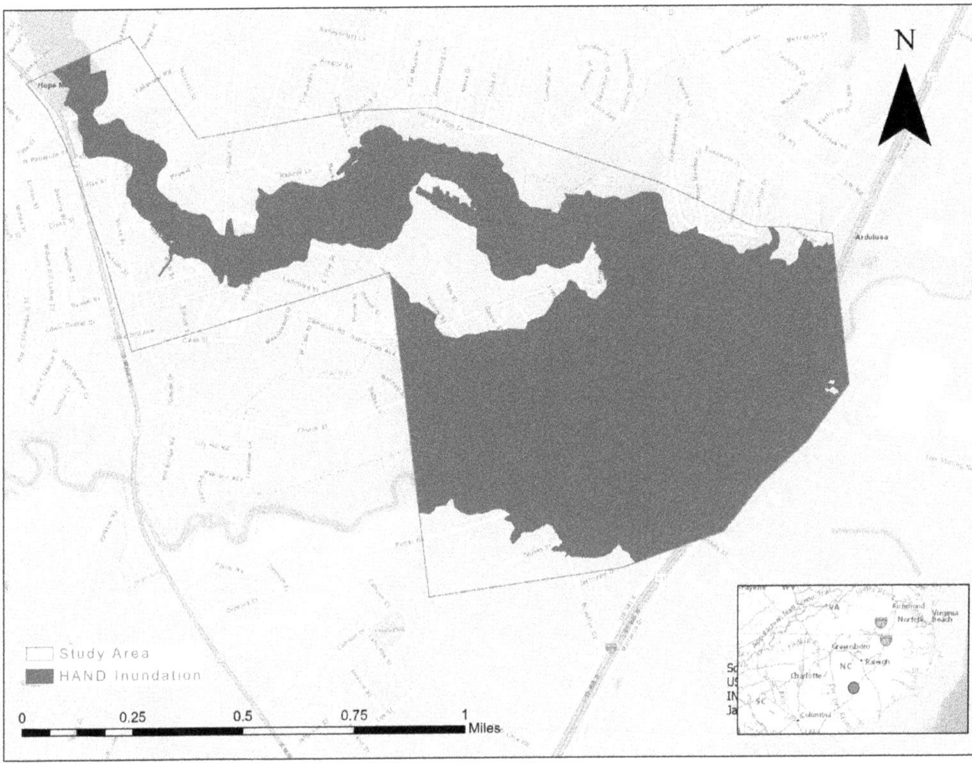

Figure 30.3 Final HAND-Derived Map of Flood Inundation Extent

Source: Original figure/figure created by authors.

inundation extent was created by first producing flood elevation raster surfaces. Seven raster surfaces were created independently for each hard-hit community in North Carolina using HWM elevation cross-sections across the direction of flow at HWMs and USGS stream gages using ArcGIS Pro Topo-to-Raster interpolation tool. The seven flood elevation surfaces were then combined with DEM raster layers. DEMs were 0.9525 m cell-sized and were derived from light detection and ranging (LiDAR) data with an 18.2 cm vertical root-mean-square-error and 0.07 m or better nominal point spacing. Inundated areas were identified where the flood elevation surface was higher than the LiDAR DEM land surface. Depth of flooding was determined as the difference between the elevation of the flood elevation surface and the LiDAR DEM land surface.

Figure 30.4 portrays the reference flood inundation extent created using USGS HWMs, while Figure 30.5 overlaps the reference inundation with the HAND method, and Table 30.2 summarizes the inundation areas extracted using both methods. Results from the models largely mirror each other, although the HAND model predicted slightly more flooding. Notable differences include greater flooding in the middle and southern parts of the study area (along Rockfish Creek) in the case of HAND map, and greater flooding in the northern part of the study area (along Little Rockfish Creek) in the case of USGS HWMs map. In general, the northern part of the study area is more residential, while the middle and southern parts are more commercial and industrial.

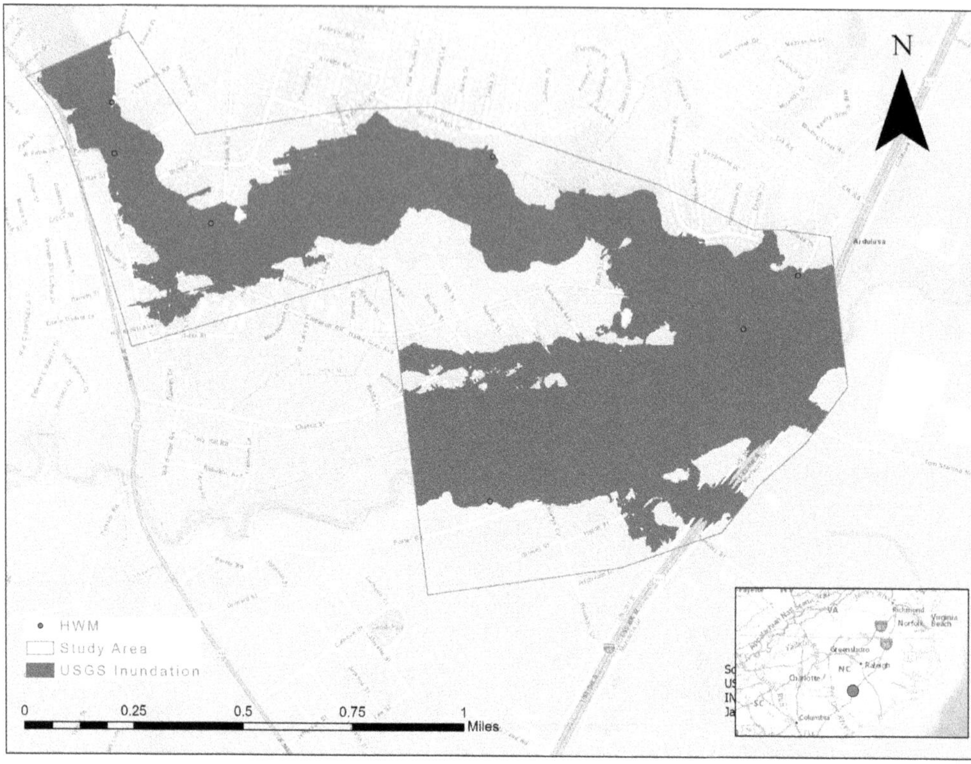

Figure 30.4 Flood Inundation Extent Based on USGS HWMs

Source: Original figure/figure created by authors.

Discussion

The flood inundation extents delineated for the same area and event using the HAND model and USGS HWMs data mirror each other but exhibit differences in flood extent and location. Some differences can be attributed to the uncertainties introduced in both methods. Uncertainty factors associated with HAND delineation include (1) inaccuracies in DEM; (2) inaccuracies in NWM discharge data; (3) selection of a constant and conservative Manning's roughness coefficient; (4) flow hydraulics simplification; and (5) misalignment of NHD and DEM-derived data (Bales and Wagner, 2009; Faghih et al., 2017; Liu and Merwade, 2019). Uncertainty factors associated with the HWM-derived delineation include the quality, quantity, and distribution of HWMs as well as simplification of hydraulics in the method used (i.e., only based on interpolation of HWM data) (Liu et al., 2014; Musser et al., 2017).

Overall, the HAND model performed well in flood inundation modeling and result indicates that the method is suitable for reconstructing historical flood inundations in this study. However, critical data requirements must be met in terms of clearly defined flowlines for the study area; close alignment between NHD flowlines (used by the NWM) and topographic-derived flowlines; and the NWM reanalysis discharge data should sufficiently represent actual river discharge spatially and temporally. The HAND model demonstrates several advantages compared to empirical and 2D and 3D hydrodynamic models such as low computational cost, low data requirements,

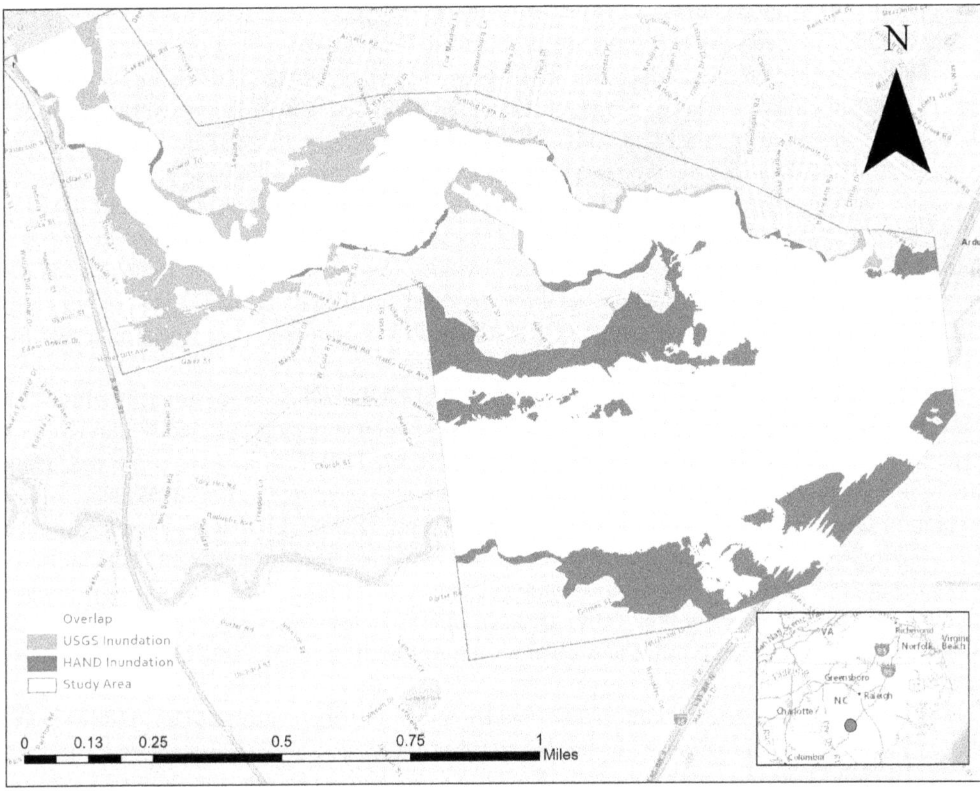

Figure 30.5 Overlay of Flood Inundation Extents Based on USGS HWMs (Medium Grey) and the HAND Model (Dark Grey)

Source: Original figure/figure created by authors.

rapid processing speed, and ability to model a larger spatial extent. Further, the method can be used for planning and predictive purposes, which is not the case for all types of models (e.g., empirical models, such as the one presented here based on USGS HWMs). These advantages are especially relevant to local planners and emergency operators focused on practical applications of disaster risk reduction. However, areas for HAND model improvement exist. The improvements can be categorized as improvements to the method, and improvements in the precision of data used.

Regarding improvements to the method, inconsistencies between the NHD flowlines and DEM-derived flowlines can introduce significant uncertainty in the flood inundation extent produced

Table 30.2 Comparison of HAND and USGS HWM-Derived Inundations Maps

	Inundated Area (km²)	*% Study Area Inundated*	*Inundation Divergence from USGS (km²)*	*% Inundation Divergence from USGS*	*Inundation Overlap of Models (km²)*	*Overlap Divergence (km²)*
HAND	1.87	63.72	+0.19	6.20	1.53	+0.34
USGS	1.69	57.52	0.00	0.00	1.53	+0.16

Source: Original table/table created by authors.

when the HAND method is used (Merwade et al., 2008; Johnson et al., 2019; Shastry et al., 2019). Inconsistencies between the flowlines are problematic because the NWM discharge data used with the HAND method are produced for NHD flowlines, whereas the synthetic rating curves in the HAND method are produced for the DEM-derived flowlines. The stream network used in the HAND method should be consistent with the DEM (i.e., topography) used to produce the HAND raster. The NHD flowlines cannot be used to produce the HAND raster because in some locations (e.g., low elevation areas with complex topographies) the NHD lines are inconsistent with the DEM-derived topographies. Therefore, the DEM-derived stream network must be used to produce the HAND raster. Inconsistencies between the flowlines from which synthetic rating curves are produced (i.e., DEM-derived flowlines) and the flowlines for which discharge data are produced (i.e., NHD flowlines) can make local/street-level inundation extent prediction uncertain using the HAND method.

Factors that contribute to inconsistencies between NHD and DEM-derived flowlines are the difference in resolution of the two flowlines, low relief in topography, and human-made structures (Li and Wong, 2010; Abood et al., 2012). NHD flowlines were produced using medium-resolution DEMs (e.g., 1 arc second) and are also used by the NWM. When a higher resolution DEM (such as 1/3 arc-second, 1/9 arc-second, or even LiDAR) is used to produce a HAND raster, inconsistencies between the two flowlines become apparent. Furthermore, inconsistencies increase in flatter topographies because determining the location of a streamline using a DEM is easier in high relief areas compared to low relief areas, and human-made structures (which physically alter and obstruct natural hydrology) are not reflected in NHD-interpolated DEMs.

The precision of data used in the HAND model constitutes another avenue for improvement. The two most important data used in the HAND method are the NWM discharge data and the DEM. The NWM systematically underpredicts discharge data, especially for low order streams (Shastry et al., 2019). This problem can be compounded when there exists a high inconsistency between the NHD/NWM flowlines and DEM-derived flowlines, which makes the assignment of spatially and temporally representative NWM discharges to the DEM-derived flowlines difficult. Moreover, the HAND model is premised on using a DEM that can sufficiently represent the river network bed elevation and its bathymetry. However, interpolated DEMs (as opposed to LiDAR DEMs) cannot adequately represent the riverbed elevation and the bathymetry of a river. This is because interpolated DEMs are derived from topographic maps with coarse resolutions that may have been created years before. This makes interpolated DEMs based on old topographic maps virtually irrelevant for flood inundation prediction at fine resolutions since they fail to reflect the detailed physical properties of a river network. Therefore, and as mentioned in other studies (see Shastry et al., 2019; Godbout et al., 2019), HAND is an approximate method that can give satisfactory results compared to hydrodynamic models' results when used over vast geographical areas and not for individual reaches with short lengths and low stream orders where the HAND method is not suitable.

A higher resolution DEM that can sufficiently represent the geometric complexities of a river network (such as a LiDAR DEM) could significantly increase accuracy of the HAND method for flood inundation modeling. However, a balance should be considered among factors such as computational and temporal costs, data requirements, project goals (e.g., is prediction an objective), and the resulting accuracy desired or required. Based on these critical model selection criteria, the HAND simplified-conceptual model for mapping flood inundation extent is a useful tool, especially for local level planners and emergency operators who may not have access to the data, computational, and time requirements demanded by other models.

Conclusions

This study investigated the use of HAND, a simplified-conceptual model, to determine flood inundation extent. The study area is Rockfish Creek at Hope Mills, NC, and NWM historical reanalysis discharge data from Hurricane Matthew in 2016 were used. The specific objectives were to (1) produce a flood inundation extent using the HAND model; (2) compare HAND-derived results to an inundation extent produced using USGS HWM reference data for validation; and (3) identify and discuss avenues for HAND model improvement.

The flood inundation extent for Rockfish Creek was delineated using the HAND model, which first required the production of a stream raster. The raster started from the same origin points as NHDPlus flowlines but then followed the site's topography (represented through a 3 m DEM). Next, the HAND raster was produced, which was followed by the calculation of hydraulic properties of each DEM-derived river reach and derivation of synthetic rating curves. Finally, NWM reanalysis discharge data were introduced to the rating curves and the inundated pixels were identified for each DEM-derived catchment.

The HAND-derived inundation map compared well to the inundation map produced using the widely used USGS HWMs in this case study. Thus, we argue that HAND is a useful tool for local planners and disaster risk reduction officials, especially considering its relative advantages compared to other available models (e.g., data requirements, computational costs, processing time, planning and predictive abilities, ability to model a large spatial area). This argument assumes that requirements are met in terms of clearly defined flowlines for the study area, alignment between NHD flowlines and topographic-derived flowlines, and the use of spatially and temporally representative NWM reanalysis discharge data. The HAND model determined 63.7% of the study area to be inundated, whereas the inundation rate was 57.5% based on USGS HWMs. The HAND model produced a greater inundation extent, especially in the middle and southern parts of the study area, while the results based on USGS HWMs produced a smaller inundation extent yet greater flooding in the northern section of the study area. Overall, the HAND method performed well in modeling flood inundation extent, although various factors exist that create uncertainties. These uncertainties are due to issues such as inaccuracies in DEMs and NWM data, using a constant and conservative Manning's roughness coefficient, flow hydraulics simplification, and misalignment of NHD and DEM-derived data.

In terms of improvements to the HAND method for flood modeling, inconsistencies between the NHD and DEM-derived flowlines likely introduce uncertainty in flood inundation extent. Factors that contribute to inconsistencies between NHD and DEM-derived flowlines are the difference in resolution of the flowlines, low relief in topography, and human-made structures not reflected in DEMs. The precision of data used in the HAND model is another area for improvement. The two most critical data used are the NWM discharge data and DEMs. The inaccuracies can be compounded when there are inconsistencies between the NHD/NWM flowlines and the DEM-derived flowlines, which makes the assignment of spatially and temporally realistic discharge data to the DEM-derived flowlines challenging. A higher resolution DEM that can sufficiently represent the complexities of a river network could significantly increase accuracy of the HAND method for inundation modeling. That being said, computational capacity, time, project objectives, and achieving a minimum level of accuracy in the output must be considered when using the HAND model or any model. Overall, while heeding limitations of the HAND method, we have demonstrated that HAND can be used effectively—even in the case of a relatively small watershed such as Rockfish Creek—thereby capitalizing on the method's many advantages. Further, we have provided a detailed outline on how to construct a HAND-based model to guide future researchers and practitioners.

Data Availability Statement

Data used for analysis can be accessed publicly at USGS (high-water marks) and National Oceanic and Atmospheric Administration (National Water Model reanalysis data).

Funding

This work was supported by a Fulbright Scholarship.

References

Abood, Sinan A., Maclean, Ann L., and Mason, Lacey A. "Modeling Riparian Zones Utilizing DEMS and Flood Height Data." *Photogrammetric Engineering & Remote Sensing* 78.3 (2012): 259–269. Print.

Aksha, Sanam K., Juran, Luke, and Resler, Lynn M. "Spatial and Temporal Analysis of Natural Hazard Mortality in Nepal." *Environmental Hazards* 17.2 (2018): 163–179. Print.

Aksha, Sanam K., Resler, Lynn M., and Juran, Luke. "A Geospatial Analysis of Multi-Hazard Risk in Dharan, Nepal." *Geomatics, Natural Hazards and Risk* 11.1 (2020): 88–111. Web.

Apel, Heiko, Thieken, Annegret H., Merz, Bruno, and Blöschl, Günter. "A Probabilistic Modelling System for Assessing Flood Risks." *Natural Hazards* 38.1–2 (2006): 79–100. Web.

Asselman, N, Ter Maat, J, De Wit, A, Verhoeven, G, and Soares, Frazão S. "Flood Inundation Modelling: Model Choice and Application." *Flood Risk Management: Research and Practice*. CRC Press, 2008. 47–47. Print.

Bales, JD, and Wagner, CR. "Sources of Uncertainty in Flood Inundation Maps." *Journal of Flood Risk Management* 2.2 (2009): 139–147. Print.

Bhuiyan, Md Jabed Abdul Naser, and Dutta, Dushmanta. "Analysis of Flood Vulnerability and Assessment of the Impacts in Coastal Zones of Bangladesh Due to Potential Sea-Level Rise." *Natural Hazards* 61.2 (2012): 729–743. Web.

Brunner, Gary W. "HEC-RAS River Analysis System User's Manual Version 5.0." *US Army Corps of Engineers* (2016): 962. Print.

Carvalho, Ana Claudia Pereira, Pejon, Osni José, and Collares, Eduardo Goulart. "Integration of Morphometric Attributes and the HAND Model for the Identification of Flood-Prone Area." *Environmental Earth Sciences* 79.14 (2020): 367. Print.

Cohen, Sagy, Praskievicz, Sarah, and Maidment, David R. "Featured Collection Introduction: National Water Model." *Journal of the American Water Resources Association* 54.4 (2018): 767–769. Web.

CRED. "The Human Cost of Natural Disasters." *Centre for Research on the Epidemiology of Disasters CRED* (2015): 58.

Di Baldassarre, Giuliano, Laio, Francesco, and Montanari, Alberto. "Design Flood Estimation Using Model Selection Criteria." *Physics and Chemistry of the Earth, Parts A/B/C* 34.10–12 (2009): 606–611. Print.

Didier, David, Bernatchez, Pascal, Boucher-Brossard, Geneviève, Lambert, Adrien, Fraser, Christian, Barnett, Robert L, and Van-Wierts, Stefanie. "Coastal Flood Assessment Based on Field Debris Measurements and Wave Runup Empirical Model." *Journal of Marine Science and Engineering* 3.3 (2015): 560–590. Web.

Dworkin, Judith. *Global Trends in Natural Disasters, 1947-1973*. Institute of Behavioral Science, University of Colorado, 1974. Print.

Faghih, Mina, Mirzaei., Majid, Adamowski, Jan F., Lee, Juneseok, and El-Shafie, A. "Uncertainty Estimation in Flood Inundation Mapping: An Application of Non-parametric Bootstrapping." *River Research and Applications* 33.4 (2017): 611–619. Print.

Gallien, Tim W. "Validated Coastal Flood Modeling at Imperial Beach, California: Comparing Total Water Level, Empirical and Numerical Overtopping Methodologies." *Coastal Engineering* 111 (2016): 95–104. Print.

Garousi-Nejad, Irene, Tarboton, David G, Aboutalebi, Mahyar, and Torres-Rau, Alfonso F. "Terrain Analysis Enhancements to the Height Above Nearest Drainage Flood Inundation Mapping Method." *Water Resources Research* 55.10 (2019): 7983–8009. Web.

Godbout, Lukas, Zheng, Jeff Y, Dey, Sayan, Eyelade, Damilola, Maidment, David, and Passalacqua, Paola. "Error Assessment for Height Above the Nearest Drainage Inundation Mapping." *Journal of the American Water Resources Association* 55.4 (2019): 952–963. Web.

Huang, Yu, and Dai, Zili. "Large Deformation and Failure Simulations for Geo-Disasters Using Smoothed Particle Hydrodynamics Method." *Engineering Geology* 168 (2014): 86–97. Print.

Johnson, Mike, Munasinghe, Dinuke, Eyelade, Dami, and Cohen, Sagy. "An Integrated Evaluation of the National Water Model (NWM)–Height Above Nearest Drainage (HAND) Flood Mapping Methodology." *Natural Hazards and Earth System Sciences* 19.11 (2019): 2405–2420. Web.

Karim, Fazlul, Marvanek, Steve, Merrin, Linda E., Nielsen, Daryl, Hughes, Justin, Stratford, Danial, and Pollino, Carmel. "Modelling Flood-Induced Wetland Connectivity and Impacts of Climate Change and Dam." *Water* 12.5 (2020): 1278. Web.

Kauffeldt, Anna, Wetterhall, Fredrik, Pappenberger, Florian., Salamon, Peter, and Thielen, J. "Technical Review of Large-Scale Hydrological Models for Implementation in Operational Flood Forecasting Schemes on Continental Level." *Environmental Modelling and Software* 75 (2016): 68–76. Web.

Kuldeep, P. K., and Garg, R. D. "Geospatial Techniques for Flood Inundation Mapping." *2016 IEEE International Geoscience and Remote Sensing Symposium (IGARSS)*. N.p., 2016. 4387–4390. *IEEE Xplore*. Web. 19 Jan. 2024.

Kvočka, Davor, Ahmadian, Reza, and Falconer, Roger A. "Flood Inundation Modelling of Flash Floods in Steep River Basins and Catchments." *Water* 9.9 (2017): 705. *www.mdpi.com*. Web.

Lhomme, Julien, Sayers, Paul, Gouldby, Ben, Samuels, Paul, Wills, Martin, and Mulet-Marti, Jonatan. *Recent Development and Application of a Rapid Flood Spreading Method*. Keble College, Oxford, UK: N.p., 2008. *eprints.hrwallingford.com*. Web. 19 Jan. 2024.

Li, Jing, David, W. S., and Wong. "Effects of DEM Sources on Hydrologic Applications." *Computers, Environment and Urban Systems* 34.3 (2010): 251–261. *ScienceDirect*. Web.

Liu, Zhu, and Merwade, Venkatesh. "Separation and Prioritization of Uncertainty Sources in a Raster Based Flood Inundation Model Using Hierarchical Bayesian Model Averaging." *Journal of Hydrology* 578 (2019): 124100. Web.

Liu, Xin, Xia, Jianhong (Cecilia), Wright, Graeme, and Arnold, Lesley. "A State of the Art Review on High Water Mark (HWM) Determination." *Ocean & Coastal Management* 102 (2014): 178–190. Web.

Madsen, Henrik, and Rosbjerg, Dan. "Generalized Least Squares and Empirical Bayes Estimation in Regional Partial Duration Series Index-Flood Modeling." *Water Resources Research* 33.4 (1997): 771–781. Web.

Maidment, David, Rajib, Adnan, Lin, Peirong, and Clark, Edward P. "National Water Center Innovators Program. Summer Institute Report 2016." January 2019 (2016): 101. Web.

Marriott, Susan. "Textural Analysis and Modelling of a Flood Deposit: River Severn, U.K." *Earth Surface Processes and Landforms* 17.7 (1992): 687–697. Web.

Merwade, Venkatesh, Olivera, Franscisco, Arabi, Mazdak, and Edleman, Scott. "Uncertainty in Flood Inundation Mapping: Current Issues and Future Directions." *Journal of Hydrologic Engineering* 13.7 (2008): 608–620. *ASCE*. Web.

Monaghan, J J. "Simulating Free Surface Flows With SPH." *Journal of Computational Physics* 110.2 (1994): 399–406. Web.

Musser, Jonathan W., Watson, Kara M., and Gotvald, Anthony J. "Characterization of Peak Streamflows and Flood Inundation at Selected Areas in North Carolina Following Hurricane Matthew." *U.S. Geological Survey Open-File Report 2017–1047* October 2016 (2017): 24. Web.

NCEM Hurricane Matthew Resilient Redevelopment Plan Southeast Region. 2017. https://files.nc.gov/rebuildnc/documents/matthew/rebuildnc_southeast_region_plan_draft.pdf

Néelz, Sylvain, and Pender, Gareth "Benchmarking the Latest Generation of 2D Hydraulic Modelling Packages." *Environment Agency: Bristol, UK* (2013): n. pag. Print.

Nobre, A. D., Cuartas, L. A., Hodnett, M., Rennó, C. D., Rodrigues, G., Silveira, A., Waterloo, M., and Saleska, S. "Height Above the Nearest Drainage – A Hydrologically Relevant New Terrain Model." *Journal of Hydrology* 404.1–2 (2011): 13–29. Web.

Panagoulia, Dionysia. "Artificial Neural Networks and High and Low Flows in Various Climate Regimes." *Hydrological Sciences Journal* 51.4 (2006): 563–587. *Taylor and Francis+NEJM*. Web.

Panagoulia, Dionysia. "Impacts of GISS-Modelled Climate Changes on Catchment Hydrology." *Hydrological Sciences Journal* 37.2 (1992): 141–163. *Taylor and Francis+NEJM*. Web.

Pappenberger, Florian, Beven, Keith, Frodsham, Kevin, Romanowicz, Renata, and Matgen, Patrick. "Grasping the Unavoidable Subjectivity in Calibration of Flood Inundation Models: A Vulnerability Weighted Approach." *Journal of Hydrology* 333.2 (2007): 275–287. *ScienceDirect*. Web.

Pender, Gareth. "Briefing: Introducing the Flood Risk Management Research Consortium." *Proceedings of the Institution of Civil Engineers - Water Management* 159.1 (2006): 3–8. *icevirtuallibrary.com (Atypon)*. Web.

Roberts, Stephen, Nielsen, Ole, Gray, Duncan, Sexton, Jane, and Davies, Gareth. *ANUGA User Manual, Release 2.0*. N.p., 2015. *ResearchGate*. Web.

Sehgal, Vinit, Sridhar, Venkataramana, Juran, Luke, and Ogejo, Jactone Arogo. "Integrating Climate Forecasts with the Soil and Water Assessment Tool (SWAT) for High-Resolution Hydrologic Simulations and Forecasts in the Southeastern U.S." *Sustainability* 10.9 (2018): 3079. *www.mdpi.com*. Web.

Shastry, Apoorva, and Durand, Michael. "Utilizing Flood Inundation Observations to Obtain Floodplain Topography in Data-Scarce Regions." *Frontiers in Earth Science* 6 (2019): n. pag. *Frontiers*. Web. 19 Jan. 2024.

Shastry, Apoorva, Egbert, Ryan, Aristizabal, Fernando, Luo, Cehong, Yu, Chen-Wei, and Praskievicz, Sarah. "Using Steady-State Backwater Analysis to Predict Inundated Area from National Water Model Streamflow Simulations." *Journal of the American Water Resources Association* 55.4 (2019): 940–951. Web.

Smith, Laurence C. "Satellite Remote Sensing of River Inundation Area, Stage, and Discharge: A Review." *Hydrological Processes* 11.10 (1997): 1427–1439. *Wiley Online Library*. Web.

Speckhann, Gustavo Andrei, Borges, Chaffe, Pedro, Luiz, Fabris, Goerl, Roberto, Abreu, Janete, Josina De, Altamirano, Flores, and Juan, Antonio. "Flood Hazard Mapping in Southern Brazil: A Combination of Flow Frequency Analysis and the HAND Model." *Hydrological Sciences Journal* 63.1 (2018): 87–100. *Taylor and Francis+NEJM*. Web.

Teng, J., Jakeman, A. J., Vaze, J., Croke, B. F. W., Dutta, D., and Dim, S. "Flood Inundation Modelling: A Review of Methods, Recent Advances and Uncertainty Analysis." *Environmental Modelling and Software* 90 (2017): 201–216. Web.

Tilford, K.A., Sene, K.J., and Khatibi, Rahman. "Flood Forecasting Model Selection." *Advances in Natural and Technological Hazards Research* 25 (2007): 401–416. *ResearchGate*. Web

Vastiall, Kaisa, KUmmu, Matti, Sangamanee, C., and Chinvanno, Suppakorn. "Modelling Climate Change Impacts on the Flood Pulse in the Lower Mekong Floodplains." *Journal of Water and Climate Change* 1.1 (2010): 67–86. Web.

Ye, J., and McCorquodale, J. A. "Simulation of Curved Open Channel Flows by 3D Hydrodynamic Model." *Journal of Hydraulic Engineering* 124.7 (1998): 687–698. Web.

Yu, Miao, Huang, Yu, Xu, Qiang, Guo, Peng, and Dai, Zili. "Application of Virtual Earth in 3D Terrain Modeling to Visual Analysis of Large-Scale Geological Disasters in Mountainous Areas." *Environmental Earth Sciences* 75.7 (2016): 563. *Springer Link*. Web.

Zheng, Xing, Tarboton, David G., Maidment, David R., Liu, Yan Y., and Passalacqua, Paola. "River Channel Geometry and Rating Curve Estimation Using Height above the Nearest Drainage." *J. of the American Water Resources Ass* 54.4 (2018): 785–806.

Section VIII

Disaster Recovery Issues and Visions

31
DISASTER RECOVERY AND BUSINESSES

Charles D. S. Chishiri and Luke Juran

Introduction

In the 2023 book *Tales from the Edge: Real-Life Risk Management Stories*, Julian Talbot travels down memory lane to 1955 when American Motors, Brown Shoe Company, Studebaker, Collins Radio, Detroit Steel, Zenith Electronics, and National Sugar Refining were all proudly listed as Fortune 500 companies. These companies were considered among the best and most powerful organizations in the United States. Talbot then fast forward to 2023 when none of these companies were on the list. The key lesson is that risk—like the invisible hand—is neutral and amoral: it simply exists and does not care what your intent is. Risk does not pick winners and losers, it merely reveals them. Thus, for organizations to survive and recover from major disruptions they must adapt to everchanging markets and manage both foreseen and unforeseen uncertainties.

According to Krell (2006), 93% of firms that lost their critical systems for more than ten days filed for bankruptcy. Therefore, institutions that fail to implement adequate response and recovery plans will in fact be defined by their ineffective management of disasters. In that vein, this chapter investigates post-disaster business recovery by first presenting a framework that helps to categorize and conceptualize the impacts of disasters on businesses. Next, the chapter discusses various ongoing and emerging issues, such as factors that commonly affect business recovery, challenges faced by small businesses, market disruptions, the significance of household and neighborhood recovery, and instruments to finance business recovery.

Background

In a recent study, Chang et al. (2022) sought to compare economic recovery from the COVID-19 pandemic to other types of disaster events at the scale of businesses. The study garnered insights by examining both COVID-19 and large-scale disasters including floods, cyclones, and earthquakes. Areas of overlap and emergent themes emanating from the COVID-19 pandemic were identified, and a disaster taxonomy and nomenclature were developed to facilitate the comparison of hazards in terms of economic disruption and business recovery. Additionally, a conceptual framework for business recovery was developed to support policymakers in anticipating business recovery needs during economically disruptive events.

 DOI: 10.4324/9781003414834-42

A key finding of Chang et al. (2022) is that recovery tends to be more difficult for smaller businesses. Those that are vulnerable to supply chain disruptions or confront disrupted markets also experience enhanced difficulties. Another theme is that locally oriented businesses (i.e., those that primarily draw consumers from a nearby geographical area) located in heavily impacted neighborhoods are particularly vulnerable. Chang et al. (2022) also found that businesses in high-income neighborhoods and central business districts can be disproportionately affected by disasters due to their reliance on foot traffic. This was the case with COVID-19 and other major disasters including earthquakes and floods. Regarding COVID-19, substantial business disruptions occurred from changes in demand as a result of the regulatory environment and associated changes in consumer behavior. Most importantly, disparities in recovery arise not only from the hazard characteristics but also the underlying vulnerabilities and conditions of the affected businesses themselves. Ultimately, business recovery is a function of the ability to continue to operate as well as the survival of its customers, suppliers, financial position, and ability to adapt to a changing environment (Stevenson et al., 2018). These common themes exist across disaster types, including the COVID-19 pandemic, and will be expanded upon throughout the chapter.

Business Recovery Conceptual Framework

Chang et al.'s (2022) comparative study led to the development of a framework for outlining dominant factors that influence business recovery. The framework focuses on the business as a unit of analysis rather than the entire sector or an economic impact approach. In proposing a taxonomy of economically disruptive events from the perspective of business recovery, disasters were broadly classified according to their spatio-temporal and economic shock profiles. Some of the characteristics under consideration include opportunities for preparedness prior to the disruptive event and resultant implications on business response, adaptation, and survival.

The *spatio-temporal profile* encompasses several characteristics (Table 31.1). Speed of onset—for example, a rapid onset earthquake or a slow onset drought—and duration of the event (e.g., a tornado or an extended heat or cold wave) have far-reaching implications on business recovery. The frequency of occurrence and predictability of a hazard's impacts are critical and should already be considered in preparedness and (financial) mitigation plans in advance of a disruptive event. The spatial extent of a disaster—for example, a neighborhood-scale industrial accident or a global pandemic—is also crucial for business recovery as the geographic scope can directly impede the mobilization and availability of needed resources.

Table 31.1 Taxonomy of Spatio-Temporal Disaster Characteristics and Implications for Business Recovery

Spatio-Temporal Profile	*Type/Examples*	*Implications for Business Recovery*
Onset Speed/Suddenness	• Rapid onset (earthquake, tsunami) • Sudden onset with advance warning (hurricane) • Gradual/slow onset (drought, fluvial flood)	• Quality and extent of preparedness • Availability of rapid assistance • Ability to mobilize funds and resources for response and recovery

(*Continued*)

Table 31.1 (Continued)

Spatio-Temporal Profile	*Type/Examples*	*Implications for Business Recovery*
Duration	• Short (flash flood) • Medium (fluvial flood) • Long (drought)	• Quality and duration of business and government responses • Ability to modify business operations • Management of business interruptions • Depletion of resources affecting long-term viability
Frequency	• 'One-time' event (industrial accident) • Repeated event over a fairly well-defined period (earthquake) • Prolonged, repetitive event (sea level rise)	• Multi-hazard preparedness and risk plans • Robust and adaptable business recovery plans • Investment in redundant and resilient infrastructure and systems • Labor and management awareness training
Predictability of Impact	• Plausible intensity known but impacts less known (earthquake) • Impacts largely known but not timing or intensity (hurricanes, wildfire) • Impacts, timing, and intensity largely unknown (solar storms, asteroids)	• Adequate planning for potential consequences • Preemptive actions (e.g., mitigation, insurance) • Robust and adaptable business recovery plans • Labor and management awareness training
Spatial Extent/Scale	• Localized (flash flood) • Regional (drought) • Global (pandemic, climate modifying volcanic eruption)	• Ease of mobilizing, accessing, and substituting resources • Overcoming governance and management challenges • Cascading impacts, supply and demand mismatches, and effects on ancillary firms
Magnitude/Severity	• Strength of the disaster event (e.g., more vs. less severe)	• Damage to power, water, transport, and communication infrastructure can severely impair business operations • Potential for road, border, and airspace closures and restrictions • Recovery plans to cope with loss of critical infrastructure are required for business continuity
Scope/Sectoral impact	• Impacts on one vs. various sectors	• Disasters impacting multiple sectors are more likely to disrupt supply chains and affect demand for goods and services by consumers and producers • Risk analysis to assess vulnerability of your firm, sector, and supply chain

Source: Original Table/Table by Authors; Adapted from Chang et al. (2022).

Table 31.2 Taxonomy of Economic Shock Characteristics of Disasters and Implications for Business Recovery

Economic Shock Profile	*Type/Examples*	*Implications for Business Recovery*
Inputs	• Damage to physical capital and infrastructure (earthquake, flood) • Damage to labor and human capital (pandemic) • Damage to financial capital (financial crisis, recession, pandemic)	• Quality and quantity of production • Possible absenteeism, poor morale, loss of workforce, and low productivity • Inability for business to generate revenue
Supply Chain	• Damage to transport networks (volcanic eruption, fluvial flood, wildfire) • Damage to crucial supply chain nodes (pandemic, conflict, disaster in a monopolistic location) • Damage to financial networks (financial crisis, depression, pandemic)	• Inaccessibility of inputs • Identification of substitutable and complementary resources • Indirect and cascading impacts • Collapse in the distribution chain and increase in failed business transactions • Risk analysis to assess vulnerability of your firm, sector, and supply chain
Demand	• Sectoral shifts in demand (natural disaster) • Behavioral change (pandemic, financial crisis) • Government mandates (pandemic, water use limits)	• Risk analysis to assess vulnerability of your consumer base and geographic market • Assessment of market vulnerability • Elasticity of demand based on product type

Source: Original Table/Table by Authors; Adapted from Chang et al. (2022).

Meanwhile, the *economic shock profile* conceptualizes how a business' ability to produce goods or services is compromised and issues of altered demand for a business' end products (Table 31.2). The supply side of a firm is affected through the damage or destruction of a business' inputs for production—including physical, financial, and human capital—and through broader supply chain disruptions (e.g., damage to bridges, ports, airports, and transportation networks). On the other side, demand for a business' goods or services can be affected by temporary or permanent regulations implemented by governments and by voluntary behavioral changes that surface as a result of the disaster. This was especially evident during the COVID-19 pandemic. Finally, it must be noted that while industries such as tourism, hospitality, and luxury goods typically observe decreased demand after a disaster, others may experience increased demand (e.g., lumber, steel, and construction). For example, extensive private property damage often generates increased household spending on durable goods, as occurred after the 1995 Kobe earthquake in Japan (Sawada and Shimizutani, 2011).

Business Recovery: Key Themes and Lessons Learned

A number of studies have surveyed the economic impacts of disruptive events, including empirical findings on business recovery after earthquakes, floods, tsunamis, hurricanes, wildfires, and tornadoes (see Webb et al., 2000; Tierney, 2007; Chang and Rose, 2012; Noy and Dupont IV,

2018; Krumel and Goodrich, 2021). This body of research has established numerous cross-cutting themes that will be explored in this section.

Defining Business Recovery

Recovery is a subjective word and represents a complex, multifaceted process. Chang and Rose (2012) define business recovery as the "process by which businesses and local economies return to conditions of stability following a disaster" (171). This implies that achieving business recovery may equate to stability or an acceptable level of stasis rather than a return to pre-disaster conditions. Other scholars have described business recovery in a dichotomous way, namely, whether a firm is closed or open. For example, LeSage et al. (2011) posit that a business that remains open/operational or that reopens after a disaster has recovered. Alternatively, Dahlhamer and Tierney (1998) classify businesses that are relatively better off or the same as the pre-disaster context as having recovered, which aligns more closely with traditional notions of recovery and the prevailing mantra of 'build back better.'

Corey and Deitch (2011) analyzed recovery through the lens of percentage change in business performance, while Marshall and Schrank (2014) proposed a framework that conceptualizes business recovery as an ongoing process. To elucidate this, Brown et al. (2015) analyzed business recovery after the 2010 Canterbury earthquake in New Zealand using various measurements. The binary measure of whether the businesses were open or closed revealed that 90% of respondent businesses had reopened with only 1% having permanently closed. However, when given self-assessment criteria to determine whether business operations were the same or better off as the pre-disaster scenario, only 66% of the same firms reported that they met this criterion. It is clear from the literature that both the definition of business recovery and the parameter(s) for assessment are crucial. Moreover, it is imperative to recognize that business recovery is an intricate process that varies across geographies, sectors, and size and type of firm.

Variables Affecting Business Recovery

The literature identifies myriad variables that influence a business recovery including: business size; economic sector; level of infrastructure disruption; impacts on stakeholders (e.g., employees, customers, and competitors); market access; product substitutability; product complementarity; and pre-disaster solvency and financial condition (Webb et al., 2000; Chang and Rose, 2012; Dahlhamer and Tierney, 1998; Alesch et al., 2001). When a disaster results in a business failing, it is often due to pre-existing vulnerabilities worsened by the disaster event (Tierney, 1997). In fact, the factors identified as having the greatest impact on a business' likelihood of failure following a disaster are the same as those that influence failure during normal times. To illustrate this, businesses that may not have been physically or directly impacted by a disaster are often profoundly affected by the disaster through disruptions to supply chains, impacts on consumers, and economy-wide impacts. Yet, all of these vulnerabilities already exist prior to a disaster (i.e., they are merely exposed or unmasked) and all can manifest in the absence of a disaster.

Cross-Cutting Themes

A range of salient themes have emerged regarding business recovery. First, small businesses typically have more difficulty recovering and/or experience a longer recovery period compared to larger businesses (as has been observed in economic downturns). Additionally, business recovery in high-income countries often begins and is reached more quickly compared to lesser-developed countries. As might be expected, businesses that are vulnerable to supply chain disruptions have more difficulty recovering as do businesses that have their markets directly disrupted (Kurmann et al., 2020). Similarly, businesses operating in severely impacted neighborhoods generally experience greater difficulty recovering. Finally, it is critical to state that faster, less restrictive funding options and financial resources tend to facilitate more rapid and effective recovery (Alesch et al., 2001; Fort et al., 2013; Basker and Miranda, 2018).

Small Businesses and Recovery

The greater vulnerability of relatively small enterprises stems from factors such as lack of disaster preparedness planning, inexperience confronting and/or less human capital knowledge on disasters, and lack of a financial position to elect for risk mitigation measures (e.g., insurance for business interruption, inventory, and property damage) (Bloom et al., 2021; Wang and Kang, 2021). Small firms also operate under conditions of higher business fragility and smaller cash buffers. In the United States, for example, it is estimated that 50% of small and medium enterprises operate with less than 27 days of cash reserves (Farrell and Wheat, 2016). Furthermore, small businesses tend to be overly represented in sectors that are more vulnerable to disruptions, including retail, restaurants, and services catering to local customers.

After the 2001 Nisqually earthquake in the United States, Chang and Falit-Baiamonte (2002) found that small businesses in the retail sector, which tend to rely on local foot traffic, experienced more significant losses than the manufacturing sector in the same area (which produced goods for customers outside of the impacted region). These findings resonate with the impacts of COVID-19, which negatively affected many small and locally oriented businesses. According to Bartik et al. (2020) and Katare et al. (2021), small enterprise recovery during COVID-19 was influenced by their higher financial fragility, narrow cash buffers, and difficulties accessing financing. Moreover, minority and female business owners were disproportionately impacted by these factors (OECD, 2021; Bloom et al., 2021).

Regarding COVID-19, tourism, restaurants, and small businesses were among the most seriously impacted (Bartik et al., 2020). However, one of the positive aspects of recovery was a surge in new business applications, which rose significantly between June 2020 and May 2023 with 2.8 million more micro-businesses opening in 2020 compared to 2019 (U.S. Census Bureau, 2021). Turkey, Chile, Australia, and the United Kingdom experienced similar increases in new business applications in the third quarter of 2020. In Japan, new store openings in mid-2021 exceeded rates for the previous 19 years before the pandemic (Saito et al., 2022). It is postulated that consumer preferences during the pandemic—notably increased online shopping and a desire for services—created unique opportunities for entrepreneurship (e.g., participating in the 'gig economy') (Djankov and Zhang, 2021). In Israel and Brazil, research found that there was an increase in the utilization of digital technologies as method for small businesses to adapt to disruptions from COVID-19 (Katare et al., 2021; Harel, 2021; Klein and Todesco, 2021).

Disruptions to Supply Chains and Hard and Soft Infrastructure

Damage to production facilities and other physical assets leads to a reduced generation of goods and services, which can trigger ripple effects when they are required as inputs by other businesses. Production shortages in one sector or even a single, crucial incapacitated business can multiply to other businesses, other sectors, and other regions through ancillary (e.g., waste processing) and inter-industry connections (e.g., semiconductors). For example, the 2011 Great East Japan earthquake resulted in shortages of intermediate goods needed for automobile production, which in turn led to a lag and sharp drop in automobile production. In the Guangdong province of China, the impacts on automobile production were felt one month after the earthquake, and the effects reverberated even longer in Thailand (Fujita and Hamaguchi, 2012).

Capacity loss in the transport and logistics sectors also produces ripple effects locally and throughout the chain of production. In Chile, a 2010 earthquake closed ports resulting in lumber shortages for the construction sector in Chicago (Altay and Ramirez, 2010). Similar disruptions were noted after the September 11, 2001 terrorist attacks in the United States, where border and airspace closures disrupted global-scale production (Sheffi and Rice, 2005). There are certain sectors that are more vulnerable to network and logistics disruptions. High technology industries are often more susceptible due to their overreliance on complex international specializations for their inputs. For example, the 1999 Taiwan earthquake led to shortages of semiconductors (Wagner and Bode, 2006), and the capacity to produce automobiles and electronics were seriously affected by the 2011 Chao Phraya River floods in Thailand (Haraguchi and Lall, 2015).

Demand Disruptions

Numerous studies have indicated that businesses have difficulty recovering when demand for their goods and services decreases (see Brown et al., 2017; Corey and Deitch, 2011; Zhang et al., 2009; Watson et al., 2020; Kroll et al., 1990). Demand disruptions in the context of disasters is often driven by population relocation (Dietch and Corey, 2011; Runyan, 2006), inaccessibility due to physical damage (Sampson et al., 2018), and changes in consumer requirements and preferences (Alesch et al., 2001; Forbes, 2017). Demand-related disruptions particularly impact locally oriented businesses such as hospitality and retail. While impacts on demand have historically been overshadowed by supply-side impacts—including direct disruptions to physical infrastructure, staff, and financial capital—the COVID-19 pandemic brought demand-side impacts to the forefront. Some of the greatest impacts were on air travel, tourism, restaurants, and retail (Donthu and Gustafsson, 2020; Sheffi, 2020). Still, there were increases in demand in many sectors, notably medical supplies, warehousing, and technological products and services (Sheth, 2020; Donthu and Gustafsson, 2020). Regulatory measures (e.g., lockdowns) and behavioral changes to limit exposure to the virus (e.g., physical distancing and self-imposed isolation) resulted in greater demand for domestic goods, technologies, and discretionary goods as opposed to discretionary services (US Bureau of Economic Analysis, 2022). Businesses operating in diverse markets and those able to identify new markets or deploy alternative methods of delivery were less affected by changes in demand.

COVID-19 and Business Recovery

The COVID-19 pandemic ushered in severe business effects that occurred unexpectedly and rapidly, including strict lockdowns, mass layoffs, and closures of production facilities, trading centers, and office and brick and mortar spaces within weeks. Border closures, government interventions in markets and social services, and structural economic changes were also implemented (Bishop, 2020; Hannon et al., 2022). In terms of predictability and onset, the pandemic allowed little time for decision makers to comprehend and prepare for the wide-ranging and long-term impacts that would unfold, let alone the short-term impacts. In comparison, it is worth noting that the 1918 influenza pandemic did not generate the drastic policy procedures introduced during COVID-19.

The geographic reach of COVID-19 far exceeded the spatial extent of any other single disaster event. In terms of economics, another distinction is that the driver of business impact was not structural imbalances in the economy. This differs from other global economic disasters such as the recent Great Recession of the 2000s. Furthermore, the pandemic did not result in physical damage as does a typical disaster, yet it still caused unprecedented disruptions to supply chains and labor availability thereby impinging the capacity of businesses to operate, generate revenue, and ultimately recover. Substantial pivots in consumer demand propelled by behavioral change to mitigate transmission of the virus—often through government restrictions and mandates—introduced additional unforeseen barriers to business recovery.

Importance of Household and Neighborhood Recovery

It has been established that business recovery is a function of household recovery in the impacted area (Xiao and Van Zandt, 2012), especially in cases of post-disaster out-migration and subsequent repopulation. As articulated by Tierney (2007), firms rely on robust business ecologies to aid their recovery. Tierney contends that 'without a critical mass of consumers, workers, and other enterprises that are back in operation, business recovery outcomes will likely be very poor' (287). Interactions among business, neighborhood, and household recovery are important and mutually reinforcing—and they all support the tax base which in turn funds services that facilitate long-term resilience and recovery. Additionally, it could be argued that a healthy household-neighborhood-business ecology supports the production of human and cultural capital (major components of capacity and resilience) due to increased social interactions, economic opportunities, and flows of ideas and resources.

Financing Business Recovery

One of the key interventions that influence the swiftness and effectiveness of business recovery is how quickly and easily it is to access post-disaster funding. The type of funding, speed at which it is made available, and eligibility criteria and information/data that will be required from firms are vital parameters when designing recovery financing programs (Furlong and Scheberle, 1998; Graham, 2007; Fischer-Smith, 2013). From a business standpoint, grants and measures such as rebates, tax and payroll credits, subsidies, protections, and employee benefits are generally more desirable compared to loans (Bartik et al., 2020; Alekseev et al., 2020). However, such programs are obviously more expensive for governments or other institutions to implement. Extensive research has indicated that loans have had both positive and negative effects on recovery, with negative results typically due to heavier indebtedness on the businesses (Dahlhamer and Tierney, 1998; McDonald et al., 2014; Cole et al., 2017).

Insurance and the insurance industry are major components of business recovery. While insurance does provide market-based funds for damages, it normally does not cover the bulk of disaster-related expenses (e.g., missed revenue from being closed) and the claims process can be difficult and lengthy to navigate (Flynn, 2007; Xiao et al., 2018; van de Lindt et al., 2020). Furthermore, many businesses lease the buildings, offices, and production spaces that they utilize. Property owners receive the payouts for damages from insurance companies, often leaving tenants with fewer recovery resources and no space to operate (at least temporarily). Gap insurance and gap financing are additional instruments that can alleviate stresses from uninsured property losses and business closure (Runyan, 2006; Resosudarmo et al., 2012). This can help businesses reopen faster and adjust to volatility in supply chains and consumer behavior that generally follows a disaster (Chang and Falit-Baiamonte, 2002; Xiao and Nilawar, 2013).

Conclusion

This chapter reviewed business disaster recovery by highlighting both practical issues and conceptual considerations. A framework and common factors that influence recovery were presented that are applicable to a range of firms, although the focus tended to be on smaller firms and the local scale. While size of business is a recovery variable, other predominant factors include the sector/market (e.g., tourism vs. manufacturing or production of goods vs. services); status of key stakeholders (e.g., labor, consumers, and competitors); and impacts on infrastructure and networks that impede the mobilization of resource inputs and production outputs.

Recovery that facilitates a robust business ecology and encourages interactions among households, the neighborhood, and the impacted area represents an effective approach to business and community redevelopment. This approach also supports the co-creation of human and cultural capital—major components of capacity and resilience—as a result of increased interactions, opportunities, and flows of ideas and resources.

Most importantly, the pre-disaster financial position of a firm and the ability to nimbly adapt to shifts in post-disaster markets and consumer behaviors are critical. Ultimately, business recovery is not merely a function of the hazard characteristics, but more so the pre-existing vulnerabilities of the business itself. Thus, the factors that often exert the greatest impact on a business' likelihood of post-disaster failure are the same factors that influence failure during normal, 'non-disaster' times.

References

Alekseev, G., Amer, S., Gopal, M., Kuchler, T., Schneider, J.W., Stroebel, J., Wernerfelt, N.C. The Effects of COVID-19 on U.S. Small Businesses: Evidence from Owners, Managers, and Employees (No. w27833), National Bureau of Economic Research, 2020.

Alesch, D.J., Holly, J.N., Mittle, E., Nagy, R. Organizations at Risk: What Happens When Small Businesses and Not-for-Profits Encounter Natural Disasters, Public Entity Risk Institute, Fairfax, VA, 2001.

Altay, N., Ramirez, A. Impact of disasters on firms in different sectors: Implications for supply chains, J. Supply Chain Manage. 46 (4) (2010) 59–80.

Bartik, A.W., Bertrand, M., Cullen, Z., Glaeser, E.L., Luca, M., Stanton, C. The impact of COVID-19 on small business outcomes and expectations, Proc. Natl. Acad. Sci. 117 (30) (2020) 17656–17666.

Basker, E., Miranda, J. Taken by storm: Business survival in the aftermath of Hurricane Katrina, J. Econ. Geogr. 18 (6) (2018) 1285–1313.

Bishop, J. (2020) Economic effects of Spanish flu. Bulletin of the Reserve Bank of Australia. pp 8–17.

Bloom, N., Fletcher, R.S., Yeh, E. The impact of COVID-19 on US firms, National Bureau of Economic Research working paper 28314 (January 2021). http://www.nberg.org/papers/w28314.

Brown, C., Seville, E., Vargo, J. Efficacy of insurance for organisational disaster recovery: Case study of the 2010 and 2011 Canterbury earthquakes, Disasters 41 (2) (2017) 388–408.

Brown, C., Stevenson, J., Giovinazzi, S., Seville, E., Vargo, J. Factors influencing impacts on and recovery trends of organisations: Evidence from the 2010/2011 Canterbury earthquakes, Int. J. Disaster Risk Reduc. 14 (2015) 56–72.

Chang, S.E., Brown, C., Handmer, J., Helgeson, J., Kajitani, Y., Keating, A., Noy, I., Watson, M., Derakhshan, S., Kim, J., Roa-Henriquez, A. Int. J. Dis. Risk Reduct. (2022), doi: 10.1016/j.ijdrr.2022.103191.

Chang, S.E., Falit-Baiamonte, A. Disaster vulnerability of businesses in the 2001 Nisqually earthquake, Environ. Hazards 4 (2/3) (2002) 59–71.

Chang, S.E., Rose, A.Z. Towards a theory of economic recovery from disasters, Int. Mass Emerg. Disasters 32 (2) (2012) 171–181.

Cole, M.A., Elliott, R.J., Okubo, T., Strobl, E. Pre-disaster planning and post-disaster aid: Examining the impact of the Great East Japan earthquake, Int. J. Disaster Risk Reduc. 21 (2017) 291–302.
Corey, C.M., Deitch, E.A. Factors affecting business recovery immediately after Hurricane Katrina, J. Contingencies Crisis Manage. 19 (2) (2011).
Dahlhamer, J.M., Tierney, K.J. Rebounding from disruptive events: Business recovery following the Northridge earthquake, Sociol. Spectr. 18 (2) (1998) 121–141.
Dietch, E.A., Corey, C.M. Predicting long-term business recovery four years after Hurricane Katrina, Manag. Res. Rev. 34 (3) (2011) 311–324.
Djankov, S., Zhang, E. Startups boom in the United States during COVID-19, PIIE Realtime Economic Watch, 2021. https://www.piie.com/blogs/realtimeeconomic-issues-watch/startups-boom-united-states-during-covid-19#_ftn1. (Accessed 18 October 2023).
Donthu, N., Gustafsson, A. Effects of COVID-19 on business and research, J. Bus. Res. 117 (2020) 284–289, https://doi.org/10.1016/j.jbusres.2020.06.008. (Accessed 14 December 2023).
Farrell, D., Wheat, C. Cash is king: flows, balances, and buffer days: Evidence from 600,000 small businesses, JP Morgan Chase & Co, 2016 Institute Report.
Fischer-Smith, R. The earthquake support subsidy for Christchurch's small and medium enterprises: Perspectives from business owners, Small Enterprise Res. 20 (1) (2013) 40–54.
Flynn, D.T. The impact of disasters on small business disaster planning: A case study, Disasters 31 (4) (2007) 508–515.
Forbes, S.L. Post-disaster consumption: Analysis from the 2011 Christchurch earthquake, Int. Rev. Retail Distrib. Consum. Res. 27 (1) (2017) 28–42.
Fort, T.C., Haltiwanger, J., Jarmin, R.S., Miranda, J. How firms respond to business cycles: The role of firm age and firm size, IMF Econ. Rev. 61 (3) (2013) 520–559.
Fujita, M., Hamaguchi, N. Japan and economic integration in East Asia: Post-disaster scenario, Ann. Reg. Sci. 48 (2012) 485–500.
Furlong, S.R., Scheberle, D. Earthquake recovery: Gaps between norms of disaster agencies and expectations of small businesses, Am. Rev. Publ. Adm. 28 (4) (1998) 367–389.
Graham, L.T. Permanently failing organizations? Small business recovery after September 11, 2001, Econ. Dev. Q. 21 (4) (2007) 299–314.
Hannon, P.A., Hammerback, K., Kava, C.M., Bravo-Acevedo, P., Strait, M., Harris, J.R. By the seat of our pants: The experience of small businesses in the COVID-19 pandemic, Washington State, March-October 2020, Prev. Chronic Dis.: Publ. Health Res. Pract. Pol 19 (2022) E14.
Haraguchi, M., Lall, U. Flood risks and impacts: A case study of Thailand's floods in 2011 and research questions for supply chain decision making, Int. J. Disaster Risk Reduc 14 (2015) 256–272.
Harel, R. The impact of COVID-19 on small businesses' performance and innovation, Global Bus. Rev. (September 2021), https://doi.org/10.1177/09721509211039145. (Accessed 7 January 2024).
Katare, B., Marshall, M.I., Valdivia, C.B. Bend or break? Small business survival and strategies during the COVID-19 shock, Int. J. Disaster Risk Reduc 61 (2021) 102332.
Klein, V.B., Todesco, J.L. COVID-19 crisis and SMEs responses: The role of digital transformation, Knowl. Process Manage. 28 (2021) 117–133, https://doi.org/10.1002/kpm.1660
Krell, E. (2006). Management Accounting Guideline: Business Continuity Management. Society of Management Accountants of Canada and the American Institute of Certified Public Accountants. ISBN 0-87051-622-1
Kroll, C.A., Landis, J.D., Shen, Q., Stryker, S. Economic impacts of the Loma Prieta Earthquake: A Focus on Small Businesses, Studies on the Loma Preta Earthquake No, 3, University of California Transportation Center, University of California, Berkeley, 1990.
Krumel, T.P. Jr., Goodrich, C. Labour demand in the time of post-COVID-19, Appl. Econ. Lett. (2021), https://doi.org/10.1080/13504851.2021.1985067.
Kurmann, A., Lalé, E., Ta, L. The impact of COVID-19 on small business employment and hours: real-time estimates with homebase data, Document de Travail, No. 2020-09, Université du Québec à Montréal, École des Sciences de la Gestion, Département des Sciences Économiques, Montréal, 2020.
LeSage, J., Pace, R., Lam, N., Campanella, R., Liu, X. New Orleans business recovery in the aftermath of Hurricane Katrina, J. Roy. Stat. Soc. A 174 (4) (2011) 1007–1027.
Marshall, M.I., Schrank, H.L. Small business disaster recovery: A research framework, Nat. Hazards 72 (2014) 597–616.
McDonald, T.M., Florax, R., Marshal, M.I. Informal and Formal Financial Resources and Small Business Resilience to Disasters (No. 329-2016-12746), 2014.

Noy, I., duPont IV, W. The long-term consequences of disasters: What do we know, and what we still don't, Int. Rev. Environ. Resource Econ. 12 (4) (2018) 325–354.
OECD, SME and Entrepreneurship Outlook 2021, OECD Publishing, Paris, 2021, https://doi.org/10.1787/97a5bbfe-en
Resosudarmo, B.P., Sugiyanto, C., Kuncoro, A. Livelihood recovery after natural disasters and the role of aid: The case of the 2006 Yogyakarta earthquake, Asian Econ. J. 26 (3) (2012) 233–259.
Runyan, R.C. Small business in the face of crisis: Identifying barriers to recovery from a natural disaster 1, J. Contingencies Crisis Manage. 14 (1) (2006) 12–26.
Saito, S., Ito, M.I., Ohnishi, T. Fluctuations in the number of stores by industry during the COVID-19 pandemic based on Japanese phone book entries, Rev. Socionetw. Strateg. 16 (2022) 545–557.
Sampson, K., Hatton, T., Brown, C. The silent assassin: Business demand changes following disaster, J. Bus. Continuity Emerg. Plan 12 (1) (2018) 79–93.
Sawada, S., Shimizutani, S. Changes in durable stocks portfolio allocation and consumption expenditure in the aftermath of the Kobe earthquake, Rev. Econ. Househ. 9 (2011) 429–443.
Sheffi, Y. The New (ab)normal: Reshaping Business and Supply Chain Strategy Beyond Covid-19, MIT CTL Media, Cambridge, MA, 2020.
Sheffi, Y., Rice, J.B.A. Jr. Supply chain view of the resilient enterprise, MIT Sloan Manage. Rev. 47 (1) (2005) 41–48.
Sheth, J. Impact of Covid-19 on consumer behavior: Will the old habits return or die? J. Bus. Res. 117 (2020) 280–283. https://doi.org/10.1016/j.jbusres.2020.05.059
Stevenson, J.R., Brown, C., Seville, E., Vargo, J. Business recovery: An assessment framework, Disasters 42 (2018) 519–540.
Talbot, J. (2023) Tales from the Edge: Real-Life Risk Management Stories. Independently published. ISBN-13: 9798852620675
Tierney, K.J. Business impacts of the Northridge earthquake, J. Contingencies Crisis Manage. 5 (2) (1997) 87–97.
Tierney, K.J. Businesses and disasters: Vulnerability, impacts, and recovery, In: Rodriguez, H., Quarantelli, E.R., Dynes, R.R. (Eds.), Handbook of Disaster. Research, Ch.16, Springer, New York, 2007, pp. 275–296.
U.S. Census Bureau, Business Applications for the United States, FRED, Federal Reserve Bank of St. Louis, 2021. https://fred.stlouisfed.org/series/BUSAPPWNSAUS. (Accessed 19 January 2024).
US Bureau of Economic Analysis, Real personal consumption expenditures for durable goods vs services, January 2002-December 2021 [Graph], 2022.
van de Lindt, J.W., Peacock, W.G., Mitrani-Reiser, J., Rosenheim, N., Deniz, D., Dillard, M., Tomiczek, T., Koliou, M., Graettinger, A., Crawford, P.S., Harrison, K., Barbosa, A., Tobin, J., Helgeson, J., Memari, M., Sutley, E.J., Hamideh, S., Gu, D., Cauffman, S., Fung, J. Community resilience-focused technical investigation of the 2016 Lumberton, North Carolina flood: An interdisciplinary approach, Nat. Hazards Rev. 21 (3) (2020) https://doi.org/10.1061/(ASCE)NH.1527-6996.0000387
Wagner, S.M., Bode, C. An empirical investigation into supply chain vulnerability, J. Purch. Supply Manage. 12 (6) (2006) 301–312.
Wang, Q., Kang, W. What are the impacts of COVID-19 on small businesses in the U.S.? Early evidence based on the largest 50 MSAs, Geogr. Rev. 111 (4) (2021) 528–557. https://doi.org/10.1080/00167428.2021.1927731
Watson, M., Xiao, Y., Helgeson, J., Dillard, M. Importance of households in business disaster recovery, Nat. Hazards Rev. 21 (4) (2020), https://doi.org/10.1061/(ASCE)NH.1527-6996.0000393
Webb, G.R., Tierney, K.R., Dahlhamer, J.M. Businesses and disasters: Empirical patterns and unanswered questions, Nat. Hazards Rev. 1 (2) (2000) 83–90.
Xiao, Y., Nilawar, U. Winners and losers: Analysing post-disaster spatial economic demand shift, Disasters 37 (4) (2013) 646–668.
Xiao, Y., Van Zandt, S. Building community resiliency: Spatial links between household and business post-disaster return, Urban Stud. 49 (11) (2012) 2523–2542.
Xiao, Y., Wu, K., Finn, D., Chandrasekhar, D. Community businesses as social units in post-disaster recovery, J. Plann. Educ. Res. 42 (1) (2018), https://doi.org/10.1177/0739456X18804328
Zhang, Y., Lindell, M.K., Prater, C.S. Vulnerability of community businesses to environmental disasters, Disasters 33 (1) (2009) 38–57.

32

DISASTER RECOVERY AND HERITAGE BUILDINGS

David A. Torres Castro

Introduction

Cultural heritage stands as a testament to the rich tapestry of human history, encompassing a diverse array of assets, including artifacts, monuments, sites, and documents. It also extends to intangible legacies, such as traditional practices, folklore, and oral histories. These hold profound significance as bridges between past generations and those yet to come through their deployment and safeguarding in the present. The preservation of cultural heritage, therefore, is of paramount importance as it stands as an act of self-definition, recognition, and cultural continuity (Smith, 2006). However, these assets and their associated principles are seldom exempt from the losses imposed by disasters, particularly in the era of climate change. This chapter delves into the critical endeavor of protecting cultural heritage in disaster situations and exploring the actors, concepts, and strategies that form the bedrock of recovery efforts.

The process of 'patrimonialization' of cultural assets is a social construction based primarily on the attribution of values by a group of people in the present, which makes cultural heritage a product of human interpretation and cultural re-significance of the past (Burón Díaz, 2020). This perspective refrains from characterizing these values—the ones carrying high cultural significance—as inherent to the objects themselves, recognizing instead that their dynamic and evolving nature is rooted in the relationships they hold with the communities that host them. Still, some values do have an intrinsic nature, as we will examine later. In any case, it is then crucial to acknowledge the pivotal role of community participation in cultural heritage recovery during disasters, particularly of so-called informal responders as they represent the foundation of locally defined heritage. While formal institutions and experts play important roles indeed, the active involvement of local communities, often the first to respond in crises, brings a depth of the understanding of affected heritage and an intimate knowledge of the cultural terrain that can make a difference in disaster recovery. Therefore, from this viewpoint, this chapter will explore how different types of response and recovery efforts arise through different stakeholders and how they interact with cultural heritage.

To understand the process of responding, recovering, and safeguarding heritage assets during disasters, we first need to briefly examine the conditions of vulnerability they face to hazards, both natural and anthropogenic. Addressing the challenges imposed by vulnerability is crucial to

DOI: 10.4324/9781003414834-43

developing effective recovery strategies, ranging from the community to the institutional scale. Next, we will delve into pre-disaster preparedness conditions, focusing on the recognition of hazards, meticulous documentation, installed capacities, as well as social appropriation and attachment to cultural assets as conditional elements that enhance the effectiveness and efficiency of response and early recovery actions. As disasters unfold, a rapid response becomes central in minimizing damage, where collaboration between responders, prioritization of highly valuable cultural assets, and employing rapid assessment techniques are decisive.

Approaching these topics from a people-based as opposed to an overly technical perspective, I intend to highlight the interconnectedness and dynamism between cultural heritage, communities, and institutions, wherein local actors play a vital role in the overall response process aimed at preserving cultural landscapes during times of crisis. In that sense, this chapter, I hope, will serve as a revision of some of the key principles needed in the construction of a good response and early recovery strategy. Furthermore, this chapter serves as an invitation to embrace a paradigm shift, transitioning from a technocentric focus to a more holistic approach when it comes to engaging stakeholders in the response and recovery phases. Working toward sustainable cultural heritage is a task that transcends borders, disciplines, and institutionalism, thus demanding a continuous commitment and collective action from communities, governments, and experts worldwide.

Cultural Heritage and the Real Causes of Disasters

Cultural heritage has been understood in the past as a simple but rather capricious cultural emanation that runs through history, static, and predefined by previous generations. From that perspective, it is often seen as a spectrum of tangible artifacts, encompassing architectural structures, buildings and sites, and—more recently—objects and archival materials that serve as a legacy of historical, cultural, and societal narratives from the past. Today's broader, somehow progressive definition, arising largely from anthropological disciplines and critical heritage studies, settles it as a construction based on the selection of assets made in the present (Burón Díaz, 2017). This alternative notion widens the definition to also encompass intangible assets such as "practices, representations, expressions, knowledge, skills—as well as the instruments, objects, artifacts and cultural spaces associated therewith—that communities, groups and, in some cases, individuals recognize [as significant]", as established in UNESCO's *Convention for the Safeguarding of the Intangible Cultural Heritage* (2003, 5). The act of acknowledging something as important within one's culture, which in turn means there are other things that are considered alien, comes from the need to satisfy political, economic, and cultural needs and discourses. Some of these, as Burón Díaz states, might be discourses of identity and self-determination but also of tradition cooptation, negotiation or confrontation, imposition or legitimization of a group or minority, even nation-building discourses, all dependent to multiple and sometimes juxtaposed contexts. In any case, despite its monumentality, imposed or adopted materiality, and often centuries-old life and tradition, cultural heritage in all its dimensions is not impervious to the dynamic and changing nature of the planet, and specially to the deliberate actions of human agency that create what we call disasters.

Disasters are very complex phenomena that arise from the confluence of several components. Some of them certainly include environmental agents, exposed systems, and triggering events, even though not exclusively. It has been said by several scholars, notably from the so-called natural sciences, that environmental phenomena such as earthquakes, characterized by ground perturbations, or floodwaters, relentless in their ingress, have demonstrated their capacity to erode the structural integrity of ancient sites, dismantle architectural grandeur, and destroy cultural relics throughout history. For instance, a seismic phenomenon disrupts the equilibrium of foundational substrates, as engineering and architectural experts state, causing structural instability and leaving behind a wave of destruction and decay. Similarly, storms cause floods, material degradation, and serious instability of a construction. In that sense, environmental phenomena have been associated with disasters and often used as interchangeable terms to refer to them. Yet, beyond the seemingly capriciousness of natural forces, this approach leaves questions that the understanding of natural events as synonyms of disasters cannot explain alone, particularly in terms of loss of heritage. Some of them, for instance, ask why do we still have centuries-old buildings in areas that are very highly prone to earthquakes, or why do some culturally invaluable objects survive in regions annually affected by storms and others do not? Furthermore, why do natural phenomena have contrasting impacts in different contexts while being similar in magnitude? And most assuredly, are environmental agents truly the cause of disasters?

The answers to these questions lie, as it was accurately stated by Blaikie et al. (1994), in the pre-disaster conditions, mostly human-induced, that have in fact been noted as baseline causes of disasters. In the realm of cultural heritage, this is not different. The creation of disaster risk is largely caused by the anthropogenic interventions, whether structural or otherwise, that render cultural heritage—as well as other associated systems—more or less susceptible to perilous circumstances. In other words, disasters are the result of vulnerable conditions mainly created or exacerbated by societies as a consequence of their own actions, decisions, and processes and not of environmental hazards (Blaikie et al., 1994; Quarantelli, 1998). But what exactly is vulnerability? How is it materialized in cultural heritage?

Vulnerabilities of Cultural Heritage

To gain a comprehensive understanding of the creation of, response to, and recovery from disaster risk in terms of cultural heritage, it is fundamental to seize the meaning of vulnerability. Nonetheless, it is worth noting that the concept of vulnerability in the realm of cultural heritage is quite an extensive topic. Here, I only intend to introduce the matter as an element of importance within the context.

According to Blaikie et al., vulnerability is associated with "the characteristics of a person or group and their situation that influence their capacity to anticipate, cope with, and recover from the impact of a natural hazard" (1994, 9). But also "the conditions determined by physical, social, economic and environmental factors or processes which increase the susceptibility of an individual, a community, *assets* or *systems* to the impacts of hazards" (UNDRR, 2023, 24, italics added). From these two broad definitions we can highlight that vulnerability is the essential agent (or group of agents) that creates the necessary conditions so that a threat, regardless of its magnitude or intensity, can become a disaster. In the realm of cultural heritage, this means a series of prevailing circumstances, dynamics, and synergies that define the scenarios in which heritage assets previously exist, defining the characteristics and scope of possible consequences. In that sense, in order to understand the wide array of responses that can take place and their capacity to prevent damage, it is imperative to unpack the different dimensions of vulnerability, where the consideration of physical, social, economic, cultural, environmental, and institutional instances is fundamental.

Broadly speaking, cultural heritage faces two types of vulnerabilities: intrinsic and extrinsic, for which several subtypes can be delineated (i.e., material, cultural, institutional, and organizational vulnerabilities). Intrinsic vulnerability, in general, hinges upon the susceptibility of materials and techniques employed in the construction or fabrication of cultural assets, and in the function for which they were made. While extant social science perspectives contend that intrinsic vulnerability is non-existent, or in any case is an epiphenomenon of social construction, here I proffer an alternative proposition. In the context of cultural heritage, I believe, a cultural facet underlies the choice of materials, techniques, and construction methods within a cultural environment determined by their availability, first, and symbolism, second, in which cultural and representational values become inexorably intertwined. The selection of those materials and techniques, as well as symbolic resources, thereby gives rise to a constellation of intrinsic characteristics—material and symbolic—that not only shape the design and production of cultural objects, yielding distinct purposes, meanings, and functions but also fundamentally constitute a set of inherent vulnerabilities.

An illustrative example can be drawn from the construction of a choir book in the 16th century, where the constituent elements, such as leather, thread, wood, and ink, are particularly susceptible to external agents such as water or fire. The concept of intrinsic vulnerability alludes to the fact that the decision to employ those specific materials is largely the result of the circumstances and resources available in that specific moment in time and space, while only partially attributable to human agency. In other words, it is inherently cultural and contextual, and not strictly intentional or a product of human deliberation. Moreover, the utilization of those materials may have a symbolic root. Thus, the use of alternative materials would yield an entirely different object, diverging from the idea of a 'choir book' as conceived in that moment in time. For instance, if stone was used instead of wood or leather in the construction of the book, the object would not become a 'choir book' but a stellae or a tablet, or would not have the same symbolic attributes as a wooden one.

Generally speaking, intrinsic vulnerability can be subdivided into two different forms: 'material vulnerability,' which pertains to the inherent properties and susceptibility to a given hazard of the materials employed, as it was explained in the above example. On the other hand, 'cultural vulnerability' is a factor closely linked to the function or usage that was conceived for the object and arises when an object is initially designed to serve a purpose that ultimately will lead to its decay. Using once again the choir book as an ongoing example, we can say that it was originally crafted for consultation or reading, albeit this practice will likely result in some degree of page wear. Should its use be discontinued to forestall deterioration (reducing the risk of observing damage), the object would forfeit its primary function. This delineates cultural vulnerability as something dependent on the interpretation of the initial circumstances that were established for the objects.

The concept of cultural vulnerability brings some ideas from the Wittgensteinian notion of things, which suggests that objects have no meaning in isolation or an independent existence. Instead, they derive their significance through their participation in events or situations. That is, solely from their interaction with other objects and the overall context in which they exist. Similarly, rather than considering elements of cultural heritage in isolation, we need to understand their existence intrinsically linked to the context and interaction with other elements and with the community that produces them. For example, a historic building is significant not only as a sheltered space but also from its relationship with the environment, its original function, and its importance in the narrative of the community in a particular moment in history. Likewise, intrinsic vulnerability is rooted in a complex network of cultural interactions in the same way cultural heritage exists in a network of connections, links, and social, cultural, and symbolic associations. Therefore, both materiality and non-materiality are essential to fully understand the nature of an object and its intrinsic vulnerability almost in an Aristotelian fashion, where the 'substance' (materiality) provides the base or substrate upon which 'qualities' (non-materiality) are expressed. Then, just as with the choir book, when examining heritage and its vulnerability it is crucial to consider how individual elements were physically and culturally made, and how they were (or may still be) connected and contribute to the cultural system of a community.

Whether material or cultural, an intrinsic susceptibility refers to a part of vulnerability that is deep-rooted to the nature of the object and thus cannot be fully eradicated. Therefore, it will exist in a residual form as long as the object endures. This does not mean that the object remains in a constant state of disaster or emergency, nor that the vulnerable conditions will inevitably evolve to become one. But it certainly delineates the prevailing circumstance in which a cultural asset exists and, most important for our purpose here, the context in which different actors could respond to protect it, as we will explore later. Luckily, these inherent conditions can indeed be counteracted by other means, for example, by mitigating or limiting the exposure—the geographic and temporal space in which a hazard can influence an asset—to certain phenomena that threatens it, thus interrupting the famous risk generation equation proposed by Blaikie et al. (1994) that 'risk = vulnerability + hazard + exposure'. Therefore, if a cultural object can remain out of the influence of a hazard, that is, it is not exposed to it, the risk of observing damage, no matter how vulnerable it may be, is considerably reduced. Thus, we must settle that intrinsic vulnerability is not necessarily deleterious, although it is an integral aspect of cultural assets, as damage materializes only when an object is exposed to a hazard to which it is vulnerable. Consequently, strategic measures, such as minimizing exposure to specific hazards, serve to mitigate intrinsic vulnerability.

Now, the existence of a vulnerability of an intrinsic nature suggests the presence of an extrinsic one. Certainly, for cultural heritage this second group of factors largely stem from environmental agents such as humidity or temperature, which can exert both direct and indirect influence on the preservation of a cultural asset. But most significantly they come from the existing socio-economic frameworks and organizational infrastructure dedicated to its protection. In this regard, extrinsic vulnerability encompasses a range of factors, distinct from the materials, techniques, and cultural purpose that we have established as intrinsic, that can worsen or improve the existing conditions of risk. These include ineffective storage and display conditions, inadequate preservation and handling practices, the absence of contingency plans, suboptimal performance of institutional structures overseeing the conservation and safeguarding of cultural assets, or insufficient maintenance afforded by a community toward its heritage, among many others. As such, the origins of this type of vulnerability can be tracked largely to agents not directly related to the object and its attributes, but to the broader context in which a particular asset exists. Furthermore, this classification can be divided into various subtypes, for instance, organizational or institutional vulnerability, among others.

Institutional vulnerability is a subtype that refers to weaknesses or inefficiencies within the institutional framework (or frameworks) established to safeguard cultural assets, whether they be of national or subnational scope. This could be originated by a poor articulation of actions assigned to individuals by policy, protocols, norms, and procedures, what is often called 'nonstructural' measures, or by a lack of physical infrastructure (i.e., fire hydrants or security surveillance systems), commonly grouped as 'structural' measures. While an institutional framework can enable the implementation of protective measures of both structural and nonstructural forms, it can also inadvertently introduce vulnerabilities through oversights, negligence, or limited capacities. For instance, if a museum does not have access to smoke or CO_2 detectors in a given room, or a training program for their use, then it is effectively increasing the vulnerability of the objects contained in that space against a fire-related hazard.

Additionally, the concept of institutional vulnerability extends to the susceptibility of the institution itself. This encompasses the entity's physical as well as operational infrastructure, be it technical or administrative, that are also exposed to hazards, meaning that in a disaster the institution may also suffer damage to its facilities. These incidents not only have the evident impact of physical damage and temporary suspension or reduction of services offered to the public, but they might have indirect effects on the management of cultural heritage as the available human, material, and financial resources are significantly reduced or become non-operational. Thus, the capacity to effectively respond or recover could also be significantly limited.

Similarly, organizational vulnerability emerges as the adverse conditions arising from the insufficient or non-existent measures that a group, whether it be formal or informal, big or small, might arrange to safeguard and protect cultural heritage. For example, a community without a local committee that could oversee heritage maintenance, which by the way could act as a liaison with institutions after a disaster, showcases organizational vulnerability, and may lead to physical deterioration due to a lack of preventive actions and increased susceptibility to hazards such as earthquakes. These factors of extrinsic vulnerability arise from a series of conditions that exist largely before a disaster, such as organizational strength, installed capacity, and the level of appropriation of the community and their willingness to adapt, which in turn determine the degree of responsiveness of that group.

Lastly, we have seen how vulnerability stems largely from pre-disaster conditions that can extend far in time, from days to several years or even decades. Even though this review of vulnerability is far from comprehensive, it illuminates the intricate interplay of intrinsic and extrinsic conditions that underlie heritage assets. These conditions, as we will explore next, are the bedrock of the types and scope of responses that may be deployed by a range of stakeholders when speaking of recovering cultural heritage. By delineating the pivotal influence of pre-existing circumstances such as cultural values and material choices, as well as organizational backgrounds and institutional strength, this section has paved the road to understand disaster response and recovery.

Response, Adaptation, and Only Then, Recovery

Disasters have been conceptualized frequently through what has been called the 'disaster cycle' model, which has long served as a foundational framework. This model envisions disaster management as a cyclical process, typically commencing with a disaster event, going through a series of distinct phases, including mitigation, preparedness, response, and recovery, and concluding with another potential disaster occurrence. The disaster cycle model gained prominence in the 1970s and 1980s as a response to the increasing frequency and impact of disasters and had emphasis primarily on humanitarian relief. However, while it has been a valuable tool in disaster management for decades, it has also faced criticism for its linear and cyclical representation. Some scholars argue that it may inadvertently downplay the crucial role of the underlying causes and root vulnerabilities that create disaster risk, as it tends to oversimplify the complex dynamics surrounding disasters. Moreover, this model commonly aims to 'build back better', without addressing the question 'better for whom?', not actually tackling the root causes that turned the hazard into a disaster, as pointed out by Chmutina and Cheek (2021). In that sense, the persistence of the typical disaster cycle model has undervalued the pre-disaster activities associated with mitigation and preparedness, and rather put excessive emphasis on post-disaster ones (i.e., response and recovery). This has led to the framing of disasters as one-off, 'extraordinary' events with impacts that can be 'managed' through technocratic solutions, rather than as processes that are neither linear nor cyclical but multidimensional and socially constructed, thereby demanding socio-political solutions.

Over the years, the predominant technocratic approach has configured disaster risk management as a merely reactive tool, even in the field of cultural heritage, framing disasters as unpredictable and 'natural' events. As such, today's risk management strategies largely rest on technological and notably financial tools that stand supposedly as preparations against disasters, when in fact they are reactive measures. This has translated into a disproportionate investment in financial devices such as the so-called risk transferring schemes, rather than in vulnerability reduction strategies.

For instance, there are cultural heritage institutions that boast of having disaster risk plans, when in reality they have only taken out insurance policies. This strategy has several drawbacks. First, there is the fact that, as with any insurance prime, financial resources are available only once a disaster has occurred and cultural assets are damaged or even completely destroyed, sometimes even without the possibility of recovery. But second and most important is the fact that these tools were created to reconstruct infrastructure such as schools, hospitals, dams, and other infrastructure, but not culturally significant assets. While budget availability is undoubtedly necessary in the recovery process of every affected sector, in terms of cultural heritage it does not offer any solution to the loss of cultural values, traditions, and practices, it does not foster social re-cohesion, nor it hastes cultural recovery at all. On the contrary, it transfers the responsibility of doing so to the communities by themselves. In this context, the disaster risk cyclical model as well as the hazard approach have been challenged by experts, for instance, Bosher et al. (2021), who assert that through alternative models with proper risk reduction measures, disasters can be significantly mitigated or even prevented, stirring from the cyclical form.

Adapting to Respond Efficiently

According to Alexander (2002), the response phase is the group of actions taken directly before, during, or immediately after a disaster to reduce its impacts and meet the needs of the affected population. Response efforts are therefore closely linked to the adaptive capacity of the affected community or group during pressing circumstances, which in turn refers to adjustments in ecological, social, economic, and most important here, cultural systems. These appear in response to actual or expected stimuli and their effects or impacts on the cultural and social spheres. But response efforts also depend on several determinants, mainly but not limited to access to information, availability of economic resources and sufficiently skilled people, installed infrastructure, existing institutions (and their own installed capacities and competencies), social capital, and, interestingly proposed by Smit and Pilifosova (2003), equity. Additionally, the physical state of the affected heritage assets (when speaking of tangible heritage, of course) must be considered, as it has been observed by Torres and García (2022), because it significantly defines its level of material vulnerability. Lack of one or several of these factors can impede or have an important influence in the downgrade of responsiveness by limiting the range of possible adaptation strategies, many of which—if not all of them—are interdependent.

In a similar way, a successful response requires recognition of the necessity to adapt, understanding of the available options, and the capacity to assess and ponder them, in addition to the proficiency to implement most suitable ones in a timely manner. The ability to respond promptly, then, is often subject to pre-disaster conditions such as the strength and effectiveness of social networks, and the robustness of infrastructure in partnership with the stability and capacities of stakeholders to manage new risks associated with emergent conditions, as suggested by Smit and Pilifosova (2003). In terms of a successful heritage-oriented response, it is crucial to consider the following key conditions in addition to the aforementioned: recognizing and assessing existing and potential hazards, including primary and secondary ones, both for the population and for heritage assets. But most importantly, it is crucial to understand that, for circumstances to become potentially threatening to culturally significant assets they must cross a conceptual threshold, posing a threat not only to the physical assets but notably to their cultural functionality, and thus to the cultural stability of the community itself. The threshold, nonetheless, is highly dynamic as well as multilayered, as it is defined by each layer of responsive stakeholders according to the discourse and concept of heritage each of them wield (Torres, 2021).

Another condition for a successful response in terms of cultural heritage is to maintain a thorough documentation of all the carried-out actions, as far as conditions permit. This assures that preservation efforts will be less likely to be derailed by rotation of personnel, rapidly changing circumstances or the long waiting periods that are usually observed during recovery due to bureaucratic processes. But also, it is important to acknowledge actual capabilities from all actors, which in turn means taking responsible decisions about the distribution of resources, prioritization, and safety.

Finally, but not less important, is the principle to foster social engagement at all stages of response and early recovery. Only by acknowledging that all stakeholders may have a different but valid understanding of what is culturally significant—and therefore worth protecting—is the satisfaction of real cultural needs genuinely achievable, as one must recognize that heritage is not only about physical objects or buildings but also about significance, functionality, and continuity, as we explored in previous pages. All these factors play a pivotal role in enhancing the effectiveness and efficiency of initial response and early recovery efforts as pre-established conditions. As disasters unfold, the swiftness of response becomes paramount in mitigating damage to cultural assets. This entails seamless collaboration among responders (formal or informal), prioritizing the safeguarding of highly valuable assets, and deploying rapid assessment techniques to make timely and informed decisions.

Preparing to Face Disasters

So far, we have explored the overall conditions that can make a response efficient and successful in terms of protecting cultural heritage. As we have seen, these shall be carried out mostly directly before, during or immediately after a hazard triggers. Nonetheless, the actions that can be implemented to strengthen the preparation of a good response can extend long before any hazard is on the horizon. This implies that various measures can and should be taken during the preparation phase. One of them that we consider key is the existence of an orderly and accessible documentation and inventory system. This is a fundamental tool in reducing the vulnerable conditions before a disaster occurs. It involves creating detailed records and inventories of cultural heritage assets, both tangible and intangible, including information regarding its condition, location, and other relevant considerations. Thus, ensuring that valuable cultural elements are cataloged and can be accounted for in the event of a disaster. Furthermore, it is important to take actions that assure that this information is accessible during a disaster.

Another vital aspect is to establish evacuation protocols. These should focus on crafting comprehensive plans and protocols to guide responses to disasters based on the relevance of the heritage assets under review (prioritization) and the potential loss that is devised (scenarios). It also includes establishing clear procedures for evacuation, communication, and coordination among relevant parties and responders, which takes us to the last point.

Lastly, the creation of training and capacity building actions is essential in enhancing preparedness. This involves providing educational and skill-building opportunities for individuals and groups involved in disaster response, including formal and informal actors. By empowering them with the necessary knowledge and skills, including the particularities and needs of the universe of heritage assets they interact with, they become more effective in implementing preparedness plans and procedures.

Different Layers of Response

In the field of cultural heritage, the range and type of actors involved in the process of responding can be quite a diverse group. Before the establishment of the Disaster Research Center (DRC) at Ohio State University in 1963, the predominant focus in social sciences disaster research was on describing responses at either the individual or community levels (Kreps and Bosworth, 2007; Quarantelli and Dynes, 1977). In the following years, as the DRC gained a deeper understanding of responses, it was found that major disasters are characterized by intricate correlations and complexities, with existing groups and organizations trying to meet the demands of the disaster by creating and using distinct types of organizational adaptations and interconnections. They also observed that new groups and organizations emerge as well, and both existing and new entities become integrated into broader social networks of collective action. This experience led to the creation of what has historically been called the 'DRC typology of organized disaster responses' (Kreps and Bosworth, 2007).

The DRC typology clusters the range of organized responses that can be observed during a disaster into a fourfold model, as shown in Table 32.1. Type I are named 'established organizations' as they exist prior to an event and much of what they do is expected. Type II are called 'expanding organizations' because while much of what they do is expected as well, their core structures change from a small nucleus of professional staff to a much larger unit, for instance, through volunteers, to stand out for the circumstances. Type III are included as 'extending organizations' because while they exist prior to a disaster, much of what they do is not predetermined and focused on disaster response. Finally, type IV are termed 'emergent organizations' because they are created for the circumstances, and both their existence and activities are ad hoc and therefore unique to the event.

Over the years, the model was used to investigate what these organizations do, how they do it, and how they relate structurally to each other as parts of a broader community response system. However, in terms of responding to protect and safeguard cultural heritage, there are slight but significant differences with this model. The first one is that we are not dealing with responses directly focused on protecting human life or life-threatening injuries, nor are they attempting to meet life-or-death needs. Even though cultural necessities are also basic needs, as they are considered basic

Table 32.1 Typology of Organized and Non-Affiliated Disaster Responders

Type	*Organizational structure*	*Tasks*	*Characteristics*
Established Organizations	Regular	Pre-existing	Already operating for emergency response or management purposes, they only focus their regular effort on coping with the situation
Expanding Organizations	Non-regular	Pre-existing	Anticipate emergency tasks and increase their structure (e.g., by taking on additional staff) in order to increase their capacity and cope with the situation
Extending Organizations	Regular	Developed	No emergency-related operations beforehand but modify their regular scope to take advantage of their already existing operational structure
Emergent Organizations	Non-regular	Developed	With no previous knowledge or experience in the field of emergency management and response, they voluntarily engage in response tasks
Non-Affiliated Responders	Non-regular	Developed	Activities come from individual initiatives and do not become part of any of the above organizations

Source: Table created by author, adapted from: Kreps and Bosworth (2007) and Torres (2021).

human rights by UNESCO, these do not necessarily have a strictly defined timeline for resolution, except in cases where cultural practices are crucial for the functioning of a community, for instance, traditional fishing or farming. The second difference is that in many cases cultural heritage can be recovered (at least at the most basic level) by the members of the community themselves and do not depend strictly on emergency services to appear. Nonetheless, these types of responders, named 'informal' or 'non-affiliated' by some scholars (see Lorenz et al., 2017 or Whittaker et al., 2015), are not included in the DRC typology, and often overlooked in typical response models, although they are often the first to come forth in a disaster situation.

The various types of actions carried out by a range of responders can be viewed as strata within a multilayered model, as shown in Figure 32.1, by combining and utilizing them as complementary components of an overarching strategy. The assertiveness of this model can only be achieved by acknowledging that heritage is defined differently by each stakeholder or actor, depending on the predominant values and principles it embodies, sometimes even implicitly. Each layer of actors may in turn become an emergency responder that will act toward the specific type of heritage it recognizes. In accordance with the UNESCO *Convention on the Protection and Promotion of the Diversity of Cultural Expressions* (2005), this means that every community has the right to consider and determine its cultural heritage, whether tangible or intangible, as significant and deserving of rescue in a disaster situation by themselves. Therefore,

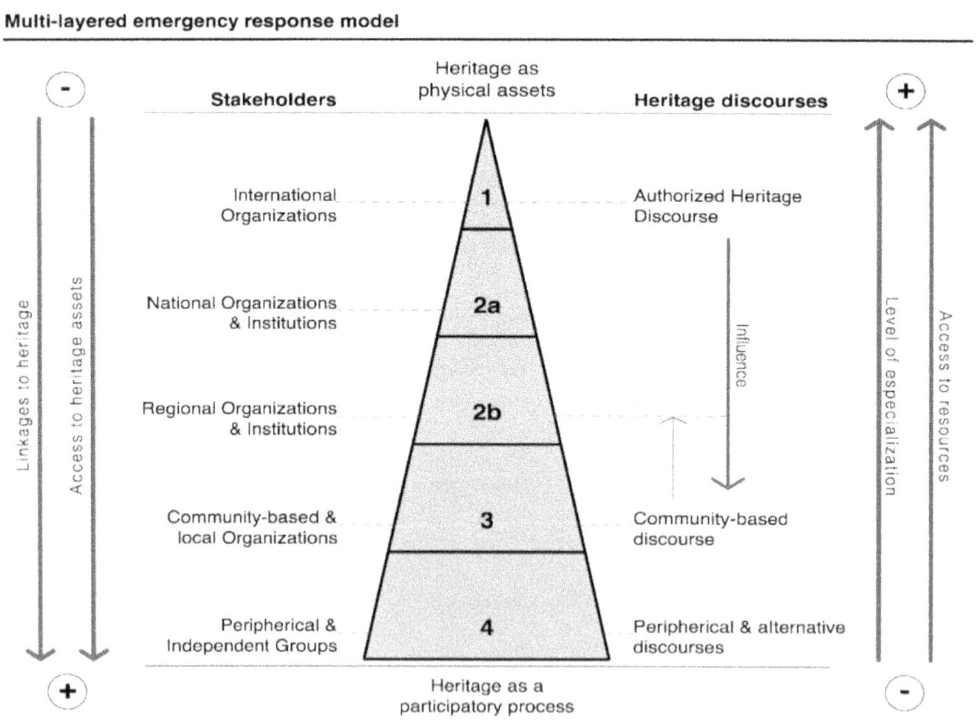

Figure 32.1 Multi-Layered Model of Disaster Responders and Their Scope

Source: Taken from Torres (2021).

a truly integrated and people-based strategy should incorporate mechanisms to ensure that all stakeholders have access to a set of recovery efforts.

Cultural Sensitivity and Ethical Considerations

As a last aspect to consider during the recovery and safeguard of cultural heritage during disasters, I would like to underline the need for cultural sensitivity and a few ethical considerations. These are pivotal components of disaster response that encompass an awareness and respect for the diverse cultural values, beliefs, and practices of communities affected by a disaster. Recognizing and valuing these differences is essential for establishing effective communication among actors and responders and to ensure that response efforts are inclusive and considerate of cultural sensitivities.

Cultural sensitivity involves understanding that affected communities may have varying cultural backgrounds, languages, traditions, and beliefs. Sometimes even these can be agents of cultural vulnerability. This knowledge enables responders to communicate clearly, accurately, and respectfully of cultural nuances, and to implement actions according to them. It also emphasizes the importance of avoiding actions or assumptions based on cultural stereotypes or biases that could unintentionally harm or offend others.

In the same way, ethical considerations in disaster response prioritize the preservation of human dignity and uphold the rights of affected individuals. This includes their right to safety, well-being, and self-determination, which becomes particularly relevant in terms of cultural heritage because, contrary to what some discourses argue, not every person has equal access to cultural resources. Ethical response efforts, therefore, should aim to be equitable and fair in terms of cultural representation, providing consistent levels of support regardless of cultural, social, or economic differences, as Seglah and Blanchard (n.d.) clearly point out. Additionally, transparency and accountability are paramount, ensuring that actions, decisions, and resource allocation are clearly communicated and that responders are held responsible for their outcomes. Furthermore, an ethical response seeks to avoid the exploitation of vulnerable populations or communities for research, media coverage, or other purposes. Instead, it places the well-being and cultural recovery of individuals and communities at the forefront of response efforts, moving beyond the limited vision of the generically called heritage 'reconstruction'.

From a practical point of view, achieving cultural competence involves engaging local leaders and community members who possess a deep understanding of the cultural context, as we have mentioned earlier. For instance, utilizing cultural liaisons or interpreters to bridge language and cultural barriers is crucial for effective communication. Additionally, it entails respecting cultural practices, rituals, and traditions, including considerations related to food, dress, religious observances, or even burial rites.

In summary, cultural sensitivity and ethical considerations during the response phase of disaster management are imperative for creating response efforts that are inclusive, respectful, and ultimately effective. They serve to build trust, facilitate communication, and promote the well-being and dignity of affected individuals and communities.

Conclusions

As we have seen, there is no doubt that heritage in all its dimensions is susceptible to the impact of disasters, which in turn represent one of the major threats for the well-being of modern societies. There are recent disaster management models emphasizing that, in terms of cultural heritage, the true causes of disasters are primarily a result of human interventions that render it more or less susceptible to hazardous circumstances. In that context, there is a need to move beyond reactive strategies that notably rely on financial mechanisms as 'preventive measures' (e.g., insurance policies) into models that truly address underlying vulnerabilities.

Response efforts during a disaster are closely tied to the adaptive capacity of affected communities, involving adjustments in various systems in response to stimuli and their impacts, including cultural and social systems. Factors such as access to information, social cohesion, access to economic resources, trained personnel, or institutional capacity among others, play a crucial role in determining the effectiveness of response strategies, as well as the physical condition of heritage assets. A successful response in terms of cultural heritage requires that actions are taken long before a disaster occurs, such as recognizing and assessing hazards, understanding the threshold at which these become threatening to cultural assets, and maintaining thorough documentation of actions taken. Additionally, prioritizing social engagement at all stages of response and early recovery is essential for effectively protecting culturally significant assets in the near future. This involves recognizing diverse perspectives on cultural significance and making informed decisions about resource allocation and safety measures.

Last, we have explored the need for cultural sensitivity and ethical considerations in response efforts, recognizing diverse cultural values, beliefs, and practices. This involves clear communication, avoiding cultural stereotypes, and upholding cultural diversity and human rights perspective, as well as engaging local leaders and community members with cultural expertise. Overall, these considerations are crucial for inclusive, respectful, and effective response efforts in safeguarding cultural heritage during disasters.

The ideas proposed here suggest that disasters should be seen as multidimensional and socially constructed events that require socio-political solutions. There are timelines within the management of a disaster that can be dynamically adjusted based on specific circumstances and needs, including the planning, display, and adjustment of response and cultural recovery efforts. As such, it is of greater relevance to understand that the actions carried out during the response phase play a fundamental role in the definition of the subsequent short- and long-term recovery phases. Furthermore, they have the potential to reconfigure the functioning and cultural cohesion of a community by selecting which cultural heritage assets will be salvaged and those that will be left behind in an act of resignation.

References

Alexander, D. (2002). *Principles of Emergency Planning and Management*. Oxford University Press.

Blaikie, P., Cannon, T., Davis, I., & Wisner, B. (1994). *At Risk*. Routledge.

Bosher, L., Chmutina, K., & van Niekerk, D. (2021). Stop going around in circles: Towards a reconceptualisation of disaster risk management phases. *Disaster Prevention and Management: An International Journal*. https://doi.org/10.1108/DPM-03-2021-0071

Burón Díaz, M. (2017). Comunidades, patrimonio y arqueólogos: Relaciones entre municipios e instituciones culturales de Oaxaca en el periodo indigenista. *Estudos Ibero-Americanos*, *43*(1), 67. https://doi.org/10.15448/1980-864X.2017.1.24196

Burón Díaz, M. (2020). Cabezas y pájaros: La construcción y restitución del patrimonio en Nueva Zelanda. *Locus: Revista de História*, *26*(2), 210–234. https://doi.org/10.34019/2594-8296.2020.v26.31022

Chmutina, K., & Cheek, W. (2021). "Building back better" is neoliberal post-disaster reconstruction. *Disasters*, 1–24.

Kreps, G. A., & Bosworth, S. L. (2007). Organizational adaptation to disaster. In *Handbook of Disaster Research*. Springer.

Lorenz, D. F., Schulze, K., & Voss, M. (2017). Emerging citizen responses to disasters in Germany. Disaster myths as an impediment for a collaboration of unaffiliated responders and professional rescue forces. *Journal of Contingencies and Crisis Management*, 358–367. https://doi.org/10.1111/1468-5973.12202

Quarantelli, E. L. (1998). *What Is a Disaster?: Perspectives on the Question* (E. L. Quarantelli, Ed.). Routledge.

Quarantelli, E. L., & Dynes, R. R. (1977). Response to social crisis and disaster. *Annual Review of Sociology*, *26*(2), 23–49.

Seglah, H. A., & Blanchard, K. (n.d.). *Invisible Again: Hyper-Marginialized Groups and Disaster Data*. DRR Dynamics. https://irp.cdn-website.com/cde3424c/files/uploaded/Invisible%20again%20report%20upload.pdf

Smith, L. (2006). *Uses of Heritage*. https://doi.org/10.4324/9780203602263

Smit, B., & Pilifosova, O. (2003). From adaptation to adaptive capacity and vulnerability reduction. In J. B. Smith, R. J. T. Klein, & S. Huq, *Climate Change, Adaptive Capacity and Development* (pp. 9–28). Imperial College Press. https://doi.org/10.1142/9781860945816_0002

Torres, D. A. (2021). Community organization for the protection of cultural heritage in the aftermath of disasters. *International Journal of Disaster Risk Reduction*, *60*(November 2020), 102321. https://doi.org/10.1016/j.ijdrr.2021.102321

Torres, D. A., & García, P. (2022). Mapeo de riesgos participativo para patrimonio cultural. El caso de Tehuantepec, Oaxaca, México. *Intervención*, *1*(25), 271–321. https://doi.org/10.30763/Intervencion.267.v1n25.46.2022

UNESCO. (2003). *Convention for the Safeguarding of the Intangible Cultural Heritage*. https://doi.org/10.1093/rsq/10.2.88

UNESCO. (2005). *Convention on the Protection and Promotion of the Diversity of Cultural Expressions* (pp. 326–342). https://doi.org/10.1163/ej.9789004164543.1-760.38

United Nations Office for Disaster Risk Reduction (UNDRR). (2023). *GAR Special Report: Mapping Resilience for the Sustainable Development Goals*. Geneva, UNDRR. https://www.undrr.org/media/88718

Whittaker, J., McLennan, B., & Handmer, J. (2015). A review of informal volunteerism in emergencies and disasters: Definition, opportunities and challenges. *International Journal of Disaster Risk Reduction*, *13*, 358–368. https://doi.org/10.1016/j.ijdrr.2015.07.010

33

DISASTER RECOVERY AND MENTAL HEALTH

BreAnne A. Danzi, Jessica T. Kelly, and Jessy L. Thomas

Introduction

Disasters have widespread impacts on individuals and communities. In addition to addressing the basic needs that emerge in the aftermath of a disaster, it is crucial to consider the mental health of people impacted by disasters. Disasters can prompt a wide range of psychological reactions across individuals. Following disasters, most people will not experience significant long-term adverse psychological or social sequelae – a trend of resiliency that is observed in humans internationally (Neria, Nandi, and Galea, 2007). Some people may experience transient distress and they may return to normal functioning within a few months of the disaster. However, for a subset of individuals, disaster initiated or exacerbated psychological disorders can lead to significant impairment in daily functioning and last for years (Bonanno et al., 2010). Furthermore, given the level of disruption caused by disasters within a localized geographic region, local mental health services can quickly become overwhelmed, creating challenges with access to appropriate treatment. In the aftermath of a disaster, a key task is to identify individuals at greatest risk for long-lasting maladaptive outcomes and to connect them with appropriate mental health services. This chapter outlines common mental health problems that may emerge after disasters, factors that predict risk versus resiliency trajectories, and prevention and intervention strategies for addressing mental health needs after disasters.

Mental Health Consequences of Disasters

Although there are many psychological disorders that could be exacerbated by exposure to a disaster, some of the most common mental health problems observed post-disaster include posttraumatic stress disorder (PTSD), depression, anxiety, and substance use problems (Forman-Hoffman, Riley, and Pici, 2005; North et al., 2004 Flory et al., 2009; Beaglehole et al., 2018; Robinson et al., 2022; Keya et al., 2023). Additionally, rates of suicide and intimate partner violence (IPV) increase in the aftermath of disasters (Safarpour et al., 2022; Rahmani et al., 2023; Cannon and Kovach, 2023). At a population level, PTSD symptoms gradually improve over time, whereas depressive and anxiety symptoms tend to remain elevated for years after the disaster (Newnham et al., 2022). However, mental health trajectories can vary substantially across individuals, from

DOI: 10.4324/9781003414834-44

resiliency to patterns of chronic distress. Further, while some people may experience distress immediately following the disaster and show gradual improvement over time, others may exhibit a delayed onset presentation where clinically significant symptoms are not immediately present but gradually worsen into a psychological disorder over time (Bonanno et al., 2010).

PTSD

PTSD is a psychological disorder that emerges in the context of life threat, including experiencing or witnessing threatened death or injury. Disasters are major sources of life threat and are among the events most likely to cause trauma exposure in large groups of people at the same time. Although feeling some distress immediately following disaster exposure is normal, people may be diagnosed with PTSD if they exhibit a pattern of distress and behavioral changes that last longer than one month (WHO, 2019; APA, 2022;).

PTSD is characterized by re-experiencing the disaster, such as in the form of intrusive memories or nightmares, and avoiding reminders of the disaster (e.g., staying away from places associated with the disaster). PTSD also involves experiencing heightened reactivity, such as sleep difficulties or being on edge, and negative changes to thought patterns and mood (WHO, 2019; APA, 2022).

PTSD is considered the most common disorder that emerges after a disaster (SAMHSA, 2019). Estimated rates of PTSD post-disaster vary and depend on several factors, such as the type, severity, unexpectedness, chronicity, and impact of the disaster as well as the amount of time that has passed since the disaster. Table 33.1 displays prevalence rates of PTSD that have been reported in adult samples following different disaster types. Although there is variability depending on the nature of the disaster and sample characteristics, typically PTSD is present in up to 30% of disaster-affected individuals (Bonanno et al., 2010).

Depression

Depressive disorders are characterized by feelings of sadness, loss of interest or pleasure in activities, irritability, changes in appetite or weight, sleep disturbances, emptiness, or fatigue that significantly affect an individual's ability to function (APA, 2022). Depressive symptoms can exist on a range of severity but can be highly debilitating in individuals with severe symptoms. Depression can also increase the risk for suicidal thoughts and behaviors (WHO, 2019).

Depression frequently co-occurs with PTSD. However, while symptoms of disaster-related PTSD must be preceded by and tied to the disaster experience, depressive symptoms could exist before, during, or after the disaster. For people with pre-existing depression, disasters can worsen symptoms due to creating conditions that increase susceptibility to depression. For example, a natural disaster might result in increased sedentary behavior due to impeded access to parks and gyms, reduced access to nutritious foods due to supply chain disruptions, or decreased participation in enjoyable activities due to community closures. People may experience significant losses and may have more difficulty accessing their usual social support networks after a disaster, which can increase risk for depression (Tang et al., 2014). Depression is considered the second most common disorder post-disaster (SAMHSA, 2019). See Table 33.1 for prevalence rates of depression by disaster type.

Anxiety

Anxiety disorders involve excessive fear or worry, increased physiological arousal, apprehension, and avoidance of feared stimuli that cause significant impairment in functioning (WHO, 2019).

Table 33.1 Representative Prevalence Rates of PTSD, Depression, and Anxiety in Disaster-Exposed Samples of Adults in Selected Studies

	PTSD		*Depression*		*Anxiety*		*Studies*
	≤6 months (%)	*≥1 year (%)*	*≤6 months (%)*	*≥1 year (%)*	*≤6 months (%)*	*≥1 year (%)*	
Hurricane/ typhoon/ tornado	2–30	14	6–43	–	2–13	–	Wang et al. (2019), Amstadter et al. (2009), Sackey et al. (2023), Chen et al. (2014)
Flood	28	10–25	35	6–11	25	8–14	Mulchandani et al. (2020), Dai et al. (2017), Jermacane et al. (2018), Mason, Andrews, and Upton (2010)
Earthquake	9–34	19–26	23–38	20–26	44	43	Zhang et al. (2011), Acharya Pandey et al. (2023), Cénat, McIntee, and Blais-Rochette (2020), Guo et al. (2014), Kun et al. (2009), Zhang et al. (2012)
Fire	24	16–40	33	13–45	20	23	Agyapong et al. (2018), Laugharne, van der Watt, and Janca (2011), Mao et al. (2022), Bryant et al. (2014), Ritchie et al. (2020)
Tsunami	3–34	16–21	10–30	16	22–37	30	Hollifield et al. (2008), Thavichachart et al. (2009), van Griensven (2006)
Human-made/ technological	7–29	5–73	52	18	45	–	Neria, Nandi, and Galea (2007), Tohme et al. (2022), Maeda et al. (2016)

Source: Table created by authors.

People with anxiety may find it difficult to control their worry and may experience symptoms such as restlessness, difficulty concentrating, irritability, muscle tension, and sleep disturbances (APA, 2022). Anxiety also frequently co-occurs with PTSD and depression. Elevated levels of anxiety following a disaster can vary depending on a variety of factors such as disaster severity, level of exposure, pre-existing mental health conditions, and available support (Agyapong et al., 2018). Increased symptoms immediately after the disaster are frequently attributed to property damage, feelings of uncertainty, separation from loved ones, lack of support, and disruption of services (Davidson and McFarlane, 2006; Agyapong et al., 2018; Keya et al., 2023). See Table 33.1 for prevalence rates of anxiety by disaster type.

Substance Use

Substance use disorders are characterized by problematic patterns of substance use, including alcohol, tobacco, and drugs. Substance use problems negatively impact occupational, social, or academic responsibilities and can lead to increased substance tolerance and symptoms of withdrawal (APA, 2022). Disaster exposure has been linked to increased substance use. For example, after Hurricane Ike, alcohol use increased from 47% to 53% in individuals most affected by the hurricane (Ma and Smith, 2017). Similarly, after the 9/11 World Trade Center attacks in New York City, 25% of residents reported increased alcohol use, 10% reported increased cigarette use, and 3% reported increased marijuana use (Vlahov et al., 2002).

While disaster exposure has been linked to increased substance use, the relationship between disasters and substance use is complex and depends on a variety of factors such as disaster type and severity, pre-existing mental health conditions, and pre-existing substance use (Goldmann and Galea, 2014). One theory regarding the observed pattern of increased substance use post-disaster suggests that individuals affected by disasters may use substances to cope with the distress and disruption associated with the experience (Alexander and Ward, 2017). For example, after the Oklahoma City bombing, 15% of survivors reported using alcohol to cope with their experiences (North et al., 2004). However, it is important to note that while some individuals may increase substance use following a disaster, it does not always result in a substance use disorder.

Other Adverse Outcomes

Globally, rates of suicidality tend to increase following disasters (Safarpour et al., 2022). A meta-analysis evaluating international suicide rates found that rates rose from 13.6% pre-disaster to 16.7% post-disaster, with Western Pacific regions of the globe experiencing the highest rates of post-disaster suicidality (i.e., 23.6%; Safarpour et al., 2022). Risk factors for increased suicidality following disasters include older age, female gender, low socioeconomic status, low social support, serious mental illness, and injury sustained during the disaster (Jafari et al., 2020). Youth are also at increased risk for suicidality post-disaster (Rahmani et al., 2023).

IPV, also commonly referred to as domestic violence, may also increase post-disaster (Cannon and Kovach, 2023). For example, following Hurricane Katrina, women reported an 11.6% increase in psychological victimization and a 4.1% increase in physical victimization, while men reported a 6.4% increase in psychological victimization and reported no change in physical victimization (Schumacher et al., 2010). Increases in IPV are associated with increases in other adverse mental health outcomes such as stress, PTSD, and depression (Cannon and Kovach, 2023). While women and individuals belonging to disadvantaged groups are more likely to experience IPV following a disaster, other factors such as disrupted social networks and low socioeconomic status are also associated with reports of IPV (Bell and Folkerth, 2016; Rao, 2020). However, living in non-urban settings, having a strong sense of community, and having high socioeconomic status appear to serve as protective factors against experiencing IPV following a disaster (Bell and Folkerth, 2016; Rao, 2020).

Mental Health in Children

Children and adolescents are a particularly vulnerable population for developing adverse mental health consequences following a disaster and may have worse post-disaster mental health outcomes compared to adults (Newnham et al., 2022; Norris et al., 2002). Although estimates vary across studies, children and adolescents show high rates of PTSD (e.g., 18%–38%), depression (e.g., 8%–45%), and anxiety (e.g., 12%–32%) following disasters (Kar and Bastia, 2006; Neria, Nandi, and Galea, 2007; Tang et al., 2014; Danzi and La Greca, 2016; Marthoenis et al., 2019). Depressive and anxiety symptoms reported by children are higher than those reported by adults and remain higher across time (Newnham et al., 2022). These disaster-related mental health consequences can negatively impact children's social development (McDermott, Berry, and Cobham, 2012) and academic performance (Gibbs et al., 2019), particularly among populations who may experience social marginalization (Peek and Domingue, 2020). Therefore, it is especially critical to appropriately assess and monitor children's mental health following a disaster so that mental health services and psychosocial support can be implemented accordingly.

Risk and Protective Factors That Predict Psychological Functioning after Disasters

Given the variability in psychological reactions experienced following a disaster, it is helpful to consider factors that may predict risk versus resiliency trajectories for mental health outcomes. Risk factors increase the likelihood that an individual may experience psychological difficulties post-disaster, whereas protective factors buffer against adverse outcomes. Specifically, factors relating to individual demographic characteristics (e.g., age, gender, race/ethnicity, socioeconomic status, and health needs), disaster exposure (e.g., perceived life threat, proximity to the disaster, loss, and life disruption), and the recovery environment (e.g., social support, community support, and management of cumulative life events) have been identified as factors that can serve as risk or protective factors for the adverse mental health consequences of disaster exposure.

Demographic Characteristics

Age is an individual characteristic that may impact risk for negative mental health outcomes following disasters. Youth represent one group that is particularly at-risk for adverse post-disaster functioning (Norris et al., 2002; Newnham et al., 2022). Due to the sensitive nature of early development, youth have been noted to exhibit greater post-disaster symptoms of anxiety, depression, posttraumatic stress, behavioral problems, isolation, and problems with learning and are at greater risk of physical harm or death during disasters (Peek, 2008). Older age is also a vulnerability factor to consider. Older age is associated with heightened symptoms of posttraumatic stress and adjustment-related symptoms following disasters (Parker et al., 2016). Moreover, elderly individuals are at greater risk of injury or death following disasters, a risk that is heightened by factors such as poor communication of disaster-related safety information to older adults, reduced physical and cognitive functioning to effectively prepare and evacuate, and increased reliance on others for care (Heagele and Pacquiao, 2019).

In considering gender, women are at an elevated risk of poor functioning following disasters compared to men. Research into gender differences in post-disaster psychological functioning have noted multiple pre-, peri-, and posttraumatic mechanisms that may contribute to this discrepancy (Bonanno et al., 2010; Juran, 2012). At the same time, the caretaking that women often hold, in addition to the risk for less financial and employment independence, place women at greater risk of harm and poor mental health functioning after disasters (Ashraf and Azad, 2015).

Low socioeconomic status and poverty are also linked with greater difficulty in disaster preparation, greater disaster-related destruction to property and long-term financial hardship, and poor mental health outcomes globally (Morganstein and Ursano, 2020). Additionally, individuals living in countries with few economic resources and high rates of poverty are at risk due to a lack of community resources needed to recover from disasters (Ferrier and Spickett, 2007).

Individuals with pre-existing mental health conditions and disabilities are also uniquely affected by disasters. Those with disabilities or other health needs are at greater risk of injury or death during disasters, experience barriers during evacuation due to the inaccessibility of evacuation resources, impaired access to medical resources, and exhibit greater rates of poor mental health outcomes following disasters (Thomas et al., 2013; Zhou et al., 2015). Of note, youth with complex physical or psychological health needs are a uniquely vulnerable group due to their reliance on caregivers, above and beyond that of other same-aged youth (Peek and Stough, 2010).

Other marginalized groups (e.g., people of color in the United States and indigenous communities in the Philippines) have historically been exposed to significant adverse experiences during natural disasters (Gaillard and Le Masson, 2007; Maurya, 2019). Globally, discriminatory ideologies and social or political practices have largely influenced the lack of opportunity for social and financial advancement among these groups in society. As such, these groups are more likely to live in disaster damage-prone areas, live in lower-quality housing, experience little employment flexibility and financial resources to prepare for and recover from disasters, poor communication of disaster information due to lack of resources or language barriers, greater loss and destruction to property, and greater long-term financial hardship (Thomas et al., 2013; Fatemi et al., 2017; Maurya, 2019). Additionally, these marginalized groups are also less likely to receive mental health support that is culturally and socially informed (Warfa et al., 2014; Maurya, 2019).

Other characteristics that serve as risk factors for adverse reactions to disasters include level of education, literacy, and employment inflexibility. Specifically, individuals with lower levels of education or literacy ability often experience complications with understanding public evacuation-related information (Thomas et al., 2013; Newnham et al., 2022). Further, individuals with inflexible work hours or an inability to take time off are often unable to effectively recover mentally and physically due to financial instability (Thomas et al., 2013; Newnham et al., 2022).

Disaster Exposure

Disaster-related exposure characteristics also pose a risk for poor post-disaster functioning. A growing body of research has investigated the role of *actual life threat*, or direct exposure to a life-threatening event, in contrast with *perceived life threat*, or the belief that one's life is in danger, in the development of psychopathology post-disaster. Research into mental health outcomes following disasters has supported the notion that perceived life threat is linked with greater posttraumatic stress symptoms in both adults and youth (Chan et al., 2016; Pittman et al., 2020) and may even better predict clinically significant adverse post-disaster functioning than actual life threat (Pittman et al., 2020).

Proximity to a traumatic event and personal loss and life disruption may also impact post-disaster functioning. Specifically, individuals who experienced a traumatic event in close physical proximity are more likely to exhibit greater mental health issues than those who were not directly exposed (Bonanno et al., 2010). Similarly, individuals who experienced greater life disruption (e.g., losing one's home in a tornado) or loss (e.g., losing a loved one in a hurricane) during a disaster are also likely to have worse mental health outcomes (Bonanno et al., 2010).

Recovery Environment

While it is important to note that all previously discussed demographic and exposure-related factors may serve to influence one's recovery following disasters, additional factors related to the recovery environment are important to address. Access to mental health care following disasters is critical, especially for those experiencing clinically elevated symptoms. However, such services are often impacted in the aftermath of disasters due to limited transportation and infrastructure damage. Access to these services are further limited for those with limited mobility, individuals and families experiencing poverty, and other marginalized groups (Peek and Stough, 2010; Fatemi et al., 2017; Morganstein and Ursano, 2020).

Social support, or positive interpersonal relationships more broadly, is also a critical factor for protecting against adverse psychological consequences of trauma and enhancing individual resiliency following disasters (Kaniasty et al., 2019). Additionally, community support and community resiliency play an important role in increasing the likelihood that a community will recover from disaster which, in turn, has an immense influence on the well-being of individuals within the community. Specifically, community resiliency in the aftermath of disasters is an important factor in the individual psychological and physical well-being of community members. However, community resilience is often highly dependent on a number of factors, including economic status and community resources (Gil-Rivas and Kilmer, 2016).

Preparation and Prevention of Mental Health Problems Post-Disaster

Many post-disaster interventions have been designed to treat mental health symptoms with the goal of preventing the development of a psychological disorder (Goldmann and Galea, 2014). Interventions may be designed for different phases of the disaster aftermath, with immediate intervention promoting safety and stability to promote healthy coping after the disaster (Bryant and Litz, 2009). In general, timely mental health intervention is critical for efficient response and prevention (Pfefferbaum et al., 2012). While psychological debriefing or critical incident stress debriefing has historically been used immediately after a disaster, it is no longer recommended as this method does not prevent psychopathology and may even contribute to worsening symptoms as it requires disaster-exposed individuals to describe the event and their emotional responses, which may strengthen the memories of the trauma (Norris et al., 2002; McNally, Bryant, and Ehlers, 2003; Bryant and Litz, 2009). Instead, preventing mental health difficulties after disasters should involve efforts related to preparedness, response, and recovery.

To prevent individuals from developing mental health problems post-disaster, it is important that communities are prepared to address the potential psychological and emotional reactions of individuals. For example, mental health services should be integrated into the healthcare emergency response plan prior to disasters so that professionals have an established framework to provide immediate support (Roudini, Khankeh, and Witruk, 2017). The framework should include a chain of command, identifying local resources, and providing disaster-specific training (Palinkas, 2015). This type of training can increase resiliency, and self-sufficiency, and enable communities to meet the mental health needs of the population (NAM, 2015). It can be administered to emergency personnel, community partners, and community members. Individuals who are trained can then implement the procedures that aid in mitigating mental health problems following a disaster (NAM, 2015). Furthermore, the response plan should include collaborating with external organizations, continuously assessing and monitoring risk, psychological support, and procedures for communicating and sharing information (Palinkas, 2015).

One existing program that is designed to prepare community leaders to respond to emergencies is the Incident Command System, a standardized approach to managing an incident that allows the integration of community resources, including personnel, equipment, and procedures, and organizes communication into a common operating structure (FEMA, 2013). It facilitates response coordination across internal and external organizations, both public and private, and establishes a shared process for planning and managing resources (FEMA, 2013). Countries with integrated community-based mental health services are better prepared for a mental health response in case of a disaster (IASC, 2014). Overall, it is critical to include mental health personnel in disaster preparedness planning to effectively prevent mental health problems post-disaster.

Additionally, it is important to establish community support programs as many adverse mental health outcomes that occur following a disaster are either directly or indirectly related to disrupted interpersonal relationships (Palinkas, 2015). Disasters often lead to individuals being separated from friends and family, sometimes temporarily while repairs are in progress or permanently if individuals are forced to relocate from their community (Gallagher et al., 2016; Kaniasty et al., 2019). While disruption in daily life may be unavoidable after a disaster, communities can develop plans to prepare for such disruption that aid in facilitating interpersonal support. For example, communities should identify socioeconomic resources, rituals, and networks, including faith-based organizations that are specific to their community's needs, to help promote a collaborative response during a disaster (Palinkas, 2015). Furthermore, communities should identify vulnerable populations and include underrepresented groups in disaster planning discussions. Incorporating input from these groups can help facilitate response actions that are designed according to the principle of 'cultural exchange' (Palinkas and Soydan, 2012). This concept promotes collaboration between local and global organizations that allow experts to identify solutions that are culturally appropriate to those being served. It allows for the option of meaningful alternatives to traditional survival activities that do not neglect or minimize a cultural value (Palinkas and Soydan, 2012). Additionally, collaborating with outside experts can help provide emotional and material support to those affected by a disaster by facilitating interactions between communities which can help aid recovery efforts.

When considering how to address the mental health needs of the community following a disaster, it is important to include vulnerable populations in those discussions. For example, children are at an increased risk of adverse mental health outcomes when exposed to a natural disaster (La Greca and Silverman, 2009; Danzi and La Greca, 2016). Poor access to services and stigma may contribute to the unmet needs of children post-disaster. Therefore, programs that target children, such as schools and community centers, can help meet these needs by providing psychosocial support (NAM, 2015). Furthermore, the stepped-care approach provides evidence of reducing symptoms and impairment in children (Marks, 2010). This approach offers services along a continuum of care that begin with less intensive interventions such as communication strategies before building to more intensive interventions (McDermott and Cobham, 2014).

Treatments for Mental Health Problems Post-Disaster

Individuals who are experiencing or are at risk for developing psychological disorders post-disaster may benefit from intervention. Different interventions may be most appropriate for different groups based on the timing post-disaster and the severity level of the psychological difficulties. Some interventions (e.g., Psychological First Aid) may be appropriate for many people and can promote coping skills that would be useful for anyone affected by a disaster, whereas other interventions (e.g., Cognitive Behavioral Therapy) are most appropriate for the minority of individuals who are exhibiting high levels of distress for an extended time post-disaster. Although there are many different inventions that may be helpful following a disaster, some commonly used inventions are discussed below.

Psychological First Aid is a widely known evidence-based treatment used to support present and future physical and mental well-being and promote individual resiliency in the immediate aftermath of disasters (Ruzek et al., 2007). Psychological First Aid is particularly helpful for providing mental health care in the immediate aftermath of a disaster. The intervention includes coping resources as well as practical assistance addressing immediate needs of those affected by disasters and connecting them with sources of support. Psychological First Aid can be provided in a variety of settings and by a variety of disaster-response personnel, which allows immediate support to reach a greater population (Ruzek et al., 2007). There is a strong emphasis on providing culturally sensitive services to diverse groups (Sim and Wang, 2021).

Skills for Psychological Recovery is another brief, empirically supported intervention that can be used with adults and youth exposed to disasters internationally (Wade et al., 2014). This intervention can follow Psychological First Aid and also can be delivered in a variety of settings. Skills for Psychological Recovery utilizes a skills-based approach to support recovery and resiliency by helping disaster-affected individuals develop skills used to enhance self-efficacy in regulating emotional and behavioral reactions following disasters (Wade et al., 2014).

For individuals experiencing higher severity of chronic distress post-disaster, Cognitive Behavioral Therapy is a structured, evidence-based intervention that is efficacious in addressing post-disaster mental health issues, such as anxiety and fear (Brymer, Reyes, and Steinberg, 2012). Specifically, the exposure-based techniques provided in Cognitive Behavioral Therapy or in more intensive exposure therapies (i.e., Prolonged Exposure) are found to be useful in addressing mental health concerns following disasters in populations globally (Lopes et al., 2014). Exposure exercises focus on exposing the individual to feared objects, situations, or people associated with a traumatic event, with the end goal of reducing the individual's distress when faced with direct contact or reminders of the traumatic experience (Lopes et al., 2014). Cognitive Processing Therapy is another well-researched and supported intervention for addressing posttraumatic stress symptoms following disasters. Cognitive Processing Therapy specifically aims to target unhelpful beliefs or thinking patterns about a traumatic event with the goal of changing how the patient thinks about their trauma to reduce distress (Tran et al., 2016).

In youth, evidence-based mental health interventions are available that are found to effectively treat post-disaster symptoms of psychopathology. Cognitive behavioral approaches have been found to be helpful for youth following disasters. Specifically, Trauma-Focused Cognitive Behavioral Therapy (TF-CBT) is an empirically supported treatment for PTSD in youth (Cohen et al., 2000). TF-CBT is structured as an individual and conjoint youth and parent intervention that utilizes a trauma narrative and in-vivo exposure approach as well as supporting the development of emotional and cognitive coping and relaxation skills in culturally diverse groups of youth (Cohen et al., 2000). Cognitive Behavioral Intervention for Trauma in Schools (CBITS) is a group-based intervention similar to TF-CBT; however, it offers support to youth in school settings (Jaycox et al., 2010). Prolonged Exposure is another evidence-based intervention for youth following disasters that utilizes exposure exercises to help build resiliency and manage fear responses following traumatic events (La Greca and Silverman, 2009). When selecting an intervention, it is important to consider the context and characteristics of the youth (Danzi and La Greca, 2021).

Conclusion

Although many people are resilient or experience only short-term psychological distress after a disaster, some people may exhibit distress reactions that can be highly impairing and last for years after the disaster. It is crucial that communities and organizations adequately prepare to address mental health needs that may arise in the wake of a disaster. Mental health problems that commonly emerge post-disaster include PTSD, depression, anxiety, and substance use problems. Awareness of the risk and protective factors that may help predict which individuals are most vulnerable to maladaptive outcomes can help with triaging resources to facilitate their access to services. Communities should prepare a disaster emergency response plan that integrates mental health services and encourage professionals to complete trainings in interventions (e.g., Psychological First Aid and Skills for Psychological Recovery) that can aid with mental health recovery after disasters.

References

Acharya Pandey, R., Chalise, P., Khadka, S., Chaulagain, B., Maharjan, B., Pandey, J., Nepal, J., & Pandey, C. (2023). Post-traumatic stress disorder and its associated factors among survivors of 2015 earthquake in Nepal. *BMC Psychiatry*, *23*(1). https://doi.org/10.1186/s12888-023-04836-3

Agyapong, V.I., Hrabok, M., Juhas, M., Omeje, J., Denga, E., Nwaka, B., Akinjise, I., Corbett, S.E., Moosavi, S., Brown, M., Chue, P., Greenshaw, A.J., & Li, X.-M. (2018). Prevalence rates and predictors of generalized anxiety disorder symptoms in residents of Fort McMurray six months after a wildfire. *Frontiers in Psychiatry*, *9*. https://doi.org/10.3389/fpsyt.2018.00345

Alexander, A.C., & Ward, K.D. (2017). Understanding postdisaster substance use and psychological distress using concepts from the self-medication hypothesis and social cognitive theory. *Journal of Psychoactive Drugs*, *50*(2), 177–186. https://doi.org/10.1080/02791072.2017.1397304

American Psychiatric Association (APA). (2022). *Diagnostic and Statistical Manual of Mental Disorders* (5th ed., text revision). https://doi.org/10.1176/appi.books.9780890425787

Amstadter, A.B., Acierno, R., Richardson, L.K., Kilpatrick, D.G., Gros, D.F., Gaboury, M.T., Tran, T.L., Trung, L.T., Tam, N.T., Tuan, T., Buoi, L.T., Ha, T.T., Thach, T.D., & Galea, S. (2009). Posttyphoon prevalence of posttraumatic stress disorder, major depressive disorder, panic disorder, and generalized anxiety disorder in a Vietnamese sample. *Journal of Traumatic Stress*, *22*(3), 180–188. https://doi.org/10.1002/jts.20404

Ashraf, M.A., & Azad, M.A. (2015). Gender issues in disaster: Understanding the relationships of vulnerability, preparedness and capacity. *Environment and Ecology Research*, *3*(5), 136–142. https://doi.org/10.13189/eer.2015.030504

Beaglehole, B., Mulder, R.T., Frampton, C.M., Boden, J.M., Newton-Howes, G., & Bell, C.J. (2018). Psychological distress and psychiatric disorder after natural disasters: Systematic review and meta-analysis. *The British Journal of Psychiatry*, *213*(6), 716–722. https://doi.org/10.1192/bjp.2018.210

Bell, S.A., & Folkerth, L.A. (2016). Women's mental health and intimate partner violence following natural disaster: A scoping review. *Prehospital and Disaster Medicine*, *31*(6), 648–657. https://doi.org/10.1017/s1049023x16000911

Bonanno, G.A., Brewin, C.R., Kaniasty, K., & Greca, A.M. (2010). Weighing the costs of disaster. *Psychological Science in the Public Interest*, *11*(1), 1–49. https://doi.org/10.1177/1529100610387086

Bryant, R.A., & Litz, B. (2009). Mental health treatments in the wake of disaster. In Y. Neria, S. Galea, & F. H. Norris (Eds.), *Mental Health and Disasters* (pp. 321–335). Cambridge: Cambridge University Press.

Bryant, R.A., Waters, E., Gibbs, L., Gallagher, H.C., Pattison, P., Lusher, D., MacDougall, C., Harms, L., Block, K., Snowdon, E., Sinnott, V., Ireton, G., Richardson, J., & Forbes, D. (2014). Psychological outcomes following the Victorian Black Saturday bushfires. *Australian & New Zealand Journal of Psychiatry*, *48*(7), 634–643. https://doi.org/10.1177/0004867414534476

Brymer, M.J., Reyes, G., & Steinberg, A.M. (2012). Disaster behavioral health for children and adolescents—Best practices for preparedness, response, and recovery. In J.L. Framingham & M.L. Teasley (Eds.), *Behavioral Health Response to Disasters* (pp. 143–158).

Cannon, C.E., & Kovach, K. (2023). Intimate partner violence and disasters: A review of the literature. *Traumatology*. https://doi.org/10.1037/trm0000491

Cénat, J.M., McIntee, S.E., & Blais-Rochette, C. (2020). Symptoms of posttraumatic stress disorder, depression, anxiety and other mental health problems following the 2010 earthquake in Haiti: A systematic review and meta-analysis. *Journal of Affective Disorders*, *273*, 55–85. https://doi.org/10.1016/j.jad.2020.04.046

Chan, C.S., Tang, K.N., Hall, B.J., Yip, S.Y., & Maggay, M. (2016). Psychological sequelae of the 2013 Super Typhoon Haiyan among survivor-responders. *Psychiatry*, *79*(3), 282–296. https://doi.org/10.1080/00332747.2015.1129874

Chen, Y., Hsu, W., Lai, C., Tang, T., Wang, P., Yeh, Y., Huang, M., Yen, C., & Chen, C. (2014). One-year follow up of PTSD and depression in elderly Aboriginal people in Taiwan after typhoon Morakot. *Psychiatry and Clinical Neurosciences*, *69*(1), 12–21. https://doi.org/10.1111/pcn.12227

Cohen, J.A., Mannarino, A.P., Berliner, L., & Deblinger, E. (2000). Trauma-focused cognitive behavioral therapy for children and adolescents: An empirical update. *Journal of Interpersonal Violence*, *15*(11), 1202–1223. https://doi.org/10.1177/088626000015011007

Dai, W., Kaminga, A.C., Tan, H., Wang, J., Lai, Z., Wu, X., Xiong, Y., Deng, J., & Liu, A. (2017). Comorbidity of post-traumatic stress disorder and anxiety in flood survivors. *Medicine*, *96*(36). https://doi.org/10.1097/md.0000000000007994

Danzi, B.A., & La Greca, A.M. (2016). DSM-IV, DSM-5, and ICD-11: Identifying children with posttraumatic stress disorder after disasters. *Journal of Child Psychology and Psychiatry*, *57*(12), 1444–1452.

Danzi, B.A., & La Greca, A.M. (2021). Treating children and adolescents with posttraumatic stress disorder: Moderators of treatment response. *Journal of Clinical Child and Adolescent Psychology*, *50*(4), 510–516.

Davidson, J.R., & McFarlane, A.C. (2006). The extent and impact of mental health problems after disaster. *The Journal of Clinical Psychiatry*, *67*(Suppl 2), 9–14.

Fatemi, F., Ardalan, A., Aguirre, B., Mansouri, N., & Mohammadfam, I. (2017). Social vulnerability indicators in disasters: Findings from a systematic review. *International Journal of Disaster Risk Reduction*, *22*, 219–227. https://doi.org/10.1016/j.ijdrr.2016.09.006

Federal Emergency Management Agency (FEMA). (2013). *Crisis Counseling Assistance and Training Program Guidance – CCP Application Toolkit*. https://dhhr.wv.gov/healthprep/programs/behavioralhealth/Documents/FEMA%20CCP%20Toolkit.pdf

Ferrier, B.A., & Spickett, J.T. (2007). Natural disasters in developing countries: An environmental health perspective. *Asia Pacific Journal of Public Health*, *19*(1_suppl), 18–24. https://doi.org/10.1177/101053950701901s04

Flory, K., Hankin, B.L., Kloos, B., Cheely, C., & Turecki, G. (2009). Alcohol and cigarette use and misuse among Hurricane Katrina Survivors: Psychosocial risk and protective factors. *Substance Use & Misuse*, *44*(12), 1711–1724. https://doi.org/10.3109/10826080902962128

Forman-Hoffman, V., Riley, W., & Pici, M. (2005). Acute impact of the September 11 tragedy on smoking and early relapse rates among smokers attempting to quit. *Psychology of Addictive Behaviors*, *19*(3), 277–283. https://doi.org/10.1037/0893-164x.19.3.277

Gaillard, J.C., & Le Masson, V. (2007). Traditional societies' response to volcanic hazards in the Philippines. *Mountain Research and Development*, *27*(4), 313–317. https://doi.org/10.1659/mrd.0949

Gallagher, H.C., Richardson, J., Forbes, D., Harms, L., Gibbs, L., Alkemade, N., MacDougall, C., Waters, E., Block, K., Lusher, D., Baker, E., & Bryant, R.A. (2016). Mental health following separation in a disaster: The role of attachment. *Journal of Traumatic Stress*, *29*(1), 56–64. https://doi.org/10.1002/jts.22071

Gibbs, L., Nursey, J., Cook, J., Ireton, G., Alkemade, N., Roberts, M., Gallagher, H.C., Bryant, R., Block, K., Molyneaux, R., & Forbes, D. (2019). Delayed disaster impacts on academic performance of primary school children. *Child Development*, *90*(4), 1402–1412. https://doi.org/10.1111/cdev.13200

Gil-Rivas, V., & Kilmer, R.P. (2016). Building community capacity and fostering disaster resilience. *Journal of Clinical Psychology*, *72*(12), 1318–1332. https://doi.org/10.1002/jclp.22281

Goldmann, E., & Galea, S. (2014). Mental health consequences of disasters. *Annual Review of Public Health*, *35*, 169–183. https://doi.org/10.1146/annurev-publhealth-032013-182435

Guo, S., Tian, D., Wang, X., Xiao, Y., He, H., Qu, Z., & Zhang, X. (2014). Protective effects of social support content and support source on depression and its prevalence 6 months after Wenchuan earthquake. *Stress and Health*, *31*(5), 382–392. https://doi.org/10.1002/smi.2563

Heagele, T., & Pacquiao, D. (2019). Disaster vulnerability of elderly and medically frail populations. *Health Emergency and Disaster Nursing*, *6*(1), 50–61. https://doi.org/10.24298/hedn.2016-0009

Hollifield, M., Hewage, C., Gunawardena, C.N., Kodituwakku, P., Bopagoda, K., & Weerarathnege, K. (2008). Symptoms and coping in Sri Lanka 20–21 months after the 2004 tsunami. *British Journal of Psychiatry*, *192*(1), 39–44. https://doi.org/10.1192/bjp.bp.107.038422

Inter-Agency Standing Committee (IASC). (2014). *Review of the Implementation of the IASC Guidelines on Mental Health and Psychosocial Support in Emergency Settings: How Are We Doing?* https://interagencystandingcommittee.org/sites/default/files/migrated/2018-10/1_iasc_guidelines_on_mhpss_review-2014-final_2_0_0.pdf

Jafari, H., Heidari, M., Heidari, S., & Sayfouri, N. (2020). Risk factors for suicidal behaviours after natural disasters: A systematic review. *The Malaysian Journal of Medical Sciences*, *27*(3), 20. https://doi.org/10.21315/mjms2020.27.3.3

Jaycox, L.H., Cohen, J.A., Mannarino, A.P., Walker, D.W., Langley, A.K., Gegenheimer, K.L., Scott, M., & Schonlau, M. (2010). Children's mental health care following Hurricane Katrina: A field trial of trauma-focused psychotherapies. *Journal of Traumatic Stress*, *23*(2), 223–231. https://doi.org/10.1002/jts.20518

Jermacane, D., Waite, T.D., Beck, C.R., Bone, A., Amlôt, R., Reacher, M., Kovats, S., Armstrong, B., Leonardi, G., James Rubin, G., & Oliver, I. (2018). The English National Cohort Study of flooding and health: The change in the prevalence of psychological morbidity at year two. *BMC Public Health*, *18*(1). https://doi.org/10.1186/s12889-018-5236-9

Juran, L. (2012). The gendered nature of disasters: Women survivors in post-tsunami Tamil Nadu. *Indian Journal of Gender Studies*, *19*(1), 1–29. https://doi.org/10.1177/097152151101900101

Kaniasty, K., de Terte, I., Guilaran, J., & Bennett, S. (2019). A scoping review of post-disaster social support investigations conducted after disasters that struck the Australia and Oceania continent. *Disasters*, *44*(2), 336–366. https://doi.org/10.1111/disa.12390

Kar, N., & Bastia, B.K. (2006). Post-traumatic stress disorder, depression and generalised anxiety disorder in adolescents after a natural disaster: A study of comorbidity. *Clinical Practice and Epidemiology in Mental Health*, *2*, 17. https://doi.org/10.1186/1745-0179-2-17

Keya, T.A., Leela, A., Habib, N., Rashid, M., & Bakthavatchalam, P. (2023). Mental health disorders due to disaster exposure: A systematic review and meta-analysis. *Cureus*. https://doi.org/10.7759/cureus.37031

Kun, P., Chen, X., Han, S., Gong, X., Chen, M., Zhang, W., & Yao, L. (2009). Prevalence of post-traumatic stress disorder in Sichuan Province, China after the 2008 Wenchuan earthquake. *Public Health*, *123*(11), 703–707. https://doi.org/10.1016/j.puhe.2009.09.017

La Greca, A.M., & Silverman, W.K. (2009). Treatment and prevention of posttraumatic stress reactions in children and adolescents exposed to disasters and terrorism: What is the evidence? *Child Development Perspectives*, *3*(1), 4–10. https://doi.org/10.1111/j.1750-8606.2008.00069.x

Laugharne, J., Van de Watt, G., & Janca, A. (2011). After the fire: The mental health consequences of fire disasters. *Current Opinion in Psychiatry*, *24*(1), 72–77. https://doi.org/10.1097/yco.0b013e32833f5e4e

Lopes, A.P., Macedo, T.F., Coutinho, E.S.F., Figueira, I., & Ventura, P.R. (2014). Systematic review of the efficacy of cognitive-behavior therapy related treatments for victims of natural disasters: A worldwide problem. *PLoS One*, *9*(10), e109013. https://doi.org/10.1371/journal.pone.0109013

Maeda, M., Ueda, Y., Nagai, M., Fujii, S., & Oe, M. (2016). Diagnostic interview study of the prevalence of depression among public employees engaged in long-term relief work in Fukushima. *Psychiatry and Clinical Neurosciences*, *70*(9), 413–420. https://doi.org/10.1111/pcn.12414

Mao, W., Adu, M., Eboreime, E., Shalaby, R., Nkire, N., Agyapong, B., Pazderka, H., Obuobi-Donkor, G., Owusu, E., Oluwasina, F., Zhang, Y., & Agyapong, V.I. (2022). Post-traumatic stress disorder, major depressive disorder, and wildfires: A fifth-year postdisaster evaluation among residents of Fort McMurray. *International Journal of Environmental Research and Public Health*, *19*(15), 9759. https://doi.org/10.3390/ijerph19159759

Marks, I. (2010). Oxford guides to cognitive behavioural therapy. In J. Bennnet-Levy (Ed.), *Oxford Guide to Low Intensity CBT Interventions*. https://doi.org/10.1093/med/9780199590117.001.0001

Marthoenis, M., Ilyas, A., Sofyan, H., & Schouler-Ocak, M. (2019). Prevalence, comorbidity and predictors of post-traumatic stress disorder, depression, and anxiety in adolescents following an earthquake. *Asian Journal of Psychiatry*, *43*, 154–159. https://doi.org/10.1016/j.ajp.2019.05.030

Ma, C., & Smith, T.E. (2017). Increased alcohol use after Hurricane Ike: The roles of perceived social cohesion and social control. *Social Science & Medicine*, *190*, 29–37. https://doi.org/10.1016/j.socscimed.2017.08.014

Mason, V., Andrews, H., & Upton, D. (2010). The psychological impact of exposure to floods. *Psychology, Health & Medicine*, *15*(1), 61–73. https://doi.org/10.1080/13548500903483478

Maurya, V. (2019). Natural disasters, psychological well-being and resilience: Concerns related to marginalized groups. *International Journal of Research and Analytical Reviews*, 6(1), 270–275.

McDermott, B., Berry, H., & Cobham, V. (2012). Social connectedness: A potential aetiological factor in the development of child post-traumatic stress disorder. *Australian & New Zealand Journal of Psychiatry*, *46*(2), 109–117. https://doi.org/10.1177/0004867411433950

McDermott, B.M., & Cobham, V.E. (2014). A stepped-care model of post-disaster child and adolescent mental health service provision. *European Journal of Psychotraumatology*, *5*(1). https://doi.org/10.3402/ejpt.v5.24294

McNally, R.J., Bryant, R.A., & Ehlers, A. (2003). Does early psychological intervention promote recovery from posttraumatic stress? *Psychological Science in the Public Interest*, *4(*2), 45–79. https://doi.org/10.1111/1529-1006.01421

Morganstein, J.C., & Ursano, R.J. (2020). Ecological disasters and mental health: Causes, consequences, and interventions. *Frontiers in Psychiatry*, *11*, 1. https://doi.org/10.3389/fpsyt.2020.00001

Mulchandani, R., Armstrong, B., Beck, C.R., Waite, T.D., Amlôt, R., Kovats, S., Leonardi, G., Rubin, G.J., & Oliver, I. (2020). The English National Cohort Study of Flooding & Health: Psychological morbidity at three years of follow up. *BMC Public Health*, *20*(1). https://doi.org/10.1186/s12889-020-8424-3

National Academy of Medicine (NAM). ((2015). *Healthy, Resilient, and Sustainable Communities after Disasters: Strategies, Opportunities, and Planning for Recovery*. The National Academies Press.

Neria, Y., Nandi, A., & Galea, S. (2007). Post-traumatic stress disorder following disasters: A systematic review. *Psychological Medicine*, *38*(4), 467–480. https://doi.org/10.1017/s0033291707001353

Newnham, E.A., Mergelsberg, E.L.P., Chen, Y., Kim, Y., Gibbs, L., Dzidic, P.L., Ishida DaSilva, M., Chan, E.Y.Y., Shimomura, K., Narita, Z., Huang, Z., & Leaning, J. (2022). Long term mental health trajectories after disasters and pandemics: A multilingual systematic review of prevalence, risk and protective factors. *Clinical Psychology Review*, *97*, 102203. https://doi.org/10.1016/j.cpr.2022.102203

Norris, F.H., Friedman, M.J., Watson, P.J., Byrne, C.M., Diaz, E., & Kaniasty, K. (2002). 60,000 disaster victims speak: Part I. An empirical review of the empirical literature, 1981–2001. *Psychiatry: Interpersonal and Biological Processes*, *65*(3), 207–239. https://doi.org/10.1521/psyc.65.3.207.20173

North, C.S., Kawasaki, A., Spitznagel, E.L., & Hong, B.A. (2004). The course of PTSD, major depression, substance abuse, and somatization after a natural disaster. *Journal of Nervous & Mental Disease*, *192*(12), 823–829. https://doi.org/10.1097/01.nmd.0000146911.52616.22

Palinkas, L.A. (2015). Behavioral health and disasters: Looking to the future. *The Journal of Behavioral Health Services & Research*, *42*(1), 86–95. https://doi.org/10.1007/s11414-013-9390-7

Palinkas, L.A., & Soydan, H. (2012). *Translation and Implementation of Evidence-Based Practice*. Oxford University Press.

Parker, G., Lie, D., Siskind, D.J., Martin-Khan, M., Raphael, B., Crompton, D., & Kisely, S. (2016). Mental health implications for older adults after natural disasters—A systematic review and meta-analysis. *International Psychogeriatrics*, *28*(1), 11–20. https://doi.org/10.1017/s1041610215001210

Peek, L. (2008). Children and disasters: Understanding vulnerability, developing capacities, and promoting resilience—An introduction. *Children Youth and Environments*, *18*(1), 1–29. https://doi.org/10.1353/cye.2008.0052

Peek, L., & Domingue, S. (2020). Recognizing vulnerability and capacity: Federal initiatives focused on children and youth across the disaster lifecycle. *Government Responses to Crisis*, 61–87. https://doi.org/10.1007/978-3-030-39309-0_5

Peek, L., & Stough, L.M. (2010). Children with disabilities in the context of disaster: A social vulnerability perspective. *Child Development*, *81*(4), 1260–1270. https://doi.org/10.1111/j.1467-8624.2010.01466.x

Pfefferbaum, B., Flynn, B.W., Schonfeld, D., Brown, L.M., Jacobs, G.A., Dodgen, D., Donato, D., Kaul, R.E., Stone, B., Norwood, A.E., Reissman, D.B., Herrmann, J., Hobfoll, S.E., Jones, R.T., Ruzek, J.I., Ursano, R.J., Taylor, R.J., & Lindley, D. (2012). The integration of mental and behavioral health into disaster preparedness, response, and recovery. *Disaster Medicine and Public Health Preparedness*, *6*(1), 60–66. https://doi.org/10.1001/dmp.2012.1

Pittman, L.D., Cutright, N.L., McNeela, L., Awada, S., Pabis, J.M., Kochanova, K., & Shelleby, E. (2020). Coping strategies and psychological symptoms among children on St. Thomas in the aftermath of Hurricanes Irma and Maria. *Journal of Traumatic Stress, 33*(2), 151–160. https://doi.org/10.1002/jts.22471

Rahmani, M., Silverman, A.L., Thompson, A., & Pumariega, A. (2023). Youth suicidality in the context of disasters. *Current Psychiatry Reports, 25*(11), 587–602. https://doi.org/10.1007/s11920-023-01454-4

Rao, S. (2020). A natural disaster and intimate partner violence: Evidence over time. *Social Science & Medicine, 247*, 112804. https://doi.org/10.1016/j.socscimed.2020.112804

Ritchie, A., Sautner, B., Omege, J., Denga, E., Nwaka, B., Akinjise, I., Corbett, S.E., Moosavi, S., Greenshaw, A., Chue, P., Li, X.M., & Agyapong, V.I. (2020). Long-term mental health effects of a devastating wildfire are amplified by sociodemographic and clinical antecedents in college students. *Disaster Medicine and Public Health Preparedness, 15*(6), 707–717. https://doi.org/10.1017/dmp.2020.87

Robinson, E., Sutin, A.R., Daly, M., & Jones, A. (2022). A systematic review and meta-analysis of longitudinal cohort studies comparing mental health before versus during the COVID-19 pandemic in 2020. *Journal of Affective Disorders, 296*, 567–576. https://doi.org/10.1016/j.jad.2021.09.098

Roudini, J., Khankeh, H.R., & Witruk, E. (2017). Disaster mental health preparedness in the community: A systematic review study. *Health Psychology Open, 4*(1), 205510291771130. https://doi.org/10.1177/2055102917711307

Ruzek, J.I., Brymer, M.J., Jacobs, A.K., Layne, C.M., Vernberg, E.M., & Watson, P.J. (2007). Psychological first aid. *Journal of Mental Health Counseling, 29*(1), 17–49. https://doi.org/10.17744/mehc.29.1.5racqxjueafabgwp

Sackey, E.T., Stewart, R.W., Young, J., & Orengo-Aguayo, R. (2023). Disaster exposure and mental health among Puerto Rican teachers after Hurricane Maria. *Journal of Traumatic Stress, 36*(6), 1066–1076. https://doi.org/10.1002/jts.22973

Safarpour, H., Sohrabizadeh, S., Malekyan, L., Safi-Keykaleh, M., Pirani, D., Daliri, S., & Bazyar, J. (2022). Suicide death rate after disasters: A meta-analysis study. *Archives of Suicide Research, 26*(1), 14–27. https://doi.org/10.1080/13811118.2020.1793045

Schumacher, J.A., Coffey, S.F., Norris, F.H., Tracy, M., Clements, K., & Galea, S. (2010). Intimate partner violence and Hurricane Katrina: Predictors and associated mental health outcomes. *Violence and Victims, 25*(5), 588–603. https://doi.org/10.1891/0886-6708.25.5.588

Sim, T., & Wang, A. (2021). Contextualization of psychological first aid: An integrative literature review. *Journal of Nursing Scholarship, 53*(2), 189–197. https://doi.org/10.1111/jnu.12613

Substance Abuse and Mental Health Services Administration (SAMHSA). (2019). *Disaster Technical Assistance Center Supplemental Research Bulletin – Disasters and People With Serious Mental Illness*. https://www.samhsa.gov/sites/default/files/disasters-people-with-serious-mental-illness.pdf

Tang, B., Liu, X., Liu, Y., Xue, C., & Zhang, L. (2014). A meta-analysis of risk factors for depression in adults and children after natural disasters. *BMC Public Health, 14*(1). https://doi.org/10.1186/1471-2458-14-623

Thavichachart, N., Tangwongchai, S., Worakul, P., Kanchanatawan, B., Suppapitiporn, S., Pattalung, A., Roomruangwong, C., & Chareonsook, O. (2009). Posttraumatic mental health establishment of the tsunami survivors in Thailand. *Clinical Practice and Epidemiology in Mental Health, 5*(1), 11. https://doi.org/10.1186/1745-0179-5-11

Thomas, D.S., Phillips, B.D., Fothergill, A., & Blinn-Pike, L. (2013). *Social Vulnerability to Disasters*. CRC press. https://doi.org/10.1201/b14854

Tohme, P., Grey, I., El-Tawil, M.T., El Maouch, M., & Abi-Habib, R. (2022). Prevalence and correlates of mental health difficulties following the Beirut Port Explosion: The roles of mentalizing and resilience. *Psychological Trauma: Theory, Research, Practice, and Policy, 16*(1), 30–38. https://doi.org/10.1037/tra0001328

Tran, K., Moulton, K., Santesso, N., & Rabb, D. (2016). *Cognitive Processing Therapy for Post-Traumatic Stress Disorder: A Systematic Review and Meta-Analysis*. https://doi.org/https://doi.org/10.1371/journal.pone.0109013

van Griensven, F. (2006). Mental health problems among adults in tsunami-affected areas in southern Thailand. *JAMA, 296*(5), 537. https://doi.org/10.1001/jama.296.5.537

Vlahov, D. (2002). Increased use of cigarettes, alcohol, and marijuana among Manhattan, New York, residents after the September 11th terrorist attacks. *American Journal of Epidemiology, 155(*11), 988–996. https://doi.org/10.1093/aje/155.11.988

Wade, D., Crompton, D., Howard, A., Stevens, N., Metcalf, O., Brymer, M., Ruzek, J., Watson, P., Bryant, R., & Forbes, D. (2014). Skills for psychological recovery: Evaluation of a post-disaster mental health training program. *Disaster Health, 2*(3–4), 138–145. https://doi.org/10.1080/21665044.2015.1085625

Wang, Z., Wu, X., Dai, W., Kaminga, A.C., Wu, X., Pan, X., Liu, Z., Wen, S., Hu, S., & Liu, A. (2019). The prevalence of posttraumatic stress disorder among survivors after a typhoon or hurricane: A systematic review and meta-analysis. *Disaster Medicine and Public Health Preparedness, 13*(5–6), 1065–1073. https://doi.org/10.1017/dmp.2019.26

Warfa, N., Chalangary, J., Amour, A.S., Mollica, R., & Bhui, K. (2014). Cultural competence in psychological interventions for psychotrauma following natural disasters: An international perspective. *Clinical Neuropsychiatry, 11*(1), 40–45.

World Health Organization (WHO). (2019). *International Statistical Classification of Diseases and Related Health Problems* (11th ed.). https://icd.who.int/

Zhang, Z., Shi, Z., Wang, L., & Liu, M. (2011). Post-traumatic stress disorder, anxiety and depression among the elderly: A survey of the hard-hit areas a year after the Wenchuan earthquake. *Stress and Health, 28*(1), 61–68. https://doi.org/10.1002/smi.1403

Zhang, Z., Wang, W., Shi, Z., Wang, L., & Zhang, J. (2012). Mental health problems among the survivors in the hard-hit areas of the Yushu earthquake. *PLoS One, 7*(10). https://doi.org/10.1371/journal.pone.0046449

Zhou, X., Song, H., Hu, M., Li, X., Cai, Y., Huang, G., Li, J., Kang, L., & Li, J. (2015). Risk factors of severity of post-traumatic stress disorder among survivors with physical disabilities one year after the Wenchuan earthquake. *Psychiatry Research, 228*(3), 468–474. https://doi.org/10.1016/j.psychres.2015.05.062

34

COVID-19 AND DISASTER RESPONSE AND RECOVERY

Sarah E. Scales, Abdulhadi A. Al Ruwaithi, and Jennifer A. Horney

COVID-19: More Than a Public Health Emergency

The COVID-19 pandemic was declared a global public health emergency by the World Health Organization on January 30, 2020. One day later, on January 31, 2020, the U.S. Secretary of Health and Human Services declared a public health emergency. While the extent and duration of the pandemic's impacts were unique, the response to and recovery from the pandemic highlighted longstanding challenges to strategic preparedness and response planning, including fragmented global and domestic governance, a fragile supply chain characterized by just-in-time procurement, and challenges related to trusted and effective communication.

Over the course of the COVID-19 response, a changing climate highlighted the intersectional risk from the public health emergency and the increasingly frequent and severe anthropogenically fueled disasters to socially and physically vulnerable communities. Public health and other aspects of our preparedness and response infrastructure began the pandemic response with workforce shortages, inadequate information technology systems, and in financial precarity. Legislation that reduced the ability of first responders to take actions to protect the public's health and the politicization of the pandemic led to exceptionally high rates of burnout and job turnover, as well as excess cases and deaths.

COVID-19 and Concurrent Disasters

Hazards and their related disasters do not occur in a vacuum; rather, unrelated but concurrent disasters have compounding impacts. From January 2020 through the end of 2022, the International Disaster Database recorded 1,658 disasters (EM-DAT, 2022). These included 34 non-COVID-19 biological disasters, 97 climatological disasters, 93 geophysical disasters, 649 hydrological disasters, 375 meteorological disasters, and 410 technological disasters. The individual, joint, and reinforcing effects of disasters and COVID-19 fomented new challenges for disaster risk reduction, preparedness, and local, national, and regional response.

DOI: 10.4324/9781003414834-45

Extreme Heat and Wildfires during the COVID-19 Pandemic: Reinforcing Negative Impacts

Health services provision and access are often impeded during emergency events, exacerbating the negative human health impacts of disasters (WHO, n.d.). The COVID-19 pandemic created novel challenges for healthcare access and utilization (Pujolar et al., 2022). Consequently, constraints on access to and utilization of healthcare, COVID-19, and simultaneous disasters created a vicious loop, negatively reinforcing the effects of one another. For example, extreme heat, wildfires, and other hazards decrease air quality and increase risk of negative health outcomes, particularly cardiopulmonary morbidity and mortality (Wettstein et al., n.d.). With limited hospital capacity—especially in the early phases and Delta and Omicron waves of the pandemic—non-COVID consults decreased appreciably, leaving non-critical illnesses and conditions untreated (French et al., 2021).

With increases in unmanaged chronic conditions due to constraints on utilization of and access to healthcare throughout the pandemic, the share of the population at-risk for negative health outcomes in many disaster scenarios increased. Populations at high risk for severe COVID-19, particularly the elderly and those with pre-existing conditions, are similarly at high risk for negative health outcomes related to poor air quality and heat-related illness (Sacks et al., 2011; Xing et al., 2016).

All-cause morbidity emergency visits typically increase during extreme heat events. Throughout the pandemic, this placed an even higher burden on strained emergency departments (EDs) and healthcare providers (Davis & Novicoff, 2018; Sun et al., 2021). In both Portland, Oregon, and Seattle, Washington, during the 2021 Pacific Northwest heatwave, heat and exertion-related emergency consults increased among insured patients (Nori-Sarma et al., 2022). There were 100 and 83 heat-attributed deaths in Washington and Oregon, respectively (R. H. White et al., 2023). In addition to the negative effects of exposure to high concentrations of particulate matter, these effects contributed to COVID-19 exacerbations (Wu et al., 2020; Hernandez Carballo et al., 2022).

Disaster Data and COVID-19

Surveillance, reporting, and communications systems should be equipped to handle multiple concurrent hazards. In the U.S., the Pandemic and All-Hazards Preparedness Act (PAHPA), which delineated objectives for improving "public health and medical preparedness and response capabilities for emergencies," was signed into law in December 2006 (Pandemic and All-Hazards Preparedness Act, 2006). PAHPA addressed eight major areas, including bolstering the core capabilities under the public health and medical response emergency support function (ESF-8).

A key action of ESF-8 is enhanced health surveillance (FEMA, n.d.). Timely, complete, and accessible data are fundamental to measuring the human health impacts of disasters, but, even in isolated events, capturing these data is extremely difficult (Saulnier et al., 2019). Therefore, mortality is frequently underestimated during and immediately following emergencies (Rivera et al., 2020). At a global level, civil registration and vital statistics systems were non-existent or slow and under-equipped prior to the pandemic (Phillips et al., 2014; WHO, 2017). The pandemic placed even more stress on these systems, exacerbating lags in reporting and complicating measuring both direct and indirect deaths throughout the course of the pandemic (Rivera et al., 2020). For example, in March 2022, estimates of global excess COVID-19-related mortality were nearly three times higher than the number of recorded deaths from 2020 to 2021 (Wang et al., 2022).

COVID-19 and Concurrent Disaster Response

Disaster response was severely impacted by the pandemic. Keeping response personnel safe and ensuring timely quarantine, isolation, and treatment when exposed to or contracting COVID-19 was a persistent challenge. More than 6,100 individuals were deployed as part of the Cameron Peak wildfire response in 2020 and 2021, bringing together personnel from across the U.S. and representing 63 distinct crews of firefights, heavy machinery operators, paramedics, and other incident management team members (Metz et al., 2022). Seventy-nine COVID-19 cases representing 17 U.S. states were identified among responders, leading to the isolation of more than 270 individuals (Metz et al., 2022). This outbreak has since been used as the impetus for implementing new screening, testing, and quarantine measures for disaster responders in an attempt to keep personnel safe while still facilitating large response operations.

The pandemic also highlighted challenges with evacuations, sheltering, and personal risk reduction practices during concurrent disasters. In China, a 1,000-year flood occurred at the height of the Delta wave for the flood-affected population (Fan et al., 2022). Lockdowns and limitations on movement, in addition to insufficient planning for concurrent biological and climatological disasters, augmented the negative impacts of both. While social distancing was a critical non-pharmaceutical public health approach for COVID-19 risk reduction, diminished social connectedness impaired important mechanisms for disaster risk reduction more broadly (Mora et al., 2014; Losee et al., 2022). Isolation and lack of social-connectedness is a predictor of negative outcomes in disaster events, including extreme heat events, especially for the elderly and disabled (White et al., 2001).

COVID-19 and Biological Hazards

In the 2022 Pakistan floods, more than 33 million people, nearly half of whom were children, were affected (UNICEF, 2023). The compounding effects of COVID-19 and the 2022 floods precipitated lapses in immunization coverage for high priority, vaccine-preventable diseases such as measles and polio, leading to spikes in annual cases (Ali & Hamid, n.d.). The pandemic itself also prompted changes in healthcare utilization that could contribute to secondary biological disasters. In the U.S., early childhood immunization rates dropped, leaving an estimated 13% of children susceptible to measles (Gambrell et al., 2022). Redoubled efforts to ensure uptake and equitable access to vaccines are critical for reducing the risk of biological disasters in a highly interconnected world.

Breakdowns in the human-environment interface due to destruction of property and inadequate shelter following disasters can increase contact rates for humans with vectors. For example, vector-borne diseases such as dengue, chikungunya, and malaria were prevalent throughout Pakistan prior to the floods. Following the floods, the incidence of cases increased, placing additional strains on the already fragile healthcare system (Vohra et al., 2023). Because of similarities in non-specific clinical presentation for COVID-19 and these high-incidence vector-borne diseases, there was notable risk of misdiagnosis and missed opportunities for timely and appropriate treatment (Rahmat et al., 2023). As the human-environment interface becomes more fragile, emerging and re-emerging infectious diseases will pose significant threats to human health (Baker et al., 2022). The example of competing hazards in Pakistan shows the challenges to be expected as climate change precipitates more frequent and severe disasters.

Opportunities to Improve Disaster Risk Reduction and Response to Concurrent Disasters

Despite the challenges presented by the pandemic for disaster risk reduction, preparedness, and response, there are ample opportunities for improving existing systems. Siloed response mechanisms—for example, public health and disaster management—complicate timely and appropriate allocation of resources, result in duplicated efforts, and decrease efficiency (Potutan & Arakida, 2021). Developing true multi-hazard disaster plans requires cooperation across sectoral silos, and, importantly, the integration of biological hazards into the surveillance systems and early warning mechanisms in place for other natural hazard types. Furthermore, building on successful single-hazard plans can ensure new information and warnings are communicated through trusted and socially acceptable channels. Integrating experiences and lessons learned from disaster response throughout the COVID-19 pandemic will be critical for reducing risks in future compounding, cascading, and concurrent disasters.

The COVID-19 Response

Many traditional and non-traditional first responders played a role in COVID-19 response. In many cases, these roles were altered from a typical response due to the early implementation of non-pharmaceutical control measures such as stay-at-home orders, physical distancing, and closures of schools and non-essential businesses and other services. As a result, early in the pandemic response there were many uncertainties and protocol changes that impacted the usual coping mechanisms for emergency responders. For example, among Emergency Medical Service (EMS) workers in Texas, a qualitative study identified increased workloads, operational changes, and fear of disease transmission as key stressors, while actions taken to limit disease transmission such as physical distancing and orders not to transport patients to overwhelmed facilities intensified the negative health impacts on EMS workers (Mendez & Horney, 2023). Among domestic violence service providers, who were not recognized as first responders by many jurisdictions, demand for services increased. There were both higher rates of abuse and violence and more severe victimization during lockdowns while the capacity to provide shelter and advocacy was dramatically reduced by social distancing and the inability to access personal protective equipment (PPE) for the workforce (Wells et al., 2024).

The Role of Traditional First Responders in COVID-19 Response

First responders typically include police, EMS clinicians, and firefighters. Traditional first responders had a large spectrum of roles during the COVID-19 response, some typical and some atypical. Although traditional first responders share similar work features, for example, working in non-defined settings and performing time-sensitive tasks, their work environment can be highly varied in terms of organizational cultures, roles, tasks, disciplines, and training. The COVID-19 pandemic increased this variation by introducing new types of responses, operational protocols, and coping mechanisms for first responders.

Policing powers, whether handled primarily by public health officials or the courts, have historically been the source of authority for control measures such as isolation, quarantine, and health inspections. While the balance of individual rights and public health police powers were altered to some extent following the terrorist attacks of September 11, 2001, and the subsequent anthrax attacks, in the U.S., law enforcement has typically worked alongside public health in a limited way. Globally, countries such as India, Italy, China, Nigeria, and the United Arab Emirates implemented strict lockdowns that often-required police permission for any travel outside the home (Koh, 2020). Beyond lockdowns early in the pandemic, law enforcement continued to work with public health around logistics and security in healthcare facilities and at testing and vaccination sites performing duties such as monitoring traffic. Like other disasters and emergencies, the COVID-19 pandemic enabled many types of fraud to occur, which would also have been investigated by law enforcement. According to the U.S. Federal Trade Commission, by October 2021, there had been nearly 300,000 reports of fraud related to stimulus payments, treatments, masks, air filters, and other products and services (FTC, 2021).

EMS clinicians focused on patient care and responding to COVID-19 patients in their homes, workplaces, and on the streets. In non-disaster and disaster settings, they provide emergency medical care and mental health support to patients and transport them to the nearest healthcare facility. However, during the COVID-19 pandemic, patient fears, severe constraints to facility capacity, and infection control protocols increased non-transport of patients by nearly 50% in April 2020 compared to the year prior (Satty et al., 2021). The COVID-19 pandemic disrupted not only the work of EMS but also the coping mechanisms paramedics and dispatchers typically relied on to deal with the stressors and vicarious trauma of their work (Mendez & Horney, 2023). This intensified the feeling that EMS workers were unable to connect directly because of the use of virtual screening protocols or extensive PPE, which also limited their ability to interact with patients and family members or perform health education around infection control and other topics.

Firefighters can also act as first responders in medical emergencies and other types of disasters. These duties can vary depending on the firefighters' scope of practice and capabilities (e.g., PPE suitable for infection control). As discussed above, wildland firefighters in Colorado were part of a large outbreak of COVID-19 that raised concerns about the inadequacy of screening, testing, vaccination coverage, and infection control protocols during responses as well as the potential for exposure to smoke to increase vulnerability to, and the severity of, COVID-19 infections (Metz et al., 2022)

The Role of Frontline Healthcare Workers

Frontline healthcare workers include physicians, nurses, ED staff, and other healthcare providers working in healthcare institutions, clinics, and departments that are directly or indirectly involved in patient care. Frontline healthcare workers also include essential, non-patient facing staff in healthcare facilities such as environmental services, technicians, and research or administrative staff. The Advisory Committee on Immunizations Practices defined essential health workers as individuals who might be exposed to infectious agents, whether these individuals are compensated for their work and regardless of whether their potential exposures are direct or indirect (CDC, 2020).

The roles of frontline workers are expansive. They may conduct research and surveillance to measure, explore, and investigate different aspects and issues of the pandemic and the healthcare capabilities designed to combat it. In addition to traditional roles of patient care and research, many frontline workers have been tasked with assuming roles in public education and outreach, advocacy, social support, and mental health and well-being (Brolan et al., 2022). Other potential roles for frontline healthcare workers included COVID-19 response coordination and planning, quarantine management, risk management, development of treatment and resource allocation guidance, policy development and advocacy, and continuity of typical emergency care services (Brolan et al., 2022). Beyond these, quaternary prevention, which aims to protect individuals from medical harm (Martins et al., 2018; Kisling & Das, 2023), became an important element of healthcare as vaccines and treatments were introduced among widespread mis- and disinformation.

Finally, the COVID-19 response required frontline healthcare workers to balance saving lives with risks to their own (and their families') safety and, in some cases, decisions about who should receive care when operating under crisis standards. While the use of triage systems is familiar to providers in disasters and emergencies, aspects of the pandemic made balancing these principles with public health strategies challenging (Bazyar et al., 2019). It is difficult to find a single perspective that would satisfy the ethical considerations given the level of resource scarcity present, particularly at the start of the COVID-19 pandemic (Robert et al., 2020). These types of conflicting decision-making processes can lead to moral stress and negative physical health impacts (Kooli, 2021).

The Role of Public Health

To understand the role of public health throughout the COVID-19 pandemic, it is important to understand the field of public health and the practitioners that comprise its workforce. Public health as a field faced many tangible challenges in fulfilling its role throughout the pandemic. In addition to the stressors of sustained emergency response, the public health workforce was operating at both funding and workforce deficits (Gary-Webb et al., 2020; de Beaumont Foundation, 2021a, 2021c, 2021b). Another significant challenge that has impeded the execution of public health's role is the general public's lack of awareness about the field and its distinction from medicine (Fineberg, 2011). Fineberg (2011) captured the latter, noting that medicine is a "well-established profession with sharp public image" while public health has "multiple professional identities with a diffuse public image." These multiple identities can include but are not limited to the fields of epidemiology, health communications, occupational and environmental health, communicable diseases, chronic diseases, laboratory sciences, and biostatistics (Otok et al., 2018).

Disease surveillance is foundational to public health. Surveillance can take on a range of forms, such as contact tracing, healthcare provider reporting, and laboratory testing. Even prior to the World Health Organization (WHO)'s declaration of the then novel coronavirus as a public health emergency of international concern, public health officials were monitoring cases of non-specific influenza-like illness (CDC, 2023a). When the WHO officially declared COVID-19 a pandemic on March 11, 2020, expanded public health surveillance systems were in place to capture cases and their contacts, many of whom were connected to international travel (Cucinotta & Vanelli, 2020; PAHO, 2020). As case numbers climbed in the early stages of the pandemic, state and local health departments used emergency funding mechanisms to hire tens of thousands of contact tracers to interview suspected and confirmed cases, identify those they might have exposed, and counsel both cases and contacts on quarantine and isolation protocols (Goldberg & Ollstein, 2020).

In addition to contact tracing, public health practitioners were granted expanded privileges for collecting and reporting COVID-19 data through the public health emergency declaration (US Department of Health and Human Services, nd). In May 2020, the U.S. Centers for Disease Control and Prevention (CDC) began the SARS-CoV-2 Sequencing for Public Health Emergency Response, Epidemiology, and Surveillance (SPHERES) program (CDC, 2023a). The Department of Health and Human Services was able to require mandatory lab reporting for all COVID-19 testing, which allowed for the calculation and sharing of key metrics for infection control such as percent positivity (DHHS ASPA, 2023). These data were subsequently used to aid in case investigation and outbreak response. Furthermore, throughout the duration of the public health emergency declaration, hospitals and healthcare providers reported COVID-19 cases daily, aiding in timely reporting of incidence at local, regional, state, and national levels. With the end of the public health emergency declaration on May 11, 2023, reporting frequency requirements decreased to weekly reports.

Public health is a central player in communicating health and safety information during and after public health emergencies. However, public distrust and uncertainty, politicization of response, and conspiracy theories made health communications exceptionally challenging throughout the COVID-19 pandemic (Cooks et al., 2022). To adapt to these challenges, public health had to embrace new communication channels to reach the populations they serve with timely and accurate information. Platforms such as Instagram and Twitter, as well as engagement with social media influencers and celebrities, proved to be important tools for health communication and debunking misinformation (Folkenflik, 2020; Sleigh et al., 2021; Mayberry et al., 2023). Typically, public health works in the background, ensuring that there is no need for the general public to regularly interface with health departments. In previous public health emergencies, specialized subgroups of the public health workforce such as health communicators and public health nurses were the primary interface between the field of public health and the public at large. However, the scale and sustained response to the COVID-19 pandemic propelled the entire field of public health into public-facing roles.

After the Food and Drug Administration granted emergency use authorization for the Pfizer-BioNTech vaccine in December 2020, state and local health departments were responsible for procurement and allocation of vaccines to the population (FDA, 2021). As part of their role in health communications, public health departments were key players in communicating the safety, effectiveness, and importance of receiving the COVID-19 vaccines. Outreach programs and engagement with local businesses (e.g., pharmacies and barber shops) were critical for delivering vaccines into rural and underserved communities (AlMahasis et al., 2023; National Association of County and City Health Officials, n.d.). Vaccination data were reported to the health departments and the CDC (CDC Vaccine Safety, 2023). Using vaccination data, these agencies monitored and reported on vaccine coverage, safety, and effectiveness, which informed updates and changes to recommended indication schedules.

Public health continued to have many other obligations for protecting community and population health during the COVID-19 pandemic. In 2022, spikes in global monkeypox (Mpox) cases—totaling over 30,500 U.S. cases by June 2023—led to enhanced surveillance, targeted risk communications, and vaccine distribution and administration (CDC, 2023c). Public health continued to collect and report routine vaccination data, finding that there were appreciable declines in coverage throughout the pandemic, especially early childhood and maternal vaccinations (Gambrell et al., 2022). Significant work is needed to promote catch-up vaccinations and address hesitancy. Health disparities by race, rural versus urban residence, gender, sexual orientation, and a range of other sociodemographic factors came to the forefront during the pandemic and simultaneous social movements in the U.S. (CDC, 2023b). Across jurisdictional levels, public health entities are responsible for quantifying these disparities, identifying areas for actionable change, and operationalizing their findings to improve the health of all in the communities they serve. These examples are nowhere near comprehensive of the routine duties of public health practitioners; however, they highlight the multifaceted, day-to-day work of the field that continued throughout the pandemic.

COVID-19 Response Impacts

First Responders and Healthcare Workers

The impacts of the COVID-19 response on traditional emergency responders and healthcare workers included higher rates of COVID-19 infection, spreading the infection to their communities and other patients, reductions in household financial security, and other impacts that resulted from increased workloads such as stress, insomnia, burnout, anxiety, panic, depression

(Torales et al., 2020), and suicide ideation and attempts (Danet Danet, 2021; Jahan et al., 2021; Brolan et al., 2022). Several studies conducted during the COVID-19 pandemic response identified factors that would better support frontline healthcare workers during an emergency response (Gibson et al., 2020; Alwidyan et al., 2022; Brolan et al., 2022). For example, ensuring adequate access to PPE and up-to-date knowledge and information about disease transmission was the most important factor for Jordanian EMS workers reporting to work during a novel respiratory infection virus (Alwidyan et al., 2022). Even in highly resourced health systems like the U.S., there was limited access to PPE early in the pandemic, making it difficult to protect healthcare workers and other first responders (Gibson et al., 2020). Forming a provider-centric, multi-disciplinary team was identified as essential for supporting providers during the pandemic in another study (Dutta et al., 2021). Much more will be needed to address the occupational, physical, and mental health impacts of the COVID-19 response among practitioners, the effects of which are likely to linger for years to come (Brolan et al., 2022).

Public Health Workers

The public health workforce was facing severe staffing shortages prior to the start of the pandemic (Gary-Webb et al., 2020; Leider et al., 2023). Both understaffed and underfunded at the beginning of the pandemic, the public health workforce at large had to take on more job roles, work longer hours, and extend work weeks with little to no surge capacity throughout the course of the pandemic (DeSalvo et al., 2021; Kintziger et al., 2021; Scales et al., 2022). The funding mechanisms that supported the on-boarding of contract employees to assist with contact tracing and COVID-19-specific projects expired before the official end of the public health emergency on May 11, 2023, leaving health departments with vast staffing gaps (Mehta et al., 2022). Furthermore, there have been large exoduses from the field, exacerbated by the circumstances faced throughout the pandemic and associated high levels of burnout and negative mental health outcomes (Stone et al., 2021; Leider et al., 2023). Substantial progress has not been made toward addressing workforce deficits, leaving the field in a precarious position for future pandemics and public health emergencies (Kenen, 2023; Leider et al., 2023).

COVID-19 Recovery

The COVID-19 pandemic highlighted what many first responders and those working in public health preparedness and emergency response and recovery have known for decades: a complex interaction of social determinants and physical risk factors play a major role in a range of negative health outcomes in general, and the role of these factors in poor health is magnified in a disaster or emergency setting. Racial and ethnic minority populations had higher risk of both infection with, and mortality from, COVID-19 (Agyemang et al., 2021; Tai et al., 2022). In addition, access to interventions such as vaccines and treatments were also highly inequitable. This was true globally, as many low- and middle-income countries could not access testing, vaccines, or therapeutics due to a range of factors ranging from a lack of funding to protectionist policies in high-income countries (Rees et al., 2022). Within high-income countries, access was limited, particularly among racial and ethnic minority groups. Between December 2021 and May 2022, dispensing rates for the oral antiviral drugs Legevrio and Paxlovid were lowest in the most socially vulnerable ZIP Codes in the U.S. (Gold, 2022), while during the same time period, Black and African American patients in the U.S. were prescribed Paxlovid 35.8% less often than White patients seeking medical care for COVID-19 infection (Boehmer, 2022).

Ensuring Equity in Disaster Recovery

Post-pandemic, many U.S. governmental agencies, including the Office of the Administration for Children and Families and the Substance Abuse and Mental Health Services Administration (SAMHSA), are focused on better defining and quantifying the disproportionate impact of disasters on individuals and communities with "less power and access to resources" (SAMHSA, 2022). It is widely acknowledged that racial and ethnic minorities and individuals living with disabilities have been disproportionately impacted by the COVID-19 pandemic (Shakespeare et al., 2021; Tai et al., 2022). Furthermore, post-pandemic it is generally recognized that some groups are less likely to have full and consistent access to disaster recovery resources, ranging from financial support, social resources, and post-disaster external investment in the community (Howell & Elliott, 2018). In effect, it is simple: exposure to disasters typically widens existing inequities. This is why developing more equitable disease recovery programs that ensure "the provision of community-specific services and resources for disaster survivors that are accessible, and culturally and linguistically tailored to mitigate disparities in health and wellbeing and support resilience" must be a priority of long-term recovery from the COVID-19 pandemic (ACF Office, 2023).

Ensuring equitable recovery is not easy even when it is prioritized. A survey conducted in Puerto Rico one year after Hurricane Maria identified sustained inclusion of communities in the governance of disaster recovery as a key priority, with nearly 60% of respondents identifying equity as the most important consideration for long-term recovery (Kim et al., 2022). While communities may often recognize the disproportionate negative impacts of a disaster event, aid and assistance is rarely able to effectively address these disproportionalities. In the case of COVID-19, long recognized health inequities led directly to negative impacts (both acute and chronic) among historically and systemically marginalized groups. We must do better to prepare for a more equitable response to the next disaster or public health emergency.

A 2021 scoping review conducted by the RAND Corporation found that baseline measures of health needs for sub-populations, namely racial minorities, and related inequities were lacking or absent from FEMA indicators and planning initiatives (Finucane et al., 2021). This lack of data complicates the processes needed to meaningfully address inequities throughout disaster recovery and into the inter-disaster period as we prepare for the next event. While we focus on billion-dollar disasters, little attention is paid to the fact that these losses are largely driven by damage to high-value properties. This can overshadow losses in communities with comparatively lower reported losses due to disproportionate challenges in recovery due to pre-existing socioeconomic circumstances. Focusing on economic measures of disaster impacts can bias not only reporting but also resource allocation. Globally, country-level income and development indices influence the reporting of disaster impact metrics, including both economic and human indicators (Jones et al., 2022). To improve equity in disaster recovery, global disaster data ecosystems must improve to better capture the true frequency and impact of disasters specific to the affected communities.

While self-efficacy is an important component of disaster preparedness, especially for individuals and communities that face disproportionate risks (Adams et al., 2019; Marceron & Rohrbeck, 2019), all individuals and communities are not equally endowed with the resources needed to plan, prepare for, respond to, and recover from disasters and emergencies. In the continuum of disaster risk, steps to improve preparedness at both individual and community levels are important for downstream improvements in response and recovery. Meaningful inclusion of marginalized populations as well as the deconstruction of barriers and systems that have historically and systemically excluded populations from mainstream healthcare and social supports are vital steps to improving equity in disaster recovery (Pearce et al., 2022). Disasters often broaden chasms in the social and financial resources available to affected populations on the lines of race, ethnicity, and disability status. By lessening these disparities prior to disaster events, recovery efforts are less likely to widen these gaps and, with deliberate engagement, can work to further close these gaps.

The Vital Conditions Framework

One way forward is through engagement with the U.S. Department of Health and Human Services' Vital Conditions for Health and Well-Being framework. These conditions include reliable transportation, a thriving natural world, basic needs for health and safety, humane housing, meaningful work and wealth, lifelong learning, and belonging and civic muscle (DHHS OASH, 2022).

While transportation is critical for day-to-day activities, it plays an even more vital role in disaster preparedness, recovery, and response (Alabbad et al., 2021). A study during the 2013–2014 academic year found that, among 558 U.S. school transportation staff, more information and resources were needed for these staff regarding the utilization of local evacuation plans and the services provided by emergency shelters (Olympia et al., 2017). Access to humane housing has far reaching social impacts broadly and in disaster preparedness and recovery specifically. Individuals experiencing homelessness are less likely to be able to evacuate before or during a disaster and less likely to be able to access recovery resources after a disaster (Wexler & Smith, 2015; SAMHSA, 2021). As seen in the example from Puerto Rico, belonging and civic muscle play an important role in disaster recovery. Feeling that one's voice is included in larger conversations and that it has an impact on the realization of baseline and improved recovery undertakings is critical for meaningful and sustainable recovery outcomes.

Conclusion

As climate change precipitates more frequent and severe hazards, public health response to concurrent disasters will become even more important. The COVID-19 pandemic highlighted the need to vastly improve support systems for first responders and the public health workforce in complex emergencies. Furthermore, equity in disaster response and recovery resources for historically marginalized communities must be prioritized across systems to reduce disproportionately negative health outcomes for these communities. The Vital Conditions Framework offers a key set of domains for identifying and improving the health and well-being for all communities. While it is important to document and acknowledge lessons learned from the COVID-19 pandemic, it is even more important to operationalize those lessons to improve health for all moving forward.

References

ACF Office. (2023, July 31). *Disaster Equity Plan Statement*. https://www.acf.hhs.gov/ohsepr/about/disaster-equity-plan-statement

Adams, R. M., Eisenman, D. P., & Glik, D. (2019). Community advantage and individual self-efficacy promote disaster preparedness: A multilevel model among persons with disabilities. *International Journal of Environmental Research and Public Health*, *16*(15), 2779. https://doi.org/10.3390/ijerph16152779

Agyemang, C., Richters, A., Jolani, S., Hendriks, S., Zalpuri, S., Yu, E., Pijls, B., Prins, M., Stronks, K., & Zeegers, M. P. (2021). Ethnic minority status as social determinant for COVID-19 infection, hospitalisation, severity, ICU admission and deaths in the early phase of the pandemic: A meta-analysis. *BMJ Global Health*, *6*(11), e007433. https://doi.org/10.1136/bmjgh-2021-007433

Alabbad, Y., Mount, J., Campbell, A. M., & Demir, I. (2021). Assessment of transportation system disruption and accessibility to critical amenities during flooding: Iowa case study. *The Science of the Total Environment*, *793*, 148476. https://doi.org/10.1016/j.scitotenv.2021.148476

Ali, I., & Hamid, S. (n.d). Implications of COVID-19 and "super floods" for routine vaccination in Pakistan: The reemergence of vaccine preventable-diseases such as polio and measles. *Human Vaccines & Immunotherapeutics*, *18*(7), 2154099. https://doi.org/10.1080/21645515.2022.2154099

AlMahasis, S. O., Fox, B., Ha, D., Qian, J., Wang, C., & Westrick, S. C. (2023). Pharmacy-based immunization in rural USA during the COVID-19 pandemic: A survey of community pharmacists from five southeastern states. *Vaccine*, *41*(15), 2503–2513. https://doi.org/10.1016/j.vaccine.2023.03.002

Alwidyan, M. T., Oteir, A. O., & Trainor, J. (2022). Working during pandemic disasters: Views and predictors of EMS providers. *Disaster Medicine and Public Health Preparedness*, *16*(1), 116–122. https://doi.org/10.1017/dmp.2020.131

Baker, R. E., Mahmud, A. S., Miller, I. F., Rajeev, M., Rasambainarivo, F., Rice, B. L., Takahashi, S., Tatem, A. J., Wagner, C. E., Wang, L.-F., Wesolowski, A., & Metcalf, C. J. E. (2022). Infectious disease in an era of global change. *Nature Reviews Microbiology*, *20*(4), Article 4. https://doi.org/10.1038/s41579-021-00639-z

Bazyar, J., Farrokhi, M., & Khankeh, H. (2019). Triage systems in mass casualty incidents and disasters: a review study with a worldwide approach. *Open Access Macedonian Journal of Medical Sciences*, *7*(3), 482–494. https://doi.org/10.3889/oamjms.2019.119

Boehmer, T. K. (2022). Racial and ethnic disparities in outpatient treatment of COVID-19 — United States, January–July 2022. *MMWR. Morbidity and Mortality Weekly Report*, *71*. https://doi.org/10.15585/mmwr.mm7143a2

Brolan, C. E., Körver, S., Phillips, G., Sharma, D., Herron, L.-M., O'Reilly, G., Mitchell, R., Kendino, M., Poloniati, P., Kafoa, B., & Cox, M. (2022). Lessons from the frontline: The COVID-19 pandemic emergency care experience from a human resource perspective in the Pacific region. *The Lancet Regional Health – Western Pacific*, *25*. https://doi.org/10.1016/j.lanwpc.2022.100514

CDC. (2020, February 11). *Healthcare Workers*. Centers for Disease Control and Prevention. https://www.cdc.gov/coronavirus/2019-ncov/hcp/guidance-risk-assesment-hcp.html

CDC. (2023a, March 15). *CDC Museum COVID-19 Timeline*. Centers for Disease Control and Prevention. https://www.cdc.gov/museum/timeline/covid19.html

CDC. (2023b, March 21). *What is Health Equity?* https://www.cdc.gov/healthequity/whatis/index.html

CDC. (2023c, June 21). *Mpox*. https://www.cdc.gov/poxvirus/mpox/index.html

CDC Vaccine Safety. (2023, June 28). *Enroll in v-Safe after Vaccination Health Checker*. Centers for Disease Control and Prevention. https://www.cdc.gov/vaccinesafety/ensuringsafety/monitoring/v-safe/index.html

Cooks, E. J., Vilaro, M. J., Dyal, B. W., Wang, S., Mertens, G., Raisa, A., Kim, B., Campbell-Salome, G., Wilkie, D. J., Odedina, F., Johnson-Mallard, V., Yao, Y., & Krieger, J. L. (2022). What did the pandemic teach us about effective health communication? Unpacking the COVID-19 infodemic. *BMC Public Health*, *22*(1), 2339. https://doi.org/10.1186/s12889-022-14707-3

Cucinotta, D., & Vanelli, M. (2020). WHO declares COVID-19 a pandemic. *Acta Bio-Medica: Atenei Parmensis*, *91*(1), 157–160. https://doi.org/10.23750/abm.v91i1.9397

Danet Danet, A. (2021). Psychological impact of COVID-19 pandemic in Western frontline healthcare professionals. A systematic review. *Medicina Clinica*, *156*(9), 449–458. https://doi.org/10.1016/j.medcli.2020.11.009

Davis, R. E., & Novicoff, W. M. (2018). The impact of heat waves on emergency department admissions in Charlottesville, Virginia, U.S.A. *International Journal of Environmental Research and Public Health*, *15*(7), 1436. https://doi.org/10.3390/ijerph15071436

de Beaumont Foundation. (2021a). *2021 Findings*. De Beaumont Foundation. https://debeaumont.org/phwins/2021-findings/

de Beaumont Foundation. (2021b). *PH WINS 2021: Rising Stress and Burnout in Public Health*. De Beaumont Foundation. https://debeaumont.org/phwins/2021-findings/stress-and-burnout/

de Beaumont Foundation. (2021c). *Staffing Up: Workforce Levels Needed to Provide Basic Public Health Services for All Americans*. De Beaumont Foundation. https://debeaumont.org/wp-content/uploads/2021/10/Staffing-Up-FINAL.pdf

DeSalvo, K., Hughes, B., Bassett, M., Benjamin, G., Fraser, M., Galea, S., Gracia, J. N., & Howard, J. (2021). Public health COVID-19 impact assessment: Lessons learned and compelling needs. *NAM Perspectives*. https://doi.org/10.31478/202104c

DHHS ASPA. (2023, May 9). *Fact Sheet: End of the COVID-19 Public Health Emergency* [Text]. HHS.Gov. https://www.hhs.gov/about/news/2023/05/09/fact-sheet-end-of-the-covid-19-public-health-emergency.html

DHHS OASH. (2022). *Vital Conditions Framework*. Federal Plan for Equitable Long-Term Recovery and Resilience. https://health.gov/our-work/national-health-initiatives/equitable-long-term-recovery-and-resilience/framework

Dutta, U., Sachan, A., Premkumar, M., Gupta, T., Sahoo, S., Grover, S., Sharma, S., Lakshmi, P. V. M., Talati, S., Biswal, M., Suri, V., Singh, M. P., Ghai, B., Chhabra, R., Bharti, B., Samanta, J., Arora, P., Mohindra, R., Malhotra, S., … & Ram, J. (2021). Multidimensional dynamic healthcare personnel (HCP)-centric model from a low-income and middle-income country to support and protect COVID-19 warriors: A large prospective cohort study. *BMJ Open*, *11*(2), e043837. https://doi.org/10.1136/bmjopen-2020-043837

EM-DAT. (2022). *EM-DAT*. https://public.emdat.be/

Fan, J., Hambly, B. D., Sun, Y., & Bao, S. (2022). Simultaneous compound disasters from COVID-19 and catastrophic flooding. *Journal of Flood Risk Management*, *15*(2), e12779. https://doi.org/10.1111/jfr3.12779

FDA. (2021, August 23). *FDA Approves First COVID-19 Vaccine*. FDA; FDA. https://www.fda.gov/news-events/press-announcements/fda-approves-first-covid-19-vaccine

FEMA. (n.d.). *National Response Framework*. Retrieved July 3, 2023, from https://www.fema.gov/emergency-managers/national-preparedness/frameworks/response

Fineberg, H. V. (2011). Public health and medicine: Where the twain shall meet. *American Journal of Preventive Medicine*, *41*(4), S149–S151. https://doi.org/10.1016/j.amepre.2011.07.013

Finucane, M. L., Warren May, L., & Chang, J. (2021). *A Scoping Literature Review on Indicators and Metrics for Assessing Racial Equity in Disaster Preparation, Response, and Recovery*. RAND Corporation. https://www.rand.org/pubs/research_reports/RRA1083-1.html

Folkenflik, D. (2020, April 1). Dr. Anthony Fauci Is Talking To Just About Anyone About The Coronavirus. *NPR*. https://www.npr.org/2020/04/01/825499536/dr-anthony-fauci-is-talking-to-just-about-anyone-about-the-coronavirus

French, G., Hulse, M., Nguyen, D., Sobotka, K., Webster, K., Corman, J., Aboagye-Nyame, B., Dion, M., Johnson, M., Zalinger, B., & Ewing, M. (2021). Impact of hospital strain on excess deaths during the COVID-19 pandemic—United States, July 2020-July 2021. *MMWR. Morbidity and Mortality Weekly Report*, *70*(46), 1613–1616. https://doi.org/10.15585/mmwr.mm7046a5

FTC. (2021, April 29). *Coronavirus (COVID-19) Pandemic: The FTC in Action*. Federal Trade Commission. https://www.ftc.gov/news-events/features/coronavirus

Gambrell, A., Sundaram, M., & Bednarczyk, R. A. (2022). Estimating the number of US children susceptible to measles resulting from COVID-19-related vaccination coverage declines. *Vaccine*, *40*(32), 4574–4579. https://doi.org/10.1016/j.vaccine.2022.06.033

Gary-Webb, R. T. W., Troisi, C. L., & Tiffany, L. (2020, April 5). A deficit of more than 250,000 public health workers is no way to fight Covid-19. *STAT*. https://www.statnews.com/2020/04/05/deficit-public-health-workers-no-way-to-fight-covid-19/

Gibson, C., Ventura, C., & Collier, G. D. (2020). Emergency medical services resource capacity and competency amid COVID-19 in the United States: Preliminary findings from a national survey. *Heliyon*, *6*(5), e03900. https://doi.org/10.1016/j.heliyon.2020.e03900

Gold, J. A. W. (2022). Dispensing of oral antiviral drugs for treatment of COVID-19 by zip code–Level social vulnerability—United States, December 23, 2021–May 21, 2022. *MMWR. Morbidity and Mortality Weekly Report*, *71*. https://doi.org/10.15585/mmwr.mm7125e1

Goldberg, D., & Ollstein, A. M. (2020, April 21). Tracking the virus may require 300,000 workers. We're nowhere close. – POLITICO. *POLITICO*. https://www.politico.com/news/2020/04/21/tracking-coronavirus-workforce-does-not-exist-197622

Hernandez Carballo, I., Bakola, M., & Stuckler, D. (2022). The impact of air pollution on COVID-19 incidence, severity, and mortality: A systematic review of studies in Europe and North America. *Environmental Research*, *215*, 114155. https://doi.org/10.1016/j.envres.2022.114155

Howell, J., & Elliott, J. R. (2018). As disaster costs rise, so does inequality. *Socius*, *4*, 2378023118816795. https://doi.org/10.1177/2378023118816795

Jahan, I., Ullah, I., Griffiths, M. D., & Mamun, M. A. (2021). COVID-19 suicide and its causative factors among the healthcare professionals: Case study evidence from press reports. *Perspectives in Psychiatric Care*, *57*(4), 1707–1711. https://doi.org/10.1111/ppc.12739

Jones, R. L., Guha-Sapir, D., & Tubeuf, S. (2022). Human and economic impacts of natural disasters: Can we trust the global data? *Scientific Data*, *9*, 572. https://doi.org/10.1038/s41597-022-01667-x

Kenen, J. (2023, March 7). *We're Not Prepared for the Next Public Health Crisis*. POLITICO. https://www.politico.com/newsletters/politico-nightly/2023/03/07/were-not-prepared-for-the-next-public-health-crisis-00085993

Kim, K., Porro, R., Ghimire, J., Durand, L. R., Serrano, R. C., Gonzalez, B., & Yamashita, E. (2022). Equity, participation, and planning for recovery in Puerto Rico. *Journal of Emergency Management (Weston, Mass.)*, *19*(8), 235–253. https://doi.org/10.5055/jem.0656

Kintziger, K. W., Stone, K. W., Jagger, M. A., & Horney, J. A. (2021). The impact of the COVID-19 response on the provision of other public health services in the U.S.: A cross sectional study. *PLoS ONE*, *16*(10), e0255844. https://doi.org/10.1371/journal.pone.0255844

Kisling, L. A., & Das, J. M. (2023). Prevention Strategies. In *StatPearls*. StatPearls Publishing. http://www.ncbi.nlm.nih.gov/books/NBK537222/

Koh, D. (2020). COVID-19 lockdowns throughout the world. *Occupational Medicine (Oxford, England)*, kqaa073. https://doi.org/10.1093/occmed/kqaa073

Kooli, C. (2021). COVID 19: Public health issues and ethical dilemmas. *Ethics, Medicine and Public Health*, *17*, 100635. https://doi.org/10.1016/j.jemep.2021.100635

Leider, J. P., Castrucci, B. C., Robins, M., Hare Bork, R., Fraser, M. R., Savoia, E., Piltch-Loeb, R., & Koh, H. K. (2023). The exodus of state and local public health employees: Separations started Before and continued throughout COVID-19. *Health Affairs*, *42*(3), 338–348. https://doi.org/10.1377/hlthaff.2022.01251

Leider, J. P., Yeager, V. A., Kirkland, C., Krasna, H., Hare Bork, R., & Resnick, B. (2023). The state of the US public health workforce: Ongoing challenges and future directions. *Annual Review of Public Health*, *44*(1), 323–341. https://doi.org/10.1146/annurev-publhealth-071421-032830

Losee, J. E., Webster, G. D., & McCarty, C. (2022). Social network connections and increased preparation intentions for a disaster. *Journal of Environmental Psychology*, *79*, 101726. https://doi.org/10.1016/j.jenvp.2021.101726

Marceron, J. E., & Rohrbeck, C. A. (2019). Disability and disasters: The role of self-efficacy in emergency preparedness. *Psychology, Health & Medicine*, *24*(1), 83–93. https://doi.org/10.1080/13548506.2018.1492730

Martins, C., Godycki-Cwirko, M., Heleno, B., & Brodersen, J. (2018). Quaternary prevention: Reviewing the concept. *The European Journal of General Practice*, *24*(1), 106–111. https://doi.org/10.1080/13814788.2017.1422177

Mayberry, K. M., Scaccia, J. P., & Mitsdarffer, M. L. (2023). Analyzing twitter for community-level public health messaging. *Journal of Public Health Management and Practice*, *29*(2), 174. https://doi.org/10.1097/PHH.0000000000001677

Mehta, J., Kenin, J., & Kelly, M. L. (2022, November 23). The pandemic isn't over yet, but thousands of public health workers lost their jobs. *NPR*. https://www.npr.org/2022/11/23/1139073592/the-pandemic-isnt-over-yet-but-thousands-of-public-health-workers-lost-their-job

Mendez, M., & Horney, J. A. (2023). Impact of the COVID-19 pandemic response on the Emergency Medical Services workforce, Texas, USA. *Journal of Emergency Management*, *21*(3), 215–222. https://doi.org/10.5055/jem.0761

Metz, A. R., Bauer, M., Epperly, C., Stringer, G., Marshall, K. E., Webb, L. M., Hetherington-Rauth, M., Matzinger, S. R., Totten, S. E., Travanty, E. A., Good, K. M., & Burakoff, A. (2022). Investigation of COVID-19 outbreak among wildland firefighters during wildfire response, Colorado, USA, 2020. *Emerging Infectious Diseases*, *28*(8), 1551–1558. https://doi.org/10.3201/eid2808.220310

Mora, K., Thomas, J., Nankivell, K., & Flude, S. (2014). *Using existing social networks to improve emergency preparedness of supported community members*. *29*, 18–24.

National Association of County and City Health Officials. (n.d.). *Shots at the Shop Initiative*. Retrieved July 5, 2023, from https://www.naccho.org/programs/our-covid-19-response/shotsinshops

Nori-Sarma, A., Milando, C., Weinberger, K. R., Hess, J. J., Errett, N. A., & Wellenius, G. A. (2022). Association between the 2021 heat wave in Portland, Oregon, and Seattle, Washington, and Emergency Department Visits. *JAMA*, *328*(23), 2360–2362. https://doi.org/10.1001/jama.2022.20665

Olympia, R. P., Weber, C., Brady, J., & Ho, S. (2017). Emergency and disaster preparedness of school transportation staff and school buses in the United States: Compliance with recommendations for school transportation safety. *Pediatric Emergency Care*, *33*(11), 718–723. https://doi.org/10.1097/PEC.0000000000000704

Otok, R., Richardson, E., Czabanowska, K., & Middleton, J. (2018). The Public Health Workforce. In *Organization and Financing of Public Health Services in Europe [Internet]*. European Observatory on Health Systems and Policies. https://www.ncbi.nlm.nih.gov/books/NBK535727/

PAHO. (2020, January 30). *WHO declares Public Health Emergency on novel coronavirus*. https://www.paho.org/en/news/30-1-2020-who-declares-public-health-emergency-novel-coronavirus

Pandemic and All-Hazards Preparedness Act, 109–417, Congress 109 (2006). chrome-extension://efaidnbmnnnibpcajpcglclefindmkaj/https://www.govinfo.gov/content/pkg/PLAW-109publ417/pdf/PLAW-109publ417.pdf

Pearce, E., Kamenov, K., Barrett, D., & Cieza, A. (2022). Promoting equity in health emergencies through health systems strengthening: Lessons learned from disability inclusion in the COVID-19 pandemic. *International Journal for Equity in Health, 21*(Suppl 3), 149. https://doi.org/10.1186/s12939-022-01766-6

Phillips, D. E., Lozano, R., Naghavi, M., Atkinson, C., Gonzalez-Medina, D., Mikkelsen, L., Murray, C. J., & Lopez, A. D. (2014). A composite metric for assessing data on mortality and causes of death: The vital statistics performance index. *Population Health Metrics, 12*, 14. https://doi.org/10.1186/1478-7954-12-14

Potutan, G., & Arakida, M. (2021). Evolving disaster response practices during COVID-19 pandemic. *International Journal of Environmental Research and Public Health, 18*(6), 3137. https://doi.org/10.3390/ijerph18063137

Pujolar, G., Oliver-Anglès, A., Vargas, I., & Vázquez, M.-L. (2022). Changes in access to health services during the COVID-19 pandemic: A scoping review. *International Journal of Environmental Research and Public Health, 19*(3), 1749. https://doi.org/10.3390/ijerph19031749

Rahmat, Z. S., Sadiq, M., Vohra, L. I., Ullah, H., & Essar, M. Y. (2023). The impact of COVID-19 followed by extreme flooding on vector borne diseases in Pakistan: A mini narrative review. *New Microbes and New Infections, 51*, 101075. https://doi.org/10.1016/j.nmni.2022.101075

Rees, H., Mihigo, R., & Gray, A. (2022). Challenges in addressing inequity in access to COVID-19 diagnostics, therapeutics and vaccines in Africa. *South African Journal of Science, 118*(5/6), Article 5/6. https://doi.org/10.17159/sajs.2022/13475

Rivera, R., Rosenbaum, J. E., & Quispe, W. (2020). Excess mortality in the United States during the first three months of the COVID-19 pandemic. *Epidemiology & Infection, 148*, e264. https://doi.org/10.1017/S0950268820002617

Robert, R., Kentish-Barnes, N., Boyer, A., Laurent, A., Azoulay, E., & Reignier, J. (2020). Ethical dilemmas due to the Covid-19 pandemic. *Annals of Intensive Care, 10*(1), 84. https://doi.org/10.1186/s13613-020-00702-7

Sacks, J. D., Stanek, L. W., Luben, T. J., Johns, D. O., Buckley, B. J., Brown, J. S., & Ross, M. (2011). Particulate matter–induced health effects: Who is susceptible? *Environmental Health Perspectives, 119*(4), 446–454. https://doi.org/10.1289/ehp.1002255

SAMHSA. (2021, July 22). *Planning for Individuals Experiencing Homelessness*. https://www.samhsa.gov/dtac/disaster-planners/homelessness

SAMHSA. (2022, July 13). *Diversity, Equity, and Inclusion in Disaster Planning and Response*. https://www.samhsa.gov/dtac/disaster-planners/diversity-equity-inclusion

Satty, T., Ramgopal, S., Elmer, J., Mosesso, V. N., & Martin-Gill, C. (2021). EMS responses and non-transports during the COVID-19 pandemic. *The American Journal of Emergency Medicine, 42*, 1–8. https://doi.org/10.1016/j.ajem.2020.12.078

Saulnier, D., et al. (2019). *Disaster Risk Reduction: Why Do We Need Accurate Disaster Mortality Data to Strengthen Policy and Practice? | PreventionWeb*. UNDRR. https://www.preventionweb.net/publication/disaster-risk-reduction-why-do-we-need-accurate-disaster-mortality-data-strengthen

Scales, S. E., Patrick, E., Stone, K. W., Kintziger, K. W., Jagger, M. A., & Horney, J. A. (2022). Lessons learned from the public health Workforce's experiences with the COVID-19 response. *Health Security, 20*(5), 387–393. https://doi.org/10.1089/hs.2022.0091

Shakespeare, T., Ndagire, F., & Seketi, Q. E. (2021). Triple jeopardy: Disabled people and the COVID-19 pandemic. *The Lancet, 397*(10282), 1331–1333. https://doi.org/10.1016/S0140-6736(21)00625-5

Sleigh, J., Amann, J., Schneider, M., & Vayena, E. (2021). Qualitative analysis of visual risk communication on twitter during the Covid-19 pandemic. *BMC Public Health, 21*(1), 810. https://doi.org/10.1186/s12889-021-10851-4

Stone, K. W., Kintziger, K. W., Jagger, M. A., & Horney, J. A. (2021). Public health workforce burnout in the COVID-19 response in the U.S. *International Journal of Environmental Research and Public Health, 18*(8), Article 8. https://doi.org/10.3390/ijerph18084369

Sun, S., Weinberger, K. R., Nori-Sarma, A., Spangler, K. R., Sun, Y., Dominici, F., & Wellenius, G. A. (2021). Ambient heat and risks of emergency department visits among adults in the United States: Time stratified case crossover study. *BMJ*, *375*, e065653. https://doi.org/10.1136/bmj-2021-065653

Tai, D. B. G., Sia, I. G., Doubeni, C. A., & Wieland, M. L. (2022). Disproportionate impact of COVID-19 on racial and ethnic minority groups in the United States: A 2021 update. *Journal of Racial and Ethnic Health Disparities*, *9*(6), 2334–2339. https://doi.org/10.1007/s40615-021-01170-w

Torales, J., O'Higgins, M., Castaldelli-Maia, J. M., & Ventriglio, A. (2020). The outbreak of COVID-19 coronavirus and its impact on global mental health. *The International Journal of Social Psychiatry*, *66*(4), 317–320. https://doi.org/10.1177/0020764020915212

UNICEF. (2023). *Devastating floods in Pakistan.* https://www.unicef.org/emergencies/devastating-floods-pakistan-2022

US Department of Health and Human Services. (nd). *Public Health Emergency Declaration Q&A.* https://www.phe.gov/Preparedness/legal/Pages/phe-qa.aspx

Vohra, L. I., Aqib, M., Jamal, H., Mehmood, Q., & Yasin, F. (2023). Rising cases of dengue and malaria in flood affected areas of Pakistan: A major threat to the Country's healthcare system. *Disaster Medicine and Public Health Preparedness*, *17*, e323. https://doi.org/10.1017/dmp.2022.293

Wang, H., Paulson, K. R., Pease, S. A., Watson, S., Comfort, H., Zheng, P., Aravkin, A. Y., Bisignano, C., Barber, R. M., Alam, T., Fuller, J. E., May, E. A., Jones, D. P., Frisch, M. E., Abbafati, C., Adolph, C., Allorant, A., Amlag, J. O., Bang-Jensen, B., …, & Murray, C. J. L. (2022). Estimating excess mortality due to the COVID-19 pandemic: A systematic analysis of COVID-19-related mortality, 2020–21. *The Lancet*, *399*(10334), 1513–1536. https://doi.org/10.1016/S0140-6736(21)02796-3

Wells, S. A., Fleury-Steiner, R. E., Miller, S. L., Camphausen, L. C., & Horney, J. A. (2024). Impacts of the COVID-19 response on the domestic violence workforce. *Journal of Interpersonal Violence*, *39*(5–6), 1190–1205. https://doi.org/10.1177/08862605231203610

Wettstein, Z. S., Hoshiko, S., Fahimi, J., Harrison, R. J., Cascio, W. E., & Rappold, A. G. (n.d). Cardiovascular and cerebrovascular emergency department visits associated with wildfire smoke exposure in California in 2015. *Journal of the American Heart Association*, *7*(8), e007492. https://doi.org/10.1161/JAHA.117.007492

Wexler, B., & Smith, M.-E. (2015). Disaster response and people experiencing homelessness: Addressing challenges of a population with limited resources. *Journal of Emergency Management (Weston, Mass.)*, *13*(3), 195–200. https://doi.org/10.5055/jem.2015.0233

White, R. H., Anderson, S., Booth, J. F., Braich, G., Draeger, C., Fei, C., Harley, C. D. G., Henderson, S. B., Jakob, M., Lau, C.-A., Mareshet Admasu, L., Narinesingh, V., Rodell, C., Roocroft, E., Weinberger, K. R., & West, G. (2023). The unprecedented Pacific Northwest heatwave of June 2021. *Nature Communications*, *14*(1), Article 1. https://doi.org/10.1038/s41467-023-36289-3

White, G. F., Kates, R. W., & Burton, I. (2001). Knowing better and losing even more: The use of knowledge in hazards management. *Global Environmental Change Part B: Environmental Hazards*, *3*(3), Article 3. https://doi.org/10.3763/ehaz.2001.0308

WHO. (n.d.). *Making hospitals safe in emergencies.* https://www.who.int/activities/making-health-facilities-safe-in-emergencies-and-disasters

WHO. (2017). *The Role of CRVS in Achieving the Sustainable Development Goals: Get Every One in the Picture*. Sustainable Development Goals. https://getinthepicture.org/resource/role-crvs-achieving-sustainable-development-goals

Wu, X., Nethery, R. C., Sabath, M. B., Braun, D., & Dominici, F. (2020). Air pollution and COVID-19 mortality in the United States: Strengths and limitations of an ecological regression analysis. *Science Advances*, *6*(45), eabd4049. https://doi.org/10.1126/sciadv.abd4049

Xing, Y.-F., Xu, Y.-H., Shi, M.-H., & Lian, Y.-X. (2016). The impact of PM2.5 on the human respiratory system. *Journal of Thoracic Disease*, *8*(1), E69–E74. https://doi.org/10.3978/j.issn.2072-1439.2016.01.19

CONCLUSIONS AND FUTURES

Bimal Kanti Paul and Luke Juran

Introduction

In this handbook, we have been confined to two stages of the traditional disaster management cycle, response and recovery, which occur during and after disaster events. Response includes all activities during and immediately following an extreme event. It seeks to saves lives, prevent further property damage, and address the short-term needs of survivors in order to reduce societal suffering. Recovery overlaps with latter parts of the response stage as well as early parts of mitigation, which is the subsequent stage of the disaster management cycle. Recovery includes numerous medium- and long-term activities, such as arranging temporary shelter and repairing and reconstructing damaged and destroyed housing, civil infrastructure, and public and business buildings (Davis and Alexander, 2016; Coppola, 2020). Historically, the primary goal of recovery is to facilitate a rapid return to the pre-disaster scenario, or preferably better (Phillips, 2009; Paul, 2022). However, recent critiques, including in this handbook, have called for recovery to more explicitly include the social and emotional recovery of affected populations and to pay particular attention to disadvantaged populations.

The other two stages of the disaster management cycle are mitigation and preparedness. These two stages, which we would like to tie to this handbook, primarily occur before disasters and will encompass the content of a subsequent companion handbook co-edited by the same authors. Mitigation aims to decrease threats to life, property, and the environment that are posed by extreme events (Paul, 2011). Mitigation involves anticipatory, sustained actions taken to reduce, avoid, or eliminate long-term risk from future disasters, such as physical measures, policies, and behavioral changes. Mitigation is proactive and widely considered the 'cornerstone of disaster management' since future disaster damage and loss of life can be avoided if appropriate mitigation is implemented. In the context of the United States, the National Institute of Building Sciences (NIBS) (2019) has estimated that \$6 USD on average are saved for every \$1 spent on federal mitigation grants.

Innumerable types of mitigation techniques can be employed to minimize the destruction and disruptions caused by disasters, and each hazard has a unique set of structural and non-structural actions from which disaster managers may choose. Structural or 'hard' mitigation approaches typically involve infrastructure and physical/engineered changes to systems and landscapes and

 DOI: 10.4324/9781003414834-46

include dams, levees, seawalls, landslide barriers, retention ponds, retrofitting buildings, and the enhanced use of nature-based solutions such as wetlands, mangroves, and sand dunes for protection. There exists a temptation to rely on structural approaches, but effective mitigation must also include robust non-structural or 'soft' measures that direct human activities away from disaster-prone areas. Non-structural mitigation seeks to nudge human behavior and decision-making though approaches such as building codes, zoning, land use planning, coastal regulation/ management zones, economic (dis)incentives, public education, data collection, and mapping. Public awareness and participation, along with their inherent ability to infuse place- and systems-based knowledge, have been recognized as integral components of disaster mitigation as the magnitude of destruction depends not only on what type of measures are introduced, but also the extent of public compliance with such measures.

Preparedness refers to the degree of readiness of an individual, household, community, government, or other unit in advance of a disaster. Among many measures or actions, preparedness includes activities such as hazard warning systems, emergency response drills, testing disaster plans, information dissemination to the public via various media, preemptive evacuation, and administering disaster training to first responders and the public. The acquisition of equipment and resources to support disaster response is also a crucial element of disaster preparedness.

Vulnerability, Risk, and Resilience

Beyond this handbook, many other topics are significant in the hazards and disaster literature. Notable among these are vulnerability, risk, and resilience, which are intricately embedded within the disaster management cycle and its four traditional stages. For example, preparedness typically begins with vulnerability and risk analyses that seek to identify which hazards (along with their anticipated frequency and range of magnitude) are likely to impact a particular location and its built and human assets. Consequently, vulnerability and risk assessment play an important role in building the resilience of human and engineered systems through informed mitigation, enhanced response, and pre-identified issues to address during the recovery process. Vulnerability is determined by physical, social, economic, and environmental factors that serve to either increase or decrease the susceptibility of an individual, household, community, or political territory to the impacts of a disaster. Thus, vulnerability is not the product of a disaster. Rather, it evolves over time and is the product of a combination of environmental and socioeconomic variables established in the pre-disaster period that can either support or suppress the ability to withstand and recover from disasters (Blaikie et al., 1994; Mileti, 1999; Juran and Trivedi, 2015).

Vulnerability and resilience are highly complementary since the former decreases as the latter increases. In fact, the Intergovernmental Panel on Climate Change (IPCC) (2001) essentially treats resilience as the opposite of vulnerability. Resilience is often described as the capacity of human and physical systems to bounce back and recover from extreme events (Tierney and Bruneau, 2007; Southwick et al., 2014), and the term gained additional prominence after Hurricane Katrina in 2005. Resilience is the capacity to remain functional during and immediately after disasters and it enables a quicker return to normality in post-disaster recovery.

Disaster risk is expressed as the likelihood of loss of life, injury, physical damage, and economic loss from an extreme event. Risk can be conceptualized as the combination of three related variables: a hazard, exposure, and vulnerability (Paul, 2011). Another definition that is preferred by many risk managers is the likelihood of an event occurring multiplied by the effects of the event (Ansel and Wharton, 1992). Disaster risk reduction (DRR) contributes to resilience and therefore decreases vulnerability. DRR aims to mitigate the effects of disasters through various protective measures that reduce hazard exposure and strengthen our engineered and human systems. DRR builds safer living environments and bolsters the coping capacity of communities. The United Nations Sendai Framework for Disaster Risk Reduction 2015–2030 (United Nations Office for Disaster Risk Reduction (UNDRR), 2015)—successor to the Hyogo Framework for Action 2005–2015 (United Nations Office for Disaster Risk Reduction (UNDRR), 2005)—explicitly emphasizes both risk reduction and resilience building as essential components of disaster management.

Ultimately, the consequences of future disasters will reflect the levels of vulnerability and risk that are already present in our living environments. Reducing the vulnerability and risk of communities while perpetually fostering resilience throughout the disaster management cycle is a way forward. For example, in the pre-disaster stages, this can be accomplished via robust preparedness plans, warning systems that reach all communities through accessible media and modes, and mitigative actions that preemptively address DRR and multi-hazard risk. Meanwhile, in the post-disaster stages, response should be informed by vulnerability and risk assessment maps/data and recovery should not only focus on infrastructure and the economy but also the social, cultural, and emotional recovery of the people. In essence, an ethos of resilience needs to be created and sustained.

Social Vulnerability and Social Inclusion

There are several dimensions of vulnerability, and different sets of variables are used to assess different types of vulnerability. A critical dimension is social vulnerability, which considers the susceptibility of particular social and demographic individuals or groups to potential losses from hazards (Blaikie et al., 1994; Hewitt, 1997; Cutter and Finch, 2008). Social vulnerability focuses on socioeconomic factors such as income, education level, race and ethnicity, insurance coverage, and proportion of elderly, children, and disabled persons. These factors, which compromise what Cutter et al. (2003) refer to as the 'social profile' of a community, operate both individually and collectively to influence social vulnerability. The assumption is that human variables such as level of affluence, minority status, disabilities, education status, and others provide advantages (or disadvantages) during normal times and that the same (dis)advantages are merely revealed in times of disaster (Tierney et al., 2001; Fothergill and Peek, 2004; Juran, 2012). Social vulnerability can be decreased in a variety of ways, including through mainstreaming the needs and experiences of vulnerable communities and increasing levels of social capital (Bourdieu, 1985).

Deeply embedded political, economic, and societal processes and norms (e.g., gender relations, androcentric systems, and environmental racism) produce inequitable outcomes and differential access to resources, networks, and power. These unnatural variables, along with the obvious dimension of poverty, represent common root causes of social vulnerability. As a group, marginalized and subaltern communities are often more socially vulnerable due to a history of exclusionary practices, systemic disadvantages, and disproportionately occupying geographic locations with substandard services, lower levels of resources, and greater hazard exposure. Other socially vulnerable populations often include women, elderly, children, migrants, and those who identify as lesbian, gay, bisexual, transgender, queer, intersex, and asexual (LGBTQIA+). Social vulnerability is complex and dynamic; it changes over time and as the conditions of vulnerable communities and their systems change. Therefore, temporal and spatial scales are important when undertaking vulnerability assessments. Analyses performed at the national, state, or even county level may not accurately reflect the social profile of a place. Ideally, social vulnerability should be examined at the local or community level. Since social vulnerability encompasses numerous variables that are highly interrelated, assessments often utilize correlations or principal component analysis (PCA) to identify the predominant factors and to derive indicators and composite indices. For example, Cutter et al. (2003) developed and deployed the Social Vulnerability Index (SoVI) for all counties in the United States. Modified versions of the SoVI have been deployed in Nepal (Aksha et al., 2019), Portugal (Guillard-Gonçalves et al., 2015), and other places through the utilization of place-based indicators that capture the social dynamics of the unique contexts.

Social inclusion can be understood as broadening participation in societal processes (e.g., in political spheres, economic markets, and decision-making) among all persons and especially communities that have been historically marginalized and underrepresented. This can be accomplished by providing opportunities, enhancing access to resources, and uplifting the voices and agency of disadvantaged groups. Mainstreaming, which seeks to achieve equality in formal and informal structures, represents another pragmatic approach. Mainstreaming assesses the implications of a policy or planned action for all groups to ensure that diverse viewpoints and concerns are infused as integral components of the design, implementation, monitoring, and evaluation of actions so that inequalities are not perpetuated and everyone benefits (Daly, 2005). Conversely, exclusionary practices rob individuals of dignity, security, and opportunities to forge a better life. Until systemic and cultural impediments to inclusion are addressed, it will be difficult to generate long-term, sustainable, and proactive disaster processes.

Conclusion

The stages of response and recovery are useful concepts that have direct implications for planning, policy, and scholarly research. They represent integral components of a continuous process through which diverse ecosystem actors communally reduce the effects of disasters across numerous synergistic activities. It is important to reiterate that, in this handbook (and the subsequent handbook on mitigation and preparedness), chapters were structured around each independent stage for purposes of organization. However, the stages do not exist in silos. Rather, they overlap, are cross-cutting, and both inform and influence each other through various feedbacks. There are also different representations of the disaster management cycle; this is explicitly addressed in the handbook and alternative conceptualizations of the cycle and its stages have been provided. Additionally, it is critical to note that the handbook excluded many topics due to space, time, and the fact that no handbook on disasters can ever be considered complete. All such handbooks are necessarily abridged as the domain is massive and new issues and considerations seem to arise after each disaster.

This concluding chapter sought to briefly review the two traditional post-disaster stages of response and recovery and link them to the pre-disaster stages of mitigation and preparedness. Again, while it is useful to categorize the stages, in actuality they constitute a perpetual, interconnected process in which an activity undertaken in any stage has the potential to affect all other stages. The importance of vulnerability, risk, and resilience was also discussed. These factors should be addressed in each stage of the disaster cycle and can be facilitated by reducing social vulnerability and including the voices and lived experiences of marginalized communities.

Unfortunately, disasters and their impacts are inevitable. Human and ecological systems will be affected and we will have to adjust and manage the impacts. Proactive, sustainable, and inclusive policies and practices are required across all institutions and at all scales if we seek to create a culture of disaster resilience.

References

Aksha, S.K., L. Juran, L.M. Resler, and Y. Zhang. 2019. An Analysis of Social Vulnerability to Natural Hazards in Nepal Using a Modified Social Vulnerability Index. *International Journal of Disaster Risk Science* 10(1): 103–116.

Ansel, J., and F. Wharton. 1992. *Risk Analysis, Assessment, and Management*. Chichester: John Wiley & Sons Ltd.

Blaikie, P., T. Cannon, I. Davies, and B. Wisner. 1994. *At Risk: Natural Hazards, People's Vulnerability, and Disasters*. London: Routledge.

Bourdieu, P. 1985. The Social Space and the Genesis of Groups. *Theory and Society* 14(6): 723–744.

Coppola, D.P. 2020. *Introduction to International Disaster Management*, 4th edition. Boston, MA: Elsevier.

Cutter, S.L., B.J. Boruff, and W.L Shirley. 2003. Social Vulnerability to Environmental Hazards. *Social Science Quarterly* 84(2): 242–261. https://doi.org/10.1111/1540-6237.8402002

Cutter, S.L., and C. Finch. 2008. Temporal and Spatial Changes to Social Vulnerability to Natural Hazards. *Proceedings of the National Academy of Sciences* 105(7): 2301–2306.

Daly, M. 2005. Gender Mainstreaming in Theory and Practice. *Social Politics: International Studies in Gender, State & Society*. 12(3): 433–450. https://doi.org/10.1093/sp/jxi023

Davis, I., and D. Alexander. 2016. *Recovery from Disaster*. London: Routledge.

Fothergill, A., and L.A. Peek. 2004. Poverty and Disasters in the United States: A Review of Recent Sociological Findings. *Natural Hazards* 32: 89–110.

Guillard-Gonçalves, C., S.L. Cutter, C.T. Emrich, and J.L. Zêzere. 2015. Application of Social Vulnerability Index (SoVI) and Delineation of Natural Risk Zones in Greater Lisbon, Portugal. *Journal of Risk Research* 18(5): 651–674.

Hewitt, K. 1997. *Regions of Risk: A Geographical Introduction to Disaster*. Essex: Longman.

Intergovernmental Panel on Climate Change (IPCC). 2001. *Climate Change 2001: The Scientific Basis*. Cambridge: Cambridge University Press. https://www.ipcc.ch/site/assets/uploads/2018/07/WG1_TAR_FM.pdf

Juran, L. 2012. The Gendered Nature of Disasters: Women Survivors in Post-Tsunami Tamil Nadu. *Indian Journal of Gender Studies* 19(1): 1–29.

Juran, L., and J. Trivedi. 2015. Women, Gender Norms, and Natural Disasters in Bangladesh. *Geographical Review* 105(4): 601–611.

Mileti, D.S. 1999. *Disaster by Design: A Reassessment of Natural Hazards in the United States*. Washington, DC: Joseph Henry.

National Institute of Building Sciences (NIBS). 2019. Mitigation Saves: Federal Grants Provide a $6 Benefit for Each $1 Invested. https://www.nibs.org/files/pdfs/ms_v3_federalgrants.pdf

Paul, B.K. 2011. *Environmental Hazards and Disasters: Context, Perspectives and Management*. Hoboken, NJ: Wiley-Blackwell.

Paul, B.K. 2022. Recovery. In *Routledge Handbook of Environmental Hazards and Society*, edited by McGee, T., and E. Penning-Rowsell, pp. 425–441. London: Routledge.

Phillips, B.D. 2009. *Disaster Recovery*. Boca Raton, FL: CRC Press.

Southwick, S.M., G.A. Bonanno, A.S. Masten, C. Panter-Brick, and R. Yehuda. 2014. Resilience Definitions, Theory, and Challenges: Interdisciplinary Perspectives. *European Journal of Psychotraumatology* 5. https://doi.org/10.3402/ejpt.v5.25338

Tierney, K.J., and M. Bruneau. 2007. Conceptualizing and Measuring Resilience: A Key to Disaster Loss Reduction. *TR News* 250(May-June): 14–17.

Tierney, K.J., M.K. Lindell, and R.W. Perry. 2001. *Facing the Unexpected: Disaster Preparedness and Response in the United States*. Washington, D.C: Joseph Henry Press.

United Nations Office for Disaster Risk Reduction (UNDRR). 2005. Hyogo Framework for Action 2005-2015: Building the Resilience of Nations and Communities to Disasters. https://www.unisdr.org/files/1037_hyogoframeworkforactionenglish.pdf

United Nations Office for Disaster Risk Reduction (UNDRR). 2015. Sendai Framework for Disaster Risk Reduction 2015-2030. https://www.unisdr.org/we/inform/publications/43291

INDEX

Note: Page references in *italics* denote figures, in **bold** tables and with "n" endnotes.

For Product Safety Concerns and Information please contact our EU
representative GPSR@taylorandfrancis.com
Taylor & Francis Verlag GmbH, Kaufingerstraße 24, 80331 München, Germany

www.ingramcontent.com/pod-product-compliance
Ingram Content Group UK Ltd.
Pitfield, Milton Keynes, MK11 3LW, UK
UKHW052056230926
475702UK00032B/2962

* 9 7 8 1 0 3 2 5 4 0 3 4 4 *